RECENT ADVANCES IN PLANT GENETIC ENGINEERING

Recent Advances in Plant Genetic Engineering

Editor:

Aleksei Anatoliyovych Stepanenko, Ph.D.

Delve Publishing

www.delvepublishing.com

Recent Advances in Plant Genetic Engineering

Editor: **Aleksei Anatoliyovych Stepanenko, Ph.D.**

708 3rd Avenue, 6th Floor
New York, NY 10017
United States of America
www.delvepublishing.com

ISBN: 978-1-68095-749-5

Library of Congress Control Number: 2016951507

About the Editor

Aleksei Anatoliyovych Stepanenko, Ph.D.

Aleksei A. Stepanenko received his Masters's degree in Biochemistry from V. N. Karazin Kharkiv National University and PhD in Molecular biology from Institute of Molecular Biology and Genetics, where he is currently working as a Research Fellow. His scope of interests is genomic and phenotypic evolution of tumor cells under chemotherapeutic stresses and viro-immunological approaches as an alternative perspective in cancer therapy.

Contents

List of Contributors

Ali Movahedi
The Cooperative Innovation Center of Southern Modern Forestry, Nanjing Forestry University, Nanjing 210037, China

Jiaxin Zhang
The Cooperative Innovation Center of Southern Modern Forestry, Nanjing Forestry University, Nanjing 210037, China

Rasoul Amirian
Department of Genomics, Agricultural Biotechnology Research Institute, Central Region of Iran (ABRICI), Najafabad Road, Isfahan 85135-487, Iran

Qiang Zhuge
The Cooperative Innovation Center of Southern Modern Forestry, Nanjing Forestry University, Nanjing 210037, China

Simona Capone
Institute of Chemical Engineering, Research Area Biochemical Engineering, Vienna University of Technology, Gumpendorfer Strasse 1a, 1060 Vienna, Austria

Lejla Ćorajević
Institute of Chemical Engineering, Research Area Biochemical Engineering, Vienna University of Technology, Gumpendorfer Strasse 1a, 1060 Vienna, Austria

Günther Bonifert
Institute of Chemical Engineering, Research Area Biochemical Engineering, Vienna University of Technology, Gumpendorfer Strasse 1a, 1060 Vienna, Austria

Patrick Murth
Institute of Chemical Engineering, Research Area Biochemical Engineering, Vienna University of Technology, Gumpendorfer Strasse 1a, 1060 Vienna, Austria

Daniel Maresch
Department of Chemistry, University of Natural Resources and Life Sciences, 1180 Vienna, Austria

Friedrich Altmann
Department of Chemistry, University of Natural Resources and Life Sciences, 1180 Vienna, Austria

Christoph Herwig
Institute of Chemical Engineering, Research Area Biochemical Engineering, Vienna University of Technology, Gumpendorfer Strasse 1a, 1060 Vienna, Austria

Oliver Spadiut
Institute of Chemical Engineering, Research Area Biochemical Engineering, Vienna University of Technology, Gumpendorfer Strasse 1a, 1060 Vienna, Austria

Surender Khatodia
Amity Institute of Biotechnology, Amity University Haryana, Gurgaon, India

Kirti Bhatotia
Amity Institute of Biotechnology, Amity University Haryana, Gurgaon, India

Nishat Passricha
Plant Molecular Biology Group, International Centre for Genetic Engineering and Biotechnology, New Delhi, India

S. M. P. Khurana
Amity Institute of Biotechnology, Amity University Haryana, Gurgaon, India

Narendra Tuteja
Plant Molecular Biology Group, International Centre for Genetic Engineering and Biotechnology, New Delhi, India
Amity Institute of Microbial Technology, Amity University, Noida, India

Zafar Iqbal
Institute of Biochemistry and Biotechnology, Quaid-i-Azam Campus, University of the Punjab, Lahore, Pakistan

Muhammad N. Sattar
Department of Environment and Natural Resources, Faculty of Agriculture and Food Science, King Faisal University, Al-Hasa, Saudi Arabia

Muhammad Shafiq
Agricultural Biotechnology Division, National Institute for Biotechnology and Genetic Engineering, Faisalabad, Pakistan

Erum Dilshad
Department of Biochemistry, Faculty of Biological Sciences, Quaid-iAzam University, Islamabad, Pakistan

Rosa Maria Cusido
Laboratorio de Fisiologia Vegetal, Facultad de Farmacia, Universidad de Barcelona, Barcelona, Spain

Javier Palazon
Laboratorio de Fisiologia Vegetal, Facultad de Farmacia, Universidad de Barcelona, Barcelona, Spain

Karla Ramirez Estrada
Laboratorio de Fisiologia Vegetal, Facultad de Farmacia, Universidad de Barcelona, Barcelona, Spain

Mercedes Bonfill
Laboratorio de Fisiologia Vegetal, Facultad de Farmacia, Universidad de Barcelona, Barcelona, Spain

Bushra Mirza
Department of Biochemistry, Faculty of Biological Sciences, Quaid-iAzam University, Islamabad, Pakistan

Sandeep Kumar
Dow AgroSciences LLC, 9330 Zionsville Road, Indianapolis, IN 46286, USA

Pierluigi Barone
Dow AgroSciences LLC, 9330 Zionsville Road, Indianapolis, IN 46286, USA

Michelle Smith
Dow AgroSciences LLC, 9330 Zionsville Road, Indianapolis, IN 46286, USA

Meizhen Wang
State Key Laboratory of Agrobiotechnology, College of Biological Sciences, China Agricultural University, No. 2 Yuanmingyuan West Road, Haidian District, Beijing 100193, China
Institute of Medicinal Plant Development, Chinese Academy of Medical Sciences & Peking Union Medical College, No. 151, Malianwa North Road, Haidian District, Beijing 100193, China

Chen Liu
State Key Laboratory of Agrobiotechnology, College of Biological Sciences, China Agricultural University, No. 2 Yuanmingyuan West Road, Haidian District, Beijing 100193, China

Shixue Li
State Key Laboratory of Agrobiotechnology, College of Biological Sciences, China Agricultural University, No. 2 Yuanmingyuan West Road, Haidian District, Beijing 100193, China

Dengyun Zhu
State Key Laboratory of Agrobiotechnology, College of Biological Sciences, China Agricultural University, No. 2 Yuanmingyuan West Road, Haidian District, Beijing 100193, China

Qian Zhao
State Key Laboratory of Agrobiotechnology, College of Biological Sciences, China Agricultural University, No. 2 Yuanmingyuan West Road, Haidian District, Beijing 100193, China

Jingjuan Yu
State Key Laboratory of Agrobiotechnology, College of Biological Sciences, China Agricultural University, No. 2 Yuanmingyuan West Road, Haidian District, Beijing 100193, China

Noemi Ruiz-Lopez
Department of Biological Chemistry, Rothamsted Research, Harpenden, Herts AL5 2JQ, UK

Sarah Usher
Department of Biological Chemistry, Rothamsted Research, Harpenden, Herts AL5 2JQ, UK

Olga V. Sayanova
Department of Biological Chemistry, Rothamsted Research, Harpenden, Herts AL5 2JQ, UK

Johnathan A. Napier
Department of Biological Chemistry, Rothamsted Research, Harpenden, Herts AL5 2JQ, UK

Richard P. Haslam
Department of Biological Chemistry, Rothamsted Research, Harpenden, Herts AL5 2JQ, UK

Philippe Jeandet
Laboratory of Stress, Defenses and Plant Reproduction, Research Unit "Vines and Wines of Champagne", UPRES EA 4707, Faculty of Sciences, University of Reims, Reims 51687, France

Christophe Clément
Laboratory of Stress, Defenses and Plant Reproduction, Research Unit "Vines and Wines of Champagne", UPRES EA 4707, Faculty of Sciences, University of Reims, Reims 51687, France

Eric Courot
Laboratory of Stress, Defenses and Plant Reproduction, Research Unit "Vines and Wines of Champagne", UPRES EA 4707, Faculty of Sciences, University of Reims, Reims 51687, France

Sylvain Cordelier
Laboratory of Stress, Defenses and Plant Reproduction, Research Unit "Vines and Wines of Champagne", UPRES EA 4707, Faculty of Sciences, University of Reims, Reims 51687, France

Hyun-Soon Kim
Plant Systems Engineering Research Center, KRIBB, 125 Gwahangno, Yuseong-gu, Daejeon 305-806, Republic of Korea

Jae-Heung Jeon
Plant Systems Engineering Research Center, KRIBB, 125 Gwahangno, Yuseong-gu, Daejeon 305-806, Republic of Korea

Kyung Jin Lee
Department of Medicine, College of Medicine, Chung-Ang University, Seoul 156-756, Republic of Korea

Kisung Ko
Department of Medicine, College of Medicine, Chung-Ang University, Seoul 156-756, Republic of Korea

Kathleen Hefferon
Cornell University, Ithaca, NY 14850, USA

Daisuke Todaka
Laboratory of Plant Molecular Physiology, Graduate School of Agricultural and Life Sciences, The University of Tokyo, Tokyo, Japan

Kazuo Shinozaki
RIKEN Center for Sustainable Resource Science, Yokohama, Japan

Kazuko Yamaguchi-Shinozaki
Laboratory of Plant Molecular Physiology, Graduate School of Agricultural and Life Sciences, The University of Tokyo, Tokyo, Japan

Hongyan Wang
Institute of Technology, Yantai Academy of China Agriculture University, Yantai, China

Honglei Wang
Institute of Technology, Yantai Academy of China Agriculture University, Yantai, China

Hongbo Shao
Jiangsu Key Laboratory for Bioresources of Saline Soils, Provincial Key Laboratory of Agrobiology, Institute of Biotechnology, Jiangsu Academy of Agricultural Sciences, Nanjing, China
Key Laboratory of Coastal Biology and Bioresources Utilization, Yantai Institute of Coastal Zone Research, Chinese Academy of Sciences, Yantai, China

Xiaoli Tang
Key Laboratory of Coastal Biology and Bioresources Utilization, Yantai Institute of Coastal Zone Research, Chinese Academy of Sciences, Yantai, China

Tetsuya Yamada
Research Faculty of Agriculture, Hokkaido University, Kita 9 Nishi 9, Kita, Sapporo, Hokkaido 060-8589, Japan

Kyoko Takagi
Soybean Applied Genomics Research Unit, National Institute of Agrobiological Sciences, 2-1-2 Kannondai, Tsukuba, Ibaraki 305- 8602, Japan

Masao Ishimoto
Soybean Applied Genomics Research Unit, National Institute of Agrobiological Sciences, 2-1-2 Kannondai, Tsukuba, Ibaraki 305- 8602, Japan

Ibrokhim Y. Abdurakhmonov
Center of Genomics and Bioinformatics, Structural and Functional Genomics, Academy of Sciences the Republic of Uzbekistan, Ministry of Agriculture and Water Resources the Republic of Uzbekistan and "Uzpakhtasanoat" Association, Kibray, Uzbekistan

Mirzakamol S. Ayubov
Center of Genomics and Bioinformatics, Structural and Functional Genomics, Academy of Sciences the Republic of Uzbekistan, Ministry of Agriculture and Water Resources the Republic of Uzbekistan and "Uzpakhtasanoat" Association, Kibray, Uzbekistan

Khurshida A. Ubaydullaeva
Center of Genomics and Bioinformatics, Structural and Functional Genomics, Academy of Sciences the Republic of Uzbekistan, Ministry of Agriculture and Water Resources the Republic of Uzbekistan and "Uzpakhtasanoat" Association, Kibray, Uzbekistan

Zabardast T. Buriev
Center of Genomics and Bioinformatics, Structural and Functional Genomics, Academy of Sciences the Republic of Uzbekistan, Ministry of Agriculture and Water Resources the Republic of Uzbekistan and "Uzpakhtasanoat" Association, Kibray, Uzbekistan

Shukhrat E. Shermatov
Center of Genomics and Bioinformatics, Structural and Functional Genomics, Academy of Sciences the Republic of Uzbekistan, Ministry of Agriculture and Water Resources the Republic of Uzbekistan and "Uzpakhtasanoat" Association, Kibray, Uzbekistan

Haydarali S. Ruziboev
Center of Genomics and Bioinformatics, Structural and Functional Genomics, Academy of Sciences the Republic of Uzbekistan, Ministry of Agriculture and Water Resources the Republic of Uzbekistan and "Uzpakhtasanoat" Association, Kibray, Uzbekistan

Umid M. Shapulatov
Center of Genomics and Bioinformatics, Structural and Functional Genomics, Academy of Sciences the Republic of Uzbekistan, Ministry of Agriculture and Water Resources the Republic of Uzbekistan and "Uzpakhtasanoat" Association, Kibray, Uzbekistan
Laboratory of Plant Physiology, Wageningen University, Wageningen, Netherlands

Sukumar Saha
Crop Science Research Laboratory, United States Department of Agriculture – Agricultural Research Service, Starkville, MS, USA

Mauricio Ulloa
Plant Stress and Germplasm Development Research, United States Department of Agriculture – Agricultural Research Service, Lubbock, TX, USA

John Z. Yu
Crop Germplasm Research Unit, United States Department of Agriculture – Agricultural Research Service, College Station, TX, USA

Richard G. Percy
Crop Germplasm Research Unit, United States Department of Agriculture – Agricultural Research Service, College Station, TX, USA

Eric J. Devor
Department of Obstetrics and Gynecology, University of Iowa Carver College of Medicine, Iowa City, IA, USA

Govind C. Sharma
Department of Biological and Environmental Sciences, Alabama A&M University, Normal, AL, USA

Venkateswara R. Sripathi
Department of Biological and Environmental Sciences, Alabama A&M University, Normal, AL, USA

Siva P. Kumpatla
Dow AgroSciences LLC, Indianapolis, IN, USA

Alexander van der Krol
Laboratory of Plant Physiology, Wageningen University, Wageningen, Netherlands

Hake D. Kater
Agricultural and Environmental Research, Cary, NC, USA

Khakimdjan Khamidov
Center of Genomics and Bioinformatics, Structural and Functional Genomics, Academy of Sciences the Republic of Uzbekistan, Ministry of Agriculture and Water Resources the Republic of Uzbekistan and "Uzpakhtasanoat" Association, Kibray, Uzbekistan

Shavkat I. Salikhov
Center of Genomics and Bioinformatics, Structural and Functional Genomics, Academy of Sciences the Republic of Uzbekistan, Ministry of Agriculture and Water Resources the Republic of Uzbekistan and "Uzpakhtasanoat" Association, Kibray, Uzbekistan

Johnie N. Jenkins
Crop Science Research Laboratory, United States Department of Agriculture – Agricultural Research Service, Starkville, MS, USA

Abdusattor Abdukarimov
Center of Genomics and Bioinformatics, Structural and Functional Genomics, Academy of Sciences the Republic of Uzbekistan, Ministry of Agriculture and Water Resources the Republic of Uzbekistan and "Uzpakhtasanoat" Association, Kibray, Uzbekistan

Alan E. Pepper
Department of Biology, Texas A&M University, Colleges Station, TX, USA

Akira Kikuchi
Faculty of Life and Environmental Sciences, Gene Research Center, University of Tsukuba, 1-1-1 Ten-noudai, Tsukuba, Ibaraki 305-8752, Japan

Huu Duc Huynh
Faculty of Life and Environmental Sciences, Gene Research Center, University of Tsukuba, 1-1-1 Ten-noudai, Tsukuba, Ibaraki 305-8752, Japan

Tsukasa Endo
Faculty of Life and Environmental Sciences, Gene Research Center, University of Tsukuba, 1-1-1 Ten-noudai, Tsukuba, Ibaraki 305-8752, Japan

Kazuo Watanabe
Faculty of Life and Environmental Sciences, Gene Research Center, University of Tsukuba, 1-1-1 Ten-noudai, Tsukuba, Ibaraki 305-8752, Japan

Sarma Rajeevkumar
Molecular Plant Biology and Biotechnology Division, Central Institute of Medicinal and Aromatic Plants Research Centre, Bangalore, India

Pushpanathan Anunanthini
Plant Genetic Engineering Laboratory, Department of Biotechnology, Bharathiar University, Coimbatore, India

Ramalingam Sathishkumar
Plant Genetic Engineering Laboratory, Department of Biotechnology, Bharathiar University, Coimbatore, India

Lucille Pourcel
Department of Botany and Plant Biology, University of Geneva, Geneva, Switzerland

Michael Moulin
Department of Botany and Plant Biology, University of Geneva, Geneva, Switzerland

Teresa B. Fitzpatrick
Department of Botany and Plant Biology, University of Geneva, Geneva, Switzerland

Thanyanan Wannathong
Present address: Department of Biology, Faculty of Science, Silpakorn University, Nakornpathom 73000, Thailand

Janet C. Waterhouse
Algal Research Group, Institute of Structural and Molecular Biology, University College London, Gower Street, London WC1E 6BT, UK

Rosanna E. B. Young
Algal Research Group, Institute of Structural and Molecular Biology, University College London, Gower Street, London WC1E 6BT, UK

Chloe K. Economou
Algal Research Group, Institute of Structural and Molecular Biology, University College London, Gower Street, London WC1E 6BT, UK

Saul Purton
Algal Research Group, Institute of Structural and Molecular Biology, University College London, Gower Street, London WC1E 6BT, UK

Preface

This book presents recent innovations in plant virus expression systems and their uses for producing biologics from plants, advances in the glycoengineering in plant to allow human-like glycomodification and optimization of desired glycan structures for enhancing safety and functionality of recombinant pharmaceutical glycoproteins, highlights progress in the development of transgenic drought-tolerant rice plants, with special attention paid to field trial investigations, reviews the recent progress in the potential utilization of transcription factors involved in plant abiotic stress responses to improve multiple stress tolerance of crops in the field conditions, describes recent advances and problems of soybean transformation, new tools for chloroplast genetic engineering to allow the synthesis of human growth hormone in the green alga, recent transgenic studies on abiotic stress tolerance and molecular breeding in potato, approaches attempting to increase the vitamin B1 in crop plants. This book also includes the recent case studies covering the diverse perspective directions of plant genetic engineering such as improving maize nutritive quality and salt tolerance, increasing the artemisinin content in Artemisia annua plants to improve the availability of this one of the most effective and much-needed drug for malaria, manipulating phytoalexins to increase the disease resistance of plants. The current status of the various applications of the CRISPR/Cas system in plants, where it is used for the successful generation of stable mutations in a steadily growing number of species, gene targeting challenges in plants and their overcoming via intra genomic homologous recombination, antisense approaches and currently applied breakthrough RNAi studies in cotton are discussed. Undesired epigenetic silencing in transgenic technology is summarized.

Editor
Aleksei Anatoliyovych Stepanenko, Ph.D.

CHAPTER 1

AN EFFICIENT AGROBACTERIUM-MEDIATED TRANSFORMATION SYSTEM FOR POPLAR

Ali Movahedi[1] Jiaxin Zhang[1] Rasoul Amirian[2] and Qiang Zhuge[1]

[1]The Cooperative Innovation Center of Southern Modern Forestry, Nanjing Forestry University, Nanjing 210037, China

[2]Department of Genomics, Agricultural Biotechnology Research Institute, Central Region of Iran (ABRICI), Najafabad Road, Isfahan 85135-487, Iran

ABSTRACT

Poplar is a model system for the regeneration and genetic transformation of woody plants. To shorten the time required for studies of transgenic poplar, efforts have been made to optimize transformation methods that use Agrobacterium tumefaciens. In this study, an Agrobacterium infective suspension was treated at 4 °C for at least 10 h before infecting explants. By transforming the Populus hybrid clone "Nanlin895" (Populus deltoides × P. euramericana) with Agrobacterium harboring the PBI121:CarNAC6 binary vector, we showed that the transformation efficiency was improved significantly by multiple independent factors, including an Agrobacterium infective suspension with an OD_{600} of 0.7, an Agrobacterium infection for 120 min, an Agrobacterium infective suspension at a pH of 5.0, an acetosyringone concentration of 200 μM, a cocultivation at 28 °C, a cocultivation for 72 h and a sucrose concentration of 30 g/L in the cocultivation medium. We also showed that preculture of wounded leaf explants for two days increased the regeneration rate. The integration of the desired gene into transgenic poplars was detected using selective medium containing kanamycin, followed by southern blot analysis. The expression of the transgene in the transgenic lines was confirmed by northern blot analysis.

INTRODUCTION

Poplar is a versatile tree species that is highly amenable to vegetative propagation, has a rapid growth rate and is a good model system for the transformation of woody plant species. These characteristics have led to its widespread use in the pulp, paper and cosmetics industries and in both highland and lowland reforestation [1,2,3,4]. Furthermore, poplar is a forest tree species, with a rather long generation cycle (seven to 10 years or more). That is why researchers usually use materials from the initial generation for the analysis of transgenic plants of forest trees, with just several months' generation cycles. Efforts to optimize regeneration and propagation methods that use leaves, petioles, internodes, stems, roots and sprouts to improve poplar through the overexpression of transgenes conferring resistance to biotic and abiotic stresses or other traits have increased markedly over the past decade [2,3,5,6]. Transformation methods using leaf segments are the most commonly used [7]. Attempts to improve poplar transformation mediated by *Agrobacterium tumefaciens* have focused on a variety of parameters, including different *Agrobacterium* strains and culture densities, various acetosyringone (AS) concentrations and other factors[1,8,9,10,11].

Citation: Ali Movahedi, Jiaxin Zhang, Rasoul Amirian, and Qiang Zhuge, An Efficient Agrobacterium-Mediated Transformation System for Poplar, doi: 10.3390/ijms150610780.

The expression of high-copy number transgenes in transgenic plants may be more or less than only a single copy of transgene, and also, sometimes, this is good for molecular analysis and genetic engineering. Furthermore, Husaini [12] reported that strawberry transgenic plants with a high transgene copy number (four copies or more) are very good for molecular analysis and genetic engineering. Moreover, Bartlett *et al.*[13] reported that an improved method in barely using *Agrobacterium*, with a high copy of transgenes, is quite good for molecular analysis and genetic engineering. The benefits of improved transformation methods include the insertion of intact full-length cDNAs into the plant genome, a reduction in the number of undesirable mutations and an increased transgene copy number [4]. For these reasons, many efforts have been made to improve *Agrobacterium*-mediated transformation efficiency [14,15]. The *CarNAC6* gene from chickpea encodes a 308-amino-acid, plant-specific regulatory transcriptional factor belonging to the NAC (*NAM*, *ATAF1,2*, and *CUC2*) domain protein family that increases plant response reactions to biotic and abiotic stresses and plays an important role in plant development. Homologues of the *CarNAC6* gene have been identified in *Arabidopsis* and rice [16,17,18,19,20]. In this study, we transformed the poplar hybrid clone "Nanlin895" (*Populus deltoides* × *P. euramericana* "Nanlin895") with the *PBI121:CarNAC6* binary vector to improve *Agrobacterium*-mediated transformation efficiency. In addition, we attempted to optimize MS culture medium supplemented with *N*-6-benzyladenine (6-BA) and thidiazuron (TDZ) hormones to minimize lateral shoot development. Other factors affecting transformation efficiency evaluated in this study included the preculture of wounded explants, *Agrobacterium* infective suspension concentration,*Agrobacterium* infection duration, *Agrobacterium* infective suspension pH, acetosyringone (AS) concentration, cold treatment of an *Agrobacterium* infective suspension, cocultivation incubation temperature and duration and sucrose concentration in cocultivation medium. All of these factors were treated independently in related experiments independent of the age of the explants. In this study, the transformation efficiency is expressed as the percentage of independently transformed explants relative to the total number of explants.

RESULTS AND DISCUSSION

Optimization of Poplar Regeneration

An overview of the steps used in the *Agrobacterium*-mediated transformation of *Populus* species is presented in Figure 1. The poplar hybrid clone "Nanlin895" (*Populus deltoides* × *P. euramericana* "Nanlin895") was cultured in medium containing various concentrations of 6-BA and TDZ (Table 1). Little or no regeneration occurs in the absence of 6-BA. Medium containing 0.5 mg/L 6-BA was used to achieve the maximum regeneration rate (85% to 90%). Increases in non-specific poplar regeneration resulted in higher numbers of lateral shoots relative to the numbers of main shoots. TDZ, an inducer of shoot regeneration, was used to increase the specific regeneration of main shoots. The proportion of main shoots to lateral shoots was increased by TDZ concentrations of 0.002 and 0.004 mg/L, but was decreased by concentrations of 0.008 and 1.0 mg/L. The optimal regeneration medium contained 0.5 mg/L 6-BA and 0.004 mg/L TDZ to maximize the frequency of main shoot regeneration. The rates of shoot elongation (Figure 2d) and root regeneration (Figure 2e) were enhanced by up to 100% when using the optimized medium. As shown in Figure 2a and Figure 3, the preculture of explants for two days caused higher transformation efficiencies than preculture for one and three days. Therefore, explants precultured for two days were used subsequently for transformation with *Agrobacterium.*

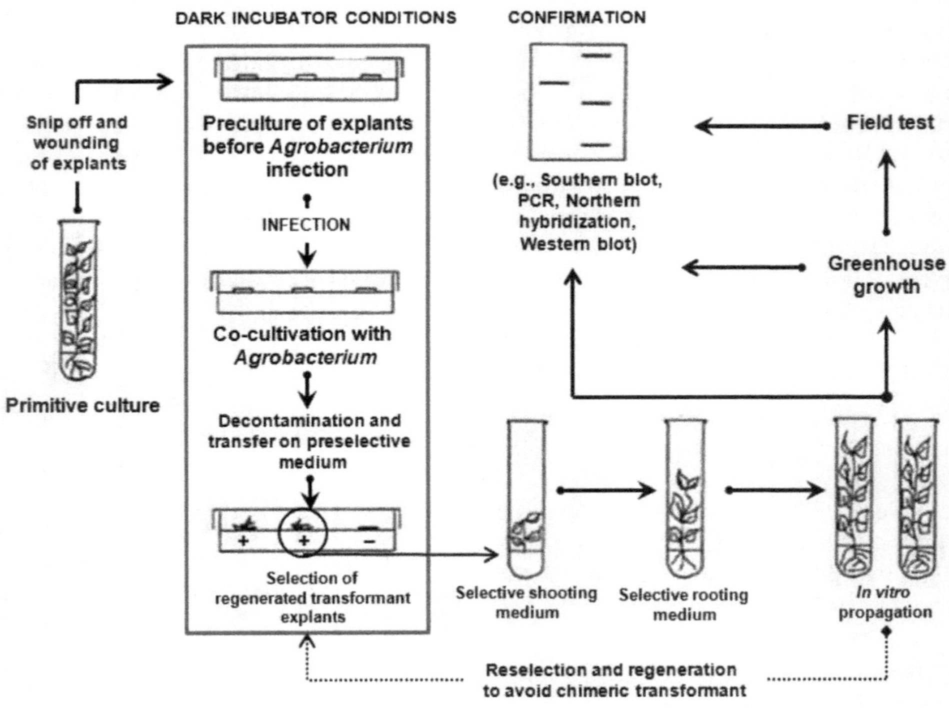

Figure 1: An overview the steps for *Agrobacterium*-mediated transformation in*Populus* species.

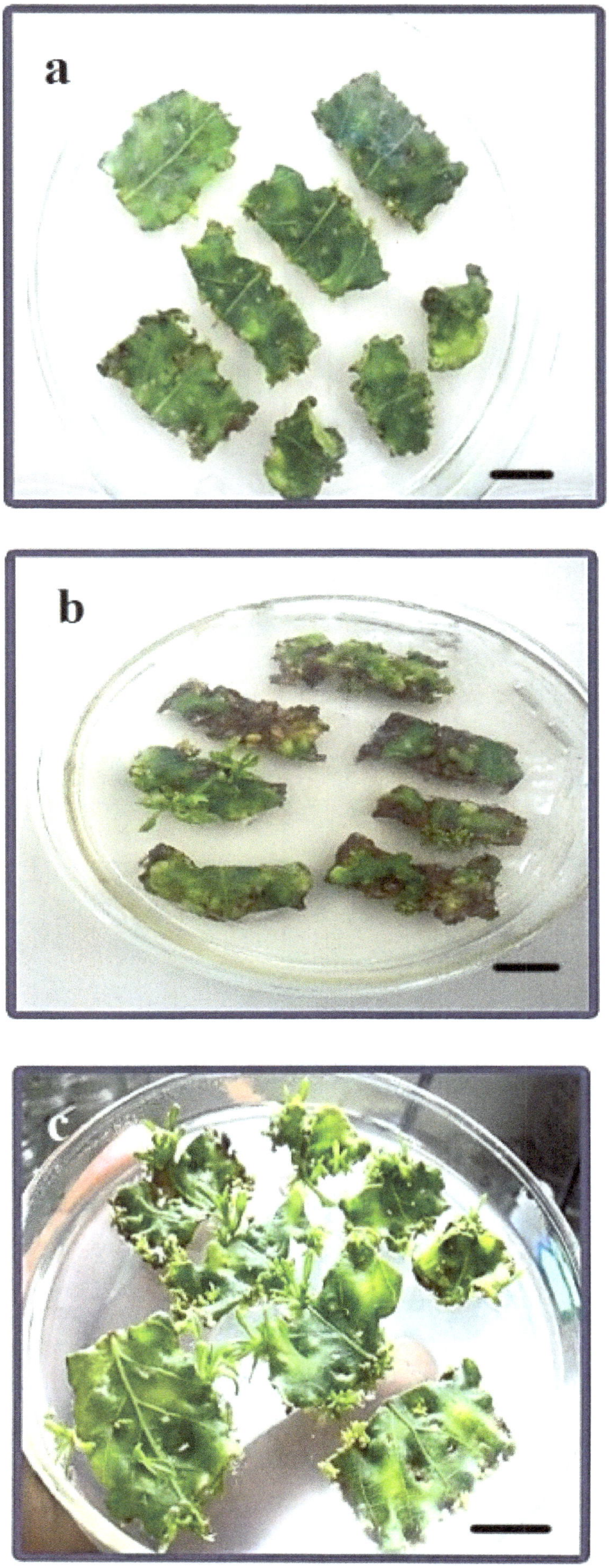
a
b
c

Figure 2: (a) Cocultivation of poplar leaf discs with *Agrobacterium* on MS medium; (b) Regenerated putative poplar transformants on MS selection medium that were generated without the optimization of transformation efficiency; (c) Regenerated putative poplar transformants .

Table 1: The effect of *N*-6-benzyladenine (6-BA) and thidiazuron (TDZ) on the main shoot and total shoot regeneration (%)

Medium Number	6-BA (mg/L)	TDZ (mg/L)	Mean Number of Main Shoots per Medium (Number of Main Shoots in Each Medium/Number of Explants in Each Medium)	Total Shoot Regeneration in Each Medium (%)
1	1.00	0.000	3.10	89.0
2	1.00	0.002	3.50	93.0
3	1.00	0.004	4.80	94.0
4	1.00	0.008	3.30	89.0
5	1.00	0.010	2.90	97.0
6	0.80	0.000	38.8	85.0
7	0.80	0.002	43.0	82.0
8	0.80	0.004	45.8	78.0
9	0.80	0.008	40.7	80.0
10	0.80	0.010	30.8	79.0
11	0.50	0.000	56.8	80.0
12	0.50	0.002	65.2	75.0

13	0.50	0.004	69.5	73.0
14	0.50	0.008	60.5	76.0
15	0.50	0.010	44.4	70.0
16	0.20	0.000	8.90	37.0
17	0.20	0.002	11.0	29.0
18	0.20	0.004	14.0	35.0
19	0.20	0.008	9.00	32.0
20	0.20	0.010	6.13	26.0
21	0.00	0.000	0.30	6.00
22	0.00	0.002	0.50	9.00
23	0.00	0.004	0.80	13.0
24	0.00	0.008	0.50	10.0
25	0.00	0.010	0.10	8.00

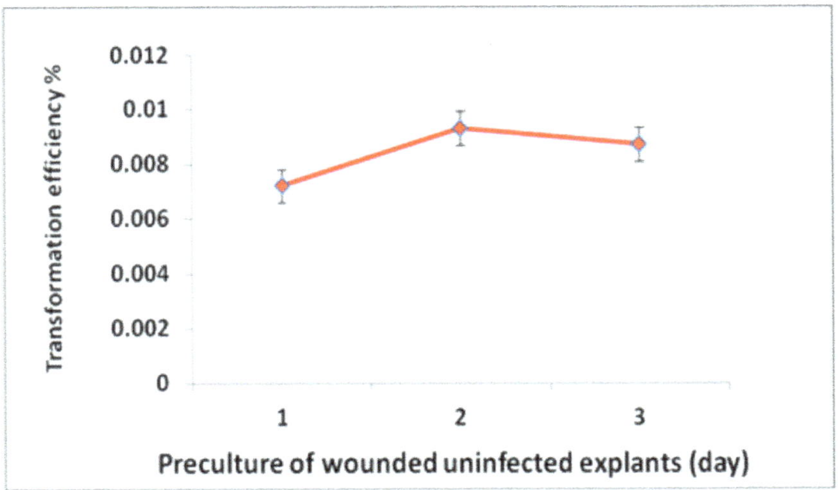

Figure 3: Precultured, wounded poplar explants before induction by*Agrobacterium*.

Improvement of Agrobacterium-Mediated Transformation Efficiency

The highest transformation efficiency was obtained from precultured explants immersed in an *Agrobacterium*infective suspension with an OD_{600} of 0.7. The use of *Agrobacterium* infective suspensions with higher densities resulted in decreased transformation efficiencies, due to increased infection (Figure 4a). We examined the effect of the duration of immersion in *Agrobacterium* infective suspensions on the transformation efficiency. The highest transformation efficiencies were obtained from immersion for 120 and 150 min and were not significantly different (Figure 4b). In addition, we examined the effect of the pH of the*Agrobacterium* infective suspension on the transformation efficiency. Low pHs increased the efficiency, but higher pHs decreased the transformation efficiency, due to decreased *Agrobacterium* growth (Figure 4c). Increasing the AS concentration in the *Agrobacterium* infective suspension to as high as 200 µM improved the transformation efficiency, probably by increasing the ability of the *Agrobacterium* to detect wounded plant cells. The transformation efficiencies obtained with the different AS concentrations were significantly different (Figure 4d).

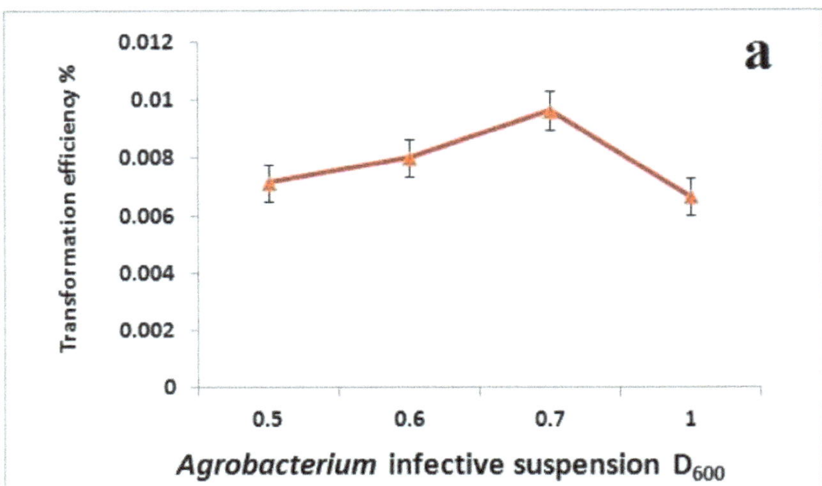

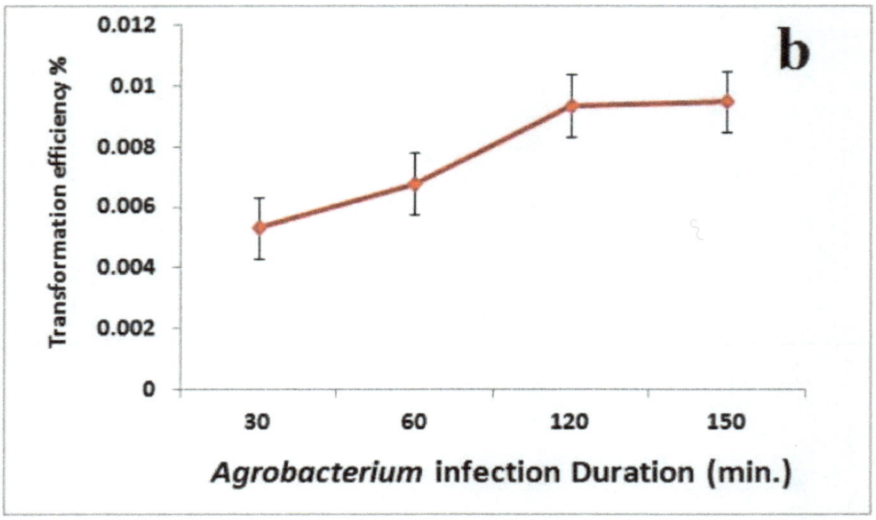

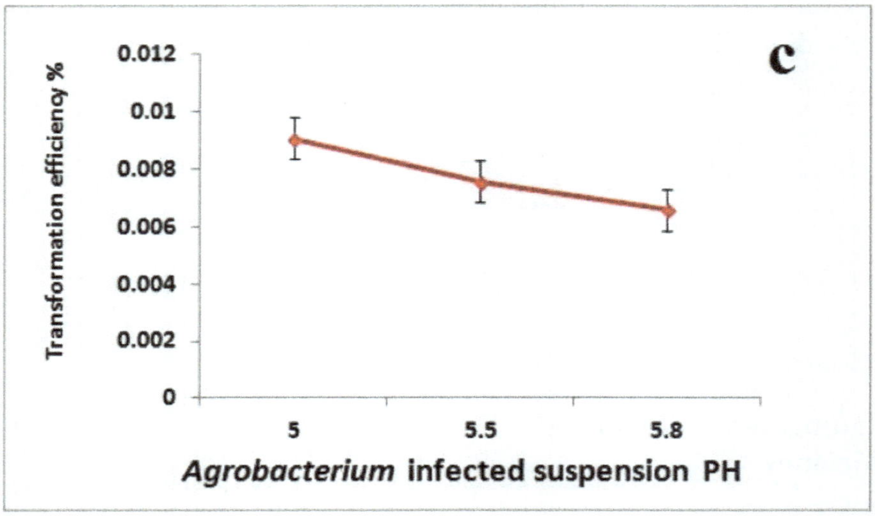

Figure 4: The effects of various factors on *Agrobacterium*-mediated transformation efficiency: (a) *Agrobacterium* concentration; (b)*Agrobacterium* infection duration; (c) *Agrobacterium* suspension pH; (d) acetosyringone (AS) concentration; (e) cold pre-treatment; ...

We developed a cold pre-treatment of the *Agrobacterium* infective suspension inspired by the freeze-thaw method [21]. The transformation efficiency was increased approximately 7% by pre-treatment of the suspension at 4 °C for at least 24 h (Figure 4e). The effect of the cold pre-treatment of the *Agrobacterium*infective suspension on the transformation efficiency was clearly demonstrated by the differences between cultures on plates for which the suspension had not received (Figure 2b) or had received (Figure 2c) the cold pre-treatment. The temperature at which the *Agrobacterium* infective suspension and the explants were co-cultivated also affected the transformation efficiency. Cocultivation at 28 °C produced the highest transformation efficiency, with significant differences in efficiency between the temperatures tested (Figure 4f). The duration of the cocultivation of the *Agrobacterium* infective suspension and explants was an important factor in the transformation efficiency. Cocultivation for up to 72 h yielded the highest transformation efficiency, with significant differences between incubation temperatures (Figure 4g)

Sucrose is a required carbon source for the growth of*Agrobacterium*. The effects of various sucrose concentrations on transformation efficiency during the cocultivation of the *Agrobacterium* infective suspension and explants were significantly different, with the highest transformation efficiency obtained using a sucrose concentration of 30 g/L (Figure 4h).

Southern and Northern Blot Analyses

Total genomic DNA isolated from transgenic plants was digested with *Sal*I, which has no restriction site in the T-DNA, and with *Eco*RI, for which one restriction site is located downstream of the *NOS* (nopaline synthase) terminator. Southern blot analysis showed that 1–2 transgene copies were present in each of the three lines generated in procedures that did not show improved transformation efficiency (Figure 5a) and that 3–4 transgene copies were present in each of the three lines generated using procedures resulting in improved transformation efficiency (Figure 5b).

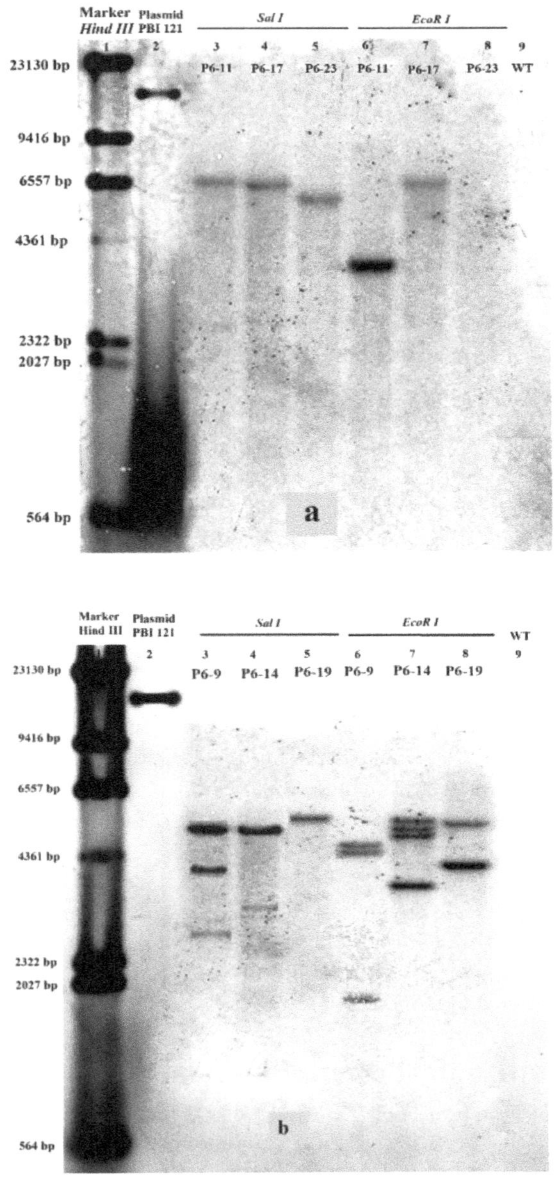

Figure 5: (a) Southern blot analysis of transgenic poplar lines generated without improved transformation efficiency. Genomic DNA extracted from three transgenic poplar lines was digested separately with *Sal*I and*Eco*RI and blotted onto a nylon membrane. The blots ...

The detection of mRNAs of 0.5–0.8 kbp by northern blot analysis confirmed that the *CarNAC6* transgene was expressed in three transgenic poplar lines generated without improved transformation efficiency (Figure 6a) and in three transgenic poplar lines generated with improved transformation efficiency (Figure 6b). The transgene expression levels in the two line types were clearly different. Supplementation of media with hormones has been investigated to determine optimal conditions for shoot regeneration with a minimum of lateral shoots and for root regeneration. High concentrations of TDZ can result in the regeneration of a high percentage of adventitious shoots [22]. Ferreira *et al.* [2] reported that the regeneration of adventitious shoots from organogenic nodules of *P. euphratica* leaf explants was achieved using a range of α-naphthaleneacetic acid (NAA) and 6-BA concentrations. Various concentrations of indoleacetic acid (IAA) were also shown to affect the frequency of regeneration from poplar explants [23]. Confalonieri *et al.* [1] reported positive effects of BA and NAA on shoot regeneration in *P. nigra*. Various media are required for shoot and root regeneration of different genotypes and are supplemented with various levels of indole-3-butyric acid (IBA) to induce rooting from aspen leaves and with zeatin to induce shooting [24]. The use of specific concentrations of shoot-inducing hormones, such as 6-BA, can increase the mean number of shoots formed per explant following regeneration, but typically causes decreased numbers of main shoots following regeneration [4]. Recently, the number of reports of successful *Agrobacterium*-mediated transformation of various plant species has increased markedly [22]. Developments in tissue culture methods and improved transformation techniques have led to the use of suitable *Agrobacterium* strains [25]. The development of techniques that increase transformation efficiency or the pathogenicity of *Agrobacterium* has been the subject of many studies. The use of appropriate AS concentrations during cocultivation has been shown to increase transformation efficiencies. Moreover, the effects of the *Agrobacterium* concentration and infection time on transformation efficiency have been demonstrated by Han *et al.* [4].

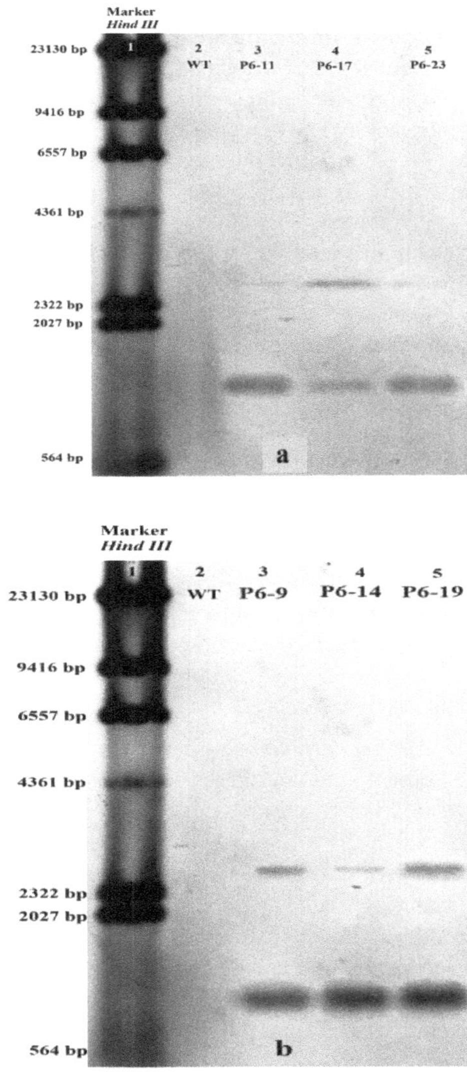

Figure 6: (a) Northern blot analysis of transgenic poplar lines generated without improved transformation efficiency. Total RNA (ribonucleic acid) was separated on a formaldehyde gel and blotted onto a nylon membrane after staining with ethidium bromide.

EXPERIMENTAL SECTION

Plant Materials and Agrobacterium Strain

Wild-type (WT) poplar leaves were collected from young *P. deltoides* × *P. euramericana* "Nanlin895" seedlings to propagate 25 lines. The binary vector, *PBI121:CARNAC6*, was generated at the State Key Laboratory of Crop Genetics and Germplasm Enhancement of the National Center for Soybean Improvement at Nanjing Agricultural University, China [26]. The pBI121 plasmid was digested with *Xba*I and *Bam*HI restriction enzymes, and a 924-bp nucleotide DNA fragment containing a full-length cDNA of the chickpea *CARNAC6* gene was inserted. *CARNAC6* expression was driven by the CaMV *35S* promoter. The plasmid was introduced into the *Agrobacterium tumefaciens* strain, EHA105, using the freeze-thaw method [21].

Optimization of Regeneration Medium for Poplar

Murashige and Skoog (MS) medium [27] was supplemented with 6-benzylaminopurine (6-BA) at five concentrations (1.0, 0.8, 0.5, 0.2 and 0 mg/L) in combination with thidiazuron (TDZ) at five concentrations (0, 0.002, 0.004, 0.008 and 0.01 mg/L) to determine the optimal combination for minimal lateral shoot development. Immature leaves were excised as 1-cm^2 discs and cultured on MS medium supplemented with the two hormones. The medium was refreshed at 2-week intervals. Regenerated shoots of 2 cm in length were cut and rooted on sterile half-strength MS medium containing 25 g/L sucrose and 6 g/L agar adjusted to pH 5.8 with no added hormones. After 6–8 weeks, the roots were examined, and leaves were collected from well-developed plantlets for gene transformation.

Agrobacterium-Mediated Transformation

Young leaves were cut into 0.5 × 0.5-cm sections and precultured on half-strength MS medium supplemented with 6-BA (0.5 mg/L) and TDZ (0.004 mg/L) for 1, 2 or 3 days. This medium was tested on 37–40 lines for each variable, with four replicates. *Agrobacterium* containing the binary vector, *PBI121:CARNAC6*, was inoculated into Luria–Bertani (LB) medium supplemented with 50 mg/L rifampicin (Rif) and 50 mg/L kanamycin (Kan) and cultured in an incubator without light at 28 °C for 72 h [28].

A single *Agrobacterium* colony was inoculated into 5 mL of LB medium containing 50 mg/L Rif and 50 mg/L Kan and incubated at 28 °C for 24 h with constant agitation (220 rpm). An additional 50 mL of LB were added and growth was continued overnight under the same conditions until an OD_{600} of 2.0 was reached. The culture was centrifuged at 3000 rpm for 10 min. The pellet was diluted in liquid MS medium with 5% sucrose to prepare an *Agrobacterium*-infective suspension. All *Agrobacterium* experiments were carried out independently using 20–25 precultured explants in triplicate.

The precultured explants were immersed in various concentrations of *Agrobacterium*-infective suspensions (OD_{600} values of 0.5, 0.6, 0.7 or 1.0). The precultured explants were also cold treated independently at 0, 4 or 10 °C. Other parameters tested independently for their effects on transformation efficiency included the duration of the cocultivation of the *Agrobacterium*-infective suspension with explants (30, 60, 120 or 150 min at 28 °C with constant agitation), the AS concentration (0, 50, 100, 150 or 200 μM) and the pH of the*Agrobacterium* infection suspension (pH 5.0, 5.5 or 5.8). Wounded leaf discs were dried on sterile paper to remove surface water. The infected explants were co-cultivated in a dark incubator at various temperatures (20, 24, 26, 28 or 30 °C) and for different durations (24, 48 or 72 h) on semi-liquid MS induction medium containing 0.5 mg/L 6-BA, 0.004 mg/L TDZ, 3 g/L agar, various sucrose concentrations (20, 25 or 30 g/L) and 200 μM AS to induce the expression of the *Agrobacterium Vir* genes [29]. The parameters used for control transformations were the preculture of explants for 1 day, an *Agrobacterium* culture OD_{600} of 0.5, 10 °C cold pre-treatment, 30 min of *Agrobacterium* cocultivation, 100 μM AS, an *Agrobacterium*-infective suspension at pH 5.8, cocultivation in darkness at 28 °C for 48 h and 25 g/L sucrose in MS induction medium. Putative transformed explants were transferred to MS selection medium supplemented with 0.5 mg/L 6-BA, 0.004 mg/L TDZ, 6 g/L agar, 25 g/L sucrose, 400 mg/L cefotaxime (Cef) and 50 mg/L kanamycin (Kan) at pH 5.8 and cultured under 16-h/8-h light/dark conditions at 23 ± 1 °C in a phytotron. Selected regenerated shoots were separated and transferred to MS shoot elongation medium supplemented with 0.25 mg/L 6-BA, 0.002 mg/L TDZ, 6 mg/L agar, 25 mg/L sucrose and 300 mg/L Cef. Shoots of approximately 2 cm in length were then transferred onto half-strength MS rooting medium supplemented with 6 mg/L agar, 25 mg/L sucrose and 300 mg/L Cef with no added hormones. Plants that were well developed were transferred to a greenhouse for experiments. Transformation efficiencies were calculated using the following equation: (number of transformed explants/total number of explants) × 100.

Southern and Northern Blot Analyses

For southern blot analysis, 10 μg of genomic DNA were extracted using the CTAB (cetyl trimethylammonium bromide) method [30,31] from WT plants, three transgenic poplar lines generated using procedures with no improvements to the transformation efficiency (P6-11, P6-17 and P6-23) and three transgenic poplar lines using procedures generated with improvements to transformation efficiency (P6-9, P6-14 and P6-19). The genomic DNA was digested with *Sal*I and *Eco*RI at 37 °C for 4 h and separated overnight on a 0.8% agarose gel at 15 V. The digested genomic DNA was transferred to a Hybond N+ nylon membrane (Amersham Biosciences BV, Eindhoven, The Netherlands) and detected using DIG (digoxigenin) reagent, according to the manufacturer's instructions (catalog number 11745832910; Roche, Basel, Switzerland). The primers, F5'-TACCGAGGATATTACACTACCAGGA-3' and R 5'-AGTCCAGTTTTGCAGCCAAG-3', were used to generate an 868-bp PCR product to use as a probe on the southern blot. Northern blot analysis was performed using 7 μg of total RNA from young leaves of WT plants and three transgenic poplar lines extracted using TRIzol reagent (Tiangen Biotech, Beijing, China), according to the manufacturer's instructions. Total RNA was separated on a 1.2% agarose formaldehyde gel, transferred to a Hybond N+ nylon membrane (Amersham Biosciences BV), and processed for northern hybridization. Random primers labelled with digoxigenin-11-dUTP using DIG-High Prime (Roche) were used to generate probes from the *CarNAC6* cDNA (GenBank Accession Number FJ477887.1). The primers, F 5›-CTGGCAAAGGTTGGAGAAAG-3› and R 5›-CGTGTTGGTCTTGTTGTTGT-3›, were used to generate a 433-bp PCR product to use as a probe for northern blotting.

Statistical Analysis

All data analyses were performed using ANOVA (analysis of variance) with a mean separation calculated using Duncan's test with SPSS version 16 (SPSS Inc., Chicago, IL, USA) and Excel 2013 software (Microsoft, Redmond, WA, USA). Confidence intervals showing no overlap of the mean values with an error value of 0.05 were considered to indicate statistically significant differences.

CONCLUSIONS

In this study, we demonstrated that the regeneration efficiency of hybrid poplar "Nanlin895" (*Populus deltoides* × *P. euramericana* "Nanlin895") could be optimized to yield an 85%–90% regeneration rate. We successfully transformed poplar with the *CarNAC6* biotic and abiotic stress resistance gene using *Agrobacterium*-mediated transformation. We showed that the transformation efficiency was increased by using a two-day preculture of wounded regeneration explants, an *Agrobacterium*-infective suspension with an OD_{600} of 0.7, cocultivation of the *Agrobacterium* infection suspension with explants for 120 min, an *Agrobacterium*-infective suspension pH of 5.0, an AS concentration of 200 μM, a 4 °C cold treatment of the *Agrobacterium*-infective suspension, a cocultivation incubation temperature of 28 °C, a cocultivation duration of 72 h and a cocultivation medium sucrose concentration of 30 g/L.

ACKNOWLEDGMENTS

We thank Hao Ma (Nanjing Agricultural University, China) for providing the *CarNAC6* plasmid and Mohammad Reza Mofid (Isfahan University of Medical Sciences, Iran) and Mahmoud Otroshi (Agricultural Biotechnology Research Institute of Iran, Isfahan) for their technical assistance. We also thank Ehsan Soleymaninejadian for his collaboration in this article. This work was supported by grants from the Forestry Public Benefit Research Program (No. 201304102), the International Science and Technology Cooperation Program of China (2014DFG32440), the National Science Foundation of China (No. 31170561), the Priority Academic Program Development of Jiangsu Higher Education Institutions and the Program for Innovative Research Team at the University of Educational Department and Jiangsu Province, China.

AUTHOR CONTRIBUTIONS

Ali Movahedi designed and performed the entire experiments. Qiang Zhuge supervised the research. Jiaxin Zhang and Rasoul Amirian were responsible for data analyses. Ali Movahedi and Qiang Zhuge wrote and edited the manuscript. All authors have read and approved the final manuscript.

REFERENCES

1. Confalonieri M., Balestrazzi A., Bisoffi S. Genetic transformation of *Populus nigra* by *Agrobacterium tumefaciens*. Plant Cell Rep. 1994;13:256–261.
2. Ferreira S., Batista D., Serrazina S., Pais M.S. Morphogenesis induction and organogenic nodule differentiation in *Populus euphratica* Oliv. leaf explants. Plant Cell Tissue Organ Cult. 2009;96:35–43. doi: 10.1007/s11240-008-9457-y.
3. Gozukirmizi N., Bajrovic K., Ipekqi Z. Genotype differences in direct plant regeneration from stem explants of *Populus tremula* in Turkey. J. For. Res. 1998;3:123–126. doi: 10.1007/BF02760313.
4. Han X., Ma S., Kong X., Takano T., Liu S. Efficient *Agrobacterium*-mediated transformation of hybrid poplar *Populus davidiana* Dode × *Populus bollena* Lauche. Int. J. Mol. Sci. 2013;14:2515–2528. doi: 10.3390/ijms14022515.
5. Kang B., Osburn L., Kopsell D., Tuskan G.A., Cheng Z.M. Micropropagation of *Populus trichocarpa* "Nisqually-1": The genotype deriving the *Populus* reference genome. Plant Cell Tissue Organ Cult. 2009;99:251–257. doi: 10.1007/s11240-009-9596-9.
6. Yadav R., Arora P., Kumar D., Katyal D., Dilbaghi N., Chaudhury A. High frequency direct plant regeneration from leaf, internode, and root segments of Eastern Cottonwood (*Populus deltoides*) Plant Biotechnol. Rep. 2009;3:175–182. doi: 10.1007/s11816-009-0088-5.
7. Horsch R.B., Fry J.E., Hoffmann N.L., Eichholtz D., Rogers S.G., Fraley R.T. A simple and general method for transferring genes into plants. Science. 1985;227:1229–1231.
8. Han K.H., Meilan R., Ma C., Strauss S.H. An *Agrobacterium tumefaciens* transformation protocol effective on a variety of cottonwood hybrids (genus *Populus*) Plant Cell Rep. 2000;19:315–320.
9. Meilan R., Ma C. Poplar (*Populus* ssp.) In: Wang K., editor. Methods in Molecular Biology, Agrobacterium Protocols. 2nd ed. Humana Press; Totowa, NJ, USA: 2006. pp. 143–151.
10. Ma C.P., Strauss S.H., Meilan R. *Agrobacterium*-mediated transformation of the genome sequenced poplar clone, Nisqually-1 (*Populus trichocarpa*) Plant Mol. Biol. Rep. 2004;22:1–9.
11. Yevtushenko D.P., Santosh M. Efficient *Agrobacterium*-mediated transformation of commercial hybrid poplar *Populus nigra* L. × *P. maximowiczii* A. Henry. Plant Cell Rep. 2010;29:211–221. doi: 10.1007/s00299-009-0806-z.
12. Husaini A.M. Pre- and post-agroinfection strategies for efficient leaf disk transformation and regeneration of transgenic strawberry plants. Plant Cell Rep. 2010;29:97–110. doi: 10.1007/s00299-009-0801-4.
13. Bartlett J.G., Alves S.C., Smedley M., Snape J.W., Harwood W.A. High-throughput *Agrobacterium*-mediated barley transformation. Plant Methods. 2008;4:1–12. doi: 10.1186/1746-4811-4-1.
14. Kondou Y., Higuchi M., Takahashi S. Systematic approaches to using the FOX hunting system to identify useful rice genes. Plant J. 2009;57:883–894. doi: 10.1111/j.1365-313X.2008.03733.x.

15. Nakamura H., Hakata M., Amano K., Miyao A. A genome-wide gain-of-function analysis of rice genes using the FOX-Hunting system. Plant Mol. Biol. 2007;65:357–371. doi: 10.1007/s11103-007-9243-y.

16. Fowler S., Thomashow M.F. Arabidopsis transcriptome profiling indicates that multiple regulatory pathways are activated during cold acclimation in addition to the CBF cold response pathway. Plant Cell.2002;14:1675–1690. doi: 10.1105/tpc.003483.

17. Nakashima K., Ito Y., Yamaguchi-Shinozaki K. Transcriptional regulatory networks in response to abiotic stresses in Arabidopsis and grasses. Plant Physiol. 2009;149:88–95. doi: 10.1104/pp.108.129791.

18. Nakashima K., Takasaki H., Mizoi J., Shinozaki K., Yamaguchi-Shinozaki K. NAC transcription factors in plant abiotic stress responses. Biochim. Biophys. Acta (BBA) Gene Regul. Mech. 2012;1819:97–103. doi: 10.1016/j.bbagrm.2011.10.005.

19. Rabbani M.A., Maruyama K., Abe H., Khan M.A., Katsura K., Ito Y., Yoshiwara K., Seki M., Shinozaki K., Yamaguchi-Shinozaki K. Monitoring expression profiles of rice genes under cold, drought, and high-salinity stresses and abscisic acid application using cDNA microarray and RNA gel-blot analyses. Plant Physiol. 2003;133:1755–1767. doi: 10.1104/pp.103.025742.

20. Yamaguchi-Shinozaki K., Shinozaki K. Transcriptional regulatory networks in cellular responses and tolerance to dehydration and cold stresses. Annu. Rev. Plant Biol. 2006;57:781–803. doi: 10.1146/annurev.arplant.57.032905.105444.

21. Holsters M., Waele D., Depicker A., Messens E., van Montagu M., Schell J. Transfection and transformation of *Agrobacterium tumefaciens*. Mol. Gen. Genet. 1978;163:181–187. doi: 10.1007/BF00267408.

22. Tzfira T., Frankmen L., Vaidya M., Citovsky V. Site-specific integration of *Agrobacterium* T-DNA via double-stranded intermediates. Plant Physiol. 2003;133:1011–1023. doi: 10.1104/pp.103.032128.

23. Thakur D.K., Srivastava D.K. High-efficiency plant regeneration from leaf explants of male Himalayan poplar (*Populus ciliata* Wall) In Vitro Cell Dev. Biol. Plant. 2006;42:144–147. doi: 10.1079/IVP2005743.

24. Zhan X.C., Kawai S., Katayama Y., Morohoshi N. A new approach based on the leaf disc method for*Agrobacterium* mediated transformation and regeneration of aspen. Plant Sci. 1997;123:105–112. doi: 10.1016/S0168-9452(96)04561-X.

25. Herrera-Estrella L., Simpson J., Martinez-Trujillo M. Transgenic plants: An historical perspective.Methods Mol. Biol. 2005;286:3–32.

26. Hui P., Cheng H.Y., Chen C., Yu X.W., Yang J.N., Gao W.R., Shi Q.H., Zhang H., Li J.G., Ma H. A NAC transcription factor gene of chickpea (*Cicer arietinum*), *CarNAC3*, is involved in drought stress response and various developmental processes. J. Plant Physiol. 2009;166:1934–1945. doi: 10.1016/j.jplph.2009.05.013.

27. Murashige T., Skoog F. A revised medium for rapid growth and bioassays with tobacco tissue culture. Plant Physiol. 1962;15:473–497. doi: 10.1111/j.1399-3054.1962.tb08052.x.

28. Horsch R.B., Fry J., Hoffmann N., Neidermeyer J., Rogers S.G., Fraley R.T. Leaf disc transformation. In: Gelvin S.B., Schilperoort R.A., editors. Plant Molecular Biology Manual. Kluwer Academic Publishers; Dordrecht, The Netherlands: 1988. pp. 1–9.

29. Satchel S.E., Messens E., van Montagu M., Zambryski P. Identification of the signal molecules produced by wounded plant cells that activate T-DNA transfer in *Agrobacterium tumefaciens*. Nature. 1985;318:624–629. doi: 10.1038/318624a0.

30. Doyle J.J., Dickson E.E. Preservation of plant samples for DNA restriction endonuclease analysis. Taxon. 1987;36:715–722. doi: 10.2307/1221122.

31. Porebski L., Bailey L., Baum B. Modification of a CTAB DNA extraction protocol for plants containing high polysaccharide and polyphenol components. Plant Mol. Biol. Rep. 1997;15:8–15. doi: 10.1007/BF02772108.

CHAPTER 2

COMBINING PROTEIN AND STRAIN ENGINEERING FOR THE PRODUCTION OF GLYCO-ENGINEERED HORSERADISH PEROXIDASE C1A IN PICHIA PASTORIS

Simona Capone[1], Lejla Ćorajević[1], Günther Bonifert[1], Patrick Murth[1], Daniel Maresch[2], Friedrich Altmann[2], Christoph Herwig[1] and Oliver Spadiut[1]

[1] Institute of Chemical Engineering, Research Area Biochemical Engineering, Vienna University of Technology, Gumpendorfer Strasse 1a, 1060 Vienna, Austria

[2] Department of Chemistry, University of Natural Resources and Life Sciences, 1180 Vienna, Austria

ABSTRACT

Horseradish peroxidase (HRP), conjugated to antibodies and lectins, is widely used in medical diagnostics. Since recombinant production of the enzyme is difficult, HRP isolated from plant is used for these applications. Production in the yeast Pichia pastoris (P. pastoris), the most promising recombinant production platform to date, causes hyperglycosylation of HRP, which in turn complicates conjugation to antibodies and lectins. In this study we combined protein and strain engineering to obtain an active and stable HRP variant with reduced surface glycosylation. We combined four mutations, each being beneficial for either catalytic activity or thermal stability, and expressed this enzyme variant as well as the unmutated wildtype enzyme in both a P. pastoris benchmark strain and a strain where the native α-1,6-mannosyltransferase (OCH1) was knocked out. Considering productivity in the bioreactor as well as enzyme activity and thermal stability, the mutated HRP variant produced in the P. pastoris benchmark strain turned out to be interesting for medical diagnostics. This variant shows considerable catalytic activity and thermal stability and is less glycosylated, which might allow more controlled and efficient conjugation to antibodies and lectins.

INTRODUCTION

The methylotrophic yeast *Pichia pastoris* is a widely used host organism for recombinant protein production. It can grow on cheap and defined media to high cell densities, is robust against stressful conditions and able to perform post-translational modifications, like glycosylation [1,2,3,4]. However, glycosylation in *Pichia pastoris* (*P. pastoris*) is characterized by a severe drawback: native glycosyltransferases recognize the amino acid motif Asn-X-Ser/Thr and link many glycan moieties to the asparagine residues of recombinant proteins yielding a hyperglycosylated product (e.g., [5,6,7]). Hyperglycosylation causes many problems as it alters the physicochemical properties of the product, hampers downstream processing [5,8] and prevents medical applications. One of the main disadvantages of yeast-derived glycosylation lies in its heterogeneous nature making subsequent conjugation to antibodies and lectins, a prerequisite for applications in medical diagnostics, extremely difficult. Two different approaches can be applied to reduce yeast-derived hyperglycosylation, namely (1) protein engineering and (2) strain engineering. The yeast *P. pastoris* is a prominent example for such efforts. In several studies detailed strain engineering approaches describing humanization of this yeast were reported (Table 1; e.g., [7,8,9,10,11,12,13,14]).

Citation: Simona Capone, Lejla Ćorajević, Günther Bonifert, Patrick Murth, Daniel Maresch, Friedrich Altmann, Christoph Herwig, and Oliver Spadiut, Combining Protein and Strain Engineering for the Production of Glyco-Engineered Horseradish Peroxidase C1A in Pichia pastoris, doi: 10.3390/ijms161023127.

Table 1: Selected studies focusing on the humanization of *N*-glycosylation in *P. pastoris*

Goal of the Study	Citation
Introduction of α-1,2-Mns and GntI, *OCH1* inactivation via a knock-in plasmid	[9]
Introduction of an UDP-GlcNAc transporter, α-1,2-MnsIA, MnsII, GntI, GntII in a *Δoch1::URA3* strain	[10]
Introduction of sialic acid biosynthesis pathway to produce sialylated glycoproteins	[11]
OCH1 knock-out and introduction of glycosidases and glycosyltransferases to produce terminally galactosylated glycoproteins	[14]

Mns, mannosidase; Gnt, β-*N*-acetylglucosaminyltransferase; UDP-GlcNAc, uridine diphosphate-*N*-acetylglucosamine; *OCH1*, outer chain elongation gene.

We used *P. pastoris* for the recombinant production of the heme-containing plant enzyme horseradish peroxidase (HRP), an enzyme widely used in medical diagnostics (e.g., [5,15]). We chose this yeast as expression host because Morawski *et al.* [16,17] had shown *P. pastoris* to produce recombinant proteins with significant shorter surface glycans than *Saccharomyces cerevisiae* (*S. cerevisiae*). We investigated various aspects of the expression of HRP in *P. pastoris*: different HRP isoenzymes were produced [18,19], production strategies were developed and optimized [20,21,22,23], media supplementation and strain engineering for increased heme-incorporation were analyzed [24] and the methanol utilization pathway of *P. pastoris* was manipulated for higher productivity [25].

In more recent studies we applied both protein and strain engineering with the goal of producing more homogenously glycosylated HRP variants in *P. pastoris*. However, we did not use fully humanized yeast strains, which are proprietary, but produced HRP in a *P. pastoris* strain with a deleted *och1* gene (Δoch1 strain; [26]). This gene codes for an α-1,6-mannosyltransferase responsible for triggering the uncontrolled addition of mannose residues to the recombinant protein. Although HRP produced in the Δoch1 strain was more homogeneously glycosylated with more than 70% of the enzyme being of the Man_8 glycosylation type, the Δoch1 strain showed a growth impaired phenotype and was hard to cultivate [26]. In a subsequent study, we analyzed the effect of the process parameters temperature, pH and dissolved oxygen concentration on strain physiology and productivity [6]. We found that the space-time-yield (STY) of the glyco-engineered strain was eight-fold lower compared to an unmodified *P. pastoris* benchmark strain, making HRP production in the Δoch1 strain unattractive.

Thus, we glyco-engineered the enzyme HRP by mutating each of the eight *N*-glycosylation sites to produce in a benchmark strain but still get HRP with reduced surface glycosylation [27]. We found different effects of the single mutations on enzyme activity and stability and demonstrated that mutation of the respective *N*-glycosylation site in fact resulted in the absence of glycans there [27]. Although the combination of all eight mutations gave a non-glycosylated enzyme, both catalytic activity and thermal stability were dramatically reduced leading to the hypothesis that some of the *N*-glycosylation sites must not be mutated to obtain active and stable HRP.

In the present study we combined both (1) protein engineering and (2) strain engineering to obtain a stable and active HRP variant with reduced and more homogeneous surface glycosylation. We combined the four mutations we had identified as being beneficial for either catalytic activity or stability in a previous study (Table 2; [27]).

Table 2: Biochemical characteristics of the unmodified horseradish peroxidase (HRP) wildtype enzyme (wt) and four different HRP variants [27]

Enzyme	ABTS			H_2O_2			Thermal Half-Life Time
Variant	K_m (mM)	v_{max} (U/mg)	v_{max}/K_m (U/mg/mM)	K_m (mM)	v_{max} (U/mg)	v_{max}/K_m (U/mg/mM)	$\tau_{½}$ (min)
wt	1.60	44.2	27.7	0.003	16.3	5433	20.6
N13D	2.90	47.2	16.3	0.005	14.7	3066	28.9
N57S	2.98	113	38.1	0.004	23.7	5378	38.5
N255D	1.72	51.5	29.9	0.005	21.6	4506	11.6
N268D	1.89	32.5	17.2	0.003	10.6	3642	61.9

ABTS, 2-2-azinobis-(3-ethylbenzothiazoline-6-sulfonic acid); H_2O_2, hydrogen peroxide.

We produced the enzyme variant, called 4/8 HRP, in both an unmodified *P. pastoris* benchmark strain and a Δoch1 strain. Summarizing, enzymes produced in the Δoch1 strain were characterized by lower catalytic activity and stability. However, 4/8 HRP produced in the *P. pastoris* benchmark strain (enzyme $wt^{4/8\ HRP}$) showed considerable catalytic activity and increased thermal stability. In combination with its significantly reduced surface glycosylation, which might allow more controlled conjugation to antibodies and lectins, this variant might be useful for applications in medical diagnostics in the future.

RESULTS AND DISCUSSION

Strain Physiology

We cultivated the four recombinant *P. pastoris* strains in dynamic batch cultivations with methanol pulses at different temperatures and analyzed the effect of temperature on strain physiology. A typical cultivation is exemplarily shown for strain $wt^{4/8\ HRP}$ in Figure 1.

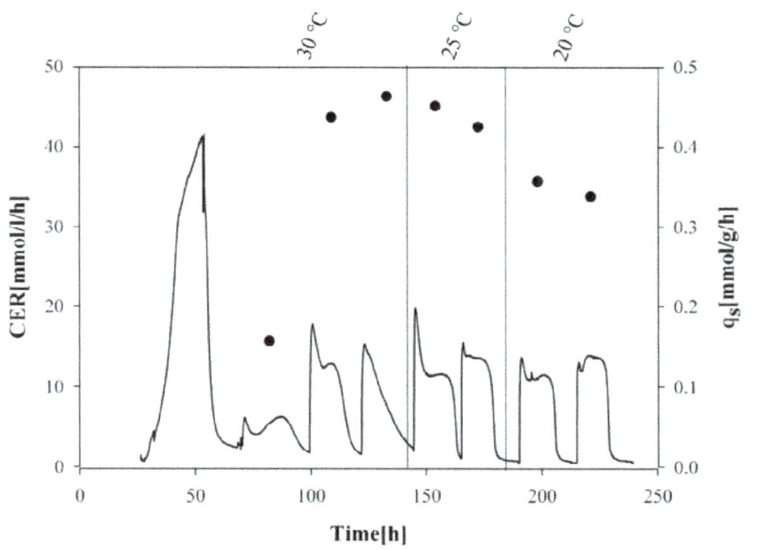

Figure 1: Schematic overview of the dynamic batch cultivation of strain $wt^{4/8\ HRP}$ with methanol pulses at different temperatures. Black continuous line, carbon dioxide evolution rate (CER); black dots, specific methanol uptake rate ($q_{s\ MeOH}$).

The most important strain physiological parameters are summarized in Table 3. Closing C-balances for the benchmark strains underline data validity. Similar to our previous studies, we observed that the Δoch1 strains lost metabolic activity over time and were affected by cell lysis [26]. Thus, C-balances did not close.

Table 3: Physiological parameters of strains $wt^{wt\ HRP}$, benchmark strain expressing the unmutated HRP enzyme; $wt^{4/8\ HRP}$, benchmark strain expressing the mutated 4/8 HRP variant; $OCH1^{wt\ HRP}$, deleted OCH1 gene strain expressing the unmutated HRP enzyme and $OCH1^{4/8\ HRP}$ delete OCH1 gene strain expressing the mutated 4/8 HRP variant were determined in dynamic batch cultivations

Strain	$\mu_{max\ gly}$ (h^{-1})	$\Delta_{time\ adapt}$ (h)	$q_{s\ adapt}$ (mmol/g/h)	$q_{s\ MeOH\ 20\ °C}$ (mmol/g/h)	$q_{s\ MeOH\ 25\ °C}$ (mmol/g/h)	$q_{s\ MeOH\ 30\ °C}$ (mmol/g/h)	C-Balance
$wt^{wt\ HRP}$	0.271	11.1	0.272	0.931	1.190	1.32	0.96
$wt^{4/8\ HRP}$	0.200	16.0	0.158	0.354	0.438	0.450	0.97
$OCH1^{wt\ HRP}$	0.199	4.5	0.370	0.891	0.780	0.632	const. Decreasing
$OCH1^{4/8\ HRP}$	0.182	3.8	0.400	1.02	0.955	0.800	const. Decreasing

When comparing the recombinant benchmark strains, great differences in specific methanol uptake rates ($q_{s\ MeOH}$) were identified. While for both strains $q_{s\ MeOH}$ increased with increasing temperature, a correlation we also had observed before [6], the strain producing the 4/8 HRP variant showed a three-fold lower $q_{s\ MeOH}$ compared to the strain producing the unmutated enzyme at the respective temperature. Apparently, production of the glyco-engineered 4/8 HRP caused a physiological burden for the yeast, decelerating metabolism and thus methanol uptake.

When comparing the recombinant Δoch1 strains, we observed that $q_{s\ MeOH}$ for both strains decreased with increasing temperature, a phenomenon we also had described before [6]. However, $q_{s\ MeOH}$ of both Δoch1 strains were comparable at each temperature indicating that the mutated product did not cause any physiological burden. We speculate that the glyco-engineered 4/8 HRP variant might cause problems in the *P. pastoris* benchmark strain during secretion as it might get stuck in the cell wall. On the contrary, the same enzyme variant can be secreted without any problems in the Δoch1 strain, which has a completely altered cell wall structure [24]. However, this remains to be elucidated in detail.

Protein Purification

After harvest, we purified the different enzyme variants by a 1-step hydrophobic charge interaction chromatography (HCIC) purification strategy [8,19]. As shown in Table 4, the majority of each HRP variant was found in the respective flow-through fraction, where purification factors (PF) between 1.34 and 1.53 were achieved. However, enzyme $OCH1^{4/8\ HRP}$ showed a different result, as nearly 20% of the enzyme was retained on the resin. The eluate fraction showed a higher PF compared to the flow-through fraction (Table 4). We ascribe this phenomenon to the fact that this HRP variant was the least glycosylated one, as the four remaining glycosylation sites mainly carried Man_8 instead of longer Man chains (respective detailed analyses had been performed in our previous study [26]). Apparently, the reduced glycosylation of enzyme OCH1$^{4/8\ HRP}$ allows physico-chemical interactions of the enzyme variant and the resin. One might speculate that the amount of enzyme variant able to interact with the resin should be even higher. We think that the rather stressful manner of cultivation with pulses and temperature shifts might have caused very heterogeneous glycosylation on the four remaining *N*-glycosylation sites and thus resulted in the still rather low fraction of only around 20% enzyme interacting with the resin. Detailed analysis of surface glycosylation by mass spectrometry, as we have done previously [24], could shed light on this speculation. However, we did not perform this analysis in this study, since variant OCH1$^{4/8\ HRP}$ did not turn out to be interesting for further applications due to low catalytic activity (see below).

Table 4: Results of the hydrophobic charged interaction chromatography (HCIC) purification and Reinheitszahl (RZ) measurements for the four different HRP enzyme variants

	HCIC				Concentrated Fraction
Enzyme Variant	**R% Total**	**R% FT**	**PF FT**	**Specific Activity (U/mg)**	**RZ (A_{404}/A_{280})**
wt$^{wt\ HRP}$	87.2	87.2	1.49	273.3	0.50
wt$^{4/8\ HRP}$	97.4	95.6	1.34	36.4	0.51
OCH1$^{wt\ HRP}$	99.8	98.7	1.53	59.8	0.63
OCH1$^{4/8\ HRP}$	93.1	76.1	1.47	11.5	0.19
Variant	**R% Total**	**R% Eluate**	**PF Eluate**	**Specific Activity (U/mg)**	**RZ (A_{404}/A_{280})**
OCH1$^{4/8\ HRP}$	93.1	17.5	2.43	19.1	0.31

After HCIC, flow-through fractions showing highest PFs were pooled and concentrated to around 1.5 mL. To assess purity of the different enzyme preparations Reinheitszahl values (RZ; A_{404}/A_{280}) were determined (Table 4). Highly pure HRP preparations are known to have RZ values of more than 3.0 [16]. Although we did not get pure enzyme preparations by the one-step HCIC purification in this study, the RZ values of the different enzyme variants were in the same range. We also analyzed the different enzyme variants on SDS-PAGE gels, to identify potential differences in apparent size (Figure 2). Furthermore, interesting protein bands were excised and analyzed by mass spectrometry (respective bands are indicated in black boxes in Figure 2).

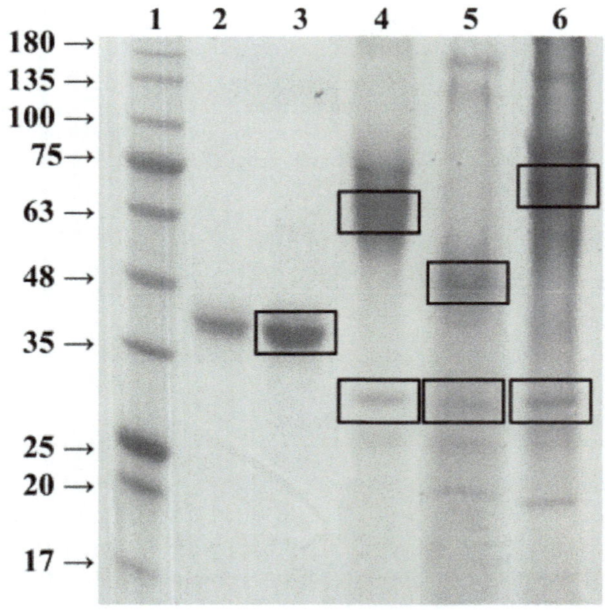

Figure 2: Sodium dodecyl sulfate polyacrylamide gel electrophoresis (SDS-PAGE) of different horseradish peroxidase (HRP) variants. Black boxes indicate protein bands analyzed by mass spectrometry. Lane 1, BLUeye Prestained Protein Ladder; lanes 2 and 3, HRP from plant in two different concentrations (4 µg protein and 8 µg protein, respectively); lane 4, wt$^{wt\ HRP}$, unmutated HRP enzyme expressed in the benchmark strain; lane 5, wt$^{4/8\ HRP}$, mutated 4/8 HRP variant expressed in the benchmark strain; lane 6, OCH1$^{4/8\ HRP}$, mutated 4/8 HRP variant expressed in the deleted OCH1 gene strain.

The enzyme preparation from plant showed a rather distinct band at an apparent size of around 40 kDa on the SDS gel (lanes 2 and 3 in Figure 2), whereas the recombinant enzyme wt$^{wt\ HRP}$ showed a smear at an apparent size of around 65 kDa (lane 4 in Figure 2). Both proteins were identified as HRP by mass spectrometry (Table 5). The apparent size difference of around 20 kDa and the smeary appearance of wt$^{wt\ HRP}$ result from the heterogeneous yeast-derived glycosylation [17,20,21]. The second prominent protein band in lane 4 at an apparent size of around 30 kDa was identified as a glucosidase from *P. pastoris* (Table 5).

Table 5: Identification of prominent protein bands by mass spectrometry

Lane	Apparent Size [kDa]	Rank	Peptides	Scores	Protein	Accession
3	45	1	11	608.9	Peroxidase C1A Organism species (OS)=Armoracia rusticana	PER1A_ARMRU
4	65	1	6	386.9	1,3-β-glucanosyltransferase OS=Komagataella pastoris	Q0QCW1_PICPA
		2	9	368.0	Peroxidase C1A OS=Armoracia rusticana	PER1A_ARMRU
	30	1	9	411.5	Glucan 1,3-β-glucosidase OS=Komagataella pastoris	F2QPL8_PICP7
5	50	1	9	454.5	Alpha-1-antichymotrypsin 2 OS=Sus scrofa	Q9GMA6_PIG
		2	9	386.6	Keratin, type II cytoskeletal 1 OS=Homo sapiens	K2C1_HUMAN
		3	9	317.4	Peroxidase C1A OS=Armoracia rusticana	PER1A_ARMRU
	30	1	9	411.5	Glucan 1,3-β-glucosidase OS=Komagataella pastoris	F2QPL8_PICP7
6	70	1	20	1136.6	Primary-amine oxidase OS=Komagataella pastoris	F2QTE6_PICP7
		2	10	685.6	1,3-β-glucanosyltransferase OS=Komagataella pastoris	Q0QCW1_PICPA
		3	9	548.8	1,3-β-glucanosyltransferase OS=Komagataella pastoris	F2QQJ2_PICP7
	30	1	9	411.5	Glucan 1,3-β-glucosidase OS=Komagataella pastoris	F2QPL8_PICP7

As shown in Figure 2, the preparation of enzyme wt$^{4/8\ HRP}$ showed a different protein pattern on the SDS gel, as the band at an apparent size of 65 kDa disappeared, whereas a prominent band at an apparent size of around 50 kDa appeared. In fact, this protein was identified as HRP (Table 5). Apparently, the mutation of four of the eight *N*-glycosylation sites resulted in the absence of glycans there and thus a size reduction of around 15 kDa. This nicely underlines the feasibility of reducing the vast and heterogeneous glycosylation of recombinant proteins from yeast by protein design. The second prominent band at an apparent size of around 30 kDa was again identified as a glucosidase (Table 5).

Lane 6 in Figure 2 shows the protein bands of enzyme preparation OCH1$^{4/8\ HRP}$. Again, we observed a different protein pattern. The prominent band at an apparent size of around 70 kDa was identified as an oxidase and the band at an apparent size of around 30 kDa again as a glucosidase (Table 5). We expected to see a protein band at an apparent size of around 45 kDa representing enzyme OCH1$^{4/8\ HRP}$. However, no respective band was visible on the SDS gel. We ascribe this absence to the extremely low protein production in the Δoch1 strain [6,26]. Furthermore, cell lysis during bioreactor cultivation resulted in a rather high impurity pattern, which is also demonstrated by the low RZ value for this enzyme variant (Table 4). However, we still measured enzymatic activity for OCH1$^{4/8\ HRP}$ and thus included this enzyme variant in the comparative biochemical characterization.

Biochemical Enzyme Characterisation

After purification, we biochemically characterized the different HRP variants. The kinetic parameters for ABTS and H_2O_2 are summarized in Table 6. We also included plant HRP for comparison. Introducing the four mutations N13D, N57S, N255D and N268D into HRP did not affect the affinity towards ABTS and H_2O_2 as K_m values of enzymes wt$^{wt\ HRP}$ and wt$^{4/8\ HRP}$ were comparable. However, the catalytic efficiency was reduced seven-fold for ABTS and six-fold for H_2O_2, respectively. Apparently, the introduced mutations did not only reduce surface glycosylation (Figure 2) but also affected the active site and reduced catalytic activity.

Table 6: Kinetic constants of four different HRP enzyme variants and plant HRP

Enzyme	ABTS			H_2O_2		
Variant	K_m (mM)	v_{max} (U/mg)	v_{max}/K_m (U/mg/mM)	K_m (mM)	v_{max} (U/mg)	v_{max}/K_m (U/mg/mM)
wt$^{wt\ HRP}$	1.50	152.9	101.9	0.009	55.2	6133
wt$^{4/8\ HRP}$	0.99	13.6	13.7	0.015	15.5	1030
OCH1$^{wt\ HRP}$	1.56	26.8	17.2	0.008	10.4	1300
OCH1$^{4/8\ HRP}$	1.34	1.28	0.96	0.016	1.29	80.7
plant HRP	1.75	567.2	324.8	0.033	377.8	11,589

ABTS, 2-2-azinobis-(3-ethylbenzothiazoline-6-sulfonic acid); H_2O_2, hydrogen peroxide.

When we produced the HRP variants in the Δoch1 strain, we found that reduced surface glycosylation, caused by the absence of the native enzyme OCH1 [26], did not alter substrate affinity either. In fact, K_m values of all four HRP variants for both ABTS and H_2O_2 were comparable. However, reducing surface glycosylation affected catalytic efficiency. Enzyme OCH1$^{wt\ HRP}$ showed a nearly six-fold reduced v_{max}/K_m compared to enzyme wt$^{wt\ HRP}$. Apparently, the benefit of having a more homogeneously glycosylated enzyme produced in the Δoch1 strain comes at cost of catalytic activity. Finally, we characterized the mutated 4/8 HRP produced in the Δoch1 strain (enzyme OCH1$^{4/8\ HRP}$). This enzyme is the least glycosylated one in this study, only providing four out of eight *N*-glycosylation sites, which are mainly occupied by

Man_8-glycan structures [26]. However, compared to enzyme wt$^{wt HRP}$ the catalytic efficiency for ABTS was reduced 119-fold and for H_2O_2 76-fold, respectively. Although, enzyme OCH1$^{4/8 HRP}$ is basically fit for medical applications, as it misses the heterogeneous yeast-derived glycan structures, the highly reduced catalytic activity as well as low productivity in the bioreactor will most likely prevent future applications.

Compared to the commercially available HRP preparation from plant, all recombinant HRP variants showed comparable K_m values but reduced catalytic activity. This might be a result from incomplete heme incorporation or different surface glycosylation. However, the plant preparation contains a mixture of different HRP isoenzymes with varying surface glycosylation, which has to be isolated from its natural source in a rather cumbersome way. Furthermore, seasonal variation in the isoenzyme content and thus the variable production scenario describe only some of the disadvantages of plant HRP. Thus, even though recombinant HRP variants show lower catalytic activity, they are still interesting for industry, as they only describe a single isoenzyme that can be produced in a predictable manner in the controlled environment of a bioreactor. Furthermore, less glycosylated enzymes might allow more controlled and efficient conjugation to antibodies and lectins, which outweighs reduced catalytic activity.

Thermal Stability

Since we had determined mutations N13D, N57S and N268D to positively affect thermal stability of HRP before (Table 2; [27]), we also analyzed thermal half-life times of the four HRP variants at 60 °C. We again included plant HRP as standard and normalized all enzyme preparations to a protein concentration of 0.1 mg/mL before incubation to guarantee comparability. In Table 7 the respective results are summarized.

Table 7: Thermal half-life times of four different HRP variants and the HRP plant preparation at 60 °C. All enzymes were normalized to a protein concentration of 0.1 mg/mL before incubation

Enzyme Variant	Protein Concentration (mg/mL)	$\tau_{1/2\ 60\ °C}$ (min)
wt$^{wt HRP}$		31.5
wt$^{4/8 HRP}$		173.2
OCH1$^{wt HRP}$	0.1	3.3
OCH1$^{4/8 HRP}$		19.3
plant HRP		53.3

As shown in Table 7, the introduction of the four mutations into HRP caused a significant increase in thermal stability. Enzyme wt$^{4/8 HRP}$ had a 5.5-fold higher $\tau_{1/2\ 60\ °C}$ than enzyme wt$^{wt HRP}$ and was even 3.3-fold more stable than the enzyme preparation from plant. The same HRP variants produced in the Δoch1 strain showed a 10-fold reduced thermal stability compared to their respective counterparts from the benchmark strain. Reducing glycosylation to Man_8 structures apparently affected stability to a greater extent than completely removing four *N*-glycosylation sites. In fact, mutating the four *N*-glycosylation sites significantly increased stability instead of decreasing it. It is remarkable that enzyme OCH1$^{4/8 HRP}$ actually showed a similar thermal half-life time as enzyme wt$^{wt HRP}$ (Table 7).

Considering both, kinetic parameters and thermal stability, enzyme wt$^{4/8 HRP}$ might be interesting for future applications in medical diagnostics. This enzyme variant can be efficiently produced in bioreactor cultivations, is less glycosylated, which might allow more controlled and efficient conjugation to antibodies and lectins, still shows considerable catalytic activity and a 5.5-fold higher thermal stability compared to enzyme wt$^{wt HRP}$. In fact, higher stability and reduced glycosylation could compensate for reduced catalytic activity.

EXPERIMENTAL SECTION

Chemicals

2,2′-Azino-bis(3-ethylbenzthiazoline-6-sulfonic acid) diammonium salt (ABTS), D(+)-biotin and hemin were purchased from Sigma-Aldrich. Difco™ yeast nitrogen base w/o amino acids (YNB), Bacto™ tryptone and Bacto™ yeast extract were obtained from Becton Dickinson (Franklin Lakes, NJ, USA). Zeocin™ was obtained from InvivoGen (San Diego, CA, USA). Other chemicals were obtained from Carl Roth (Karlsruhe, Germany).

Strain Generation

All strains in this study are based on the *P. pastoris* wildtype strain CBS7435. Generation of the Δoch1 strain and the single HRP variants are described in detail in our previous studies [26,27]. In short, the four Asn, representing glycosylation sites of HRP, were mutated by site directed mutagenesis and subsequent digestion with *Dpn*I. The mutagenic PCR was performed as: 98 °C for 30 s; then 10 cycles of 98 °C for 10 s, 57 °C for 20 s, 72 °C for 1 min-10 cycles of 98 °C for 10 s, 60 °C for 20 s, 72 °C for 1 min-10 cycles of 98 °C for 10 s, 63 °C for 20 s, 72 °C for 1 min; with a final incubation at 72 °C for 10 min. Each reaction contained 1× HF buffer (Fermentas, Waltham, MA, USA), 0.01 μg of plasmid DNA, 2.5 U *Phusion* DNA polymerase (Fermentas), 10 μM of each dNTP and 5 pmol of each primer in a total volume of 50 μL.

After PCR, the methylated template DNA was degraded by digestion with 10 U of *Dpn*I at 37 °C for at least three hours. The remaining PCR products were purified using the QIAquick PCR purification kit (QIAGEN, Hilden, Germany) and 5 μL of each purified PCR product were transformed into electro-competent *E. coli*TOP10 F′ cells. The successful introduction of the desired mutation and the absence of further mutations were confirmed by DNA sequencing (Microsynth). Transformation of approximately 2 μg *Swa*I-linearized pPpT4_S plasmid DNA harbouring the respective mutated HRP gene into the *P. pastoris* benchmark and the Δoch1 strain was done by electroporation. Stable transformants were generated via homologous recombination between the linearized plasmid DNA and genomic yeast DNA. Selection of successfully transformed clones was performed on Yeast Extract Peptone Dextrose medium (YPD; 10 g/L yeast extract, 20 g/L peptone, 20 g/L glucose, 20 g/L agar) supplemented with 100 mg/L Zeocin. In total, we generated and compared four recombinant *P. pastoris* strains producing four different HRP enzyme variants (Table 8).

Table 8: Recombinant *P. pastoris* strains and HRP enzyme variants in this study

P. pastoris Chassis Strain	HRP Variant	Name of Recombinant *P. pastoris* Strain and Enzyme
benchmark strain	wt HRP	$wt^{wt\ HRP}$
	4/8 HRP	$wt^{4/8\ HRP}$
Δoch1 strain	wt HRP	$OCH1^{wt\ HRP}$
	4/8 HRP	$OCH1^{4/8\ HRP}$

Bioreactor Cultivations

Preculture

Precultures were grown in 100 mL YNB_Zeo medium (0.1 M potassium phosphate buffer, pH 6.0; 3.4 g/L YNB w/o amino acids and ammonium sulfate, 10 g/L $(NH_4)_2SO_4$, 400 mg/L biotin, 20 g/L glucose, 100 mg/L Zeocin) in 1000 mL baffled shake flasks at 30 °C and 230 rpm for 24 h.

Dynamic Batch Cultivation

For dynamic batch cultivations, 3582.6 mL 2-fold concentrated basal salt medium (BSM; 26.7 mL/L 85% phosphoric acid, 1.17 g/L $CaSO_4 \cdot 2H_2O$, 18.2 g/L K_2SO_4, 14.9 g/L $MgSO_4 \cdot 7H_2O$, 4.13 g/L KOH, 0.3 mL/L Antifoam Struktol J650, 60 g/L glycerol) were sterilized in a 5 L working volume glass bioreactor (Infors, Molndal, Sweden). After sterilization, 4.35 mL PTM1 per litre medium were added (*i.e.*, 17.4 mL for 4 litres) and pH was set to 5.0 with concentrated ammonia solution. Precultures were aseptically transferred to the respective vessel (10% inoculation volume) and the batch phase on glycerol was carried out at 30 °C with the stirrer fixed at 1300 rpm. Aeration with compressed dry air was set to 1 volume per volume per minute (vvm). Dissolved oxygen (dO_2) was measured with a sterilizable VisiFerm DO 225 probe (Hamilton, ON, Canada). The pH was measured with a sterilizable electrode (Mettler Toledo, Greifensee, Switzerland) and maintained constant at pH 5.0. Reactor weight was continuously recorded by a precision balance (Sartorius, Göttingen, Germany). Batch and methanol adaptation were performed at 30 °C. After the complete consumption of glycerol, indicated by an increase of dO_2 and a drop in off-gas activity, the first methanol pulse at a final concentration of 0.5% volume per volume (*v*/*v*) was conducted with methanol supplemented with 12 mL/L PTM1. Following pulses were performed with 1% methanol/PTM1 (*v*/*v*) at different temperatures. At 30, 25 and 20 °C at least two pulses were applied, respectively. For each pulse, two samples were taken to determine the concentrations of substrate and product, as well as dry cell weight to calculate specific rates and yields.

Analysis of Growth and Expression Parameters

Dry cell weight (DCW) was determined by centrifugation of 2 × 5 mL fermentation broth (4800 rpm, 10 min, 4 °C), washing the pellet with 5 mL water and subsequent drying to a constant weight at 105 °C. Optical density at 600 nm (OD_{600}) was determined in a spectrophotometer (Thermo Scientific, Waltham, MA, USA). Activity of HRP in the cell-free supernatant was determined using a previously described assay in a CuBiAn-XC enzymatic robot (Innovatis, Bielefeld, Germany) [27]. Protein concentration was determined using a Bradford protein assay kit (Thermo Scientific). All growth and protein expression parameters were determined in duplicates.

Analysis of Substrate Concentration

Concentrations of methanol were determined in cell free samples by HPLC (Agilent Technologies, Santa Clara, CA, USA) equipped with an ion-exchange column (Supelcogel C-610H, Sigma-Aldrich, St. Louis, MO, USA) and a refractive index detector (Agilent Technologies). The mobile phase was 0.1% H_3PO_4 with a constant flow rate of 0.5 mL/min and the system was run isocratic at 30 °C. Calibration was done by measuring standard points in the range from 0.1 to 10 g/L methanol. Measurements of biomass, product and substrate concentration were executed in duplicates.

Calculation of Strain Physiological Parameters

The relevant parameters to physiologically characterize the recombinant yeast strains were: carbon dioxide evolution rate (CER; mmol/L/h), adaptation time to methanol (h), specific methanol uptake rate (q_{MeOH}; mmol/g/h), biomass yield ($Y_{X/MeOH}$; C-mol/C-mol), carbon dioxide yield ($Y_{CO2/MeOH}$; C-mol/C-mol) and C-balance. Details concerning the calculation of these parameters have been published before [20,21,27].

Protein Purification

Cell-free cultivation broth was diafiltrated with a Centramate 500S TFF system (Pall) using a 10 kDa MWCO membrane. The buffer was HCIC-A (20 mM NaOAc, 1.0 M NaCl, pH 6.0) and the protein solution was concentrated to a final volume of 100–120 mL. The HCIC resin MEP HyperCel™ was obtained from Pall (Port Washington, NY, USA) and HCIC was performed in flow-through mode [8,19]: a column containing 25 mL of MEP HyperCel™ resin was equilibrated with at least four column volumes (CV) buffer HCIC-A. The HRP solution in HCIC-A was loaded onto the column that was then washed with at least five CV of HCIC-A at a flow rate of 20 cm/h. Then a step elution to 100% buffer HCIC-B (50 mM Tris, pH 8.0) was performed. After elution, the column was washed with five CV 0.8 M NaOH before it was stored in EtOH 20%, 1.0 M NaCl. During all the steps, fractions were collected and analyzed for protein content and catalytic activity. Fractions showing HRP activity were pooled and concentrated to around 1.5 mL using Amicon Ultra-15 Centrifugal Filter Units with 10 kDa MWCO (Merck Millipore, Darmstadt, Germany).

The efficiency of the purification was evaluated by determining the purification factor (PF; Equation (1)) and the recovery yield of HRP activity in percentage (R%; Equation (2)). The suffixes "pre" and "post" indicate the respective values before and after the HCIC step.

$$PF = \frac{\text{specific activity}_{post}}{\text{specific activity}_{pre}} \quad (1)$$

$$R\% = \frac{\text{volumetric activity}_{post} \cdot \text{volume}_{post}}{\text{volumetric activity}_{pre} \cdot \text{volume}_{pre}} \quad (2)$$

Additionally, Reinheitszahl values (RZ; A_{404}/A_{280}) of the concentrated enzyme preparations were measured. Absorbance at 280 and 404 nm were determined in a quartz cuvette in a spectrophotometer (UV-1601; Shimadzu, Long Beach, CA, USA).

SDS-PAGE

Apparent sizes and purities of produced HRP variants were followed by SDS-PAGE according to the Laemmli protocol [28]. Electrophoresis was done using an Amersham ECL Gel 8%–16% gel (GE Healthcare, Buckinghamshire, UK) in 1× Tris-glycine buffer. Before loading, the gel had to be pre-run at 160 V for 12 min. Protein separation was performed at 140 V for about 2 h. BLUeye Prestained Protein Ladder (GeneDirex, Taoyuan County, Taiwan) was used as protein mass standard. Gels were stained with Coomassie Blue sensitive stain.

Protein Identification and Peptide Analysis Using LC-ESI-MS

The relevant protein bands were cut out from the SDS gel and digested in gel. S-alkylation with iodoacetamide and digestion with sequencing grade modified trypsin (Promega, Madison, WI, USA) were performed. The peptide mixture was analysed using a Dionex Ultimate 3000 system directly linked to an ion trap instrument (amaZon speed ETD, Bruker, Billerica, NA, USA) equipped with the standard ESI source in the positive ion, DDA mode (=switching to MSMS mode for eluting peaks). MS-scans were recorded (range: 400–1600 *m/z*; icc target: 100,000; max. accu time: 200 ms) and the 12 highest peaks were selected for fragmentation. Instrument calibration was performed using ESIcalibration mixture (Agilent). For separation of the peptides a Thermo BioBasic C18 separation column (5 μm particle size, 150 × 0.360 mm) was used. A gradient from 97% solvent A and 3% solvent B (Solvent A: 65 mM ammonium formiate buffer, B: 100% ACCN) to 32% B in 45 min was applied, followed by a 15 min gradient from 32% B to 75% B, at a flow rate of 6 μL/min. The analysis files were converted using Data Analysis 4.0 (Bruker) to XML files, which are suitable to perform MS/MS ion searches with MASCOT (embedded in ProteinScape 3.0, Bruker) for protein identification. Only proteins identified with at least 2 peptides with a protein score higher than 80 were accepted. For the searches the SwissProt database was used. Peptide MS/MS data were evaluated against the target sequence using X! Tandem (www.thegpm.org/tandem/) with the following settings: reversed sequences no; check parent ions for charges 1, 2 and 3 yes; models found with peptide $\log_e$ lower −1 and proteins $\log_e$ lower −1; residue modifications: oxidation M, W and deamidation N, Q; isotope error was considered; fragment type was set to monoisotopic; refinement was used with standard parameters; fragment mass error of 0.3 Da and ± 50 ppm parent mass error; fragment types b and y ions; maximum parent ion charge of 3; missed cleavage sites allowed was set to 1; semi-cleavage yes.

Biochemical Enzyme Characterization

We determined the basic kinetic parameters K_m and v_{max} for the substrates ABTS and H_2O_2 for the different HRP variants in a spectrophotometer UV-1601 (Shimadzu). The reaction mixture with a final volume of 1.0 mL contained 20 μL of HRP variant, 50 mM potassium phosphate buffer, pH 6.5, and either varying concentrations of ABTS (0.01–10 mM) and a saturating concentration of H_2O_2 of 1.0 mM or varying concentrations of H_2O_2 (0.001–1.0 mM) and a saturating concentration of ABTS of 10.0 mM, respectively. The increase in absorption was followed at 420 nm at 30 °C for 180 s. Absorption curves were recorded with a software program (UVPC Optional Kinetics software; Shimadzu). The maximum reaction rate (v_{max}) and the Michaelis constant (K_m) were calculated with the software Sigma Plot (Systat Software Inc., Chicago, IL, USA).

The thermal stability of individual HRP variants was tested at 60 °C. The residual activity with ABTS was measured after 1, 5, 10, 15, 30, 45, 60, 90 and 120 min of incubation at 60 °C in a PCR thermoblock. Protein concentrations were normalized to 0.1 mg/mL in Bis-Tris buffer (50 mM Bis-Tris, pH 6.5) to guarantee comparability. Residual activities were plotted *versus* the incubation time and the half-life times of thermal inactivation at 60 °C (τ½) were calculated using Equation (3):

$$\tau_{1/2} = \frac{\ln 2}{k_{in}} \quad (3)$$

k_{in}: rate of inactivation (slope of the logarithmic residual activity).

CONCLUSIONS

In this study we combined protein and strain engineering to obtain an active and stable recombinant HRP variant with reduced surface glycosylation. We combined four mutations, which individually had shown beneficial effects on either catalytic activity or thermal stability before, and expressed this enzyme variant as well as the unmutated wildtype enzyme in both an unmodified *P. pastoris* benchmark strain and a Δoch1 strain. Our results can be summarized as:

- Production of the 4/8 HRP variant caused a physiological burden for the *P. pastoris* benchmark strain slowing down its metabolism. In contrast, production of the same enzyme variant did not affect physiology of the Δoch1 strain. While $q_{s\ MeOH}$ increased with increasing temperature for the benchmark strain, the opposite was true for the Δoch1 strain. Based on strain physiological parameters identified in dynamic batch experiments, fed-batch strategies can be easily designed for future production of these enzyme variants.
- Reduced and missing surface glycosylation did not affect substrate affinity of HRP, but significantly reduced catalytic activity.
- Introducing the four mutations into HRP significantly boosted thermal stability. In fact, the 4/8 HRP variant produced in the *P. pastoris* benchmark strain even showed a 3.3-fold increased thermal half-life time compared to the HRP preparation from plant.
- Considering both, enzyme activity and thermal stability, the 4/8 HRP variant produced in the *P. pastoris* benchmark strain might be interesting for future applications in medical diagnostics. This enzyme variant can be efficiently produced in the bioreactor, shows considerable catalytic activity and high thermal stability and is less glycosylated, which might allow more controlled and efficient conjugation to antibodies and lectins.

In this study we show that enzymes can be modified not only by mutation, but that a combination of both strain and protein engineering is a useful way to obtain enzyme variants tailored to specific needs.

ACKNOWLEDGMENTS

The authors are grateful to the Austrian Science Fund FWF: project P24861-B19 for financial support. We further thank Alexander Pekarsky for assistance in the laboratory.

AUTHOR CONTRIBUTIONS

Simona Capone, Lejla Ćorajević, Günther Bonifert and Patrick Murth performed experiments. Oliver Spadiut was responsible for experimental design. Daniel Maresch and Friedrich Altmann conducted mass spectrometry analyses. Oliver Spadiut and Christoph Herwig initiated, planned and supervised the study. Simona Capone and Oliver Spadiut wrote the manuscript. All authors read and approved the manuscript.

REFERENCES

1. Mattanovich D., Sauer M., Gasser B. Yeast biotechnology: Teaching the old dog new tricks. Microb. Cell Factories. 2014;13:34. doi: 10.1186/1475-2859-13-34.
2. Cregg J.M., Cereghino J.L., Shi J.Y., Higgins D.R. Recombinant protein expression in *Pichia pastoris*.Mol. Biotechnol. 2000;16:23–52. doi: 10.1385/MB:16:1:23.
3. Cregg J.M. *Pichia* protocols, second edition. In: Cregg J.M., editor. Methods in Molecular Biology.Humana Press; Totowa, NJ, USA: 2007. p. 268.
4. Cregg J.M., Tolstorukov I., Kusari A., Sunga J., Madden K., Chappell T. Expression in the yeast *Pichia pastoris*. Methods Enzymol. 2009;463:169–189.
5. Spadiut O., Herwig C. Production and purification of the multifunctional enzyme horseradish peroxidase: A review. Pharm. Bioproc. 2013;1:283–295.
6. Gmeiner C., Saadati A., Maresch D., Krasteva S., Frank M., Altmann F., Herwig C., Spadiut O. Development of a fed-batch process for a recombinant *Pichia pastoris* Δoch1 strain expressing a plant peroxidase. Microb. Cell Factories. 2015;14:1. doi: 10.1186/s12934-014-0183-3.
7. Hamilton S.R., Gerngross T.U. Glycosylation engineering in yeast: The advent of fully humanized yeast.Curr. Opin. Biotechnol. 2007;18:387–392. doi: 10.1016/j.copbio.2007.09.001.
8. Spadiut O., Rossetti L., Dietzsch C., Herwig C. Purification of a recombinant plant peroxidase produced in *Pichia pastoris* by a simple 2-step strategy. Protein Expr. Purif. 2012;86:89–97. doi: 10.1016/j.pep.2012.09.008.
9. Vervecken W., Kaigorodov V., Callewaert N., Geysens S., de Vusser K., Contreras R. *In vivo* synthesis of mammalian-like, hybrid-type *N*-glycans in *Pichia pastoris*. Appl. Environ. Microbiol. 2004;70:2639–2646. doi: 10.1128/AEM.70.5.2639-2646.2004.
10. Hamilton S.R., Bobrowicz P., Bobrowicz B., Davidson R.C., Li H.J., Mitchell T., Nett J.H., Rausch S., Stadheim T.A., Wischnewski H., et al. Production of complex human glycoproteins in yeast. Science.2003;301:1244–1246. doi: 10.1126/science.1088166.
11. Hamilton S.R., Davidson R.C., Sethuraman N., Nett J.H., Jiang Y., Rios S., Bobrowicz P., Stadheim T.A., Li H., Choi B.K., et al. Humanization of yeast to produce complex terminally sialylated glycoproteins. Science. 2006;313:1441–1443. doi: 10.1126/science.1130256.
12. Li H., Miele R.G., Mitchell T.I., Gerngross T.U. *N*-linked glycan characterization of heterologous proteins. Methods Mol. Biol. 2007;389:139–150.
13. Li H., Sethuraman N., Stadheim T.A., Zha D., Prinz B., Ballew N., Bobrowicz P., Choi B.K., Cook W.J., Cukan M., et al. Optimization of humanized IgGs in glycoengineered *Pichia pastoris*. Nat. Biotechnol.2006;24:210–215. doi: 10.1038/nbt1178.
14. Jacobs P.P., Geysens S., Vervecken W., Contreras R., Callewaert N. Engineering complex-type *N*-glycosylation in *Pichia pastoris* using glycoswitch technology. Nat. Protoc. 2009;4:58–70. doi: 10.1038/nprot.2008.213.
15. Krainer F.W., Glieder A. An updated view on horseradish peroxidases: Recombinant production and biotechnological applications. Appl. Microbiol. Biotechnol. 2015;99:1611–1625. doi: 10.1007/s00253-014-6346-7.
16. Morawski B., Lin Z.L., Cirino P.C., Joo H., Bandara G., Arnold F.H. Functional expression of horseradish peroxidase in *Saccharomyces cerevisiae* and *Pichia pastoris*. Protein Eng. 2000;13:377–384. doi: 10.1093/protein/13.5.377.
17. Morawski B., Lin Z.L., Bandara G., Arnold F.H. Expression of horseradish peroxidase (HRP) in *S. cerevisiae* and *P. pastoris*. Abstr. Pap. Am. Chem. Soc. 2000;219:U165–U165.
18. Naatsaari L., Krainer F.W., Schubert M., Glieder A., Thallinger G.G. Peroxidase gene discovery from the horseradish transcriptome. BMC Genom. 2014;15:227. doi: 10.1186/1471-2164-15-227
19. Krainer F.W., Pletzenauer R., Rossetti L., Herwig C., Glieder A., Spadiut O. Purification and basic biochemical characterization of 19 recombinant plant peroxidase isoenzymes produced in *Pichia pastoris*. Protein Expr. Purif. 2013;95C:104–112. doi: 10.1016/j.pep.2013.12.003.
20. Dietzsch C., Spadiut O., Herwig C. A fast approach to determine a fed batch feeding profile for recombinant *Pichia pastoris* strains. Microb. Cell Factories. 2011;10:85. doi: 10.1186/1475-2859-10-85
21. Dietzsch C., Spadiut O., Herwig C. A dynamic method based on the specific substrate uptake rate to set up a feeding strategy for *Pichia pastoris*. Microb. Cell Factories. 2011;10:14. doi: 10.1186/1475-2859-10-14
22. Spadiut O., Dietzsch C., Herwig C. Determination of a dynamic feeding strategy for recombinant *Pichia pastoris* strains. Methods Mol. Biol. 2014;1152:185–194.
23. Spadiut O., Zalai D., Dietzsch C., Herwig C. Quantitative comparison of dynamic physiological feeding profiles

for recombinant protein production with *Pichia pastoris*. Bioprocess Biosyst. Eng. 2013;37:1163–1172. doi: 10.1007/s00449-013-1087-z.

24. Krainer F.W., Capone S., Jager M., Vogl T., Gerstmann M., Glieder A., Herwig C., Spadiut O. Optimizing cofactor availability for the production of recombinant heme peroxidase in *Pichia pastoris*.Microb. Cell Factories. 2015;14:4. doi: 10.1186/s12934-014-0187-z.

25. Krainer F.W., Dietzsch C., Hajek T., Herwig C., Spadiut O., Glieder A. Recombinant protein expression in *Pichia pastoris* strains with an engineered methanol utilization pathway. Microb. Cell Factories.2012;11:22. doi: 10.1186/1475-2859-11-22.

26. Krainer F.W., Gmeiner C., Neutsch L., Windwarder M., Pletzenauer R., Herwig C., Altmann F., Glieder A., Spadiut O. Knockout of an endogenous mannosyltransferase increases the homogeneity of glycoproteins produced in *Pichia pastoris*. Sci. Rep. 2013;3:3279. doi: 10.1038/srep03279.

27. Capone S., Pletzenauer R., Maresch D., Metzger K., Altmann F., Herwig C., Spadiut O. Glyco-variant library of the versatile enzyme horseradish peroxidase. Glycobiology. 2014;24:852–863. doi: 10.1093/glycob/cwu047.

28. Laemmli U.K. Cleavage of structural proteins during the assembly of the head of bacteriophage T4.Nature. 1970;227:680–685. doi: 10.1038/227680a0.

CHAPTER 3

THE CRISPR/CAS GENOME-EDITING TOOL: APPLICATION IN IMPROVEMENT OF CROPS

Surender Khatodia[1], Kirti Bhatotia[1], Nishat Passricha[2], S. M. P. Khurana[1] and Narendra Tuteja[2,3]

[1] Amity Institute of Biotechnology, Amity University Haryana, Gurgaon, India

[2] Plant Molecular Biology Group, International Centre for Genetic Engineering and Biotechnology, New Delhi, India

[3] Amity Institute of Microbial Technology, Amity University, Noida, India

ABSTRACT

The Clustered Regularly Interspaced Short Palindromic Repeats associated Cas9/sgRNA system is a novel targeted genome-editing technique derived from bacterial immune system. It is an inexpensive, easy, most user friendly and rapidly adopted genome editing tool transforming to revolutionary paradigm. This technique enables precise genomic modifications in many different organisms and tissues. Cas9 protein is an RNA guided endonuclease utilized for creating targeted double-stranded breaks with only a short RNA sequence to confer recognition of the target in animals and plants. Development of genetically edited (GE) crops similar to those developed by conventional or mutation breeding using this potential technique makes it a promising and extremely versatile tool for providing sustainable productive agriculture for better feeding of rapidly growing population in a changing climate. The emerging areas of research for the genome editing in plants include interrogating gene function, rewiring the regulatory signaling networks and sgRNA library for high-throughput loss-of-function screening. In this review, we have described the broad applicability of the Cas9 nuclease mediated targeted plant genome editing for development of designer crops. The regulatory uncertainty and social acceptance of plant breeding by Cas9 genome editing have also been described. With this powerful and innovative technique the designer GE non-GM plants could further advance climate resilient and sustainable agriculture in the future and maximizing yield by combating abiotic and biotic stresses.

CRISPR/CAS TECHNOLOGY – AN OVERVIEW

Clustered regularly interspaced short palindromic repeat/Cas system was discovered in bacteria as an adaptive immune system which helps the bacteria in protecting itself against invading foreign DNA, such as that of a bacteriophage. This system comprises of CRISPR loci in the genome and a Cas9 protein. CRISPR, i.e., Clustered Regularly Interspaced Short Palindromic Repeats (CRISPRs) – is a genomic locus of tandem direct repeat sequences and protospacers, the spaces in between repeat sequences, both of which are derived from the invading elements (Kim and Kim, 2014). The CRISPR loci contains a combination of Cas9 genes; sequences for non-coding RNA elements called CRISPR RNA (crRNA) and sequences for small trans-encoded CRISPR RNA, i.e., trans-activating crRNA (tracrRNA). The two RNA sequences crRNA and tracrRNA forms a complex known as guide RNA, which determines the specificity of the cleavage of the target sequence in the nucleic acid along with the Protosapcer Adjacent Motif (PAM), a 5′-NGG sequence (Barrangou, 2013; Jinek et al., 2013). The cleavage of the double stranded target DNA occurs

Citation: Surender Khatodia, Kirti Bhatotia, Nishat Passricha, S. M. P. Khurana and Narendra Tuteja, The CRISPR/Cas Genome-Editing Tool: Application in Improvement of Crops, doi: 10.3389/fpls.2016.00506.

within the limits of protospacer region. The Cas9 protein is an endonuclease associated with CRISPR loci, which is responsible for the double-stranded breaks (DSBs) at the site, when targeted by a guide RNA (Cong et al., 2013; Mali et al., 2013). This revealing of molecular mechanism of the CRISPR/Cas system in 2012 opened up its vast area of applications as a promising component of genome editing termed as RNA-guided engineered nucleases (RGENs), which were used as sequence specific nucleases for precise genetic modifications (Doudna and Charpentier, 2014; Fichtner et al., 2014; Liang et al., 2015). RGENs are developed as programmable nucleases composed of two components, which must be expressed in cells to perform genome editing; the Cas9 nuclease and an engineered single guide RNA (sgRNA). The sgRNA has 20 nucleotides at the 5′ end that directs Cas9 to the complementary target site. Any DNA sequence of the form N20-NGG can be targeted by altering the first 20 nucleotides of the gRNA for novel genome editing applications (Sander and Joung, 2014) (Figure1).

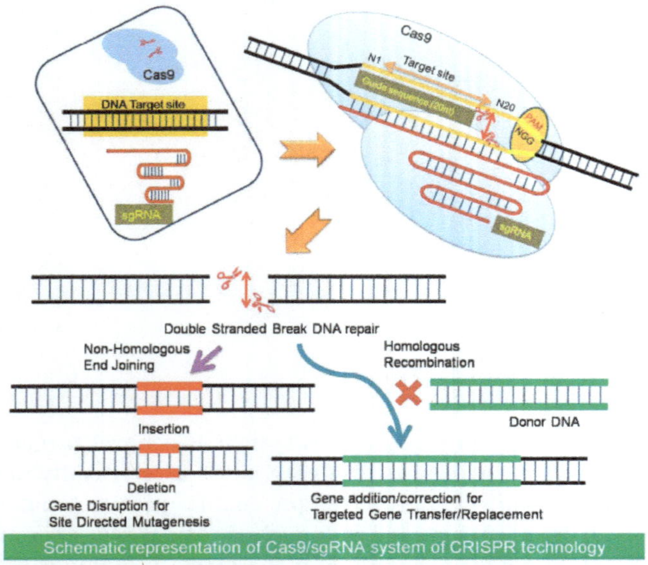

Figure 1: The basic strategy of Cas9/sgRNA system. The Cas9 is a RNA guided endonuclease consists of two nuclease domains namely HNH and RuvC. The target specificity of Cas9 depends upon the guide sequence (20 nt) short guide RNA (sgRNA). The target sites must lay immediately 50of a PAM (Protospacer Adjacent Motif) sequence of the form N20-NGG (or N20-NAG). The Cas9 nuclease induces double stranded breaks (DSB) at the target site which can be repaired either by Non- Homologous End Joining method or Homologous Recombination by cellular system which results in gene disruption by indels or gene addition/correction, respectively.

The targeted genome editing is utilized for the generation of desired endogenous modifications like gene disruption, addition, or correction at one or more specific genome site by introduction of nuclease mediated DNA-break using customized engineered nucleases (Kim and Kim, 2014). After DSB the cellular recombination repair mechanism can do the desired modification in a broad range of organisms and cell types. The elucidation of the CRISPR-Cas9 mechanism have resulted in many fundamental discoveries in biology (Doudna and Charpentier, 2014). CRISPR/Cas genome editing as a fledgling technology has reinvented the genetic and molecular biology research due to its simplicity and ease of design. There are many Cas9 and gRNA variants available, which could be utilized for further novel applications particularly in the field of plant biotechnology.

THE CAS9 NUCLEASE VARIANTS

The Cas9 endonuclease consists of two different domains, which includes a large globular recognition (REC) specific functional domain, connected to a smaller nuclease (NUC) domain. The NUC domain further accommodates two nuclease sites, RuvC and HNH, and also a PAM-interacting site (Doudna and Charpentier, 2014; Jinek et al., 2014). Cas9 protein is activated upon the loading of guide RNA, which further undergoes a conformational rearrangement to form a central channel for RNA-DNA heteroduplex binding and canonical PAM motifs recognition (Anders et al., 2014). The mechanism of Cas9 nuclease action was recently elucidated by crystal structure studies in a complex with partially duplexed target DNA containing PAM motif and sgRNA. This provides an insight into how Cas9 may be engineered to create variants with novel PAM specificities (Belhaj et al., 2013). There are many Cas9 variants available today, which can be utilized for high-throughput genome editing, silencing and transcriptional control with improved specificity and reduced off target effects in various systems, from yeast, *Drosophila*, bacteria, monkey, zebrafish, human, and plants.

The Native Cas9

The DSB created by a native Cas9 can be repaired by either HR or NHEJ method (Wyman and Kanaar, 2006; Shuman and Glickman, 2007). HR-mediated repair can be used to introduce specific point mutations like nucleotide substitutions or to insert desired sequences through recombination of the target locus with exogenously supplied DNA templates (Jiang et al., 2013) (Figure Figure11). NHEJ can, however, lead to the efficient introduction of insertion/deletion mutations (*indels*), which can disrupt the translational reading frame of a coding sequence or the transcription factors binding sites in promoters or enhancers (Cong et al., 2013). The high rate of alterations after DSBs created by Cas9, makes easy identification of the desired mutations without drug-resistance marker selection (Shimatani et al., 2015). So far, the applicability of the native Cas9/sgRNA system has been demonstrated in 10 plant species including model crops for targeted mutagenesis to gene knockouts and replacement as well as multiplex plant gene editing (Table1).

Table 1: List of applications of Cas9/sgRNA system in various plant species for single or multiplex genome editing and gene insertion or replacement

Plant species	Target genes	Reference
Gene knockout or editing with Cas9/sgRNA		
Arabidopsis thaliana	*AtPDS3, AtFLS2, TT4,BRI1, JAZ1, GAI,ADH1, CHLI, AP1, FT, SPL4, AtCRU3, At1g56650*	Feng et al., 2013; Jiang et al., 2013; Li et al., 2013; Mao et al., 2013; Fauser et al., 2014; Feng et al., 2014; Jiang et al., 2014; Hyun et al., 2015; Johnson et al., 2015; Ma et al., 2015
Nicotiana benthamiana	*NbPDS, PDS, NbPDS, NbPDS3, NbIspH*	Jiang et al., 2013; Li et al., 2013; Nekrasov et al., 2013; Upadhyay et al., 2013; Yin et al., 2015
Nicotiana tabacum	*NtPDS, NtPDR6, ALS*	Baltes et al., 2014; Gao et al., 2014
Oryza sativa	*OsPDS,OsBADH2,Os02g23823,OsMPK2, OsSWEET11, OsSWEET14, CAO1, LAZY1, OsMPK5, OsMYB1, ROC5, SPP, YSA.BEL,SWEET13/1a/1b, PMS3, EPSPS, DERF1, MSH1, MYB5, CDKB2, OsGSTU, OsMRP15, OsAnP, OsAOX1a, OsAOX1b, OsAOX1c, OsBEL*	Feng et al., 2013; Jiang et al., 2013; Mao et al., 2013; Miao et al., 2013; Shan et al., 2013; Xie and Yang, 2013; Endo et al., 2014; Xu et al., 2014, 2015; Zhang et al., 2014; Zhou et al., 2014;
Triticum Aestivum	*TaMLO,INOX, PDS,TaMLO-A1*	Shan et al., 2013; Upadhyay et al., 2013; Wang et al., 2014
Sorghum bicolor	*DsRED2*	Jiang et al., 2013
Marchantia polymorpha	*MpARF1*	Sugano et al., 2014
Citrus sinensis	*CsPDS*	Jia and Wang, 2014
Solanum lycopersicum	*SlAGO7,mGFP5, eGFP, RIN, ANT1*	Brooks et al., 2014; Ron et al., 2014; Cermak et al., 2015; Ito et al., 2015
Zea mays	*ZmIPK, LIG1, Ms26, Ms45, ALS2*	Liang et al., 2014; Svitashev et al., 2015
Glycine max	*Glyma07g14530, Glyma06g14180, Glyma08g02290, Glyma12g37050, Glyma18g04660, Glyma20g38560; GmFEI2, GmSHR*	Cai et al., 2015; Jacobs et al., 2015; Michno et al., 2015; Sun et al., 2015
Medicago trancatula	*GUS*	Michno et al., 2015
Populus tomentosa	*PtoPDS, 4CL, PtPDS*	Fan et al., 2015; Tingting et al., 2015; Zhou et al., 2015
Solanum tubersum	*StIAA2, StALS1*	Butler et al., 2015; Wang et al., 2015
Gene knockout or editing with Cas9 paired nickase/sgRNA		
Arabidopsis thaliana	*RTEL1, ADH1, TT4*	Fauser et al., 2014; Schiml et al., 2014
Multiplex genome editing with Cas9/sgRNA		
Arabidopsis thaliana	*AtRACK1b+AtRACK1c, CHLI1+CHLI2, ETC2, CPC, TRY, and PYL1-6, At5g55580*	Li et al., 2013; Mao et al., 2013; Xing et al., 2014; Ma et al., 2015; Zhang et al., 2015
Nicotiana tabacum	*NtPDS+NtPDR6*	Gao et al., 2014
Oryza sativa	*CDKB1, CDKA1, MPK1/2/5/6, PDSOsFTL*	Endo et al., 2014; Ma et al., 2015; Xie et al., 2015
Solanum lycopersicum	*Solyc07g021170+Solyc12g044760*	Brooks et al., 2014
Zea mays	*ZmHKT1*	Xing et al., 2014
Triticum aestivum	*TaMLO-A1+TaMLO-B1+TaMLO-D1*	Wang et al., 2014
Glycine max	*01gDDM1, 11gDDM1, 01g+11gDDM1; GmFEI2, and GmSHR*	Cai et al., 2015; Jacobs et al., 2015
Populus tomentosa	*PtPDS1 and PtPDS2*	Tingting et al., 2015
Gene insertion or replacement by HR with Cas9/sgRNA and donor template		
Arabidopsis thaliana	*YFFP,GU.US, DGU.US → GUS functional, nptII → ADH1*	Feng et al., 2013; Mao et al., 2013; Fauser et al., 2014; Schiml et al., 2014
Nicotiana benthamiana	*AvrII site → NbPDS*	Li et al., 2013
Oryza sativa	*OsPDS,OsBADH2*	Shan et al., 2013
Zea mays	*UBI:MoPAT*	Svitashev et al., 2015

The Cas9 Nickase

A Cas9 nickase (Cas9n) was first developed by Cong et al. (2013) through a mutation in native Cas9 (D10A, aspartate to alanine substitution). The Cas9 nickase with a RuvC or HNH mutation has the ability to create a nick, instead of a DSB at the target site. The individual nicks in the genome can be typically repaired with high fidelity homology-directed repair (HDR). The Cas9n has been used into a paired nickase system with two different gRNA to extend the number of specifically recognized bases for target cleavage, which has improved the specificity and helped mitigate the off-target phenomena (Cong et al., 2013; Mali et al., 2013;Ran et al., 2013; Fauser et al., 2014). The paired-nicking strategy has high-efficiency of HDR with reduced off-target cleavages by 50- to 1,500-fold in human cells, without sacrificing on-target cleavage efficiency (Ran et al., 2013) (Figure Figure2A).

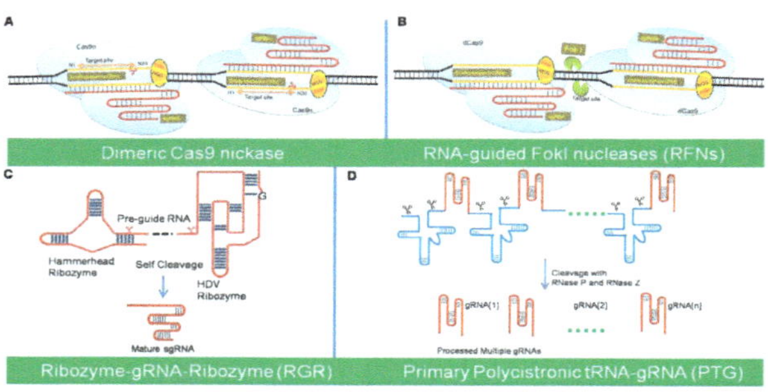

Figure 2: The different variants of Cas9/sgRNA system of genome editing.(A) The Cas9 nickase (Cas9n) with a RuvC or HNH mutation create a nick instead of a DSB at the target site. The dimeric Cas9n can be used for enhances specificity and reduced off target effects. (B) Dimeric RNA-guided FokI nucleases (RFNs) are the fusion of a catalytically inactive dCas9 protein with the FokI nuclease domain. Dimerization of two RFNs used for high genome editing frequencies and reduced off-target mutations. (C) Ribozyme-gRNA-Ribozyme (RGR) is an artificial gene, which generates self-catalyzed desired gRNA after transcribed from any promoter for tissue-specific genome editing. (D) Primary Polycistronic tRNA-gRNA (PTG) is tandemly arrayed tRNA-gRNA units, which is cleaved by the endogenous tRNA-processing system for simultaneously targeting multiple sites.

The Inactive dCas9

The nuclease deficient catalytically inactive mutant version of Cas9 (dCas9) has been used for RNA-guided transcription regulation, instead of genome editing (Gilbert et al., 2013; Qi et al., 2013). This modified system has been used for CRISPR interference (CRISPRi) and CRISPR activator (CRISPRa) for highly efficient and precise gene silencing and activation, respectively, using dCas9 with an effector and sgRNA. The dCas9 has an ability to incorporate gRNA and binding to the target (Xu et al., 2014). In comparison to the RNAi concept of transcript degradation and/or translation blocking, the CRISPRi system blocks the transcription initiation and elongation when dCas9/sgRNA is fused with a repressor. Therefore, dCas9/

sgRNA system offers a general platform for RNA-guided DNA targeting for stable and efficient modulation of transcription. The dCas9 has been fused to effector domains with distinct regulatory functions for functional mapping of promoters and other genomic regulatory modules (Gilbert et al., 2013; Qi et al., 2013; Figure Figure33).

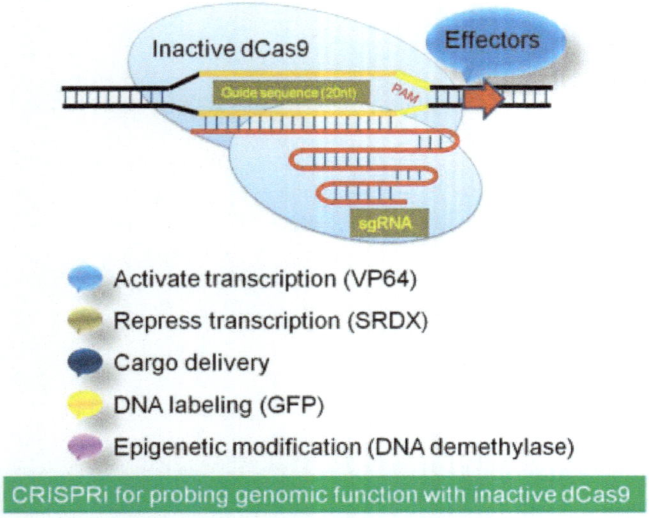

Figure 3: The strategic demonstration of inactive Cas9 system for Regulation of transcription and effector delivery. The inactive dCas9 is a catalytically inactive mutant repurposed for RNA-guided transcription regulation for high efficiency and specific CRISPR nterference (CRISPRi). It has been used by fusion of dCas9 to effector domains like activator or repressor for RNA-guided DNA modulation of transcription. The dCas9 has also been used to deliver specific effectors to targeted genomic locations like GFP and DNA demethylase.

Lowder et al. (2015) developed a CRISPR/Cas9 toolbox of applications in plants by transcriptional regulation. They have examined the applicability of the fusing the deactivated Cas9 with the transcriptional activator VP64 or the transcriptional repressor domain SRDX for transcriptional activation and repression of multiple endogenous genes (Lowder et al., 2015). The dCas9 has also been used to deliver GFP to targeted genomic locations (Anton et al., 2014). The novel light and chemical inducible dCas9 system have been developed, which could be utilized for light and agrochemical mediated transcription activation using Cas9 system. These will further, widen the applicability of Cas9/sgRNA system in crop improvement and functional genomics (Cong et al., 2013; Polstein and Gersbach, 2015; Zetsche et al., 2015).

Dimeric RNA-Guided FokI Nucleases (RFNs)

Dimeric RNA-guided FokI Nucleases (RFNs) are a fusion of a catalytically inactive dCas9 protein with the FokI nuclease domain. Dimerization of two RFNs rather than co-localization is required for efficient genome editing activity, which gives a plus point over the Cas9 nickase for high genome editing frequencies and reduced off-target mutations (Tsai et al., 2014; Bortesi and Fischer, 2015). Their cleavage activity depends strictly on the binding of two gRNAs with a defined spacing and orientation, which reduces the likelihood of target site occurring more than once in the genome (Tsai et al., 2014) (Figure Figure2B).

GUIDE RNA VARIANTS

There are some other modifications in the guide RNA of the Cas9/sgRNA system, which have provided other improvements in the native system for utilization in broader applicability of this significant technology.

Truncated Guide RNAs (truRNA)

Truncated guide RNAs (truRNA) are the sgRNA variants with shorter regions of target complementarities that is 17 nucleotides in length, which offers a simple, effective strategy to improve the specificities of Cas9 nucleases or paired nickases by reducing the off-target effects (Fu et al., 2014; Bortesi and Fischer, 2015). The truncated gRNAs, with shorter regions of target complementarities can decrease undesired mutagenesis at some off-target sites without sacrificing on-target genome editing efficiencies (Fu et al., 2014).

Ribozyme-gRNA-Ribozyme (RGR)

Ribozyme-gRNA-Ribozyme (RGR) is an artificial gene which generates RNA molecule with ribozyme sequences, which undergo self-catalyzed cleavage to generate the desired gRNA both *in vitro* and *in vivo*(Gao and Zhao, 2014b). RGR can be transcribed from any type of promoter and thus allow tissue-specific genome editing and efficient detection of mutations (Figure Figure2C2C).

Polycistronic tRNA-gRNA (PTG/Cas9)

Polycistronic tRNA-gRNA (PTG/Cas9) consists of array of tandem tRNA-gRNA units, with each gRNA containing a target-specific spacer for simultaneously targeting multiple sites (Xie et al., 2015). The primary transcript of PTG is cleaved after precise processing via the endogenous tRNA-processing system by RNase P and RNase Z, which releases numerous mature gRNAs *in vivo* from a synthetic polycistronic gene. The excised mature gRNAs direct Cas9 to multiple targets which significantly increase CRISPR/Cas9 multiplex editing efficiency in plants. Xie et al. (2015) demonstrated that targeting one gene with two gRNAs using PTG would greatly increase the efficiency of complete gene knock-out in comparison to the sgRNA (**Figure2D**).

THE CAS9/SGRNA SYSTEM FOR PLANT GENOME EDITING

There are broadly three categories of applications of the RNA guided endonuclease particularly in plants. First, in which DSBs created by Cas9 were repaired by non-homologous end joining (NHEJ) method for generation of *indels*, which leads to frame-shift mutations similar to natural variants, or those produced by physical or chemical mutagenesis as in mutation breeding (Chen and Gao, 2014; Saika et al., 2014). In second category, a short DNA repair template or a transgene has been used with Cas9 to repair DSB by homologous recombination (HR) for generation of the point mutations or targeted transgene insertion, gene replacement and gene stacking at predetermined sites. This avoids the position effects associated with random insertion of genes into plant genomes using genetic engineering. Third category uses, the multiplex genome editing for targeting multiple different sites with multiple sgRNAs along with the

Cas9 nuclease. The multiplex genome editing in plants can be used for dissecting the functions of gene family members with redundant functions and for analyzing epistatic relationships in genetic pathways (Xing et al., 2014). Here, we have briefly reviewed the achievements of Cas9 mediated genome editing in plants (Table 1).

CRISPR Achievements in Plants

The CRISPR/Cas system generates stable and heritable mutations, which can easily segregate from the Cas9/sgRNA construct to avoid further modifications by CRISPR/Cas. This results in development of homozygous modified transgene free plants in only a generation (Brooks et al., 2014; Fauser et al., 2014;Feng et al., 2014; Gao and Zhao, 2014a; Jiang et al., 2014; Schiml et al., 2014; Zhang et al., 2014; Zhou et al., 2014). Xu et al. (2015) successfully developed transgene-free rice with desired gene mutation by segregating out the transgene with self-fertilization in the T1 generation. The relative cleavage efficiency of Cas9 nucleases has been found better in comparison to previously- described TALENs and ZFNs against the same target sites (Gaj et al., 2013; Johnson et al., 2015). Xing et al. (2014) have developed a toolkit for multiplex genome editing in plants using CRISPR/Cas9-based binary vector set and a gRNA module vector set. This will facilitate transient or stable expression of CRISPR/Cas9 in a variety of plant systems and is especially useful for high-efficiency multiplex plant genome editing (Xing et al., 2014). Hence, the delivery of just two components, i.e., Cas9 and sgRNA to the host cell by genetic transformation methods is the only requirement for plant genome editing.

Baltes et al. (2014) have suggested that Gemini virus replicons (GVRs) can be used to deliver Cas9/sgRNA to plant cells with enhanced mutagenesis, in case when replication initiation protein gene (REP) was co-transformed with the Cas9/sgRNA construct. Further, to exploit the usefulness of this technology for trait discovery and development, efficient delivery methods like viruses based DNA replicons could be used for delivery of genome engineering reagents to all plant parts with higher repair frequencies and getting seeds with the desired modifications, without transformation (Baltes et al., 2014; Ali et al., 2015b). Two recent reports of direct delivery using the *Tobacco rattle virus* (TRV) (Ali et al., 2015b) and *Cabbage Leaf Curl virus* (CaLCuV) (Yin et al., 2015) have clearly demonstrated the feasibility of different virus medited Cas9/sgRNA delivery for efficient plant genome editing.

CRISPRi in Plants

The CRISPRi (CRISPR interference) has been demonstrated for RNA-guided, stable and efficient modulation of transcription of target genes in plants by fusion of inactivated dCas9 to effector domains (Larson et al., 2013). dCas9 has been used for functional genetics for regulation of gene expression and novel synthetic biology applications. It has been recruited to specific DNA sequences by gRNAs as a fusion protein with the activation or repression domain of a transcription factor (Gilbert et al., 2013; Piatek et al., 2014). The transcription of both a reporter construct and the endogenous PDS gene in *Nicotiana benthamiana*, have been modulated by fusing the dCas9 C-terminus to the EDLL domain as transcriptional activators, and to the SRDX domain as a repressor (Piatek et al., 2014). The recognition complex of dCas9/sgRNA/effector interferes with transcriptional regulation for efficient sgRNA dependent inducible and reversible inhibition of gene expression (Qi et al., 2013; Laganà et al., 2014). Gilbert et al. (2015) identified that the target site for effective CRISPRi should lie from -50 to +300 bp relative to the Transcription Start Site (TSS) of a gene.

CRISPR activator (CRISPRa) system modulates gene expression over a ~1,000-fold range by expression of a single sgRNA with one binding site (Gilbert et al., 2015). Development of genome-scale CRISPRi and CRISPRa libraries, would prove to be powerful tools for mapping complex stress-related signaling pathways in plants for functional genomics analysis (Laganà et al., 2014; Tingting et al., 2015; Gilbert et al., 2015). The available online resources of the CRISPR/Cas system's materials and tools have made the wide adoption and applications of this technique very simple (Table2). These include the web resources for CRISPR/Cas system and software tools for sgRNA designing with minimized off-target effects.

Table 2: Available online resources for CRISPR/Cas system

Name	Remarks	Reference
Addgene	Reagents and resources	https://www.addgene.org/crispr/
sgRNA Designer	Guide RNA Design tool	http://broadinstitute.org/rnai/public/analysis-tools/sgrna-design
Cas9 Design	Guide RNA DesignTool	http://cas9.cbi.pku.edu.cn
CHOPCHOP	Target sites finding tool	https://chopchop.rc.fas.harvard.edu
CRISPR Design	Design and analysis of Guide RNA	http://crispr.mit.edu
CRISPR Genome Analyzer	Genome editing experiment analysis plateform	http://crispr-ga.net
CRISPR-PLANT	Genome-wide gRNAs prediction tool in plants	http://genome.arizona.edu/crispr
CRISPRseek	Target-specific guide RNAs design tool	http://bioconductor.org/packages/release/bioc/html/CRISPRseek.html
DNA 2.0 gRNA Design Tool	gRNA Design tool	https://dna20.com/eCommerce/cas9/input
E-CRISP	Target sites design tool	http://e-crisp-test.dkfz.de/E-CRISP
RGEN Tools	Potential off-target sites prediction tool	http://rgenome.net/cas-offinder
sgRNAcas9	sgRNA design and potential off-target sites prediction tool	http://biootools.com
CRISPR MultiTargeter	Multiplex design tool	http://multicrispr.net/
CRISPR-P	Guide RNA design in plants	http://cbi.hzau.edu.cn/crispr/
AGEseq	Analysis of Genome Editing by Sequencing	https://github.com/liangjiaoxue/AGEseq
Stupar Lab's CRISPR Design	Target sites identifier	http://stuparcrispr.cfans.umn.edu/CRISPR/

Virus Interference in Plants: the CRISPR Approach

The CRISPR/Cas system of genome editing has been used as a tool for imparting resistance to viruses in plants (Chaparro-Garcia et al., 2015). Three recent reports have described the CRISPR/Cas approach for protection to plants against geminiviruses (Ali et al., 2015a; Baltes et al., 2015; Ji et al., 2015). This system conferred enhanced resistance to the plants against the geminiviruses species including BCTV (Beet curly top virus), TYLCV (Tomato yellow leaf curl virus), and MeMV (Merremia mosaic virus) at the same time with single gRNA (Ali et al., 2015a); BeYDV (Bean yellow dwarf virus) (Baltes et al., 2015), and BSCTV (Beet severe curly top virus) (Ji et al., 2015).

REGULATIONS AND SOCIAL ACCEPTANCE OF CRISPR EDITED CROPS

Routine targeted mutagenesis in plants by the CRISPR/Cas9 system will definitely help open up new dimensions in plant biology research. The conventional random mutagenesis approaches were not able to access every gene for inactivation study because of the random nature of the gene integration. CRISPR-Cas9 technology would be of great help in generating mutants for inaccessible genes, mutate multiple loci and generate large deletions, which can, therefore, accelerate plant breeding without actually introducing a transgene (Figure4). Although, genetically modified crops could have been a solution for crop improvement, if the controversies concerning the probable environmental and health implications of GM crops could be avoided (Hilbeck et al., 2011). Using the technique of genetic engineering a DNA construct can be directly inserted into one or more chromosomes in a random manner for manipulation of the genome. However, the random gene insertions can have undesirable effects and are not favorable for making large intensive changes, such as in case of adding an entire metabolic pathway into a plant (Lau et al., 2014). The plant breeding techniques have been used regularly to introduce new traits into cultivated varieties using existing natural genetic variation and random mutagenesis. The CRISPR/Cas system introduces precise modifications into a plant genome, which inherits stably and transgene region could be removed simply after the target gene editing to make transgene free plants during crop variety improvement (Mahfouz et al., 2014;Gao and Zhao, 2014b; Kanchiswamy et al., 2015; Xu et al., 2015; Figure5). The Cas9/sgRNA system, is now one of the new plant breeding techniques (NPBT) along with the ZFNs (Zinc Finger Nucleases) and TALENs (Transcription activator like effector nucleases) like genome editing technologies. The NPBTs are faster than traditional breeding methods and can produce a null segregant line that notably lacks the transgenic insert (Podevin et al., 2012; Araki and Ishii, 2015; Schaart et al., 2015; Woo et al., 2015). The plants developed by NPBTs are identical to the classically bred plants and these should be evaluated according to the resulting end product rather than the process of creation (Giddings et al., 2012; Hartung and Schiemann, 2014; Woo et al., 2015). Thus, in such a case CRISPR edited plants might be out of the current GMO regulations, but still this pose new challenges for the regulation and social acceptance of genetically edited (GE) crops (Voytas and Gao, 2014; Araki and Ishii, 2015; Jones, 2015; Kanchiswamy et al., 2015). NPBTs are currently in debate by advisory and regulatory authorities in relation to the GMO legislation to classify GE crop varieties produced using genome editing as non-GM for the social acceptance (Lusser et al., 2012; Voytas and Gao, 2014; Araki and Ishii, 2015).

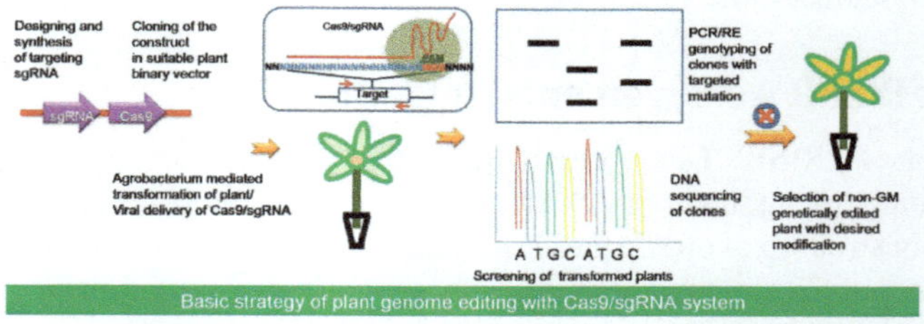

Figure 4: The strategy of using plant genome editing by Cas9/sgRNA system. Starting from the selection of the target gene, the available online resources has been utilized for designing and synthesis of sgRNA. The target sgRNA along with the suitable Cas9 variant have been cloned into a plant binary vector for transformation of the target plant species with Agrobacterium generally. After transformation the putative transformed plants would be selected for the presence of the Cas9 and sgRNA. Then screening of the plants with the desired mutation or editing would be done using PCR/RE genotyping and DNA sequencing.

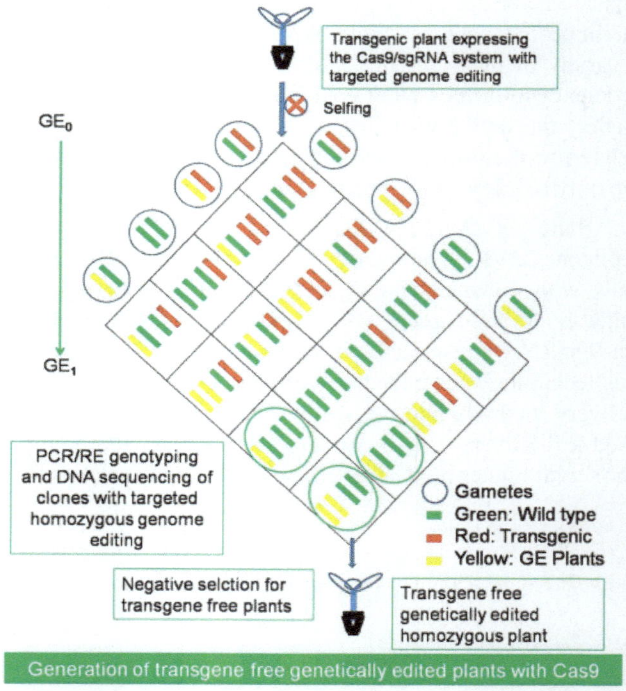

Figure 5: The generation of transgene free genetically edited (GE) crops. The transgene free homozygous mutants with desired genetic modifications at the targeted loci and without RGEN transgene construct could be selected by selfing of GE_0 generation plants and after segregation of the transgene in the next GE1 generation. The GE plants could be selected by PCR/RE genotyping and DNA sequencing of clones and negatively selecting for the transgene free plants with desired modification in the first generation only.

Araki and Ishii (2015) suggested a regulatory concept for GE crops based on the regulatory responses from the authorities in the world by categorizing the genome edited organisms under product- based and process-based GMO regulations. They recommend that those crops which have been engineered at gene/genome level using CRISPR-Cas9 technology should not be considered GMO under product based GMO regulation, because the final products are not genetically modified organism. While more work is needed to optimize the CRISPR/Cas system in plants as this ultimate plant genome editing toolkit shows encouraging potential for producing more complex useful agronomic traits in plant varieties (Quétier, 2016). This technology can better supplement the classical breeding techniques for understanding the complex quantitative traits, whose combinations of genes are selected by traditional breeding methods (Fichtner et al., 2014). The GE crops with the appropriate regulatory structures in place might prove to be more acceptable than the plants that carry foreign DNA in their genomes (Lozano-Juste and Cutler, 2014; Osakabe and Osakabe, 2015). The US Department of Agriculture has indicated that GE plants without any foreign DNA will not be considered as GMOs, however, the European Commission is expected to publish in the near future about the regulatory uncertainty of genome editing (Jones, 2015). The extent of the potential of CRISPR technology in applied plant research and crop breeding benefit for world's food security will depend upon the performance and public perception of the GE crop varieties (Belhaj et al., 2015; Jones, 2015; Wolt et al., 2015).

FUTURE PROSPECTS

The potential future crops for sustainable productive agriculture by genome editing are those which have better pest resistance, with enhanced nutritional value, and that are able to survive in changing climate. Climate resilient agriculture for combating abiotic and biotic stress is the future of crop improvement using genome editing for both the targeted mutagenesis mediated manipulation and study of transcriptional control by dissection of physiological and molecular cross talk under combined stress (Kissoudis et al., 2014; Jain, 2015). Genome editing will play very important role in developing new bio-energy crops, which could give maximum yield on wastelands and changing climate (Bosch and Hazen, 2013). This technology could offer any possible novel genome-editing concept for plants in order to improve crops for better nutrition and food security. We here suggest some of the possible concepts, which could be utilized for crop improvement and plant biotechnology applications (Liu and Fan, 2014) (Figure6). Further, direct delivery methods of Cas9 and gRNA using *Agrobacterium* and Viral replicons by using nanoparticles can be very useful for simplifying the genome editing technology (Hiei et al., 2014; Khatodia et al., 2014; Nonaka and Ezura, 2014). Inducible Cas9 system for transcription modulation like split-Cas9 for chemically inducible system and light activated Cas9 effector (LACE) could be utilized for crop improvement in future (Polstein and Gersbach, 2015; Zetsche et al., 2015). The generation of large-scale whole-genome targeted sgRNA library for high-throughput loss-of-function screening applications based on the CRISPRi system like that of RNAi is particularly feasible for model plants in future (Heintze et al., 2013).

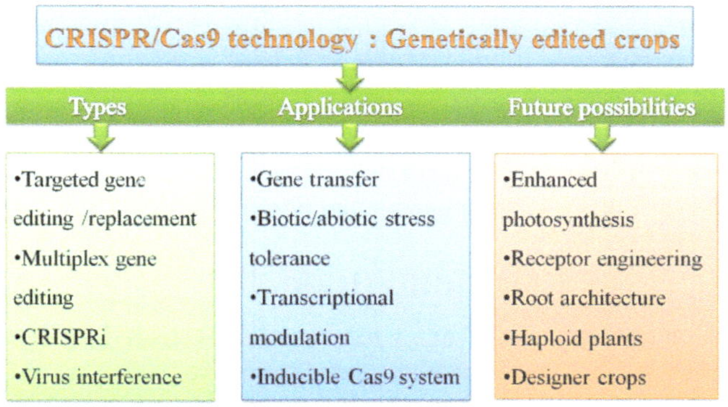

Figure 6: The types, applications and future possibilities of CRISPR/Cas9 system for development of GE crops for crop improvement.

Root trait is very important target for crop improvement using the CRISPR/Cas9 genome editing by studying the regulation of stress responses at cellular level in roots, allele replacement for QTL validation, and the epigenetic regulation of roots (Ahmadi et al., 2014). A major breeding target using CRISPR/Cas9 genome editing is to build cereal crops harboring root systems that can capture unevenly distributed water and nutrient resources in climate instability and resource scarcity. Targeted mutagenesis using the CRISPR-Cas9 system can bring advances in functional genomic research of legumes especially by generating the target mutants of the genes involved in roots and nodules (Sun et al., 2015). A recently published report of naturally transgenic sweet potato having T-DNAs has raised a few questions (Kyndt et al., 2015). The CRISPR/Cas9 editing methods can be used to delete T-DNAs for 'non-transgenic' sweet potato generation to test whether the T-DNAs confer a useful phenotype in naturally transgenic sweet potato (Jones, 2015). This technology can also be utilized for the production of haploid plants by alterations in the histone proteins found in centromeric nucleosomes (Kumar and Jain, 2014). Centromeric histone H3 protein (CENH3) is a member of the kinetochore complex, which is required for kinetochore formation and for chromosome segregation. Using the CRISPR/Cas9 technology, the expression of endogenous CENH3 gene can be suppressed by targeted mutagenesis in a plant, which can lead to selective loss of one set of chromosomes for haploid plants production (Maheswari et al., 2014). The ease and multiplicity of CRISPR/Cas system have shown the potential in three dimensions of plant functional genomics, i.e., genomics, transcriptomics, and epigenomics (Puchta, 2016). This can allow simultaneous induction as well as repression of certain sets of genes and at the same time aid in reprogramming the epigenome (Puchta, 2016).

In conclusion, this technique is getting more precise and efficient day by day, as the future opportunities like inducible Cas9 expression and direct delivery of Cas9 protein, are being explored in different organisms and cell types among the biologists all over the world (Ramakrishna et al., 2014; Polstein and Gersbach, 2015). These novel improvements can help avoid the boosting of off-target effects by expressing the Cas9/sgRNA only when required. Plant biotechnological applications of Cas9 technology not only lies in engineering non-food crops but also have the potential of generating altogether new plant varieties/species for production of specialized chemicals and biomaterials. The swiftness at which Cas9/sgRNA system has developed and improved, it has actually provided human control over genetic information with a huge genomic revolution. This will revolutionize both basic and applied research to improve a wide variety of agronomic traits in crop plants. Another remarkable worth of this CRISPR/cas9 system is that it is by far the most user friendly technique among all the currently available genome editing techniques.

AUTHOR CONTRIBUTIONS

SK and KB conceived and wrote the manuscript. NP reviewed and helped in figures. SMPK and NT reviewed the manuscript. All authors listed, have made substantial, direct, and intellectual contribution to the work, and approved it for publication.

CONFLICT OF INTEREST STATEMENT

The authors declare that the research was conducted in the absence of any commercial or financial relationships that could be construed as a potential conflict of interest.

ACKNOWLEDGMENTS

We thank the reviewers for their many informative comments and suggestions, which have improved the manuscript a lot.

REFERENCES

1. Ahmadi N., Audebert A., Bennett M. J., Bishopp A., de Oliveira A. C., Courtois B., et al. (2014). The roots of future rice harvests. *Rice (NY)* 7 29 10.1186/s12284-014-0029-y
2. Ali Z., Abulfaraj A., Idris A., Ali S., Tashkandi M., Mahfouz M. M. (2015a). CRISPR/Cas9-mediated viral interference in plants. *Genome Biol.* 16 238 10.1186/s13059-015-0799-6
3. Ali Z., Abul-faraj A., Piatek M., Mahfouz M. M. (2015b). Activity and specificity of TRV-mediated gene editing in plants. *Plant Signal. Behav.* 10:e1044191 10.1080/15592324.2015.1044191
4. Anders C., Niewoehner O., Duerst A., Jinek M. (2014). Structural basis of PAM-dependent target DNA recognition by the Cas9 endonuclease. *Nature* 513 569–573. 10.1038/nature13579
5. Anton T., Bultmann S., Leonhardt H., Markaki Y. (2014). Visualization of specific DNA sequences in living mouse embryonic stem cells with a programmable fluorescent CRISPR/Cas system. *Nucleus* 5163–172. 10.4161/nucl.28488
6. Araki M., Ishii T. (2015). Towards social acceptance of plant breeding by genome editing. *Trends Plant Sci.* 20 145–149. 10.1016/j.tplants.2015.01.010
7. Baltes N. J., Gil-Humanes J., Cermak T., Atkins P. A., Voytas D. F. (2014). DNA replicons for plant genome engineering. *Plant Cell* 26 151–163. 10.1105/tpc.113.119792
8. Baltes N. J., Hummel A. W., Konecna E., Cegan R., Bruns A. N., Bisaro D. M., et al. (2015).Conferring resistance to geminiviruses with the CRISPR-Cas prokaryotic immune system. *Nat. Plants*1 15145 10.1038/nplants.2015.145
9. Barrangou R. (2013). CRISPR-Cas systems and RNA-guided interference. *Wiley Interdiscip. Rev. RNA* 4 267–278. 10.1002/wrna.1159
10. Belhaj K., Chaparro-Garcia A., Kamoun S., Nekrasov V. (2013). Plant genome editing made easy: targeted mutagenesis in model and crop plants using the CRISPR/Cas system. *Plant Methods* 9 3910.1186/1746-4811-9-39
11. Belhaj K., Chaparro-Garcia A., Kamoun S., Patron N. J., Nekrasov V. (2015). Editing plant genomes with CRISPR/Cas9. *Curr. Opin. Biotechnol.* 32 76–84. 10.1016/j.copbio.2014.11.007
12. Bortesi L., Fischer R. (2015). The CRISPR/Cas9 system for plant genome editing and beyond.*Biotechnol. Adv.* 33 41–52. 10.1016/j.biotechadv.2014.12.006
13. Bosch M., Hazen S. P. (2013). Lignocellulosic feedstocks: research progress and challenges in optimizing biomass quality and yield. *Front. Plant Sci.* 4:474 10.3389/fpls.2013.00474
14. Brooks C., Nekrasov V., Lippman Z., Van Eck J. (2014). Efficient gene editing in tomato in the first generation using the CRISPR/Cas9 system. *Plant Physiol.* 166 1292–1297. 10.1104/pp.114.247577
15. Butler N. M., Atkins P. A., Voytas D. F., Douches D. S. (2015). Generation and inheritance of targeted mutations in potato (*Solanum tuberosum* L.) Using the CRISPR/Cas System. *PLoS ONE* 10:e014459110.1371/journal.pone.0144591
16. Cai Y., Chen L., Liu X., Sun S., Wu C., Jiang B., et al. (2015). CRISPR/Cas9-mediated genome editing in soybean hairy roots. *PLoS ONE* 10:e0136064 10.1371/journal.pone.0136064
17. Cermak T., Baltes N. J., Cegan R., Zhang Y., Voytas D. F. (2015). High-frequency, precise modification of the tomato genome. *Genome Biol.* 16 232 10.1186/s13059-015-0796-9
18. Chaparro-Garcia A., Kamoun S., Nekrasov V. (2015). Boosting plant immunity with CRISPR/Cas. *Genome Biol.* 16 254 10.1186/s13059-015-0829-4
19. Chen K., Gao C. (2014). Targeted genome modification technologies and their applications in crop improvements. *Plant Cell Rep.* 33 575–583. 10.1007/s00299-013-1539-6
20. Cong L., Ran F. A., Cox D., Lin S., Barretto R., Habib N., et al. (2013). Multiplex genome engineering using CRISPR/Cas systems. *Science* 339 819–823. 10.1126/science.1231143
21. Doudna J. A., Charpentier E. (2014). Genome editing. The new frontier of genome engineering with CRISPR-Cas9. *Science* 346 1258096 10.1126/science.1258096
22. Endo M., Mikami M., Toki S. (2014). Multigene knockout utilizing off-target mutations of the CRISPR/Cas9 system in rice. *Plant Cell Physiol.* 1 1–7. 10.1093/pcp/pcu154
23. Fan D., Liu T., Li C., Jiao B., Li S., Hou Y., et al. (2015). Efficient CRISPR/Cas9-mediated targeted mutagenesis in *Populus* in the first generation. *Sci. Rep.* 5 12217 10.1038/srep12217

24. Fauser F., Schiml S., Puchta H. (2014). Both CRISPR/Cas-based nucleases and nickases can be used efficiently for genome engineering in *Arabidopsis thaliana. Plant J.* 79 348–359. 10.1111/tpj.12554

25. Feng Z., Mao Y., Xu N., Zhang B., Wei P., Yang D.-L., et al. (2014). Multigeneration analysis reveals the inheritance, specificity, and patterns of CRISPR/Cas-induced gene modifications in *Arabidopsis.Proc. Natl. Acad. Sci. U.S.A.* 111 4632–4637. 10.1073/pnas.1400822111

26. Feng Z., Zhang B., Ding W., Liu X., Yang D.-L., Wei P., et al. (2013). Efficient genome editing in plants using a CRISPR/Cas system. *Cell Res.* 23 1229–1232. 10.1038/cr.2013.114

27. FichtnerF.,UrreaCastellanosR.,ÜlkerB.(2014).Precision genetic modifications: a new era in molecular biology and crop improvement. *Planta* 239 921–939. 10.1007/s00425-014-2029-y

28. Fu Y., Sander J. D., Reyon D., Cascio V. M., Joung J. K. (2014). Improving CRISPR-Cas nuclease specificity using truncated guide RNAs. *Nat. Biotechnol.* 32 279–284. 10.1038/nbt.2808

29. Gaj T., Gersbach C. A., Barbas C. F. (2013). ZFN, TALEN, and CRISPR/Cas-based methods for genome engineering. *Trends Biotechnol.* 31 397–405. 10.1016/j.tibtech.2013.04.004

30. Gao J., Wang G., Ma S., Xie X., Wu X., Zhang X., et al. (2014). CRISPR/Cas9-mediated targeted mutagenesis in *Nicotiana tabacum. Plant Mol. Biol.* 87 99–110. 10.1007/s11103-014-0263-0

31. Gao Y., Zhao Y. (2014a). Self-processing of ribozyme-flanked RNAs into guide RNAs in vitro and in vivo for CRISPR-mediated genome editing. *J. Integr. Plant Biol.* 56 343–349. 10.1111/jipb.12152

32. Gao Y., Zhao Y. (2014b). Specific and heritable gene editing in *Arabidopsis*. *Proc. Natl. Acad. Sci. U.S.A.* 111 4357–4358. 10.1073/pnas.1402295111

33. Giddings L. V., Potrykus I., Ammann K., Fedoroff N. V. (2012). Confronting the gordian knot. *Nat. Biotechnol.* 30 208–209. 10.1038/nbt.2145

34. Gilbert L. A., Horlbeck M. A., Adamson B., Villalta J. E., Chen Y., Whitehead E. H., et al. (2015).Genome-scale CRISPR-mediated control of gene repression and activation. *Cell* 159 647–661. 10.1016/j.cell.2014.09.029

35. Gilbert L. A., Larson M. H., Morsut L., Liu Z., Brar G. A., Torres S. E., et al. (2013). CRISPR-mediated modular RNA-guided regulation of transcription in eukaryotes. *Cell* 154 442–451. 10.1016/j.cell.2013.06.044

36. Hartung F., Schiemann J. (2014). Precise plant breeding using new genome editing techniques: opportunities, safety and regulation in the EU. *Plant J.* 78 742–752. 10.1111/tpj.12413

37. Heintze J., Luft C., Ketteler R. (2013). A CRISPR/Cas for high-throughput silencing. *Front. Genet.*4:193 10.3389/fgene.2013.00193

38. Hiei Y., Ishida Y., Komari T. (2014). Progress of cereal transformation technology mediated by*Agrobacterium tumefaciens. Front. Plant Sci.* 5:628 10.3389/fpls.2014.00628

39. Hilbeck A., Meier M., Römbke J., Jänsch S., Teichmann H., Tappeser B. (2011). Environmental risk assessment of genetically modified plants – concepts and controversies. *Environ. Sci. Eur.* 23 1–12. 10.1186/2190-4715-23-13

40. Hyun Y., Kim J., Cho S. W., Choi Y., Kim J.-S., Coupland G. (2015). Site-directed mutagenesis in*Arabidopsis thaliana* using dividing tissue-targeted RGEN of the CRISPR/Cas system to generate heritable null alleles. *Planta* 241 271–284. 10.1007/s00425-014-2180-5

41. Ito Y., Nishizawa-Yokoi A., Endo M., Mikami M., Toki S. (2015). CRISPR/Cas9-mediated mutagenesis of the RIN locus that regulates tomato fruit ripening. *Biochem. Biophy. Res. Commun.*467 76–82. 10.1016/j.bbrc.2015.09.117

42. Jacobs T. B., LaFayette P. R., Schmitz R. J., Parrott W. A. (2015). Targeted genome modifications in soybean with CRISPR/Cas9. *BMC Biotechnol.* 15:16 10.1186/s12896-015-0131-2

43. Jain M. (2015). Function genomics of abiotic stress tolerance in plants: a CRISPR approach. *Front. Plant Sci.* 6:375 10.3389/fpls.2015.00375

44. Ji X., Zhang H., Zhang Y., Wang Y., Gao C. (2015). Establishing a CRISPR-Cas-like immune system conferring DNA virus resistance in plants. *Nat. Plants* 1 15144 10.1038/nplants.2015.144

45. Jia H., Wang N. (2014). Targeted genome editing of sweet orange using Cas9/sgRNA. *PLoS ONE*9:e93806 10.1371/journal.pone.0093806

46. Jiang W., Yang B., Weeks D. P. (2014). Efficient CRISPR/Cas9-mediated gene editing in *Arabidopsis thaliana* and inheritance of modified genes in the T2 and T3 generations. *PLoS ONE* 9:e9922510.1371/journal.pone.0099225

47. Jiang W., Zhou H., Bi H., Fromm M., Yang B., Weeks D. P. (2013). Demonstration of CRISPR/Cas9/sgRNA-mediated targeted gene modification in *Arabidopsis*, tobacco, sorghum and rice. *Nucleic Acids Res.* 41:e188 10.1093/nar/gkt780

48. Jinek M., East A., Cheng A., Li S. (2013). RNA-programmed genome editing in human cells. *Elife*2:e00471 10.7554/eLife.00471

49. Jinek M., Jiang F., Taylor D. W., Sternberg S. H., Kaya E., Ma E., et al. (2014). Structures of Cas9 endonucleases reveal RNA-mediated conformational activation. *Science* 343:124799710.1126/science.1247997

50. Johnson R. A., Gurevich V., Filler S., Samach A., Levy A. (2015). Comparative assessments of CRISPR-Cas nucleases' cleavage efficiency in planta. *Plant Mol. Biol.* 87 143–156. 10.1007/s11103-014-0266-x

51. Jones H. D. (2015). Regulatory uncertainty over genome editing. *Nat Plants* 1 1401110.1038/nplants.2014.11

52. Kanchiswamy C. N., Sargent D. J., Velasco R., Maffei M. E., Malnoy M. (2015). Looking forward to

genetically edited fruit crops. *Trends Biotechnol.* 33 62–64. 10.1016/j.tibtech.2014.07.003

53. Khatodia S., Kharb P., Batra P., Chowdhury V. K. (2014). Development and characterization of transgenic chickpea (*Cicer arietinum* L.) plants with cry1Ac gene using tissue culture independent protocol. *Int. J. Adv. Res.* 2323–331.

54. Khatodia S., Khurana S. M. P. (2014). Trending: the Cas nuclease mediated genome editing technique.*Biotech Today* 4 46–49. 10.5958/2322-0996.2014.00019.2

55. Kim H., Kim J. S. (2014). A guide to genome engineering with programmable nucleases. *Nat. Rev. Genet.* 15 321–334. 10.1038/nrg3686

56. Kissoudis C., van de Wiel C., Visser R. G. F., van der Linden G. (2014). Enhancing crop resilience to combined abiotic and biotic stress through the dissection of physiological and molecular crosstalk.*Front. Plant Sci.* 5:207 10.3389/fpls.2014.00207

57. Kumar V., Jain M. (2014). The CRISPR–Cas system for plant genome editing: advances and opportunities. *J. Exp. Bot.* 66 47–57. 10.1093/jxb/eru429

58. Kyndt T., Quispe D., Zhai H., Jarret R., Ghislain M., Liu Q., et al. (2015). The genome of cultivated sweet potato contains *Agrobacterium* T-DNAs with expressed genes: an example of a naturally transgenic food crop. *Proc. Natl. Acad. Sci. U.S.A.* 112 5844–5849. 10.1073/pnas.1419685112

59. LaganàA., ShashaD., CroceC. M. (2014). Synthetic RNAs for gene regulation: design principles and computational tools. *Front. Bioeng. Biotechnol.* 2:65 10.3389/fbioe.2014.00065

60. Larson M. H., Gilbert L. A., Wang X., Lim W. A., Weissman J. S., Qi L. S. (2013). CRISPR interference (CRISPRi) for sequence-specific control of gene expression. *Nat. Protoc.* 8 2180–2196. 10.1038/nprot.2013.132

61. Lau W., Fischbach M. A., Osbourn A., Sattely E. S. (2014). Key applications of plant metabolic engineering. *PLoS Biol.* 12:e1001879 10.1371/journal.pbio.1001879

62. Li J. F., Norville J. E., Aach J., McCormack M., Zhang D., Bush J., et al. (2013). Multiplex and homologous recombination-mediated genome editing in *Arabidopsis* and *Nicotiana benthamiana*using guide RNA and Cas9. *Nat. Biotechnol.* 31 688–691. 10.1038/nbt.2654

63. Liang Q., Huashan L., Yunhan J., Chunsheng D. (2015). The molecular mechanism of CRISPR/Cas9 system and its application in gene therapy of human diseases. *Yi Chuan* 37 974–982. 10.16288/j.yczz.15-109

64. Liang Z., Zhang K., Chen K., Gao C. (2014). Targeted mutagenesis in *Zea mays* using TALENs and the CRISPR/Cas system. *J. Genet. Genomics* 41 63–68. 10.1016/j.jgg.2013.12.001

65. Liu L., Fan X. D. (2014). CRISPR-Cas system: a powerful tool for genome engineering. *Plant Mol. Biol.* 85 209–218. 10.1007/s11103-014-0188-7

66. Liu W., Yuan J. S., Stewart C. N. (2013). Advanced genetic tools for plant biotechnology. *Nat. Rev. Genet.* 14 781–793. 10.1038/nrg3583

67. Lowder L. G., Zhang D., Baltes N. J., Paul J. W., III, Tang X., Zheng X., et al. (2015). A CRISPR/Cas9 toolbox for multiplexed plant genome editing and transcriptional regulation. *Plant Physiol.* 169 971–985. 10.1104/pp.15.00636

68. Lozano-Juste J., Cutler S. R. (2014). Plant genome engineering in full bloom. *Trends Plant Sci.* 19284–287. 10.1016/j.tplants.2014.02.014

69. Lusser M., Parisi C., Plan D., Rodríguez-Cerezo E. (2012). Deployment of new biotechnologies in plant breeding. *Nat. Biotechnol.* 30 231–239. 10.1038/nbt.2142

70. Ma X., Zhang Q., Zhu Q., Liu W., Chen Y., Qiu R., et al. (2015). A robust CRISPR/Cas9 system for convenient, high-efficiency multiplex genome editing in monocot and dicot plants. *Mol. Plant* 8 1274–1284. 10.1016/j.molp.2015.04.007

71. Maheswari A., Comai L., Chan S. (2014). *Generation of Haploid Plants. WO2014110274A2*. Washington, DC: U.S. Patent and Trademark Office.

72. Mahfouz M. M., Piatek A., Stewart C. N. (2014). Genome engineering via TALENs and CRISPR/Cas9 systems: challenges and perspectives. *Plant Biotechnol. J.* 12 1006–1014. 10.1111/pbi.12256

73. Mali P., Aach J., Stranges P. B., Esvelt K. M., Moosburner M., Kosuri S., et al. (2013). CAS9 transcriptional activators for target specificity screening and paired nickases for cooperative genome engineering. *Nat. Biotechnol.* 31 833–838. 10.1038/nbt.2675

74. Mao Y., Zhang H., Xu N., Zhang B., Gou F., Zhu J.-K. (2013). Application of the CRISPR-Cas system for efficient genome engineering in plants. *Mol. Plant* 6 2008–2011. 10.1093/mp/sst121

75. Miao J., Guo D., Zhang J., Huang Q., Qin G., Zhang X., et al. (2013). Targeted mutagenesis in rice using CRISPR-Cas system. *Cell Res.* 23 1233–1236. 10.1038/cr.2013.123

76. Michno J. M., Wang X., Liu J., Curtin S. J., Kono T. J., Stupar R. M. (2015). CRISPR/Cas mutagenesis of soybean and *Medicago truncatula* using a new web-tool and a modified Cas9 enzyme.*GM Crops Food* 6 243–252. 10.1080/21645698.2015.1106063

77. Nekrasov V., Staskawicz B., Weigel D., Jones J. D. G., Kamoun S. (2013). Targeted mutagenesis in the model plant *Nicotiana benthamiana* using Cas9 RNA-guided endonuclease. *Nat. Biotechnol.* 31691–693. 10.1038/nbt.2655

78. Nonaka S., Ezura H. (2014). Plant-*Agrobacterium* interaction mediated by ethylene and super-*Agrobacterium* conferring efficient gene transfer. *Front. Plant Sci.* 5:681 10.3389/fpls.2014.00681

79. Osakabe Y., Osakabe K. (2015). Genome editing with engineered nucleases in plants. *Plant Cell Physiol.* 56 389–400. 10.1093/pcp/pcu170

80. Piatek A., Ali Z., Baazim H., Li L., Abulfaraj A., Al-Shareef S., et al. (2014). RNA-guided transcriptional regulation in planta via synthetic dCas9-based transcription factors. *Plant Biotechnol. J.*13 578–589. 10.1111/pbi.12284

81. Podevin N., Devos Y., Davies H. V., Nielsen K. M. (2012). Transgenic or not? No simple answer! New biotechnology-based plant breeding techniques and the regulatory landscape. *EMBO Rep.* 131057–1061. 10.1038/embor.2012.168

82. Polstein L. R., Gersbach C. A. (2015). A light-inducible CRISPR-Cas9 system for control of endogenous gene activation. *Nat. Chem. Biol.* 11 198–200. 10.1038/nchembio.1753

83. Puchta H. (2016). Using CRISPR/Cas in three dimensions: towards synthetic plant genomes, transcriptomes and epigenomes. *Plant J.* [Epub ahead of print].10.1111/tpj.13100

84. Qi L. S., Larson M. H., Gilbert L. A., Doudna J. A., Weissman J. S., Arkin A. P., et al. (2013).Repurposing CRISPR as an RNA-guided platform for sequence-specific control of gene expression.*Cell* 152 1173–1183. 10.1016/j.cell.2013.02.022

85. Quétier F. (2016). The CRISPR-Cas9 technology: closer to the ultimate toolkit for targeted genome editing. *Plant Sci.* 242 65–76. 10.1016/j.plantsci.2015.09.003

86. Ramakrishna S., Kwaku Dad A. B., Beloor J., Gopalappa R., Lee S. K., Kim H. (2014). Gene disruption by cell-penetrating peptide-mediated delivery of Cas9 protein and guide RNA. *Genome Res.* 24 1020–1027. 10.1101/gr.171264.113

87. Ran F. A., Hsu P. D., Lin C.-Y., Gootenberg J. S., Konermann S., Trevino A. E., et al. (2013). Double nicking by RNA-guided CRISPR Cas9 for enhanced genome editing specificity. *Cell* 154 1380–1389. 10.1016/j.cell.2013.08.021

88. Ron M., Kajala K., Pauluzzi G., Wang D., Reynoso M. A., Zumstein K., et al. (2014). Hairy root transformation using *Agrobacterium rhizogenes* as a tool for exploring cell type-specific gene expression and function using tomato as a model. *Plant Physiol.* 166 455–469. 10.1104/pp.114.239392

89. Saika H., Nishizawa-Yokoi A., Toki S. (2014). The non-homologous end-joining pathway is involved in stable transformation in rice. *Front. Plant Sci.* 5:560 10.3389/fpls.2014.00560

90. Sander J. D., Joung J. K. (2014). CRISPR-Cas systems for editing, regulating and targeting genomes.*Nat. Biotechnol.* 32 347–355. 10.1038/nbt.2842

91. Schaart J. G., van de Wiel C. C. M., Lotz L. A. P., Smulders M. J. M. (2015). Opportunities for products of new plant breeding techniques. *Trends Plant Sci.* [Epub ahead of print].10.1016/j.tplants.2015.11.006

92. Schiml S., Fauser F., Puchta H. (2014). The CRISPR/Cas system can be used as nuclease for in planta gene targeting and as paired nickases for directed mutagenesis in *Arabidopsis* resulting in heritable progeny. *Plant J.* 80 1139–1150. 10.1111/tpj.12704

93. Shan Q., Wang Y., Li J., Zhang Y., Chen K., Liang Z., et al. (2013). Targeted genome modification of crop plants using a CRISPR-Cas system. *Nat. Biotechnol.* 31 686–688. 10.1038/nbt.2650

94. Shimatani Z., Nishizawa-Yokoi A., Endo M., Toki S., Terada R. (2015). Positive-negative-selection-mediated gene targeting in rice. *Front. Plant Sci.* 5:748 10.3389/fpls.2014.00748

95. Shuman S., Glickman M. S. (2007). Bacterial DNA repair by non-homologous end joining. *Nat. Rev. Microbiol.* 5 852–861. 10.1038/nrmicro1768

96. Sugano S. S., Shirakawa M., Takagi J., Matsuda Y., Shimada T., Hara-Nishimura I., et al. (2014). CRISPR/Cas9-mediated targeted mutagenesis in the liverwort *Marchantia polymorpha* L. *Plant Cell Physiol.* 55 475–481. 10.1093/pcp/pcu014

97. Sun X., Hu Z., Chen R., Jiang Q., Song G., Zhang H., et al. (2015). Targeted mutagenesis in soybean using the CRISPR-Cas9 system. *Sci. Rep.* 5 10342 10.1038/srep10342

98. Svitashev S., Young J. K., Schwartz C., Gao H., Falco S. C., Cigan A. M. (2015). Targeted mutagenesis, precise gene editing, and site-specific gene insertion in maize using Cas9 and guide RNA. *Plant Physiol.* 169 931–945. 10.1104/pp.15.00793

99. Tanenbaum M. E., Gilbert L. A., Qi L. S., Weissman J. S., Vale R. D. (2014). A protein-tagging system for signal amplification in gene expression and fluorescence imaging. *Cell* 159 635–646. 10.1016/j.cell.2014.09.039

100. Tingting L., Di F., Lingyu R., Yuanzhong J., Rui L., Keming L. (2015). Highly efficient CRISPR/Cas9–mediated targeted mutagenesis of multiple genes in *Populus*. *Yi Chuan* 37 1044–1052. 10.16288/j.yczz.15-303

101. Tsai S. Q., Wyvekens N., Khayter C., Foden J. A., Thapar V., Reyon D., et al. (2014). Dimeric CRISPR RNA-guided FokI nucleases for highly specific genome editing. *Nat. Biotechnol.* 32 569–576. 10.1038/nbt.2908

102. Upadhyay S. K., Kumar J., Alok A., Tuli R. (2013). RNA-guided genome editing for target gene mutations in wheat. *G3 (Bethesda)* 3 2233–2238. 10.1534/g3.113.008847

103. Voytas D. F., Gao C. (2014). Precision genome engineering and agriculture: opportunities and regulatory challenges. *PLoS Biol.* 12:e1001877 10.1371/journal.pbio.1001877

104. Wang S., Zhang S., Wang W., Xiong X., Meng F., Cui X. (2015). Efficient targeted mutagenesis in potato by the CRISPR/Cas9 system. *Plant Cell Rep.* 34 1473–1476. 10.1007/s00299-015-1816-7

105. Wang Y., Cheng X., Shan Q., Zhang Y., Liu J., Gao C., et al. (2014). Simultaneous editing of three homoeoalleles in hexaploid bread wheat confers heritable resistance to powdery mildew. *Nat. Biotechnol.* 32 947–951. 10.1038/nbt.2969

106. Wolt J. D., Wang K., Yang B. (2015). The regulatory status of genome-edited crops. *Plant Biotechnol. J.* 14 510–518. 10.1111/pbi.12444

107. Woo J. W., Kim J., Kwon S. I., Corvalan C., Cho S. W., Kim H., et al. (2015). DNA-free genome editing in plants with preassembled CRISPR-Cas9 ribonucleoproteins. *Nat. Biotechnol.* 33 1162–1164. 10.1038/nbt.3389

108. Wyman C., Kanaar R. (2006). DNA double-strand break repair: all's well that ends well. *Annu. Rev. Genet.* 40 363–383. 10.1146/annurev.genet.40.110405.090451

109. Xie K., Minkenberg B., Yang Y. (2015). Boosting CRISPR/Cas9 multiplex editing capability with the endogenous tRNA-processing system. *Proc. Natl. Acad. Sci. U.S.A.* 112 3570–3575. 10.1073/pnas.1420294112

110. Xie K., Yang Y. (2013). RNA-guided genome editing in plants using a CRISPR-Cas system. *Mol. Plant* 6 1975–1983. 10.1093/mp/sst119

111. Xing H.-L., Dong L., Wang Z.-P., Zhang H.-Y., Han C.-Y., Liu B., et al. (2014). A CRISPR/Cas9 toolkit for multiplex genome editing in plants. *BMC Plant Biol.* 14:327 10.1186/s12870-014-0327-y

112. Xu R. F., Li H., Qin R. Y., Li J., Qiu C. H., Yang Y. C., et al. (2015). Generation of inheritable and "transgene clean" targeted genome-modified rice in later generations using the CRISPR/Cas9 system.*Sci. Rep.* 5 11491 10.1038/srep11491

113. Xu T., Li Y., Van Nostrand J. D., He Z., Zhou J. (2014). Cas9-based tools for targeted genome editing and transcriptional control. *Appl. Environ. Microbiol.* 80 1544–1552. 10.1128/AEM.03786-13

114. Yin K., Han T., Liu G., Chen T., Wang Y., Yu A. Y., et al. (2015). A geminivirus-based guide RNA delivery system for CRISPR/Cas9 mediated plant genome editing. *Sci. Rep.* 5 1492610.1038/srep14926

115. Zetsche B., Volz S. E., Zhang F. (2015). A split-Cas9 architecture for inducible genome editing and transcription modulation. *Nat. Biotechnol.* 33 139–142. 10.1038/nbt.3149

116. Zhang H., Zhang J., Wei P., Zhang B., Gou F., Feng Z., et al. (2014). The CRISPR/Cas9 system produces specific and homozygous targeted gene editing in rice in one generation. *Plant Biotechnol. J.*12 797–807. 10.1111/pbi.12200

117. Zhang L., Zhou Q. (2014). CRISPR/Cas technology: a revolutionary approach for genome engineering. *Sci. China Life Sci.* 57 639–640. 10.1007/s11427-014-4670-x

118. Zhang Z., Mao Y., Ha S., Liu W., Botella J. R., Zhu J. K. (2015). A multiplex CRISPR/Cas9 platform for fast and efficient editing of multiple genes in *Arabidopsis*. *Plant Cell Rep.* 1–5. 10.1007/s00299-015-1900-z

119. Zhou H., Liu B., Weeks D. P., Spalding M. H., Yang B. (2014). Large chromosomal deletions and heritable small genetic changes induced by CRISPR/Cas9 in rice. *Nucleic Acids Res.* 42 10903–10914. 10.1093/nar/gku806

120. Zhou X., Jacobs T. B., Xue L.-J., Harding S. A., Tsai C. J. (2015). Exploiting SNPs for biallelic CRISPR mutations in the outcrossing woody perennial *Populus* reveals 4-coumarate: CoA ligase specificity and redundancy. *New Phytol.* 208 298–301. 10.1111/nph.13470

CHAPTER 4

CRISPR/CAS9: A TOOL TO CIRCUMSCRIBE COTTON LEAF CURL DISEASE

Zafar Iqbal[1], Muhammad N. Sattar[2] and Muhammad Shafiq[3]

[1] Institute of Biochemistry and Biotechnology, Quaid-i-Azam Campus, University of the Punjab, Lahore, Pakistan

[2] Department of Environment and Natural Resources, Faculty of Agriculture and Food Science, King Faisal University, Al-Hasa, Saudi Arabia

[3] Agricultural Biotechnology Division, National Institute for Biotechnology and Genetic Engineering, Faisalabad, Pakistan

ABSTRACT

The begomoviruses (family Geminiviridae) associated with cotton leaf curl disease (CLCuD) pose a major threat to cotton productivity in South–East Asia including Pakistan and India. These viruses have single-stranded, circular DNA genome, of ~2800 nt in size, encapsidated in twinned icosa-hedera, transmitted by ubiquitous whitefly and are associated with satellite molecules referred to as alpha- and betasatellite. To circumvent the proliferation of these viruses numerous techniques, ranging from conventional breeding to molecular approaches have been applied. Such devised strategies worked perfectly well for a short time period and then viruses relapse due to various reasons including multiple infections, where related viruses synergistically interact with each other, virus proliferation and evolution. Another shortcoming is, until now, that all molecular biology approaches are devised to control only helper begomoviruses but not to control associated satellites. Despite the fact that satellites could add various functions to helper begomoviruses, they remain ignored. Such conditions necessitate a very comprehensive technique that can offer best controlling strategy not only against helper begomoviruses but also their associated DNA-satellites. In the current scenario clustered regulatory interspaced short palindromic repeats (CRISPR)/CRISPR associated nuclease 9 (Cas9) has proved to be versatile technique that has very recently been deployed successfully to control different geminiviruses. The CRISPR/Cas9 system has been proved to be a comprehensive technique to control different geminiviruses, however, like previously used techniques, only a single virus is targeted and hitherto it has not been deployed to control begomovirus complexes associated with DNA-satellites. Here in this article, we proposed an inimitable, unique, and broad spectrum controlling method based on multiplexed CRISPR/Cas9 system where a cassette of sgRNA is designed to target not only the whole CLCuD-associated begomovirus complex but also the associated satellite molecules.

INTRODUCTION

Cotton leaf curl disease (CLCuD) is a top ranked endemic disease to cotton in Pakistan, northwestern India and Africa, and causes a severe short fall in the economy thus, it is detrimental to the socio-economic values of the people (Mansoor et al., 2003; Varma and Malathi, 2003; Sattar et al., 2013). CLCuD on the Indian subcontinent is caused by a complex of begomoviruses in association with certain satellite molecules (alpha- and betasatellite). Viruses included in this complex are *Cotton leaf curl Alabad Virus* (CLCuAlV), *Cotton leaf curl Bangalore virus* (CLCuBaV), *Cotton leaf curl Kokhran virus* (CLCuKoV), CLCuKoV-Bu (Burewala strain), *Cotton leaf curl Multan virus* (CLCuMuV),

Citation: Zafar Iqbal, Muhammad N. Sattar and Muhammad Shafiq, CRISPR/Cas9: A Tool to Circumscribe Cotton Leaf Curl Disease, doi: 10.3389/fpls.2016.00475.

and *Cotton leaf curl Rajasthan virus*(CLCuRaV) (Zhou et al., 1998; Kirthi et al., 2004; Yousaf et al., 2013). All CLCuD-associated begomoviruses (CABs) induce typical symptoms of leaf curling (upward and downward), vein swelling, enation, and stunted growth (Figure 1). CABs belonging to genus begomovirus (Family *Geminiviridae*) are exclusively transmitted by ubiquitous whitefly (*Bemisia tabaci*) and have a circular single-stranded DNA genome of ~2800 nt in size (Figure 2A) encapsidated in twinned quasi-icosahedera (Briddon and Markham, 2001; Böttcher et al., 2004). The genome of CABs encodes six genes in bi-directional manner separated by non-coding intergenic regions (IR), containing promoter elements and the origin of replication (Figure 2A). Two virion sense genes, the coat protein (*CP*) and *V2* (pre-coat protein), are involved in insect transmission, encapsidation, *in planta* movement and suppression of host defense (Rojas et al., 2001). The remaining four complementary sense genes include the replication-associated protein [(*Rep*; sole protein involved in viral replication; (Hanley-Bowdoin et al., 2004)], the *C2/TrAP* protein [(transcriptional activator protein; involved in suppression of host defense, activation of late genes and overcomes virus induced hypersensitive cell death; Hussain et al., 2007; Yang et al., 2007; Mubin et al., 2010)], the replication enhancer protein *C3/REn*; (Settlage et al., 2005) and the *C4* protein [(a pathogenicity/symptoms determinant and a suppressor of PTGS;Vanitharani et al., 2004; Saeed et al., 2008; Iqbal et al., 2012; Sunitha et al., 2013)]. Contrary to helper begomovirus, beta- and alphsatellites are half the size (~1350 nt) of their helper begomoviruses, each encoding a single protein β*C1* and *Rep*, respectively (Mansoor et al., 2001; Briddon et al., 2004) (Figure 2B). Of these two DNA-satellites, betasatellites are more diverse and their encoded single protein, β*C1*, adds more functions towards the viral pathogenesis than the alphasatellites. This protein is a suppressor of host defense, a pathogenicity/symptom determinant (Saeed et al., 2005; Qazi et al., 2007), may be involved in movement (Saeed et al., 2007), increases the helper begomovirus titer (Briddon et al., 2001), complements the missing functions of helper begomovirus genes (Iqbal et al., 2012), modulates the levels of developmental microRNAs (Amin et al., 2011), binds to DNA/RNA in sequence independent manner (Cui et al., 2005), interacts with various host-encoded factors, forms multimers, and also interacts with *CP* (Cheng et al., 2011), and suppresses jasmonic acid production in plants (Zhang et al., 2012).

Figure 1: Cotton (*Gossypium hirsutum* **L.) plants showing typical symptoms of** CLCuD caused by CAB (s) alone (a), associated with alpha- and betasatellite (b,c), respectively. Presence of betasatellite incurred severe vein thickening and leaf enations on the underside of the leaves.

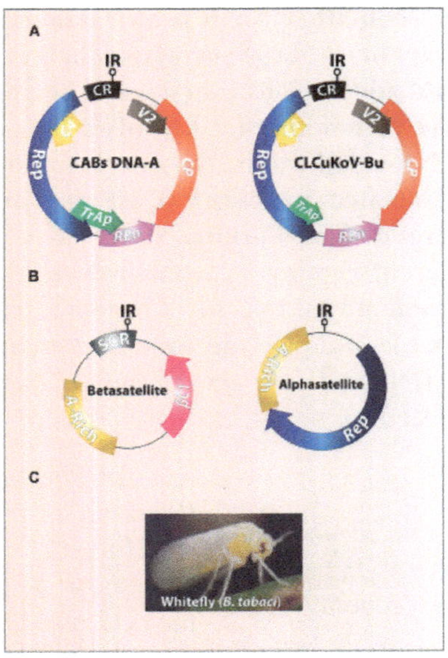

Figure 2: Genome organization of CABs (A) and associated DNA-satellites (B). The difference between genomes of resistance breaking CLCuKoV-Bu and other CABs is of a truncated *TrAP* **(A)**. The intergenic region (IR) contains stem loop structure and the origin of replication (ori). The CABs and associated DNA-satellites are vectored by ubiquitous whitefly (Bemisia tabaci) (C).

Although the exact history of CLCuD is difficult to map, CLCuD was, however, established in the sub-continent due to introgression of high yielding but highly susceptible varieties of cotton (*Gossypium hirsutum*) in place of local resistant cotton species (*G. arboreum*) (Ali, 1997). Symptoms of this disease were discerned in 1967 near Multan, Pakistan (Hussain and Ali, 1975). Later in 1980, its first outbreak linked with dramatic decrease in cotton productivity gave scientists and policy makers cause for serious concern. In 1992–1997, it surged again and about 0.2 million hectares (mha) out of 2.5 mha were affected resulting in 29% reduction in cotton productivity and the country had to suffer a short fall in economy of about $US 5 billion (Briddon and Markham, 2001). Subsequently, CLCuD spread across the borders and reported from Sriganganagar, India in 1993 (Kumar et al., 2010). In the last decade of the previous century, continuous conventional breeding efforts and introduction of resistant/tolerant cotton varieties in Pakistan heaved the economy. However, in 2001–2004, the second outbreak was observed and the disease spread to southern parts of Pakistan, the Sindh province. Later, investigations showed that this outbreak was associated with a recombinant begomovirus, *Cotton leaf curl Burewala virus* associated with recombinant Cotton leaf curl Multan betasatellite (CLCuMuB) (Briddon, 2003; Mahmood et al., 2003). *Cotton leaf curl Burewala virus*, now referred as CLCuKoV-Bu (Brown et al., 2015), has also been established widely in India (Rajagopalan et al., 2012; Zaffalon et al., 2012). Recently, different variants of CLCuKoV-Bu have been identified from Pakistan and India (Shuja et al., 2014; Kumar et al., 2015), hence posing the possibility of another alarming situation that can lead to the third outbreak in sub-continent. Until now, despite all efforts (both conventional and molecular approaches) the almost the whole available germplasm is susceptible, except few lines which are claimed to be tolerant, against this menace (Rahman and Zafar, 2007).

The failure of controlling strategies against these begomoviruses has urged the scientists to explore better measures to control CLCuD. In the current scenario, CRISPR/Cas9, as an extremely versatile, multifaceted and emerging technology, has gained the attention of research communities across the globe. This system has recently been exploited successfully against *Bean yellow dwarf virus* (BeYDV) (Baltes et al., 2015), *Beet severe curly top virus* (BSCTV) (Ji et al., 2015), and *Tomato yellow leaf curl virus* (TYLCV) (Ali et al., 2015), where reduced viral load up to 97% has been achieved linked to attenuated disease symptoms and apparently with no off-target affects. Keeping in mind the successful implementation of CRIPSR/Cas9, here in this study, a tentative methodology based on multiplex CRIPSR/Cas9 system is proposed, where universal sgRNAs are designed to target the most conserved regions of CABs and their associated DNA-satellite molecules that can potentially interfere with replication and movement of CABs and their cognate satellites, hence can resuscitate the cotton production by curbing CLCuD.

WHY CRIPSR/CAS9 AGAINST CLCUD?

In the past few decades, CABs have proliferated extensively. Their evolution and dispersion may likely be exacerbated by recombination, selection pressure imposed by introgression of resistant cultivars, change in virus complex and rise in temperature due to global warming. Despite concerted efforts made against CLCuD (see Techniques Employed to Control Cabs), nothing has been entirely successful in yielding resistance to this disease (Farooq et al., 2011; Sattar et al., 2013). The prevalent situation of unfruitful controlling strategies against CLCuD necessitates devising more strict controlling mechanisms that can curtail the CABs and their associated satellites. Under such circumstances, CRISPR/Cas9 system offers substantial advantage. The CRISPR/Cas9 system, apart from Zinc finger (ZFN) and Transcription activator-like effector nucleases (TALENs), has gained the attention of scientists across all major fields of science, especially plant biologists, as a promising programmable genome editing tool with higher level of specificity. Moreover, robustness, wide adaptability, and easy engineering of this system have proved its potential as a tool to control viruses – especially CABs. Very recently, this has also been successfully used to control plant viruses including different geminiviruses (Ali et al., 2015; Baltes et al., 2015; Ilardi and Tavazza, 2015; Ji et al., 2015). At the advent of CRISPR/Cas9 system, the major limitation observed was the degree to which off-target mutations take place. Until now, only few studies showed very negligible off-target activities. Improved versions of CRISPR/Cas9 DNA editing system and bioinformatics tools (Table1) have increased the specificity to next level. New versions of sgRNAs can direct endonuclease Cas9 to induce precise cleavage at a target site (Jinek et al., 2012). In short, this system has validated the capability across a wide range of living organism to accomplish gene interference, where it has demonstrated its applicability and efficacy, thus rendering it as a final choice to curb viral diseases (Feng et al., 2013; Li et al., 2013), especially against CABs.

Table 1: Available bioinformatics tools for selecting optimal CRISPR/Cas9 target sites and predicting off-targets

Purpose	Web address	Reference
CRISPR/Cas9 design tool to find target sites within an input sequence.	http://www.genome-engineering.org	Hsu et al., 2013
Designed sgRNA can be checked for off-targets and specificity against different genomes including *Arabidopsis thaliana*.	http://crispr.mit.edu/	Ran et al., 2013
Online tool for designing highly active sgRNAs.	http://www.broadinstitute.org/mai/public/analysis-tools/sgma-design	Doench et al., 2014
Open sourced tool that is used locally, designed to identify potential off-target sites in any user specified genome.	http://eendb.zfgenetics.org/casot/	Xiao et al., 2014
Download link to access 38 plant genomes.	http://plants.ensembl.org/info/website/ftp/index.html	
Web-based tool to design sgRNA sequences for genome library projects or individual sequences. Target site homology is also evaluated to predict off targets. Five plant genomes are available.	http://www.e-crisp.org/E-CRISP/designcrispr.html	Heigwer et al., 2014
Eight representative plant genomes are available to predict sgRNAs with low chance of off-target sites.	www.genome.arizona.edu/crispr	Xie et al., 2014
Online tool for accurate target sequence selection and prediction of off-target binding of sgRNAs. Includes the design of target specific primers for PCR genotyping. The only plant genome available is *A. thaliana*.	https://chopchop.rc.fas.harvard.edu/	Montague et al., 2014
Cas-OFFinder.	http://www.rgenome.net/cas-offinder/	Kleinstiver et al., 2016
Addgene service of non-profit plasmid distribution.	http://www.addgene.org/crispr/church/ http://www.addgene.org/crispr-cas	Mali et al., 2013 Kleinstiver et al., 2016
Software for the analysis of GUIDE-seq.	http://www.jounglab.org/guideseq	Kleinstiver et al., 2016
Software for the analysis of GUIDE-seq.	http://www.jounglab.org/guideseq	Kleinstiver et al., 2016
Database that identifies Cas proteins and CRISPRs; includes many features complementary to CRISPR.	http://crispi.genouest.org/	Bhaya et al., 2011
CRISPR database; includes several tools to identify and analyze CRISPRs. Maintained by Universite Paris Sud.	http://crispr.upsud.fr/	Bhaya et al., 2011
To obtain Cas genes sequences.	http://mbgd.genome.ad.jp http://ergo.integratedgenomics.com/ERGO	Bolotin et al., 2005
Cas9 online designer.	http://cas9.wicp.net/	Wang et al., 2016

TECHNIQUES EMPLOYED TO CONTROL CABS

A diverse group of techniques ranging from conventional breeding to advance molecular approaches have been implicated to control CLCuD, however, over time, resistance or tolerance was broken down by the virus (Aragão and Faria, 2009). Such techniques include accustom breeding approaches (Siddig, 1970; Ali, 1997; Rahman, 1997; Iqbal et al., 2001; Ahmad et al., 2010, 2011; Akhtar et al., 2010), pathogen derived resistance (PDR) including protein and non-protein mediated resistance (Amudha et al., 2010, 2011; Yousaf et al., 2013) (Asad et al., 2003; Sanjaya et al., 2005; Mubin et al., 2011; Ali et al., 2013; Tahir, 2014), DNA interference (Liu et al., 1998; Tahir, 2014), and non-PDR including expression of *cry1Ac* and *GroEL* (Akad et al., 2007; Edelbaum et al., 2009; Shepherd et al., 2009; Rana et al., 2012). Moreover, controlling the whitefly vector has also become a major focus of world renowned laboratories to control CLCuD through RNAi (author's personal communication). These techniques are lagging in "fair control" of CLCuD. All of these techniques were deployed directly or indirectly to control only CABs but not their cognate DNA-satellites. The role of CABs associated DNA-satellites to salvage CABs has been totally ignored during course of these efforts. This scenario has put the scientist in a situation where they plan to deploy more than one strategies (pyramiding) at once, which itself is a very laborious job.

SUCCESSFUL IMPLICATIONS OF CRISPR/CAS9 SYSTEM AGAINST GEMINIVIRUSES

CRISPR/Cas9 systems have superiority over other nucleases such as ZFN and TALEN for targeting multiple genes at the same time. There is variety of multiplex models available for genome engineering using CRISPR/Cas9 system. Either these models involve single RNA Pol-III promotor (Guo et al., 2015) or several different promotors at the same time (Kabadi et al., 2014; Sakuma et al., 2014; Cobb et al., 2015) to drive sgRNA/s co-expression using a single expression vector. Using multiple promoters for sets of sgRNAs in a single vector is very laborious and inconvenient (Guo et al., 2015). The ultimate strategy is to use a single promoter to drive multiple sgRNAs with each of them following a direct repeat (DR) sequence (Cong et al., 2013; Jiang et al., 2013; Guo et al., 2015), which is very similar to the native strategy in *Streptococcus thermophilus* involving crRNA interspacers followed by almost similar sized conserved DR sequences (Guo et al., 2015).

The CRISPR/Cas9 system has been successfully employed in controlling BeYDV (Baltes et al., 2015), BSCTV (Ji et al., 2015), and TYLCV (Ali et al., 2015). Ji et al. (2015) found that viral load was significantly reduced (up to 97%) using different BSCTV constructs in *Nicotiana benthamiana* and *Arabidopsis thaliana*plants. Whereas, Baltes et al. (2015) reported 87% reduction in targeted viral load in *N. benthamiana* using different sgRNAs from BeYDV, which was further improved by a combination of two sgRNAs together. The strategy involving use of CRISPR/Cas9 mediated resistance against geminiviruses was further extended to TYLCV – a monopartite begomovirus causing yellow leaf curl disease in tomatoes in most of the tomato growing areas of the world (Ali et al., 2015). During their study Ali et al. (2015) found that sgRNA constructs targeting stem loop structure in IR conferred better resistance against TYLCV in *N. benthamiana*plants. These successful implications suggest that there is huge potential in employing CRISPR/Cas9 system against geminiviruses, thereby targeting several sites in the virus genome (Zlotorynski, 2015).

PROPOSED METHODOLOGY FOR DESIGNING MULTIPLEX SGRNA TO CIRCUMVENT CLCUD

Till now, the CRISPR/Cas9 has been employed against just single species of geminiviruses, as previously mentioned, which limits this technology from reaching its full potential. To maximize the output of this technology, a multiplex type sgRNA could be designed that allow simultaneous editing of multiple begomoviruses in just one go and this is a noteworthy advantage of CRISPR/Cas9 system in comparison to other virus controlling strategies. For effective, rapid and wide-scale control over CABs, here we addressed key considerations for designing multiplex CRISPR/Cas9 system whose applications include, but not circumscribed to, simultaneous targeting of all types of CABs along with their cognate DNA-satellites (present in south east Asia). To achieve an absolute control over such a huge complex demands a multiplex genome editing strategy based on multiplex sgRNA that can potentially target whole CLCuD complex. For the said purpose, full length nucleotide sequences of CABs and their cognate DNA-satellites were retrieved from NCBI databank, aligned in Mega6 (Thompson et al., 1994) and most conserved regions, bearing protospacer adjacent motif (PAM) sequences, were selected to design multiplex sgRNA cassette (Figures3and 4). All the designed sgRNAs (16 for CABs, 08 for alphsatellites and 12 for CLCuMuB) were analyzed on the basis of available *Arabidopsis* genome for specificity, off-targets and quality score. Analyses showed that the majority of designed sgRNAs have off-target effects, though in non-coding regions, thus posing a major challenge in the course of their successful exploitation. Several strategies have been reported to enhance Cas9 specificity, including use of Cas9 nickase mutant (Mali et al., 2013; Ran et al., 2013), catalytically inactive Cas9 nucleases (Guilinger et al., 2014; Tsai et al., 2014), and few more strategies but with limited applications. Very recently, high-fidelity Cas9 nucleases were developed on the basis of structure-guided protein engineering strategy where a few amino acids were replaced successfully to reduce off-target effects and maintain robust on-target cleavage (Kleinstiver et al., 2016; Slaymaker et al., 2016). Importantly, Baltes et al. (2015) showed that designed sgRNA coupled with catalytically inactive Cas9 (dCas9) could control the BeYDV genome without cleaving the genome, hence, Cas9-HF (Kleinstiver et al., 2016; Slaymaker et al., 2016) or dCas9 (Baltes et al., 2015) could be used to eliminate the off-targets effects. We hereby proposed that same strategy could be employed here to reduce the off-target effects.

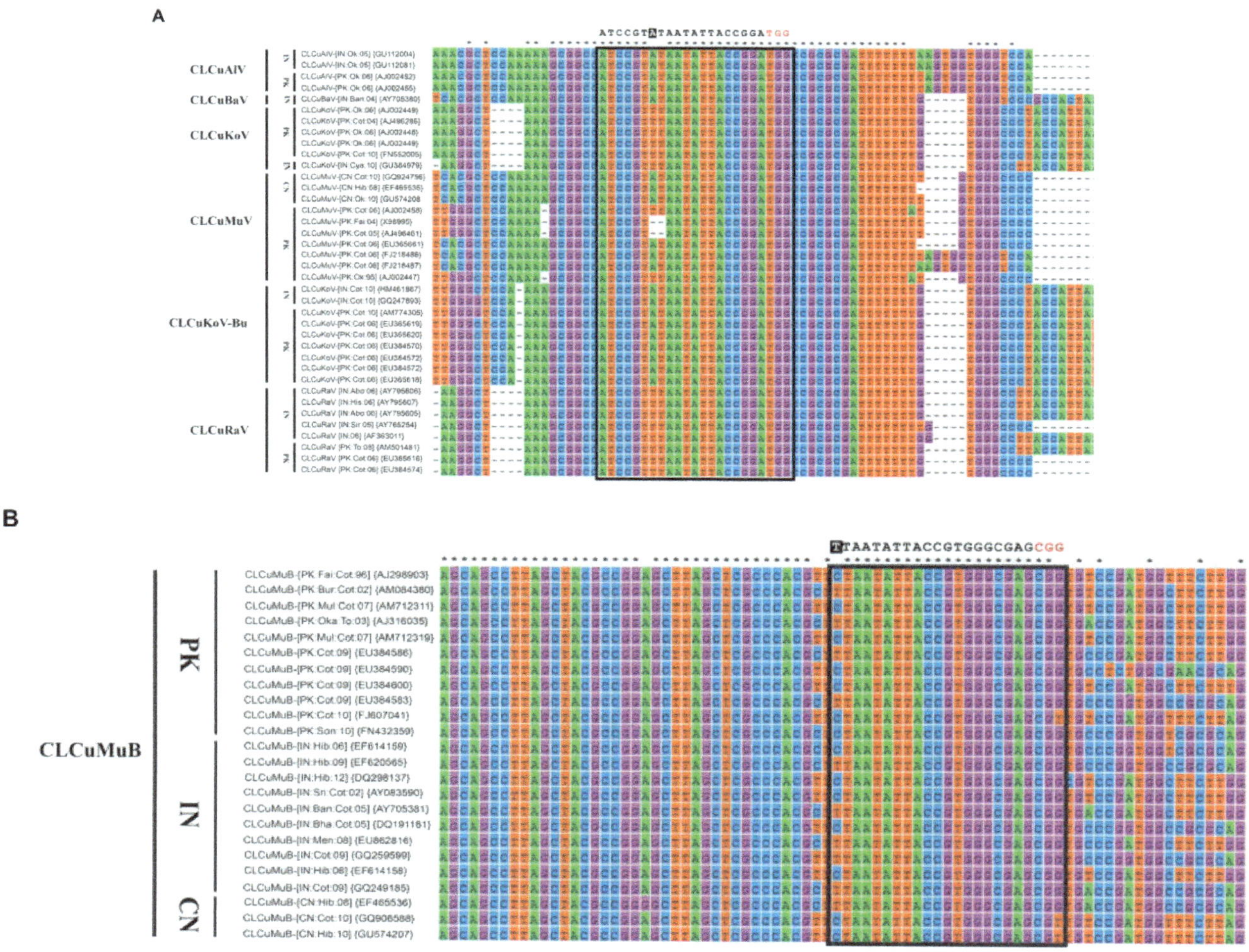

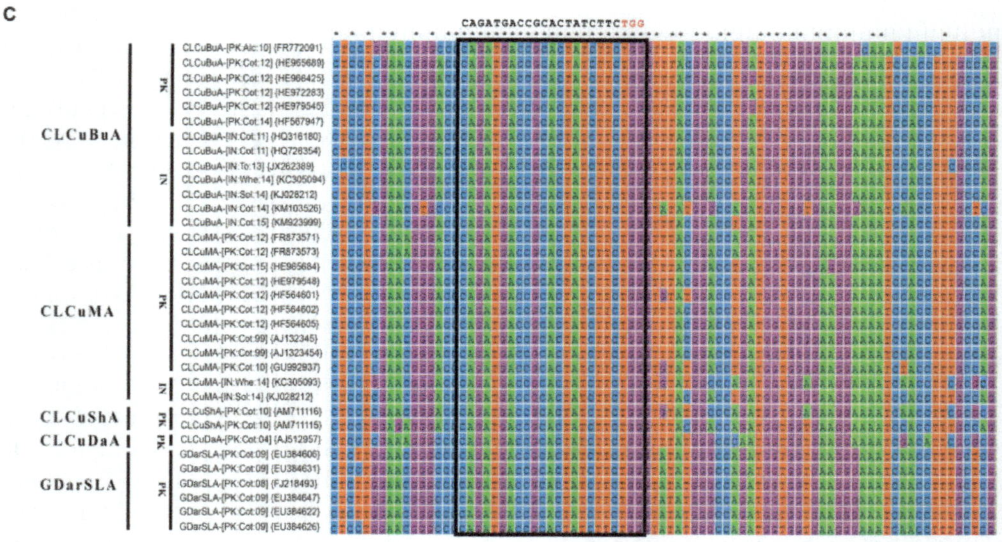

Figure 3: Nucleotide sequence alignments of CABs (A) and their associated DNA-satellites (B,C). All the sequences were retrieved from NCBI GenBank database and aligned using ClustalW in Mega6. The isolate descriptors and accession numbers are given on the left side of each alignment. Only the recommended sgRNA sequence is given at the top of each alignment as black text. The protospacer adjacent motif (PAM) sequence is indicated in red text. Any mismatch outside of the seed sequence is highlighted as white text with black background.

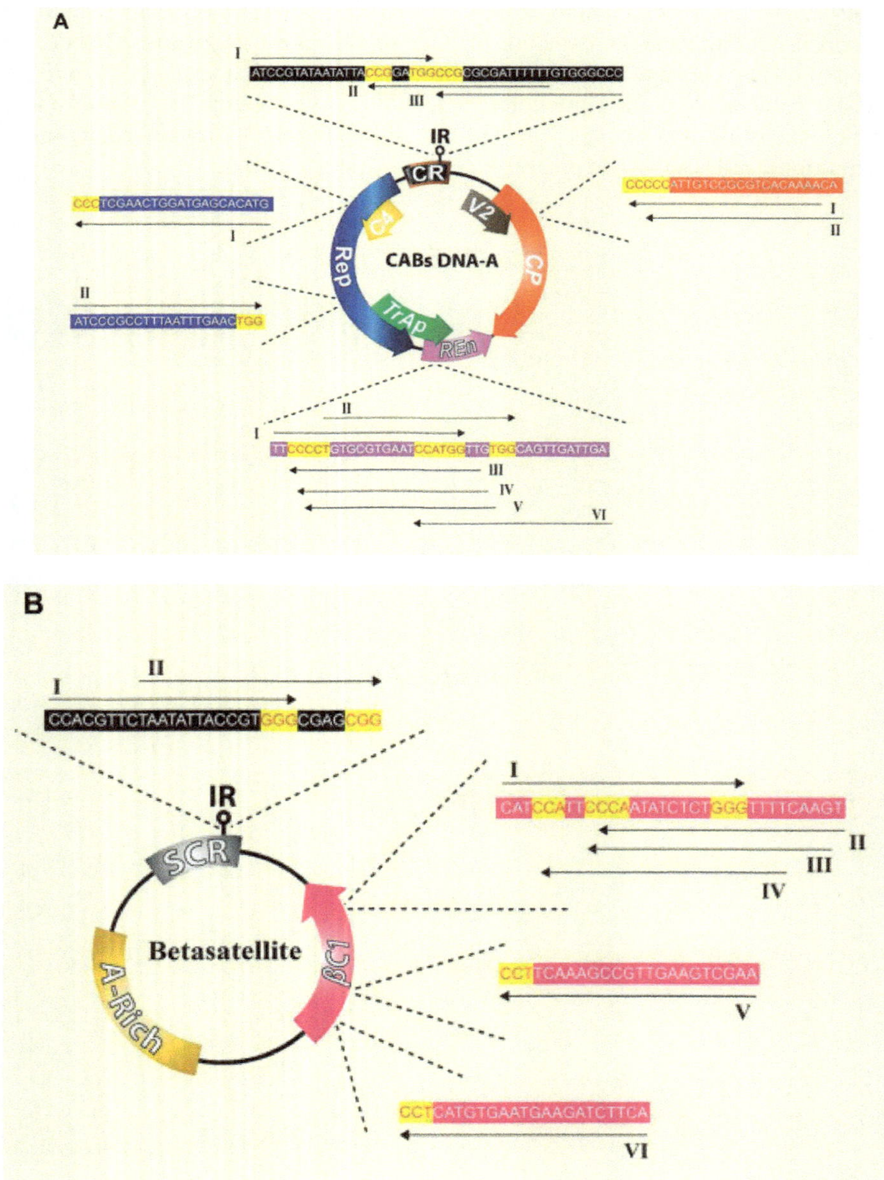

C

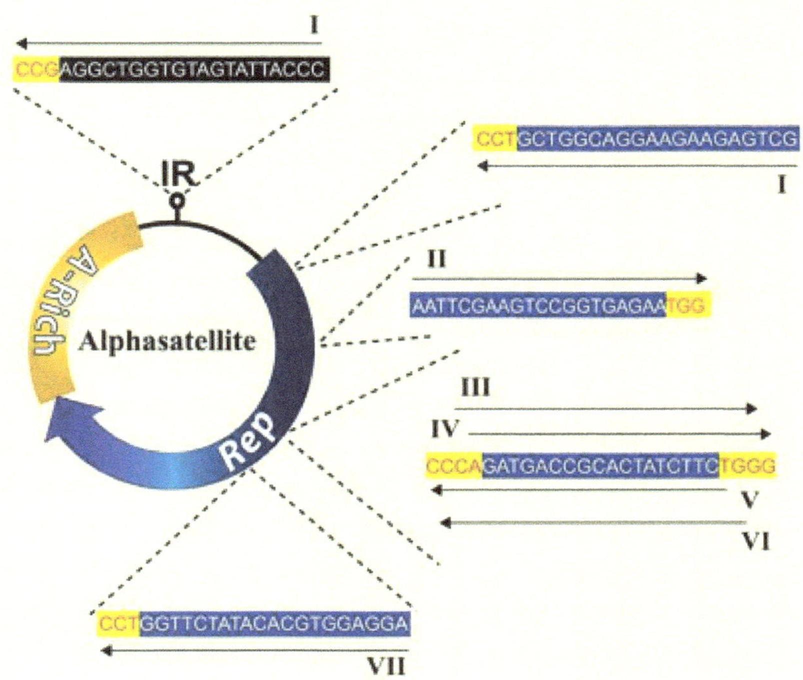

Figure 4: Wild type genome of CABs (A) and associated DNA-satellites(B,C) illustrating the suggested sgRNAs sequences and their corresponding potential targets in each molecule, respectively. The PAM sequences are indicated in red text with yellow background whereas, rest of the sequences are indicated in black text with colored background similar to the respective target gene. All sgRNAs are indicated as unbroken black arrows showing the direction of expression of each sgRNA.

Our analyses showed that designed sgRNAs targeting IR regions have less off-targets in *Arabidopsis* genome. Moreover, it has been proved that sgRNAs targeting IR of geminiviruses confer better resistance as compared to other targeted sgRNAs in a geminivirus genome (Ali et al., 2015; Baltes et al., 2015). Therefore, we proposed a multiplex sgRNA cassette targeting IR and Rep of the CABs, IR of CLCuMuB and *Rep* of diverse alphasatellites, respectively. Targeting IR could provide broad-spectrum resistance against these viruses under natural conditions because it is conserved among all members of the plant virus family *geminiviridae*, hence chance of resistance breaking would be minimized. To assemble such a cassette bearing multiple sgRNAs, many different strategies could be used such as golden gate cloning (Engler et al., 2008; Xing et al., 2014), array assembly (Guo et al., 2015), tRNA processing (Xie et al., 2015), and use of chimeric constructs (Cong et al., 2013). The expression of the gRNA cassette should be driven by single RNA Pol-III promoter and Cas9 may be expressed from CaMV-35S promoter followed by a NOS terminator sequence. Moreover, each sgRNA should be followed by a DR sequence to facilitate homologous recombination (Figure5B). Finally, the designed cassette could be cloned directly into a binary vector to facilitate *Agrobacterium*-mediated stable genetic transformation in the model plant *N. benthamiana* (Figure 5C). The transgenic plants harboring the sgRNA multiplex cassette along with all necessary controls would be challenged by different CABs with and without their cognate DNA-satellites using agroinfiltration (Figure 5D). The control plants will start showing typical CLCuD symptoms after 2–3 weeks of post inoculation. Whereas, the transgenic plants harboring sgRNA multiplex cassette would be symptom free depending upon the level of expression of each sgRNA (Figure 5E). Once proof-of-concept established in *N. benthamiana* plants then whole system could be tested in the local cotton germplasm such as 'Cocker', which is not only susceptible to CLCuD but also easy to transform (Ikram-Ul-Haq, 2004; Sohrab et al., 2014). The plant transformation will be done by*Agrobacterium*-mediated transformation and the fully established transgenic plants will be challenged by viruliferous whiteflies either in the cages or directly exposed to the viral inoculum in the field. The multiplex CRISPR/Cas9 strategy proposed here, could provide a comprehensive approach to control CABs and their associated DNA-satellites simultaneously.

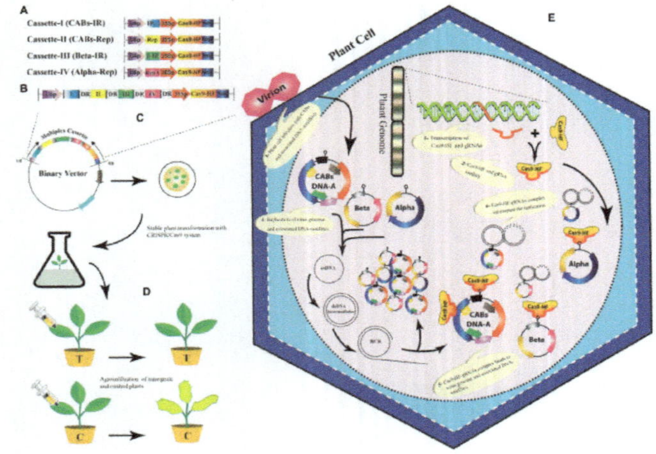

Figure 5: A tentative approach to engineer resistance against CABs and their associated DNA-satellites using CRISPR/Cas9 system. Graphical representation to construct sgRNA cassettes: A uniplex approach to construct individual sgRNAs for IR and Rep of CABs, βC1 and Rep-A to target CABs, betasatellite and alphasatellite, respectively, (B). All of the sgRNAs are shown to be expressed from common RNA polymerase-III promoter (U6p). Each sgRNA is followed by a DR sequence (rounded rectangle) to ensure homologous recombination. Cas9 may be expressed from CaMV-35S promoter followed by a NOS terminator sequence. The multiplex cassette would be inserted between left- and right border of a suitable binary vector to carry out stable genetic transformation in Nicotiana benthamiana plants (C). The control plants will be transformed with empty binary vector to ensure no resistance against CABs and associated DNA-satellites. The transgenic N. benthamiana plants will be agroinfiltrated with the CABs alone and/or with the associated DNA-satellites (D). The transgenic plants successfully expressing CRISPR/Cas9 would show resistance against the viral complex whereas, the negative control plants will start showing typical CLCuD symptoms. A simultaneous in planta graphical representation of CRISPR/Cas9 system (E): 1- Principle components of CRISPR/Cas9 system are transcribed from the plant genome in transgenic plants. 2- Assemblage of Cas9-HF and sgRNA. 3- Virus infection starts inside the cell after agroinfiltration using Agrobacterium tumefacience. 4- The single stranded DNA (ssDNA) of begomovirus and the associated DNA-satellites start proliferating by rolling circle replication (RCR) via a double stranded (ds) DNA intermediates in an infected plant cell. 5- The sgRNA and Cas9-HF complex bind at the complementary target sites along the dsDNA intermediates. 6- The activated sgRNA complex target the viral genome (s) and produce double-strand break (DSB) that can lead to degradation of viral and satellite genome. However, DSB can be repaired by non-homologous end joining (NHEJ) repair mechanism thus leaving the mutation in viral genome.

CONCLUSION

Development of resistance against CLCuD is an intricate practice amenable to be overcome by counter strategies adopted by either the CABs or associated DNA-satellites. Offering resistance through conventional as well as other synthetic biology approaches is genetically tractable through molecular genetics but the history showed that the resistance, once achieved, was not long-lasting. Alternatively, we have to devise a multifaceted strategy, which could provide unusually precise and specific approach to counter CLCuD-complex. CRISPR/Cas9 system offers a flexible approach for stacking multiple nucleases as one transgene, thereby offering targeted cleavage of mixed infections by multiple viruses and associated DNA-satellites, such as CLCuD-complex. The CRISPR/Cas9 system could be an answer to open up unprecedented possibilities to develop CLCuD resistant cotton plants. Thus, it could be beneficial to the scientific community in devising future resistance strategies against whole CLCuD complex and this approach will be an outstanding example.

AUTHOR CONTRIBUTIONS

ZI conceived the idea, performed the bioinformatics analysis for designed sgRNAs and drafted the first draft of the manuscript. MNS designed all the sgRNAs, draw figures and helped in preparation of first draft of the manuscript. MS drew all tables, searched all the web based tools and helped in preparation of first draft of the manuscript. First draft of the manuscript was edited and approved by all authors.

ACKNOWLEDGMENTS

Authors would like to thank Dr. Thomas Martin, Uppsala University, Sweden for valuable suggestions, and critically reading this manuscript.

REFERENCES

1. Ahmad S., Islam N., Mahmood A., Ashraf F., Hayat K., Hanif M. (2010). Screening of cotton germplasm against *Cotton leaf curl virus*. *Pak. J. Bot.* 42 3327–3342.
2. Ahmad S., Mahmood K., Hanif M., Nazeer W., Malik W., Qayyum A., et al. (2011). Introgression of cotton leaf curl virus-resistant genes from Asiatic cotton (*Gossypium arboreum*) into upland cotton (*G. hirsutum*). *Genet. Mol. Res.* 10 2404–2414. 10.4238/2011.October.7.2
3. Akad F., Eybishtz A., Edelbaum D., Gorovits R., Dar-Issa O., Iraki N., et al. (2007). Making a friend from a foe: expressing a GroEL gene from the whitefly *Bemisia tabaci* in the phloem of tomato plants confers resistance to Tomato yellow leaf curl virus. *Arch. Virol.* 152 1323–1339. 10.1007/s00705-007-0942-0
4. Akhtar K. P., Khan M. K. R., Ahmad M., Sarwar N., Ditta A. (2010). Partial resistance of a cotton mutant to *Cotton leaf curl Burewala virus*. *Span. J. Agric. Res.* 8 1098–1104. 10.1007/s00705-012-1225-y
5. Ali I., Amin I., Briddon R., Mansoor S. (2013). Artificial microRNA-mediated resistance against the monopartite *Begomovirus Cotton leaf curl Burewala virus*. *Virol. J.* 10 231 10.1186/1743-422X-10-231
6. Ali M. (1997). Breeding of cotton varieties for resistance to cotton leaf curl virus. *Pak. J. Phytopathol.*9 1–7. 10.1016/j.virusres.2013.12.008
7. Ali Z., Abulfaraj A., Idris A., Ali S., Tashkandi M., Mahfouz M. (2015). CRISPR/Cas9-mediated viral interference in plants. *Genome Biol.* 16 238 10.1186/s13059-015-0799-6
8. Amin I., Patil B. L., Briddon R. W., Mansoor S., Fauquet C. M. (2011). Comparison of phenotypes produced in response to transient expression of genes encoded by four distinct *Begomovirus*es in*Nicotiana benthamiana* and their correlation with the levels of developmental miRNAs. *Virol. J.* 8 23810.1186/1743-422X-8-238
9. Amudha J., Balasubramani G., Malathi V., Monga D., Bansal K., Kranthi K. (2010). Cotton transgenics with

Antisense AC1 gene for resistance against cotton leaf curl virus. *Electronic J. Plant Breed.* 1 360–369.

10. Amudha J., Balasubramani G., Malathi V. G., Monga D., Kranthi K. R. (2011). Cotton leaf curl virus resistance transgenics with antisense coat protein gene (AV1). *Curr. Sci.* 101 300–307.
11. Aragão F. J. L., Faria J. C. (2009). First transgenic geminivirus-resistant plant in the field. *Nat. Biotechnol.* 27 1086–1088. 10.1038/nbt1209-1086
12. Asad S., Haris W., Bashir A., Zafar Y., Malik K., Malik N., et al. (2003). Transgenic tobacco expressing geminiviral RNAs are resistant to the serious viral pathogen causing cotton leaf curl disease. *Arch. Virol.* 148 2341–2352. 10.1007/s00705-003-0179-5
13. Baltes N. J., Hummel A. W., Konecna E., Cegan R., Bruns A. N., Bisaro D. M., et al. (2015).Conferring resistance to geminiviruses with the CRISPR–Cas prokaryotic immune system. *Nat. Plants*1 15145 10.1038/nplants.2015.145
14. Bhaya D., Davison M., Barrangou R. (2011). CRISPR-Cas systems in bacteria and archaea: versatile small RNAs for adaptive defense and regulation. *Annu. Rev. Genet.* 45 273–297. 10.1146/annurev-genet-110410-132430
15. Bolotin A., Quinquis B., Sorokin A., Ehrlich S. D. (2005). Clustered regularly interspaced short palindrome repeats (CRISPRs) have spacers of extrachromosomal origin. *Microbiology* 151 2551–2561. 10.1099/mic.0.28048-0
16. Böttcher B., Unseld S., Ceulemans H., Russell R. B., Jeske H. (2004). Geminate structures of African cassava mosaic virus. *J. Virol.* 78 6758–6765. 10.1128/JVI.78.13.6758-6765.2004
17. Briddon R. W. (2003). Cotton leaf curl disease, a multicomponent *Begomovirus* complex. *Mol. Plant Pathol.* 4 427–434. 10.1046/j.1364-3703.2003.00188.x
18. Briddon R. W., Bull S. E., Amin I., Mansoor S., Bedford I. D., Rishi N., et al. (2004). Diversity of DNA 1; a satellite-like molecule associated with monopartite *Begomovirus*-DNA β complexes.*Virology* 324 462–474. 10.1016/j.virol.2004.03.041
19. Briddon R. W., Mansoor S., Bedford I. D., Pinner M. S., Saunders K., Stanley J., et al. (2001).Identification of DNA components required for induction of cotton leaf curl disease. *Virology* 285234–243. 10.1006/viro.2001.0949
20. Briddon R. W., Markham P. G. (2001). Cotton leaf curl virus disease. *Virus Res.* 71 151–159. 10.1016/S0168-1702(00)00195-7
21. Brown J., Zerbini F. M., Navas-Castillo J., Moriones E., Ramos-Sobrinho R., Silva J. F., et al. (2015).Revision of *Begomovirus* taxonomy based on pairwise sequence comparisons. *Arch. Virol.* 160 1593–1619. 10.1007/s00705-015-2398-y
22. Cheng X., Wang X., Wu J., Briddon R. W., Zhou X. (2011). βC1 encoded by tomato yellow leaf curl China betasatellite forms multimeric complexes in vitro and in vivo. *Virology* 409 156–162. 10.1016/j.virol.2010.10.007
23. Cobb R. E., Wang Y., Zhao H. (2015). High-efficiency multiplex genome editing of *Streptomyces*species using an engineered CRISPR/Cas system. *ACS Synth. Biol.* 4 723–728. 10.1021/sb500351f
24. Cong L., Ran F. A., Cox D., Lin S., Barretto R., Habib N., et al. (2013). Multiplex genome engineering using CRISPR/Cas systems. *Science* 339 819–823. 10.1126/science.1231143
25. Cui X., Li G., Wang D., Hu D., Zhou X. (2005). A *Begomovirus* DNAβ-encoded protein binds DNA, functions as a suppressor of RNA silencing, and targets the cell nucleus. *J. Virol.* 79 10764–10775. 10.1128/JVI.79.16.10764-10775.2005
26. Doench J. G., Hartenian E., Graham D. B., Tothova Z., Hegde M., Smith I., et al. (2014). Rational design of highly active sgRNAs for CRISPR-Cas9-mediated gene inactivation. *Nat. Biotechnol.* 321262–1267. 10.1038/nbt.3026
27. Edelbaum D., Gorovits R., Sasaki S., Ikegami M., Czosnek H. (2009). Expressing a whitefly GroEL protein in *Nicotiana benthamiana* plants confers tolerance to tomato yellow leaf curl virus and cucumber mosaic virus, but not to grapevine virus A or tobacco mosaic virus. *Arch. Virol.* 154 399–407. 10.1007/s00705-009-0317-9
28. Engler C., Kandzia R., Marillonnet S. (2008). A one pot, one step, precision cloning method with high throughput capability. *PLoS ONE* 3:e3647 10.1371/journal.pone.0003647
29. Farooq A., Farooq J., Mahmood A., Batool A., Rehman A., Shakeel A., et al. (2011). An overview of cotton leaf curl virus disease (CLCuD) a serious threat to cotton productivity. *Aust. J. Crop Sci.* 51823–1831.
30. Feng Z., Zhang B., Ding W., Liu X., Yang D.-L., Wei P., et al. (2013). Efficient genome editing in plants using a CRISPR/Cas system. *Cell Res.* 23 1229–1232. 10.1038/cr.2013.114
31. Guilinger J. P., Thompson D. B., Liu D. R. (2014). Fusion of catalytically inactive Cas9 to FokI nuclease improves the specificity of genome modification. *Nat. Biotechnol.* 32 577–582. 10.1038/nbt.2909
32. Guo L., Xu K., Liu Z., Zhang C., Xin Y., Zhang Z. (2015). Assembling the *Streptococcus thermophilus* clustered regularly interspaced short palindromic repeats (CRISPR) array for multiplex DNA targeting. *Anal. Biochem.* 478 131–133. 10.1016/j.ab.2015.02.028
33. Hanley-Bowdoin L., Settlage S. B., Robertson D. (2004). Reprogramming plant gene expression: a prerequisite to geminivirus DNA replication. *Mol. Plant Pathol.* 5 149–156. 10.1111/j.1364-3703.2004.00214.x
34. Heigwer F., Kerr G., Boutros M. (2014). E-CRISP: fast CRISPR target site identification. *Nat. Methods* 11 122–123. 10.1038/nmeth.2812
35. Hsu P. D., Scott D. A., Weinstein J. A., Ran F. A., Konermann S., Agarwala V., et al. (2013). DNA targeting specificity of RNA-guided Cas9 nucleases. *Nat. Biotechnol.* 31 827–832. 10.1038/nbt.2647

36. Hussain M., Mansoor S., Iram S., Zafar Y., Briddon R. W. (2007). The hypersensitive response to tomato leaf curl New Delhi virus nuclear shuttle protein is inhibited by transcriptional activator protein.*Mol. Plant Microbe Interact.* 20 1581–1588. 10.1094/MPMI-20-12-1581

37. Hussain T., Ali M. (1975). A review of cotton diseases of Pakistan. *Pak. Cottons* 19 71–86.

38. Ikram-Ul-Haq (2004). Agrobacterium-mediated transformation of cotton (*Gossypium hirsutum* L.) via vacuum infiltration. *Plant Mol. Biol. Rep.* 22 279–288. 10.1007/BF02773138

39. Ilardi V., Tavazza M. (2015). Biotechnological strategies and tools for Plum pox virus resistance: trans-, intra-, cis-genesis and beyond. *Front. Plant Sci.* 6:379 10.3389/fpls.2015.00379

40. Iqbal M. J., Reddy O. U. K., El-Zik K. M., Pepper A. E. (2001). A genetic bottleneck in the 'evolution under domestication' of upland cotton *Gossypium hirsutum* L. examined using DNA fingerprinting. *Theor. Appl. Genet.* 103 547–554. 10.1007/PL00002908

41. Iqbal Z., Sattar M. N., Kvarnheden A., Mansoor S., Briddon R. W. (2012). Effects of the mutation of selected genes of *Cotton leaf curl Kokhran virus* on infectivity, symptoms and the maintenance of Cotton leaf curl Multan betasatellite. *Virus Res.* 169 107–116. 10.1016/j.virusres.2012.07.016

42. Ji X., Zhang H., Zhang Y., Wang Y., Gao C. (2015). Establishing a CRISPR–Cas-like immune system conferring DNA virus resistance in plants. *Nat. Plants* 1 15144 10.1038/nplants.2015.144

43. Jiang W., Bikard D., Cox D., Zhang F., Marraffini L. A. (2013). RNA-guided editing of bacterial genomes using CRISPR-Cas systems. *Nat. Biotechnol.* 31 233–239. 10.1038/nbt.2508

44. Jinek M., Chylinski K., Fonfara I., Hauer M., Doudna J. A., Charpentier E. (2012). A programmable dual-RNA–guided DNA endonuclease in adaptive bacterial immunity. *Science* 337 816–821. 10.1126/science.1225829

45. Kabadi A. M., Ousterout D. G., Hilton I. B., Gersbach C. A. (2014). Multiplex CRISPR/Cas9-based genome engineering from a single lentiviral vector. *Nucleic Acids Res.* 42:e147 10.1093/nar/gku749

46. Kirthi N., Priyadarshini C. G. P., Sharma P., Maiya S. P., Hemalatha V., Sivaraman P., et al. (2004). Genetic variability of *Begomovirus*es associated with cotton leaf curl disease originating from India.*Arch. Virol.* 149 2047–2057.

47. Kleinstiver B. P., Pattanayak V., Prew M. S., Tsai S. Q., Nguyen N. T., Zheng Z., et al. (2016). High-fidelity CRISPR–Cas9 nucleases with no detectable genome-wide off-target effects. *Nature* 529 490–495. 10.1038/nature16526

48. Kumar A., Kumar J., Khan J. (2010). Sequence characterization of cotton leaf curl virus from Rajasthan: phylogenetic relationship with other members of geminiviruses and detection of recombination. *Virus Gen.* 40 282–289. 10.1007/s11262-009-0439-x

49. Kumar J., Gunapati S., Alok A., Lalit A., Gadre R., Sharma N., et al. (2015). *Cotton leaf curl Burewala virus* with intact or mutant transcriptional activator proteins: complexity of cotton leaf curl disease. *Arch. Virol.* 160 1219–1228. 10.1007/s00705-015-2384-4

50. Li J.-F., Norville J. E., Aach J., McCormack M., Zhang D., Bush J., et al. (2013). Multiplex and homologous recombination-mediated genome editing in *Arabidopsis* and *Nicotiana benthamiana*using guide RNA and Cas9. *Nat. Biotechnol.* 31 688–691. 10.1038/nbt.2654

51. Liu Y., Robinson D. J., Harrison B. D. (1998). Defective forms of cotton leaf curl virus DNA-A that have different combinations of sequence deletion, duplication, inversion and rearrangement. *J. Gen. Virol.* 79 1501–1508. 10.1099/0022-1317-79-6-1501

52. Mahmood T., Arshad M., Gill M. I., Mahmood H. T. (2003). Burewala strain of cotton leaf curl virus: a threat to CLCuV cotton resistant varieties. *Asian J. Plant Sci.* 2 968–970. 10.3923/ajps.2003.968.970

53. Mali P., Aach J., Stranges P. B., Esvelt K. M., Moosburner M., Kosuri S., et al. (2013). CAS9 transcriptional activators for target specificity screening and paired nickases for cooperative genome engineering. *Nat. Biotechnol.* 31 833–838. 10.1038/nbt.2675

54. Mansoor S., Amin I., Hussain M., Zafar Y., Bull S., Briddon R. W., et al. (2001). Association of a disease complex involving a *Begomovirus*, DNA 1 and a distinct DNA beta with leaf curl disease of okra in Pakistan. *Plant Dis.* 85 922 10.3390/v5092116

55. Mansoor S., Amin I., Iram S., Hussain M., Zafar Y., Malik K. A., et al. (2003). Breakdown of resistance in cotton to cotton leaf curl disease in Pakistan. *Plant Pathol.* 52 784 10.1111/j.1365-3059.2003.00893.x

56. Montague T. G., Cruz J. M., Gagnon J. A., Church G. M., Valen E. (2014). CHOPCHOP: a CRISPR/Cas9 and TALEN web tool for genome editing. *Nucleic Acids Res.* 42 W401–W407. 10.1093/nar/gku410

57. Mubin M., Amin I., Amrao L., Briddon R. W., Mansoor S. (2010). The hypersensitive response induced by the V2 protein of a monopartite *Begomovirus* is countered by the C2 protein. *Mol. Plant Pathol.* 11 245–254. 10.1111/j.1364-3703.2009.00601.x

58. Mubin M., Hussain M., Briddon R. W., Mansoor S. (2011). Selection of target sequences as well as sequence identity determine the outcome of RNAi approach for resistance against cotton leaf curl geminivirus complex. *Virol. J.* 8 122 10.1186/1743-422X-8-122

59. Qazi J., Amin I., Mansoor S., Iqbal J., Briddon R. W. (2007). Contribution of the satellite encoded gene βC1 to cotton leaf curl disease symptoms. *Virus Res.* 128 135–139. 10.1016/j.virusres.2007.04.002

60. Rahman H. (1997). *Breeding Approaches to Cotton Leaf Curl Resistance. The ICAC RECORDER,*Vol. XV Washington, DC: International Cotton Advisory Committee, 11–14.

61. Rahman M., Zafar Y. (2007). Registration of NIBGE-115 cotton. *J. Plant Regist.* 1 51–52. 10.3198/jpr2006.12.0778crg

62. Rajagopalan P., Naik A., Katturi P., Kurulekar M., Kankanallu R., Anandalakshmi R. (2012).Dominance of resistance-breaking *Cotton leaf curl Burewala virus* (CLCuBuV) in northwestern India.*Arch. Virol.* 157 855–868. 10.1007/s00705-012-1225-y

63. Ran F. A., Hsu P. D., Lin C.-Y., Gootenberg J. S., Konermann S., Trevino A. E., et al. (2013). Double nicking by RNA-Guided CRISPR Cas9 for enhanced genome editing specificity. *Cell* 154 1380–1389. 10.1016/j.cell.2013.08.021

64. Rana V. S., Singh S. T., Priya N. G., Kumar J., Rajagopal R. (2012). Arsenophonus GroEL interacts with CLCuV and is localized in midgut and salivary gland of whitefly *B. tabaci*. *PLoS ONE* 7:e4216810.1371/journal.pone.0042168

65. Rojas M. R., Jiang H., Salati R., Xoconostle-Cázares B., Sudarshana M. R., Lucas W. J., et al. (2001). Functional analysis of proteins involved in movement of the monopartite *Begomovirus*, Tomato yellow leaf curl virus. *Virology* 291 110–125. 10.1006/viro.2001.1194

66. Saeed M., Behjatnia S. A. A., Mansoor S., Zafar Y., Hasnain S., Rezaian M. A. (2005). A single complementary-sense transcript of a geminiviral DNA β satellite is determinant of pathogenicity. *Mol. Plant Microbe Interact.* 18 7–14. 10.1094/MPMI-18-0007

67. Saeed M., Mansoor S., Rezaian M. A., Briddon R. W., Randles J. W. (2008). Satellite DNA β overrides the pathogenicity phenotype of the C4 gene of Tomato leaf curl virus, but does not compensate for loss of function of the coat protein and V2 genes. *Arch. Virol.* 153 1367–1372. 10.1007/s00705-008-0124-8

68. Saeed M., Zafar Y., Randles J. W., Rezaian M. A. (2007). A monopartite begomovirus-associated DNA β satellite substitutes for the DNA B of a bipartite begomovirus to permit systemic infection *J. Gen. Virol.* 88 2881–2889. 10.1099/vir.0.83049-0

69. Sakuma T., Nishikawa A., Kume S., Chayama K., Yamamoto T. (2014). Multiplex genome engineering in human cells using all-in-one CRISPR/Cas9 vector system. *Sci. Rep.* 4 1–6. 10.1038/srep05400

70. Sanjaya, Satyavathi V. V., Prasad V., Kirthi N., Maiya S. P., Savithri H. S., et al. (2005). Development of cotton transgenics with antisense AV2 gene for resistance against cotton leaf curl virus (CLCuD) via*Agrobacterium tumefaciens*. *Plant Cell Tiss. Organ Cult.* 81 55–63. 10.1007/s11240-004-2777-7

71. Sattar M. N., Kvarnheden A., Saeed M., Briddon R. W. (2013). Cotton leaf curl disease – an emerging threat to cotton production worldwide. *J. Gen. Virol.* 94 695–710. 10.1099/vir.0.049627-0

72. Settlage S. B., See R. G., Hanley-Bowdoin L. (2005). Geminivirus C3 protein: replication enhancement and protein interactions. *J. Virol.* 79 9885–9895. 10.1128/JVI.79.15.9885-9895.2005

73. Shepherd D. N., Martin D. P., Thomson J. A. (2009). Transgenic strategies for developing crops resistant to geminiviruses. *Plant Sci.* 176 1–11. 10.1016/j.plantsci.2008.08.011

74. Shuja M. N., Briddon R. W., Tahir M. (2014). Identification of a distinct strain of *Cotton leaf curl Burewala virus*. *Arch. Virol.* 159 2787–2790. 10.1007/s00705-014-2097-0

75. Siddig M. A. (1970). "Breeding for leaf curl resistance in Sakel cotton," in *Cotton Growth in the Gezira Environment*, eds Siddig M. A., Hughes L. C., editors. (Wad Madani: Agricultural Research Cororation;), 153.

76. Slaymaker I. M., Gao L., Zetsche B., Scott D. A., Yan W. X., Zhang F. (2016). Rationally engineered Cas9 nucleases with improved specificity. *Science* 351 84–88. 10.1126/science.aad5227

77. Sohrab S. S., Kamal M. A., Ilah A., Husen A., Bhattacharya P. S., Rana D. (2014). Development of*Cotton leaf curl virus* resistant transgenic cotton using antisense ßC1 gene. *Saudi J. Biol. Sci.* 22 279–288.

78. Sunitha S., Shanmugapriya G., Balamani V., Veluthambi K. (2013). Mungbean yellow mosaic virus (MYMV) AC4 suppresses post-transcriptional gene silencing and an AC4 hairpin RNA gene reduces MYMV DNA accumulation in transgenic tobacco. *Virus Genes* 46 496–504. 10.1007/s11262-013-0889-z

79. Tahir M. N. (2014). *Use of Amplicon-Based System for Enhanced Resistance against Cotton Leaf Curl Disease.* Ph.D. thesis, Quaid-i-Azam University, Islamabad.

80. Thompson J. D., Higgins D. G., Gibson T. J. (1994). CLUSTAL W: improving the sensitivity of progressive multiple sequence alignment through sequence weighting, positions-specific gap penalties and weight matrix choice. *Nucleic Acids Res.* 22 4673–4680. 10.1093/nar/22.22.4673

81. Tsai S. Q., Wyvekens N., Khayter C., Foden J. A., Thapar V., Reyon D., et al. (2014). Dimeric CRISPR RNA-guided FokI nucleases for highly specific genome editing. *Nat. Biotechnol.* 32 569–576. 10.1038/nbt.2908

82. Vanitharani R., Chellappan P., Pita J. S., Fauquet C. M. (2004). Differential roles of AC2 and AC4 of cassava geminiviruses in mediating synergism and suppression of posttranscriptional gene silencing. *J. Virol.* 78 9487–9498. 10.1128/JVI.78.17.9487-9498.2004

83. Varma A., Malathi V. G. (2003). Emerging geminivirus problems: a serious threat to crop production.*Ann. Appl. Biol.* 142 145–164. 10.1111/j.1744-7348.2003.tb00240.x

84. Wang X., Tang Y., Lu J., Shao Y., Qin X., Li Y., et al. (2016). Characterization of novel cytochrome P450 2E1 knockout rat model generated by CRISPR/Cas9. *Biochem. Pharmacol.* 105 80–90. 10.1016/j.bcp.2016.03.001

85. Xiao A., Cheng Z., Kong L., Zhu Z., Lin S., Gao G., et al. (2014). CasOT: a genome-wide Cas9/gRNA off-target searching tool. *Bioinformatics* 30 1180–1182. 10.1093/bioinformatics/btt764

86. Xie K., Minkenberg B., Yang Y. (2015). Boosting CRISPR/Cas9 multiplex editing capability with the endogenous tRNA-processing system. *Proc. Natl. Acad. Sci. U.S.A.* 112 3570–3575. 10.1073/pnas.1420294112

87. Xie K., Zhang J., Yang Y. (2014). Genome-wide prediction of highly specific guide RNA spacers for CRISPR–Cas9-mediated genome editing in model plants and major crops. *Mol. Plant* 7 923–926. 10.1093/mp/ssu009

88. Xing H.-L., Dong L., Wang Z.-P., Zhang H.-Y., Han C.-Y., Liu B., et al. (2014). A CRISPR/Cas9 toolkit for multiplex genome editing in plants. *BMC Plant Biol.* 14:327 10.1186/s12870-014-0327-y

89. Yang X., Baliji S., Buchmann R. C., Wang H., Lindbo J. A., Sunter G., et al. (2007). Functional modulation of the geminivirus AL2 transcription factor and silencing suppressor by self-interaction. *J. Virol.* 81 11972–11981. 10.1128/JVI.00617-07

90. Yousaf S., Rasool G., Amin I., Mansoor S., Saeed M. (2013). Interference of a synthetic rep protein to develop resistance against cotton leaf curl disease. *Int. J. Agric. Biol.* 15 1140–1144.

91. Zaffalon V., Mukherjee S., Reddy V., Thompson J., Tepfer M. (2012). A survey of geminiviruses and associated satellite DNAs in the cotton-growing areas of northwestern India. *Arch. Virol.* 157 483–495. 10.1007/s00705-011-1201-y

92. Zhang T., Luan J. B., Qi J. F., Huang C. J., Li M., Zhou X. P., et al. (2012). *Begomovirus*–whitefly mutualism is achieved through repression of plant defences by a virus pathogenicity factor. *Mol. Ecol.*21 1294–1304. 10.1111/j.1365-294X.2012.05457.x

93. Zhou X., Liu Y., Robinson D. J., Harrison B. D. (1998). Four DNA-A variants among Pakistani isolates of cotton leaf curl virus and their affinities to DNA-A of geminivirus isolates from okra. *J. Gen. Virol.* 79 915–923. 10.1099/0022-1317-79-4-915

94. Zlotorynski E. (2015). Plant cell biology: CRISPR-Cas protection from plant viruses. *Nat. Rev. Mol. Cell Biol.* 16 642–642. 10.1038/nrm4079

CHAPTER 5

ENHANCED ARTEMISININ YIELD BY EXPRESSION OF ROL GENES IN ARTEMISIA ANNUA

Erum Dilshad[1], Rosa Maria Cusido[2], Javier Palazon[2], Karla Ramirez Estrada[2], Mercedes Bonfill[2] and Bushra Mirza[1]

[1]Department of Biochemistry, Faculty of Biological Sciences, Quaid-iAzam University, Islamabad, Pakistan

[2] Laboratorio de Fisiologia Vegetal, Facultad de Farmacia, Universidad de Barcelona, Barcelona, Spain

ABSTRACT

Background

Despite of many advances in the treatment of malaria, it is still the fifth most prevalent disease worldwide and is one of the major causes of death in the developing countries which accounted for 584,000 deaths in 2013, as estimated by World Health Organization. Artemisinin from Artemisia annua is still one of the most effective treatments for malaria. Increasing the artemisinin content of A. annua plants by genetic engineering would improve the availability of this much-needed drug.

Methods

In this regard, a high artemisinin-yielding hybrid of A. annua produced by the centre for novel agricultural products of the University of York, UK, was selected (artemisinin maximally 1.4 %). As rol genes are potential candidates of biochemical engineering, genetic transformation of A. annua with Agrobacterium tumefaciens GV3101 harbouring vectors with rol B and rol C genes was carried out with the objective of enhancement of artemisinin content. Transgenic lines produced were analysed by the LC–MS for quantitative analysis of artemisinin and analogues. These high artemisinin yielding transgenics were also analysed by real time quantitative PCR to find the molecular dynamics of artemisinin enhancement. Genes of artemisinin biosynthetic pathway were studied including amorphadiene synthase (ADS), cytochrome P450, (CYP71AV1) and aldehyde dehydrogenase 1 (ALDH1). Trichome-specific fatty acyl-CoA reductase 1(TAFR1) is an enzyme involved in both trichome development and sesquiterpenoid biosynthesis and both processes are important for artemisinin biosynthesis. Thus, real time qPCR analysis of the TAFR1 gene was carried out, and trichome density was determined.

Citation: Erum Dilshad, Rosa Maria Cusido, Javier Palazon, Karla Ramirez Estrada, Mercedes Bonfill and Bushra Mirza, DOI 10.1186/s12936-015-0951-5.

Results

Transgenics of rolB gene showed two- to ninefold (the decimal adds nothing in the abstract, please simplify to two- to ninefold) increase in artemisinin, 4–12-fold increase in artesunate and 1.2–3-fold increase in dihydroartemisinin. Whereas in the case of rol C gene transformants, a fourfold increase in artemisinin, four to ninefold increase in artesunate and one- to twofold increase in dihydroartemisinin concentration was observed. Transformants with the rol B gene had higher expression of these genes than rol C transformants. TAFR1 was also found to be more expressed in rol gene transgenics than wild type A. annua, which was also in accordance with the trichome density of the respective plant.

Conclusion

Thus it was proved that rol B and rol C genes are effective in the enhancement of artemisinin content of A. annua, rol B gene being more active to play part in this enhancement than rol C gene.

BACKGROUND

Artemisinin, a highly potent anti-malarial agent found in the medicinal plant *Artemisia annua*, is part of the artemisinin-based combination therapy recommended by the World Health Organization for the treatment of malaria [1, 2]. Besides malaria, artemisinin is also effective against some viral and parasitic diseases i.e. hepatitis B, schistosomiasis, chagas disease and African sleeping sickness. It is also effective in the reduction of angiogenesis and treatment of some cancers [3]. Owing to the fact that availability of artemisinin is insufficient to meet worldwide demand, efforts are needed to increase its quantity. In this regard, some improvement has taken place by the Artemisia research project of CNAP (Centre for Novel Agricultural Products, University of York, UK). Their breeding programme has developed some high artemisinin-producing hybrids by making crosses between parental lines with consistent field performance, high artemisinin content and high seed productivity. Two hybrids which have been released commercially include Hyb1209r (Shennong, artemisinin 1.5 %) and Hyb8001r (Zenith, artemisinin 1.4 %) [4].

Artemisinin synthesis occurs in the glandular trichomes (GLTs) present on flowers, floral buds and leaves. Trichome-specific fatty acyl-CoA reductase 1 (TAFR1) is thought to be involved in GLT development and sesquiterpenoid biosynthesis; which are important for artemisinin production [5, 6]. Other enzymes like amorphadiene synthase (ADS), cytochrome P450, CYP71AV1 (CYP), double bond reductase 2 (DBR2) and aldehyde dehydrogenase 1 (ALDH1) catalyze different steps of artemisinin biosynthesis [7–11]. Overexpression of these enzymes may lead to increased artemisinin accumulation and also the derivatives. Some studies provide evidence that metabolic engineering of the biosynthetic pathway that leads to artemisinin by the insertion of different genes can increase artemisinin content *in planta* [12, 13]. In this regard, *rol* genes of *A. rhizogenes* are considered to be effective inducers of secondary metabolites production in plants [14]. *Rol A* is a DNA binding protein and stimulator of growth, *rol B* having tyrosine phosphatase activity regulates signal transduction pathway of auxin [15, 16], and *rol C* has cytokinin glucosidase activity. However, each gene seems to affect plant morphology and stimulate production of different secondary metabolites [17–20]. However, there is no report about production of transgenic *A. annua* plants with individual *rol* genes for enhancement of artemisinin and its derivatives. Previous work on transformation of other *Artemisia* sp. with *rol* genes showed not only over expression of artemisinin biosynthetic genes, but also artemisinin in the plant [21].

In the continuation of that work, genetic transformation of *A. annua* (Hyb8001r, Zenith) with *rol* genes (*rol B*and *rol C*) was carried out and consequences on the content of artemisinin were evaluated by LC–MS. The real time qPCR analysis of the genes involved in biosynthetic pathway of artemisinin (ADS, CYP71AV1 and ALDH1) provided an understanding of the molecular mechanism through which *rol* genes can enhance the artemisinin production. Expression of the gene for TAFR1 was also found to be relevant in this respect, providing evidence for molecular dynamics of artemisinin accumulation pattern and correlating it with the expression of genes involved in its biosynthesis.

METHODS

Seeds germination and DNA barcoding

Seed of *A. annua* were purchased from East West seeds international company (Thailand). After collection and surface sterilization with 70 % ethanol, they were germinated on half strength MS medium. Genomic DNA of germinated plantlets was extracted according to reported protocol [22]. Non-coding spacer region between the *psbA* and *trnH* genes of chloroplast DNA was amplified by PCR for the identification of *A. annua*, using primers of *psbAtrnH* region given in Table 1. The PCR reaction was carried out according to the reaction conditions reported [23]. PCR products were purified by Rapid PCR Purification System 9700 (Marligen Biosciences, Ijamsville, MD, USA), which was then sequenced following dideoxy-chain termination method using an ABI Prism 310 Automated DNA Sequencer (PE, Applied Biosystems, Foster City, CA, USA). Sequences were identified and aligned via BioEdit sequence alignment tool (editor version 7.2.5.0).

Table 1: Accession numbers and primer sequences of the genes studied

S. no	Gene name	Primer sequences	Accession no
1	psbA-trnH	Forward: GTTATGCATGAACGTAATGCTC-3 Reverse: CGCGCATGGTGGATTCACAATC	FJ418749
2	Rol B	Forward: GCTCTTGCAGTGCTAGATTT Reverse: GAAGGTGCAAGCTACCTCTC	X15952.1
3	Rol C	Forward: GAAGACGACCTGTGTTCTC Reverse: CGTTCAAACGTTAGCCGATT	X64255.1
4	ADS	Forward: ATTACTGGCGGTGCTAAC Reverse: GTGCAGAGACAGCCCATT	JQ319661
5	CYP71AV1	Forward: ATTTTGGATATGTTTGGAGCAGGC Reverse: TCCGCTTGTACTTTCTCCATTGCT	HQ315834
6	ALDH1	Forward: CAGGAGCTAATGGAAGTTCTAAGTCAG Reverse: TTTCTTCCTTCGGCCACTGTTG	FJ809784
7	TFAR1	Forward: CCTTGGAGATCCTGAAGCTG Reverse: CGTTGGATTGTGCTGAACTG	GU733320
8	Actin	Forward: ATCAGCAATACCAGGGAACATAGT Reverse: AGGTGCCCTGAGGTCTTGTTCC	EU531837
9	GADPH	Forward: TGGGATCGTTGAGGGTCTT Reverse: GCTGCTGGGAATAATGTTGAA	GQ870632.1

Bacterial Strains and Plasmids

Agrobacterium tumefaciens strain GV3101 containing plasmid pPCV002-CaMVBT and pPCV002-CaMVC kindly provided by Dr. A. Spena, Max-Planck-Institut fur Zuchtungsforschung, 5000 Koin 30, FRG [24] were used for transformation purpose. The T-DNA region of the plasmid pPCV002-CaMVBT contained the coding sequence of *rol* B gene and that of pPCV002-CaMVC contained coding region of *rol* C gene, which were under the control of CaMV35S promoter. T-DNA of pPCV002-CaMVBT and pPCV002-CaMVC also contained the *neomycin phosphotransferase* (*NPTII*) gene with nopaline synthase (NOS) promoter and NOS terminator sequences (Fig. 1). *A. tumefaciens* containing the plasmids pPCV002-CaMVBT and pPCV002-CaMVC were grown overnight in Luria broth (Sigma Cat # L-1900). After inoculation, bacterial cultures were maintained at 28 °C and 120 rpm in a shaking incubator. Growth was obtained in 24 h and OD was checked by spectrophotometer at 600 nm, when in the range of 0.2–0.8, the bacterial suspension was used for the transformation purpose.

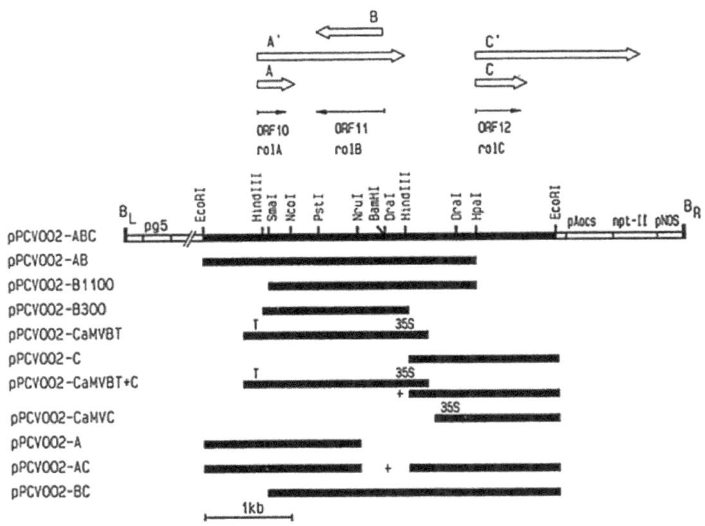

Figure 1: Detail of vectors used for transformation of *Artemisia annua*: Vectors (pPCV002-CaMVBT and pPCV002-CaMVC) used in the transformation of *Artemisia annua* (Fig. taken from Spena et al. [24]).

Transformation and Confirmation of Transgene Integration

For transformation purpose 1 month-old in vitro grown plants were used. Nodal explants were prepared and after preculturing of 2 days on shooting media (2.5 mg/l BAP and 0.25 mg/l NAA) supplemented with 200 μM acetosyringone, they were given the infection with bacterial strains containing the desired constructs. Explants were given bacterial infection for 10–15 min afterwards explants were dried on autoclaved filter paper for 2–3 min and then placed on MS shooting media supplemented with 200 μM acetosyringone. After 2 days of incubation in dark at 28 °C, explants were washed with antibiotics and placed on selection medium (2.5 mg/l BAP, 0.25 mg/l NAA, 50 mg/l kanamycin, 300 mg/l cefotaxime). Regeneration occurred within 1 month, explants were shifted to fresh selection media after every 20 days. After getting the shoots of considerable, length they were shifted to the rooting media (0.1 mg/l NAA, 50 mg/l kanamycin) Three to four cycles of selection were completed until complete plant was regenerated on selection media.

Molecular analysis was performed after extraction of genomic DNA from shoots of 2 month-old transformed and wild type plants by the CTAB method [22]. The plasmid from GV3101 was isolated by alkaline lysis method. PCR analysis was performed using a programmed DNA thermal cycler (Biometra, USA). The primers of *rol B* and *rol C* gene are given in Table 1. Conditions applied for PCR were according to a previous report [21].

Analysis of Artemisinin and its Derivatives by HPLC-ESI/MS

Extraction of artemisinin and its derivatives from leaves of 4–5 month old transformed and wild type plants was carried out with a previously published method [25]. HPLC–MS analysis was performed with Varian 500 Ion Trap Mass spectrometer (Varian, Spain) coupled to a HPLC system equipped with auto sampler (Varian, Spain). Separation of the analytes was achieved by C 18 (5 μm) column (150 mm × 4.6 mm) (AKDAY Chromatografica, Spain) using mobile phase consisting of water with 0.1 % formic acid (A) and acetonitrile (B) in a gradient as follows: [t (min), % B] (0, 50) (10, 50) (15, 99) (23, 50) (26, 50). The flow rate was 1 ml/min. Electrospray ionization (ESI) in positive mode was applied with a capillary voltage of 4000 V and plate voltage of 600 V. Other parameters include: spray chamber temperature 50 °C, nebulizing gas (N_2) pressure: 35 psi, dry gas (N_2) 10 L/min at 35 psi and 400 °C. The mass analyzer scanned from 66 to 500 u. Fragmentation amplitude was 1.0 V.

Expression analysis of *rol* genes and the genes related to artemisinin biosynthesis

Total RNA from leaves of transformed and wild type plants (4–5 months old) was extracted according to a previously published procedure [26]. Turbo DNAse (Ambion) was used to ensure complete removal of DNA from RNA after extraction. The purity and quantity was checked by taking absorbance at 260 and 280 nm on Nanodrop ND-2000 spectrophotometer (Thermoscientific). The quality of RNA was also assessed by running RNA samples on 1.2 % agarose gel. Reverse transcription of 1 μg of RNA was carried by using 1st strand cDNA synthesis kit (Invitrogen) following the manufacturer's instructions. In order to check the expression of *rol* B and *rol* C gene, semi-quantitative reverse transcriptase-polymerase chain reaction was performed with *rol* B and *rol* C gene primers as done previously by using 1 μl of cDNA reaction mixture as template. The gel image of PCR products were scanned by Kodak Molecular Imaging software (version 4.2) and integrated density values were calculated which were found to be different for each band.

To evaluate the possible effects of *rol* genes on the expression of artemisinin biosynthetic genes, quantitative real time PCR of four selected genes was performed, namely those encoding amorpha-4, 11 diene synthase (ADS), cytochrome P450 (CYP71AV1) and aldehyde dehydrogenase 1 (ALDH1). The gene involved in trichome development and sesquiterpenoid biosynthesis (TFAR1) was also studied. The β-actin gene was used as reference gene [27]. For real time qPCR, a 1:4 dilution of cDNA was used. The amplification reaction was performed by gene specific primers as given in Table 1; primer amplification efficiency was determined as previously described [28]. The qPCR was performed using iTAqTM Universal SYBR Green Supermix (BioRad, Hercules, CA, USA) in a 384-well platform system (ABI Prism® 7900HT sequence detection system, Applied Biosystems, Foster, CA, USA). The reaction conditions for real time qPCR were as follows: denaturation for 5 min at 95 °C, followed by 45 cycles each of denaturation for 10 s at 95 °C, annealing for 10 s at 60 °C, followed by elongation for 10 s at 72 °C. For each gene, the relative expression levels were normalized with respect to the wild-type plant (reference value = 1).

Calculation of trichome density

The leaves of wild type and transformed plants of *A. annua* were selected for the calculation of trichome density. For that purpose the ventral surfaces of the leaves were adhered to microscope slides with the help of glue stick and placed overnight at 4 °C for dehydration. The next day, slides were examined under the fluorescent microscope (Leica) with a FITC green filter. Slides were studied under 10× resolution and photographed.

Statistical analysis

All the experiments including, PCR, LC–MS and trichomes analysis were run in triplicate. Results of metabolites content, real-time quantitative PCR analysis and calculation of trichome density are presented as mean values ± SE. All the data obtained for the qualitative and quantitative analysis of artemisinin and derivatives, were statistically analysed by ANOVA and Duncan's multiple range test using Mstat C software. Statistical significance of trichomes was determined by t test (**, $P < 0.01$; *, $P < 0.05$).

RESULTS AND DISCUSSION

Plant identification through DNA barcoding

Amplification of 500 bp *psbA-trnH* region of the chloroplast genome was successfully carried out. DNA samples were sequenced in triplicate to confirm the authenticity of species-specific nucleotides and got the same results. Gene bank accession number [NCBI: FJ418749] (*A. annuapsbA-trnH* sequence) are reference sequence to confirm the plant species. After performing the CLUSTAL W in BioEdit software and the BLAST in NCBI, *psbA-trnH* sequence of *A. annua* was confirmed. There was 99.8 % similarity between standard and analysed sequence with a single base pair difference at 330 bp position where cytosine was observed instead of adenine. Similar results were obtained by another group that amplified the *psbA-trnH*region of *A. annua* under same conditions [23].

Transformation and confirmation of transgene integration

Successful transformation of *A. annua* with *A. tumefaciens* GV3101 harbouring the *rol* B and *rol* C gene was carried out. Single transformation experiment was carried out, and 300 explants were used for each gene. Transformation efficiency was 30 %, calculated on the basis of plants showing PCR positive results with respect to total no of infected explants. Due to contaminating factors like fungal and bacterial, only three transformants of *rol* B and three transformants of *rol* C survived until maturity on selection media. PCR performed for the *rol* gene transformants of *A. annua* showed the amplified products of 779 bp for *rol* B, 549 bp for *rol* C gene, as shown in Fig. 2a, b. Similar amplified products were obtained from plasmid DNA of GV3101-CaMVBT and GV3101-CaMVC respectively. The genome of wild type plants did not contain these genes.

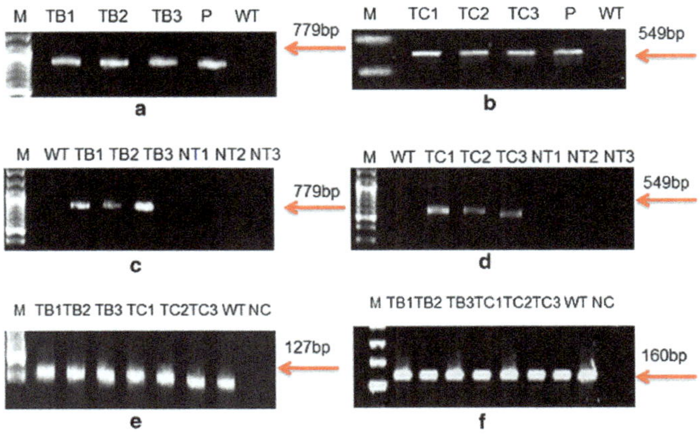

Figure 2: Gel pictures showing results of normal and semiquantitative reverse transcriptase PCR: PCR amplified products of rol B (a), rol C (b) gene are shown in the figure. Semiquantitative RT-PCR showing the relative expression of rol B (c) and rol C gene (d). GADPH and β-actin used as housekeeping gene (e, f). TB1–TB3 represents rol b transgenics whereas TC1–TC3 represent rol C transgenics. WT stands for wild type plant of A. annua, lane P refers to the plasmid DNA and lane M corresponds to the marker DNA (Fermentas). NT indicates the negative or "no RT control" for each transgenic line.

Morphological variability was distinct among transgenics and wild type *A. annua*. *Rol* B transgenics grew faster on the selection media with altered morphology showing increased inflorescence, while transgenics of*rol* C gene had totally different morphology with an altered leaf phenotype showing narrow leaf blade and decreased internode length (Fig. 3). It is reported that morphological variations can be due to the presence of*rol* genes transcripts and can be attributed to the position effect [29]. Several papers report variation in morphological characteristics of various plants species due to *rol* gene integration [30–32]. In fact, enhanced sensitivity of *rol B* transformed cells toward auxin caused by increased tyrosine phosphatase activity of *rol B*gene can result in altered phenotype of *rol B* transgenics [33, 34]. Apart from this, H^{+}ATPase and ionic imbalance [33] also synchronize the metabolic pathways leading to phenotypic variations [35]. *Rol C* is involved in functional imbalance of polyamines, gibberellins, auxins, and cytokinins [36, 37]. Such imbalances in plant growth hormones can alter plant root and shoot phenotypes [35]. However, there was no effect on the overall growth rate of shoots as confirmed by measuring the fresh and dry weight of the wild type and transformed plants.

Figure 3: Vegetative propagation of Artemisia annua: seeds germination (a), explants after co-cultivation with bacteria (b), regeneration on selection media (c), tranformant of rol B (d), transformant of rol C gene (e) and wild type A. annua (f).

Analysis of artemisinin and derivatives by LC–MS

Differences were observed in the content of artemisinin and derivatives in transformed and untransformed plants. All of the transgenic lines of *rolB* and *rol C* gene showed enhancement of artemisinin and derivatives (Fig. 4). Transgenics of *rolB* gene showed 2.7–9.2-fold increase in artemisinin, 4–12.6-fold increase in artesunate and 1.2–3-fold increase in D.H.A. Whereas in case of *rol C* gene transformants 4–4.6-fold increase in artemisinin, 4.4–9.1-fold increase in artesunate and 1.5–2-fold increase in D.H.A concentration was observed. Previously it is reported that *rol* genes (*rol A, rol B* and *rol C*) when expressed individually or combined result in increased plant secondary metabolism by transcriptional activation of defense genes through an unknown mechanism [38]. The effects of individual *rol* genes of TL-DNA of *Agrobacterium rhizogenes*, A4 strain on production of ginsenoside of *Panax ginseng* cell cultures has been reported [17]. In this report, *rolC* cultures accumulated 1.8–3 times more ginsenoside than control plant [17]. *Rol B* transgenics of *Rubia cordifolia* showed enhanced production of anthraquinones [38]. Further, *rolB* gene from *A. rhizogenes* when expressed in tomato improved foliar tolerance against fungal pathogens [39]. However, in the current study differences in artemisinin content of *rol B* and *rol C* transgenics were observed. *Rol B* gene was found to be more active in the enhancement of artemisinin content than *rol C* gene. As previously reported, high expression of the *rol B* gene dramatically increased the biosynthesis of secondary metabolites in transformed plant cells [38]. Compared to the *rol B* gene, the *rol C* gene activated the biosynthesis of secondary metabolites to a lesser extent. Evidence indicates that each of the *rol* genes has its own role in plant metabolic processes [14]. Transgenic plants showing increased artemisinin concentration and altered morphology were analysed by semi quantitative RT-PCR to check that whether the results correlate with *rol*gene expression. Results confirmed the findings to be the result of presence of *rol* gene transcripts. Also it was observed that the lines TB3 and TC1 had higher artemisinin content and more transcripts of *rol* B and *rol*C gene, respectively (Fig. 2c, d). As these genes are under the control of CaMV35S promoter, their effect on the induction of secondary metabolism become more clear due to systemic high level of expression [40].

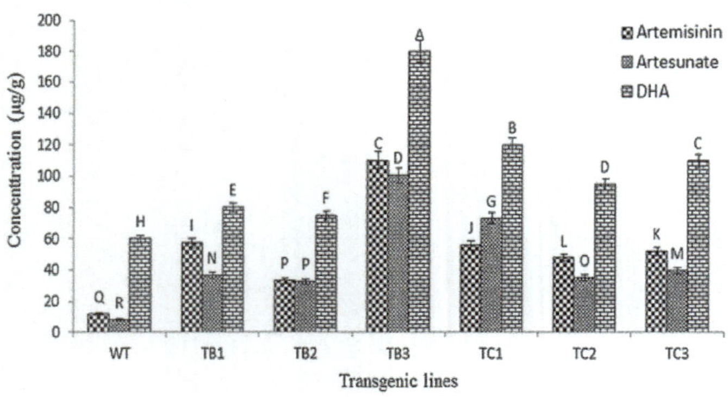

Figure 4: Statistical and quantitative analysis of artemisinin and derivatives: comparative and statistical analysis of artemisinin content and its derivatives in wild type *A. annua* and transgenics of *rol* B and *rol* C gen. *Letters* on the *bars* show LSD ranking.

Expression analysis of artemisinin and trichome biosynthesis genes

Relatively higher transcripts of all studied genes of artemisinin biosynthetic pathway were found in the transgenics of *rol* B and *rol* C gene although in variable fashion (Fig. 5), which also correlates the accumulation pattern of artemisinin and analogues. Regarding the biosynthesis of artemisinin, ADS, CYP71AV1 and ALDH1 are key genes involved in the conversion of FDP (Farnesyl diphosphate) to artemisinic acid which is a late precursor of artemisinin, as discussed in the background section [7–11]. Transgenics of *rolB* and *rolC* gene showed overexpression of studied genes with CYP71AV1 found to be most expressed in transgenics with 8–20.9-fold increase in *rol* B tranformants and 6–15.5-fold more expression in *rol* C transformants. ALDH1 gene showed eight- to tenfold and ADS gene showed 5.9–8.5-fold more expression in *rol* B transformants. Similarly Transgenics of *rol C* gene showed 6.4–9.5-fold increase in ALDH1 and 5–7.5-fold increase in ADS gene expression. Transgenic line TB3 and TC1 were found with highest expression of these genes. This relative expression of all the genes is in accordance with the accumulation of artemisinin and derivatives in respective transgenic lines. These results are supported by a previous report which demonstrates that overexpression of farnesyl pyrophosphate synthase (FPS) in *A. annua* increases the accumulation of artemisinin through conversion of isopentenyl diphosphate (IPP) and dimethylallyl diphosphate (DMADP) into farnesyl diphosphate (FDP) [41]. In another report, increased amount of artemisinin in *A. annua* transformed with *A. tumefaciens* wild type nopaline strains was observed [42]. Similarly, another group transferred the *Ipt* gene into *A. annua* and observed increased amount of artemisinin in transformed plants [13].

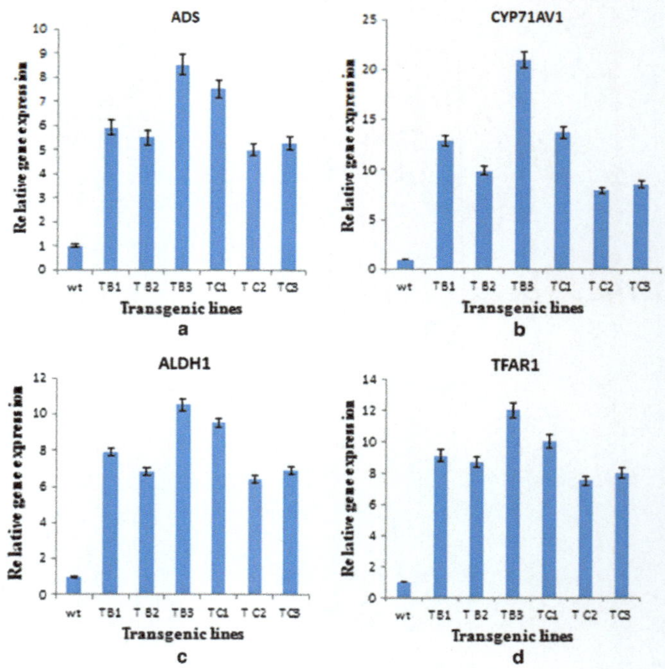

Figure 5: Expression analysis of artemisinin biosynthetic pathway genes and TAFR1 gene: expression level of artemisinin biosynthetic genes, ADS (a), CYP71AV1 (b) and ALDH1 (c) in wild type plant of A. annua and transgenics of rol B and rol C gene. TFAR1 is the gene involved in trichome development and sesquiterpenoid biosynthesis (d).

Trichome density and expression of TFAR1 gene

The number of glandular trichomes was determined at the adaxial side of three random pieces of fresh leaf material from each sample with an accurately determined leaf area of approximately 1 cm^2 per leaf and significant differences were observed as shown by Fig. 6. The leaves of *A. annua* transformed with *rol B* and*rol C* gene produced more trichomes (20–30 trichomes/cm^2) as compared to untransformed leaves (5–7 trichomes/cm^2).

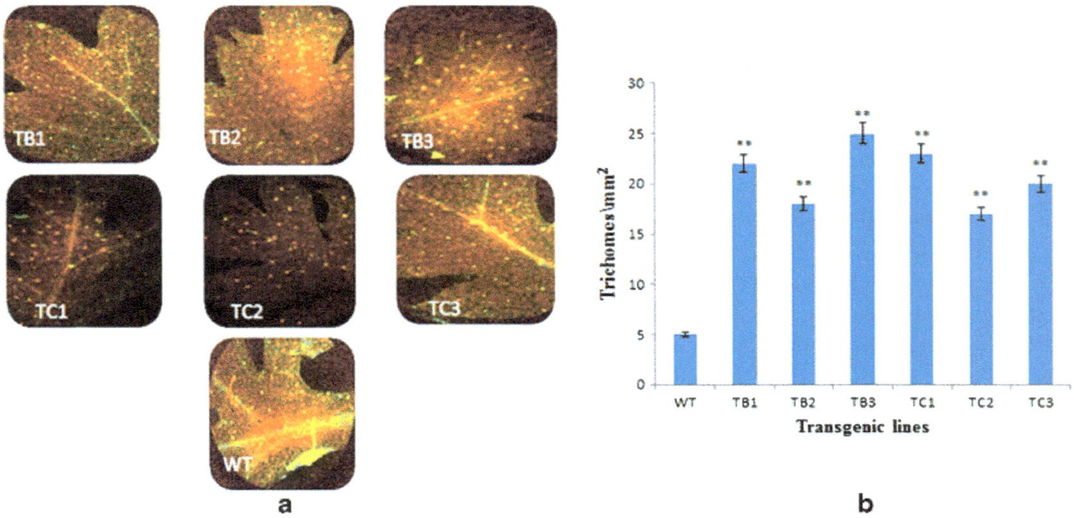

Figure 6: **a** Trichome density: comparison of trichome density of transformed and untransformed plants of A. annua. TB1–TB3 represent rol B transgenics whereas TC1–TC3 represent transgenics of rol C gene. WT represents the control plant of A. annua. **b** Graphical representation of trichomes density of wild type A. annua and transformants of rol B and rol C gene.

TFAR1, which stimulates trichome development and catalyzes sesquiterpenoid biosynthesis [41] also showed increased expression in all transgenic lines. *Rol* B transformants with 2.3–3-fold increase and *rol* C gene transformants with 1.6–2-fold increase in TFAR1 gene expression were found. As mentioned in a previous report that glandular trichomes are the sites of production of important phytochemicals including artemisinin and a mixture of chemicals that have been found to have an enormous array of uses in the pesticides, pharmaceutical and flavor industries [43, 44]. Artemisinin content was found to be directly related to the trichome index in previous reports [45, 46].

CONCLUSION

Transformation of *A. annua* with *rol* B and *rol* C gene results in the enhancement of its secondary metabolites particularly the artemisinin and derivatives, *rol B* gene being more active to play part in the enhancement of artemisinin than *rol C* gene. Further the level of transcripts of the *rolB* and *rolC* gene found in transgenics also correlate with their artemisinin accumulation pattern. Altered expression of genes involved in biosynthesis of artemisinin and trichomes was observed.

AUTHORS' CONTRIBUTIONS

ED carried out the practical work regarding all experiments i.e. DNA barcoding, genetic transformation, molecular analysis, LC–MS and trichome study. RMC & JP supervised all the work performed in the department of pharmacy, University of Barcelona Spain. They also reviewed the manuscript intellectually and critically to give it final form for publication. KR provided technical as well as intellectual support in doing real time quantitative PCR analysis. She also made efforts in making experimental design. MB supervised the quantitative analysis of artemisinin and derivatives by HPLC–MS by helping in experimental design and interpretation of results. BM supervised all the work performed in plant molecular Biology laboratory of QAU Islamabad Pakistan. She provided the intellectual support in making the experimental design of DNA barcoding, genetic transformation and trichome study. She also helped in writing the manuscript and its proofreading. All authors read and approved the final manuscript.

ACKNOWLEDGEMENTS

We are grateful to Dr. A. Spena, Max-Planck-Institut fur Zuchtungsforschung, 5000 Koin 30, FRG for providing the vectors. We are thankful to the Higher Education Commission (HEC) of Pakistan for financially supporting this research project. Part of this work was financially supported by the Spanish MEC (BIO2014-51861-R) and the Generalitat de Catalunya (2014SGR215).

REFERENCES

1. Duffy PE, Mutabingwa TK. Artemisinin combination therapies. Lancet. 2006;367:2037–2039. doi: 10.1016/S0140-6736(06)68900-9.
2. White NJ. Qinghaosu (artemisinin): the price of success. Science. 2008;320:330–334. doi: 10.1126/science.1155165.
3. Efferth TW. Schwabe Award 2006: antiplasmodial and antitumor activity of artemisinin—from bench to bedside. Planta Med. 2007;73:299–309. doi: 10.1055/s-2007-967138.
4. CNAP *Artemisia annua* F1 hybrids. 2015.http://www.york.ac.uk/org/cnap/artemisiaproject/artemisiaF1seed/index.html. Accessed 15 May 2015.

5. Ferreira JFS, Simon JE, Janick J. Developmental studies of *Artemisia annua*: flowering and artemisinin production under greenhouse and field conditions. Planta Med. 1995;61:167–170. doi: 10.1055/s-2006-958040.

6. Olsson ME, Olofsson LM, Lindahl AL, Lundgren A, Brodelius M, Brodelius PE. Localization of enzymes of artemisinin biosynthesis to the apical cells of glandular trichomes of *Artemisia annua* L. Phytochemistry. 2009;70:1123–1128. doi: 10.1016/j.phytochem.2009.07.009.

7. Brown GD, Sy LK. In vivo transformations of artemisinic acid in *Artemisia annua* plants. Tetrahedron.2007;63:9548–9566. doi: 10.1016/j.tet.2007.06.062.

8. Croteau R, Kutchan T, Lewis N. Natural products (secondary metabolites) In: Buchanan B, Gruissem W, Jones R, editors. Biochemistry and molecular biology of plants. Rockville: American Society of Plant Physiology; 2000. pp. 1250–1318.

9. Teoh KH, Polichuk DR, Reed DW, Nowak G, Covello PS. *Artemisia annua* L. (Asteraceae) trichome-specific cDNAs reveal CYP71AV1, a cytochrome P450 with a key role in the biosynthesis of the antimalarial sesquiterpene lactone artemisinin. FEBS Lett. 2006;580:1411–1416. doi: 10.1016/j.febslet.2006.01.065.

10. Wallaart TE, Pras N, Beekman AC, Quax WJ. Seasonal variation of artemisinin and its biosynthetic precursors in plants of *Artemisia annua* of different geographical origin: proof for the existence of chemotypes. Planta Med. 2000;66:57–62. doi: 10.1055/s-2000-11115.

11. Zhang Y, Teoh KH, Reed DW, Maes L, Goossens A, Olson DJH, et al. The molecular cloning of artemisinic aldehyde D11 (13) reductase and its role in glandular trichome-dependent biosynthesis of artemisinin in *Artemisia annua*. J Biol Chem. 2008;283:21501–21508. doi: 10.1074/jbc.M803090200.

12. Chen D, Ye H, Li G. Expression of a chimeric farnesyl diphosphate synthase gene in *Artemisia annua* L. transgenic plants via *Agrobacterium tumefaciens*-mediated transformation. Plant Sci. 2000;155:179–185. doi: 10.1016/S0168-9452(00)00217-X.

13. Sa G, Ma M, Ye HC, Liu BY, Li GF, Chong K. Effects of *ipt* gene expression on the physiological and chemical characteristics of *Artemisia annua* L. Plant Sci. 2001;160:69198. doi: 10.1016/S0168-9452(00)00453-2.

14. Bulgakov VP. Functions of *ROL* genes in plant secondary metabolism. Biotechnol Adv. 2008;26:31824.

15. Filippini F, Lo Schiavo F, Terzi M, Constantino P, Trovato M. The plant oncogene *rol* B alters binding of auxin to plant membranes. Plant Cell Physiol. 1994;35:767–771.

16. Filippini F, Rasi V, Marin O, Trovato M, Downey PM, Lo SF, et al. A plant oncogene as a phosphatase. Nature. 1994;379:499–500. doi: 10.1038/379499a0.

17. Bulgakov VP, Khodakovskaya MV, Labetskaya NV, Tchernoded GK, Zhuravlev YN. The impact of plant *rol* C oncogene on ginsenoside production by *ginseng* hairy root cultures. Phytochemistry.1998;49:1929–1934. doi: 10.1016/S0031-9422(98)00351-3.

18. Casanova E, Trillas MI, Moysset L, Vainstein A. Influence of *rol* genes in floriculture. Biotechnol Adv.2005;23:3–39. doi: 10.1016/j.biotechadv.2004.06.002.

19. Palazon J, Cusido RM, Roig C, Pinol MT. Expression of the *rol* C gene and nicotine production in transgenic roots and their regenerated plants. Plant Cell Rep. 1998;17:384–390. doi: 10.1007/s002990050411.

20. Palazon J, Cusido RM, Gonzalo J, Bonfill M, Morales S, Pinol MT. Relation between the amount the *rol*C gene product and indole alkaloid accumulation in *Catharanthus roseus* transformed root cultures. J Plant Physiol. 1998;153:712–718. doi: 10.1016/S0176-1617(98)80225-3.

21. Kiani BH, Safdar N, Mannan A, Mirza B. Comparative Artemisinin analysis in *Artemisia dubia*transformed with two different *Agrobacteria* harbouring *rol* ABC genes. Plant Omics J. 2012;5:386–391.

22. Ahmed I, Islam M, Arshad W, Mannan A, Ahmad W, Mirza B. High-quality plant DNA extraction for PCR: an easy approach. J Appl Genet. 2009;50:105–107. doi: 10.1007/BF03195661.

23. Tao L, Yunheng J. *psbA-trnH* sequence analysis from chloroplast on medicinal plants of *Artemisia*. Chin Agric Sci Bull. 2009;25:46–49.

24. Spena A, Schmiilling T, Koncz C, Schell JS. Independent and synergistic activity of *rol A*, *B* and *C* loci in stimulating abnormal growth in plants. EMBO J. 1987;6:3891–3899.

25. Towler MJ, Weathers PJ. Evidence of artemisinin production from IPP stemming from both the mevalonate and the non-mevalonate pathways. Plant Cell Rep. 2007;26:2129–2136. doi: 10.1007/s00299-007-0420-x.

26. Safdar N. *Agrobacterium*-mediated transformation of tomato with yeast *halotolerance* genes. Dissertation. Quaid-iAzam University Islamabad; 2012. p. 36–7.

27. Olofsson L, Engstrom A, Lundgren A, Brodelius PE. Relative expression of genes of terpene metabolism in different tissues of *Artemisia annua* L. BMC Plant Biol. 2011;11:45–47. doi: 10.1186/1471-2229-11-45.

28. Sabater-Jara AB, Onrubia M, Moyano E, Bonfill M, Palazon J, Pedreno MA, et al. Synergistic effect of cyclodextrins and methyl jasmonate on taxane production in Taxus × media cell cultures. Plant Biotech J.2014

29. Schmulling T, Schell J, Spena A. Single genes from *Agrobacterium rhizogenes* influence plant development. EMBO J. 1988;7:2621–2629.

30. Cardarelli M, Mariotti D, Pomponi M, Spano L, Capone I, Costantino P. *Agrobacterium rhizogenes* T-DNA genes capable of inducing hairy root phenotype. Mol Gen Genet. 1987;209:475–480. doi: 10.1007/BF00331152.

31. Van Altvorst AC, Bino RJ, van Dijk AJ, Lamers AMJ, Lindhout WH, Van der Mark F, et al. Effects of the introduction of *Agrobacterium rhizogenesROL* genes on tomato plant and flower development. Plant Sci.1992;83:77–85. doi: 10.1016/0168-9452(92)90064-S.

32. Spena A, Esttruch JJ, Prinsen E, Nacken W, Vanonckelen H, Sommer H. Anther-specific expression of

the *rolB* gene of *Agrobacterium-rhizogenes* increases IAA content in anthers and alters anther development and whole flower growth. Theor Appl Genet. 1992;84:520–527. doi: 10.1007/BF00224147.

33. Maurel C, Barbier-Brygoo H, Spena A, Tempe J, Guern J. Single *ROL* gene from the *Agrobacterium rhizogenes* TL-DNA alters some of the cellular responses to auxin in *Nicotiana tabacum*. Plant Physiol.1991;97:212–216. doi: 10.1104/pp.97.1.212.

34. Maurel C, Leblanc N, Barbier-Brygoo H, Perrot-Rechenmann C, Bouvier-Durand M, Guren J. Alterations of auxin perception in *rolB* transformed tobacco protoplast. Plant Physiol. 1994;105:1209–1215. doi: 10.1104/pp.105.4.1209.

35. Zia M, Mirza B, Malik SA, Chaudhary MF. Expression of ROL genes in transgenic soybean (*Glycine max* L.) leads to changes in plant phenotype, leaf morphology, and flowering time. Plant Cell Tissue Organ Cult. 2010;103:227–236. doi: 10.1007/s11240-010-9771-z.

36. Cui M, Takayanagi K, Kamada H, Nishimura S, Handa T. Efficient shoot regeneration from hairy roots of *Antirrhinum majus* L. transformed by the *rol* type MAT vector system. Plant Cell Rep. 2001;20:55–59. doi: 10.1007/s002990000277.

37. García-Sogo B B, Pineda B, Castelblanque L, Antón T, Medina M, Roque E, et al. Efficient transformation of *Kalanchoe blossfeldiana* and production of male-sterile plants by engineered anther ablation. Plant Cell Rep. 2010;29:61–77. doi: 10.1007/s00299-009-0798-8.

38. Shkryl YN, Veremeichik GN, Bulgakov VP, Tchernoded GK, Mischenko NP, Fedorevev SA, et al. Individual and combined effects of the *rol* A, B and C genes on anthraquinone production in *Rubia cordifolia*transformed calli. Biotechnol Bioeng. 2008;100:118–125. doi: 10.1002/bit.21727.

39. Arshad W, Haq IU, Waheed MT, Mysore KS, Mirza B. *Agrobacterium*-mediated transformation of tomato with *rol B* gene results in enhancement of fruit quality and foliar resistance against fungal pathogens. PLoS One. 2014;9:e96979. doi: 10.1371/journal.pone.0096979.

40. Odell JT, Nagy F, Chua NH. Identification of DNA sequences required for activity of the cauliflower mosaic virus 35S promoter. Nature. 1985;313:810–812. doi: 10.1038/313810a0.

41. Lies M, Filip CW, Van N, Yansheng Z, Darwin WR, Jacob P. Dissection of the phytohormonal regulation of trichome formation and biosynthesis of the antimalarial compound artemisinin in *Artemisia annua* plants. New Phytol. 2011;189:176–189. doi: 10.1111/j.1469-8137.2010.03466.x.

42. Biswajit G, Swapna M, Sumita J. Genetic transformation of *Artemisia annua* by *Agrobacterium tumefaciens* and artemisinin synthesis in transformed cultures. Plant Sci. 1997;122:193–199. doi: 10.1016/S0168-9452(96)04558-X.

43. Duke MV, Paul RN, Elsohly HN, Sturtz G, Duke SO. Localization of artemisinin and artemisitene in foliar tissues of glanded and glandless biotypes of *Artemisia annua* L. Int J Plant Sci. 1994;155:365–372. doi: 10.1086/297173.

44. Ferreira JF, Simon JE, Janick J. *Artemisia annua*: botany, horticulture, pharmacology. In: Janick J, editor. Horticultural reviews. New York: John Wiley and Sons, Inc.; 1997. pp. 319–371.

45. Dangash A, Pandya N, Bharillya A, Jhala A, Jain D. Impact of exogenous elicitors on artemisinin production and trichome density in *Artemisia annua* L. under subtropical conditions. Not Sci Biol.2014;6:349–353. doi: 10.15835/nsb.6.3.9109.

46. Kiani BH. Transfer of *rol* genes and evaluation of artemisinin synthesis in transformed *Artemisia annua*L. and *Artemisia dubia* WALL. Dissertation. Quaid-i-Azam University Islamabad; 2013.

CHAPTER 6

GENE TARGETING AND TRANSGENE STACKING USING INTRA GENOMIC HOMOLOGOUS RECOMBINATION IN PLANTS

Sandeep Kumar, Pierluigi Barone and Michelle Smith

Dow AgroSciences LLC, 9330 Zionsville Road, Indianapolis, IN 46286, USA

ABSTRACT

Modern agriculture has created a demand for plant biotechnology products that provide durable resistance to insect pests, tolerance of herbicide applications for weed control, and agronomic traits tailored for specific geographies. These transgenic trait products require a modular and sequential multigene stacking platform that is supported by precise genome engineering technology. Designed nucleases have emerged as potent tools for creating targeted DNA double strand breaks (DSBs). Exogenously supplied donor DNA can repair the targeted DSB by a process known as gene targeting (GT), resulting in a desired modification of the target genome. The potential of GT technology has not been fully realized for trait deployment in agriculture, mainly because of inefficient transformation and plant regeneration systems in a majority of crop plants and genotypes. This challenge of transgene stacking in plants could be overcome by Intra-Genomic Homologous Recombination (IGHR) that converts independently segregating unlinked donor and target transgenic loci into a genetically linked molecular stack. The method requires stable integration of the donor DNA into the plant genome followed by intra-genomic mobilization. IGHR complements conventional breeding with genetic transformation and designed nucleases to provide a flexible transgene stacking and trait deployment platform.

BACKGROUND

The Green Revolution in the 1960s combined advances in breeding and agricultural practice, and provided food security to millions of people [1]. Given an increasing global population, there is a projected need to increase world food production by 40 % in the next 20 years [2]. In addition to a growing population, climate change, degrading natural resources and changing food preferences have raised food and nutritional security to the level of the biggest challenge of the twenty-first century [3].

Citation: Sandeep Kumar, Pierluigi Barone and Michelle Smith, Gene targeting and transgene stacking using intra genomic homologous recombination in plants, DOI 10.1186/s13007-016-0111-0.

Genetically modified (GM) trait technology in the mid-1990s made a major impact in meeting the world food demand and there has been a rapid adoption of the technology. These first generation trait products involved simple herbicide and insect traits that required introduction of a single gene. Control of the broad range of insect pests and weeds desired today requires multiple insect and herbicide tolerance genes [4]. In addition, modern genomics and gene networking tools have revealed that many agronomic traits depend on different genes and complex interactions of proteins reacting to various external stimuli [1]. The next generation trait products, therefore, require integration of multiple transgenes and would also benefit from a flexible and modular trait stacking platform that would accommodate development of increasingly complex future products. Conventional breeding has been successfully employed for trait stacking, but this method requires substantial time and resources for sorting and deregulation of multiple unlinked transgenes [4, 5], and a limited number of independent loci can practically be stacked.

Designed nucleases have become a powerful gene targeting (GT) tool to create targeted DNA double strand breaks (DSBs) at specified genomic locations, which stimulate the cell's DNA repair machinery leading to integration of exogenously supplied transgenes into a specified genomic site. While designed nuclease-mediated targeted mutagenesis is becoming routine in plants [6–9], site-directed transgene integration remains elusive, mainly due to low transformation and regeneration efficiencies in the majority of plant species and genotypes. A GT method that requires minimal transformation effort would be very attractive to address this challenge. This review focuses on conventional Intra-chromosomal somatic homologous recombination work in plants and its recent application using designed nucleases that can provide solutions to some of the challenges associated with the deployment of GT technology for transgene stacking in crop plants.

GENE TARGETING: A BYPRODUCT OF GENOMIC DOUBLE-STRAND BREAK

DSBs can arise spontaneously, may be induced by ionizing radiation and chemicals, or recently by designed nucleases (For review, see references [10–15]). Genomic DSBs could be negatively mutagenic or lethal to cells if not repaired efficiently. In plants, DSBs are repaired by homologous recombination (HR) or non-homologous end joining (NHEJ). HR and NHEJ mechanisms are conserved in eukaryotes; however, the efficiency of these pathways differs not only between species but also between cell types [16]. HR is a precise DSB repair pathway that requires sequences homologous (almost identical) to those flanking the DSB site [12, 13]. HR is the predominant DNA recombination pathway during meiosis in higher eukaryotes including plants [17]. NHEJ mainly involves ligation to unrelated sequences or to sequences with micro-homologies, resulting mostly in non-precise repair with small insertions or deletions at the DSB site. NHEJ is the primary DNA repair pathway in the somatic cells, while HR mainly occurs during S and G2 phases of the cell cycle [18].

Targeted DSB-induced NHEJ has been previously described for mutagenesis, deletions or imprecise insertions [6–9, 13, 19, 20]. In contrast, HR, a more precise mode of DNA repair, is preferred for GT [12,13]. Gene targeting through HR requires simultaneous introduction of the nuclease to create targeted DSB at desired genomic location, and donor DNA containing flanking homologies, acting as a template for repair of the DSB [21].

GENE TARGETING CHALLENGES IN PLANTS

Targeted DSBs stimulate the cell's DNA repair machinery making the DSB site accessible to a donor transgene for site-specific integration. The DSBs however do not preclude the ectopic integration of a donor transgene elsewhere in the genome. In addition, the GT process requires efficient delivery of the donor molecule to the DSB site and the ability to regenerate whole plants from the cells with a precisely repaired targeted genomic site. Random integration of the donor transgene and an inefficient transformation method for the donor delivery therefore are two major challenges for the routine deployment of GT technology in crop plants. Positive selection for GT such that precise insertion of the donor complements the non-functional selectable marker in the target locus has been used to avoid random integration of the donor [22, 23, 24] genes in the target locus. A positive–negative selection approach has also been used very successfully for GT in rice [25, 26]. A sequential GT method providing flexibility of incremental modifications of the target locus with new trait genes was recently developed [27]. That method exploited positive GT selection using an intron sequence homology between the donor and target that allowed sequential swapping of selection markers, providing a multi-generational GT method (Fig. 1) for trait product deployment [28].

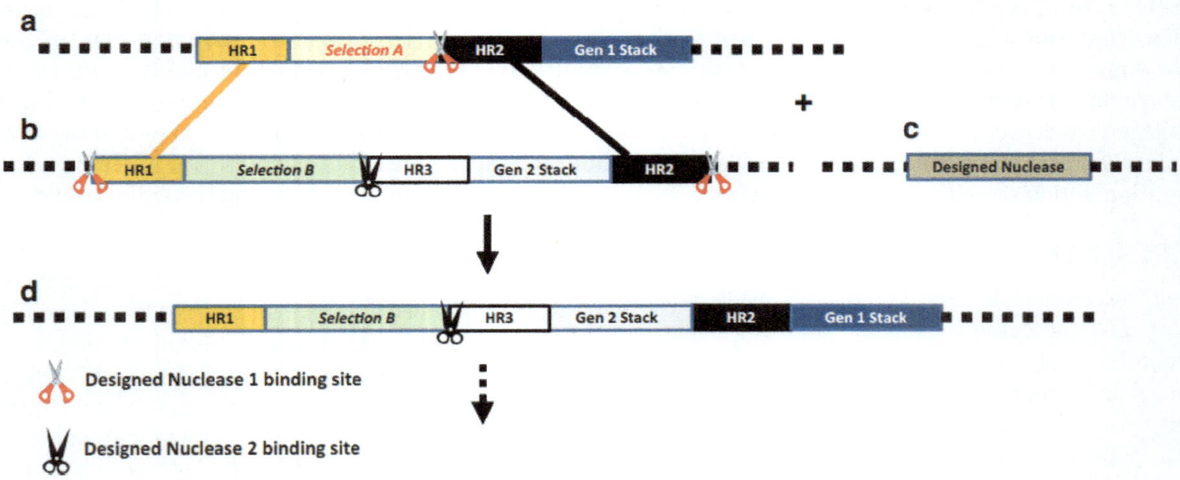

Figure 1: Basic design of constructs used for gene targeting via intra genomic homologous recombination. a Target construct contains a generation 1 (Gen 1) stack and a selection marker A flanked by unique homology sequences (HR1 and 2). A designed nuclease 1 binding site is inserted between selection marker A and HR2 sequence. b The donor construct contains selection marker B, downstream HR3 sequence, generation 2 (Gen 2) stack, and HR1 and 2 homology sequences matching the target. A designed nuclease 2 binding site is inserted between the selection marker and HR3 sequence for future targeting. The donor is flanked by a designed nuclease 1 binding site on each end. c The designed nuclease construct contains designed nuclease 1 coding sequence driven by appropriate promoter. d Target locus containing functional selection marker B gene and generation 2 (Gen 2) stack precisely inserted after gene targeting.

The accessibility of the donor transgene to the DSB site is another key bottleneck for efficient GT process. Donor DNA is exogenously supplied either via direct DNA delivery [29], mostly using the microparticle bombardment, or via indirect DNA delivery, mainly mediated by *Agrobacterium* [30]. The production of a transgenic plant is the result of a sequence of events: a) transfer of exogenous DNA into the plant cell nucleus, b) integration of the foreign DNA in a transcriptionally active region of the host genome, and c) regeneration into a fully developed plant, either via organogenesis or somatic embryogenesis, of the original cell where the transgene integrated. Regardless of the gene transfer method used (direct or indirect) cell competence for foreign DNA integration and regeneration varies with cell type and developmental stage making the recovery of transgenic events a challenging task in most crop plants.

The nuclear targeting of the exogenous DNA is hindered by physical (e.g. cell wall), cellular (e.g. proteases, nucleases) and biological barriers (e.g. plant defense) and our understanding of how to best overcome these barriers is still limited [31, 32]. Actively dividing cells are the most amenable targets for DNA insertion and it has been shown that higher transformation is obtained in cells with nuclei at the S and G2 phases [33, 34] where chromatin remodeling takes place. A localized and temporary decondensation of the chromatin is believed to be necessary for efficient transgene integration in gene-rich euchromatic regions [35, 36].

Plant biology parameters associated with species, genotype, and explant type play an important role in the efficiency of transformation and regeneration. In rice, for example, between the two subspecies indica and japonica, indica is generally more recalcitrant to tissue culture and transformation [37]. Similarly for maize transformation and regeneration, the most responsive type of explant is the immature embryo where scutellum cells are induced to proliferate and undergo somatic embryogenesis [38], but this process is highly genotype-dependent and still limited mainly to crosses and derivatives of the maize inbred lines A188 [37]. Also in soybean, the ability to regenerate transgenic plants has been limited to a few soybean model genotypes (Jack and Williams 82) with some successful examples of competence for somatic embryogenesis transferred and combined in other cultivars via introgression [39].

GENE TARGETING VIA INTRA GENOMIC HOMOLOGOUS RECOMBINATION

The challenges of inefficient transformation systems in crop plants could be overcome by intra-genomic homologous recombination (IGHR), which utilizes a cell's recombinational machinery to replicate and supply donor DNA for IGHR mediated insertion of a donor within the target site. Intra-chromosomal HR in somatic cells of the whole plant was reported more than two decades ago (Reviewed in [40, 41, 42]). Two overlapping, non-functional pieces of a chimeric *beta-glucuronidase* (*uidA*) gene were used as recombination substrates, which upon HR led to a restoration of the functional *uidA* gene that was detected by histochemical staining of the encoded functional *uidA* protein. HR was reported in different organs and tissues during different stages of the plant development, including meristematic recombination events that revealed cell lineage patterns. The system was later used to demonstrate that an induced DSB in the target site resulted in twofold increase the HR frequency [22, 43]. The germline in plants is formed during later developmental stages, and any HR occurring during the life cycle of the plant could be germinally transmitted to the next generation. The demonstration of HR between linked overlapping DNA sequences within somatic cells of the whole plant was an important milestone in the GT field. The work paved the way for HR between unlinked DNA sequences in the genome of somatic cells, and regeneration of whole plants from these cells (see below).

The next significant development in the field was the application of designed nucleases for excision of the stably integrated transgene. In tobacco, a transgenic line containing a single copy of the *codA* gene flanked by cleavage sites specific to I-*Sce*I nuclease was created. After induction of DSBs by transient expression of I-*Sce*I, the *codA* gene was successfully removed from the calli, and plants lacking the *codA* gene were regenerated [44]. Similarly, tobacco plants containing a stably integrated *uidA* gene cassette flanked by designed nuclease sites were crossed with plants expressing the corresponding nuclease. The complete deletion of a 4.3 kb sequence comprising the *uidA* gene cassette was obtained in F_1 progenies [45]. These reports were later followed by deletions of large endogenous genomic sequences in different plant species using designed nucleases [46–48].

Researchers in mammalian GT field were first to exploit cells' recombination machinery to catalyze HR between a target locus and an in vivo-liberated donor [49]. In this system, the donor transgene is first inserted stably into the genome. The randomly inserted donor molecule is later released intragenomically within the genome of intact tissue. The IGHR based method was demonstrated using a site-specific recombinase (FLP) and a site-specific endonuclease (I-*Sce*I) for the modification of the *yellow* locus in the *Drosophila* genome [49–51]. The method has been successfully applied for the modification of more than 20 loci in *Drosophila*[52].

A similar IGHR approach was also proposed for plant GT [53]; the first proof-of-principle in plants came several years later in *Arabidopsis* [54] using a single site-specific endonuclease (I-*Sce*I). The GT system was designed using a non-functional truncated *uidA* target transgene containing cleavage sites for I-*Sce*I nuclease, a donor transgene containing a complementary *uidA* GT cassette flanked by I-*Sce*I sites, and a transgene containing a I-*Sce*I expressing cassette that upon expression would generate in vivo release of linear donor after I-*Sce*I expression. Single copy target and donor lines were crossed and lines homozygous for both the transgenes were obtained. The homozygous target/donor lines were then crossed with an I-*Sce*I line and F_1progenies were screened for IGHR-mediated GT using *uidA* histochemical staining. Some F_1 progenies revealed chimeric blue spots indicating GT in somatic cells during the plant development. The F_1 lines were self-pollinated and F_2 progenies were scored for the blue seedlings indicating germinal transmittance of GT. Targeted events were obtained up to one per 100 seeds. A similar approach was later attempted with some success in a native genomic target site in *Arabidopsis* using the CRISPR/Cas system [55].

After initial work on IGHR-mediated GT in a model system, the method was successfully demonstrated in maize by somatic ectopic recombination and tissue culture selection [56]. Similar to a previous effort in*Arabidopsis*, the target construct contained a non-functional partial *neomycin phophotransferase II* (*nptII*) gene and a cleavage site for I-*Sce*I nuclease. The donor construct contained dexamethasone-inducible I-*Sce*I, and an excisable *nptII* sequence complementing the partial sequence at the target locus such that GT would constitute the functional *nptII* gene. The target and donor plants were crossed and F_1 progenies were selfed. No fully kanamycin-resistant plants were obtained from the dexamethasone-induced F_2 progeny for target and donor. However kanamycin-resistant leaf sectors were observed indicating IGHR occurred in some somatic cells during plant development. The embryos isolated from immature kernels of F_2 plants were subjected to callus induction on medium with and without dexamethasone. Kanamycin resistant GT events were recovered and the repair of the *nptII* gene was confirmed by molecular analyses. GT frequencies ranging from 0.13 to 0.55 % (per immature embryo treated) were obtained. The authors also made an interesting observation of GT at a cleaved target locus without excision of the donor molecule.

The demonstration of GT via IGHR in *Arabidopsis* and maize has created potential for the application of GT technology in transformation-inefficient crop plant species. Unlike direct transformation methods that limit donor molecules to a small number of treated cells, IGHR utilizes the plant system to replicate donor DNA in every cell throughout the life cycle. The extra-chromosomal donor molecule could be liberated and used by the target site as a template in plant tissues or stages that favor HR over NHEJ. The previous GT approaches relied on efficient transformation systems to produce a large number of events to obtain a few targeted plants. Since most economically important crop plants remain recalcitrant to transformation, GT technology has so far been practical in only a small number of crop plants. Additionally, IGHR releases only one to two copies of the donor, leading to high quality targeted events, in contrast to previous GT methods that require additional segregation work to remove randomly integrated undesired truncated donor molecules.

The IGHR method reviewed here generates tremendous opportunity for biotechnological application of GT in commercial transgenic trait deployment. This approach when combined with a sequential GT method (Fig. 1) [27] would provide the modular and flexible transgenic trait stacking platform (Fig. 2) currently needed for complex product needs in the agriculture industry. The strategy provides flexibility to stably integrate 1st generation or geography-specific traits in the target plant, while new traits are placed in the donor plant. Donor and target plants are crossed to create a breeding stack, which is then crossed with appropriate designed nuclease expressing plants. The F_1 progenies are then subjected to tissue culture selection and targeted plant regeneration. Multiple donor lines containing different traits could strategically be made to keep modularity required for creating on-demand stacked transgenic traits. The additional tissue culture selection step restricts the use of this method to crop plants that are amenable to tissue culture techniques. Precise tissue-specific expression of designed nuclease in reproductive cells [57–59] can circumvent the need for a tissue culture regeneration process, providing broader application of this approach across different crops.

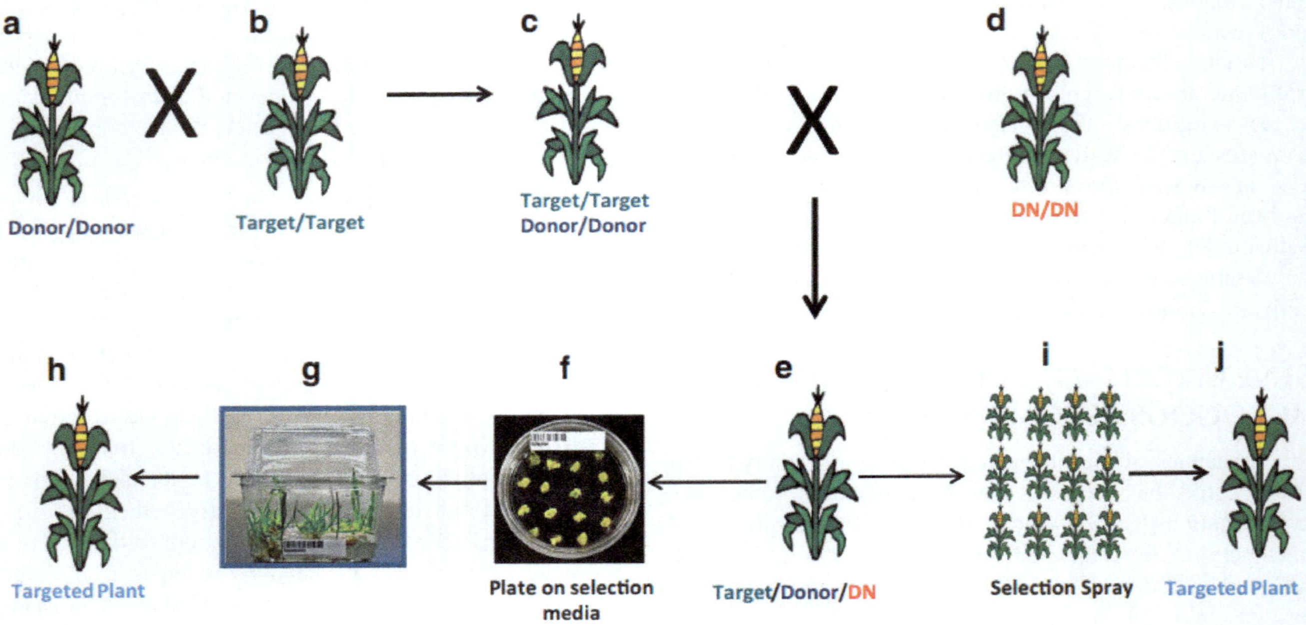

Figure 2: Crossing and targeted plant production strategy in maize using intra-genomic homologous recombination. Plants homozygous to donor (a) and target (b) are crossed and self-pollinated to obtain progenies that are homozygous to target and donor loci (c). The homozygous target donor plants are crossed with plants containing designed nuclease (DN) transgene (d) to obtain F1 progenies transgenic to target, donor and DN (e). The F1 immature embryos are treated on appropriate selection media (f) and targeted plants are regenerated on selection (g and h). Alternatively, the F1 plants could be selection sprayed (i) to obtain targeted plants (j).

CONCLUSIONS

Future biotech crops are projected to require multiple transgenes to confer resistance to a broad spectrum of insect pests and provide herbicide tolerance with different modes of action. Insects and weeds will eventually develop resistance, new target pests will emerge and new traits will inevitably be needed and desired, so designing those future products to be further modified and developing capabilities to accomplish the modifications are wise investments. It is clear that producing and modifying transgenic events through GT has many advantages over random integration, and technology continues to develop to make GT increasingly efficient and flexible. Intra genomic homologous recombination using designed nucleases has good potential to overcome limitations in plant transformation and breeding to achieve targeted and highly complex stacked trait crops.

ACKNOWLEDGEMENTS

We are grateful to Otto Folkerts, Lakshmi Sastry-Dent, and Rodrigo Sarria for reviewing the manuscript and providing valuable suggestions.

COMPETING INTERESTS

The authors in this publication certify that they are employees of Dow AgroSciences LLC, a subsidiary of The Dow Chemical Company. As employees, they authors may have some company stock (The Dow Chemical Company), but no other personal financial interest in the subject matter of the materials discussed in this manuscript.

REFERENCES

1. Dockter C, Hansson M. Improving barley culm robustness for secured crop yield in a changing climate. J Exp Bot. 2015;66(12):3499–3509. doi: 10.1093/jxb/eru521.
2. Parry MA, Hawkesford MJ. An integrated approach to crop genetic improvement. J Integr Plant Biol.2012;54(4):250–259. doi: 10.1111/j.1744-7909.2012.01109.x.
3. Bohra A, Sahrawat KL, Kumar S, Joshi R, Parihar AK, Singh U, et al. Genetics- and genomics-based interventions for nutritional enhancement of grain legume crops: status and outlook. J Appl Genet. 2015
4. Que Q, Chilton MD, de Fontes CM, He C, Nuccio M, Zhu T, et al. Trait stacking in transgenic crops: challenges and opportunities. GM Crops. 2010;1(4):220–229. doi: 10.4161/gmcr.1.4.13439.
5. Siebert MW, Nolting SP, Hendrix W, Dhavala S, Craig C, Leonard BR, et al. Evaluation of corn hybrids expressing Cry1F, Cry1A.105, Cry2Ab2, Cry34Ab1/Cry35Ab1, and Cry3Bb1 against southern United States insect pests. J Econ Entomol. 2012;105(5):1825–1834. doi: 10.1603/EC12155.
6. Zhang F, Maeder ML, Unger-Wallace E, Hoshaw JP, Reyon D, Christian M, et al. High frequency targeted mutagenesis in Arabidopsis thaliana using zinc finger nucleases. Proc Natl Acad Sci U S A. 2010;107(26):12028–12033. doi: 10.1073/pnas.0914991107.
7. Liang Z, Zhang K, Chen K, Gao C. Targeted mutagenesis in Zea mays using TALENs and the CRISPR/Cas system. J Genet Genomics. 2014;41(2):63–68. doi: 10.1016/j.jgg.2013.12.001.
8. Gao J, Wang G, Ma S, Xie X, Wu X, Zhang X, et al. CRISPR/Cas9-mediated targeted mutagenesis in Nicotiana tabacum. Plant Mol Biol. 2015;87(1–2):99–110. doi: 10.1007/s11103-014-0263-0.
9. Gao HR, Smith J, Yang MZ, Jones S, Djukanovic V, Nicholson MG, et al. Heritable targeted mutagenesis in maize using a designed endonuclease. Plant J. 2010;61(1):176–187. doi: 10.1111/j.1365-313X.2009.04041.x.
10. Camerini-Otero RD, Hsieh P. Homologous recombination proteins in prokaryotes and eukaryotes. Annu Rev Genet. 1995;29:509–552. doi: 10.1146/annurev.ge.29.120195.002453.
11. Wood RD. DNA repair in eukaryotes. Annu Rev Biochem. 1996;65:135–167. doi: 10.1146/annurev.bi.65.070196.001031.
12. Lieberman-Lazarovich M, Levy AA. Homologous recombination in plants: an antireview. Methods Mol Biol. 2011;701:51–65. doi: 10.1007/978-1-61737-957-4_3.
13. Waterworth WM, Drury GE, Bray CM, West CE. Repairing breaks in the plant genome: the importance of keeping it together. New Phytol. 2011;192(4):805–822. doi: 10.1111/j.1469-8137.2011.03926.x.
14. Puchta H, Fauser F. Synthetic nucleases for genome engineering in plants: prospects for a bright future.Plant J. 2014;78(5):727–741. doi: 10.1111/tpj.12338.
15. Weeks DP, Spalding MH, Yang B. Use of designer nucleases for targeted gene and genome editing in plants. Plant Biotechnol J. 2015
16. Puchta H, Fauser F. Synthetic nucleases for genome engineering in plants: prospects for a bright future.Plant J. 2013
17. Osman K, Higgins JD, Sanchez-Moran E, Armstrong SJ, Franklin FC. Pathways to meiotic recombination in Arabidopsis thaliana. New Phytol. 2011;190(3):523–544. doi: 10.1111/j.1469-8137.2011.03665.x.
18. Puchta H. The repair of double-strand breaks in plants: mechanisms and consequences for genome evolution. J Exp Bot. 2005;56(409):1–14.
19. Weinthal DM, Taylor RA, Tzfira T. Nonhomologous end joining-mediated gene replacement in plant cells. Plant Physiol. 2013;162(1):390–400. doi: 10.1104/pp.112.212910.
20. Chilton MD, Que Q. Targeted integration of T-DNA into the tobacco genome at double-stranded breaks: new insights on the mechanism of T-DNA integration. Plant Physiol. 2003;133(3):956–965. doi: 10.1104/pp.103.026104.
21. D'Halluin K, Ruiter R. Directed genome engineering for genome optimization. Int J Dev Biol.2013;57(6–8):621–627. doi: 10.1387/ijdb.130217kd.
22. Puchta H, Dujon B, Hohn B. Two different but related mechanisms are used in plants for the repair of genomic

double-strand breaks by homologous recombination. Proc Natl Acad Sci U S A.1996;93(10):5055–5060. doi: 10.1073/pnas.93.10.5055.

23. Paszkowski J, Baur M, Bogucki A, Potrykus I. Gene targeting in plants. EMBO J. 1988;7(13):4021–4026.

24. Cai CQ, Doyon Y, Ainley WM, Miller JC, Dekelver RC, Moehle EA, et al. Targeted transgene integration in plant cells using designed zinc finger nucleases. Plant Mol Biol. 2009;69(6):699–709. doi: 10.1007/s11103-008-9449-7.

25. Nishizawa-Yokoi A, Nonaka S, Osakabe K, Saika H, Toki S. A universal positive-negative selection system for gene targeting in plants combining an antibiotic resistance gene and its antisense rna. Plant Physiol. 2015;169(1):362–370. doi: 10.1104/pp.15.00638.

26. Shimatani Z, Nishizawa-Yokoi A, Endo M, Toki S, Terada R. Positive-negative-selection-mediated gene targeting in rice. Front Plant Sci. 2014;5:748.

27. Kumar S, AlAbed D, Worden A, Novak S, Wu H, Ausmus C, et al. A modular gene targeting system for sequential transgene stacking in plants. J Biotechnol. 2015;207:12–20. doi: 10.1016/j.jbiotec.2015.04.006.

28. Petolino JF, Kumar S. Transgenic trait deployment using designed nucleases. Plant Biotechnol J. 2015

29. Rivera AL, Gomez-Lim M, Fernandez F, Loske AM. Physical methods for genetic plant transformation. Phys Life Rev. 2012;9(3):308–345. doi: 10.1016/j.plrev.2012.06.002.

30. Tzfira T, Citovsky V. Agrobacterium-mediated genetic transformation of plants: biology and biotechnology. Curr Opin Biotech. 2006;17(2):147–154. doi: 10.1016/j.copbio.2006.01.009.

31. Taylor NJ, Fauquet CM. Microparticle bombardment as a tool in plant science and agricultural biotechnology. DNA Cell Biol. 2002;21(12):963–977. doi: 10.1089/104454902762053891.

32. Pitzschke A. Agrobacterium infection and plant defense—transformation success hangs by a thread.Front Plant Sci. 2013;4:519. doi: 10.3389/fpls.2013.00519.

33. Arias RS, Filichkin SA, Strauss SH. Divide and conquer: development and cell cycle genes in plant transformation. Trends Biotechnol. 2006;24(6):267–273. doi: 10.1016/j.tibtech.2006.04.007.

34. Delporte F, Jacquemin JM, Masson P, Watillon B. Insights into the regenerative property of plant cells and their receptivity to transgenesis: wheat as a research case study. Plant Signal Behav. 2012;7(12):1608–1620. doi: 10.4161/psb.22424.

35. Szabados L, Kovács I, Oberschall A, Ábrahám E, Kerekes I, Zsigmond L, et al. Distribution of 1000 sequenced T-DNA tags in the Arabidopsis genome. Plant J. 2002;32(2):233–242. doi: 10.1046/j.1365-313X.2002.01417.x.

36. Gelvin SB, Kim S-I. Effect of chromatin upon agrobacterium T-DNA integration and transgene expression. Biochimica et Biophysica Acta (BBA)—Gene Structure and Expression. 2007;1769(5–6):410–21. doi:http://dx.doi.org/10.1016/j.bbaexp.2007.04.005.

37. Hiei Y, Ishida Y, Komari T. Progress of cereal transformation technology mediated by Agrobacterium tumefaciens. Front Plant Sci. 2014;5:628. doi: 10.3389/fpls.2014.00628.

38. Frame B, Main M, Schick R, Wang K. Genetic transformation using maize immature zygotic embryos. Methods Mol Biol. 2011;710:327–341. doi: 10.1007/978-1-61737-988-8_22.

39. Kita Y, Nishizawa K, Takahashi M, Kitayama M, Ishimoto M. Genetic improvement of the somatic embryogenesis and regeneration in soybean and transformation of the improved breeding lines. Plant Cell Rep. 2007;26(4):439–447. doi: 10.1007/s00299-006-0245-z.

40. Puchta H, Hohn B. In planta somatic homologous recombination assay revisited: a successful and versatile, but delicate tool. Plant Cell. 2012;24(11):4324–4331. doi: 10.1105/tpc.112.101824.

41. Swoboda P, Hohn B, Gal S. Somatic homologous recombination in planta—the recombination frequency is dependent on the allelic state of recombining sequences and may be influenced by genomic position effects. Mol Gen Genet. 1993;237(1–2):33–40.

42. Swoboda P, Gal S, Hohn B, Puchta H. Intrachromosomal homologous recombination in whole plants.EMBO J. 1994;13(2):484–489.

43. Chiurazzi M, Ray A, Viret JF, Perera R, Wang XH, Lloyd AM, et al. Enhancement of somatic intrachromosomal homologous recombination in Arabidopsis by the HO endonuclease. Plant Cell.1996;8(11):2057–2066. doi: 10.1105/tpc.8.11.2057.

44. Siebert R, Puchta H. Efficient repair of genomic double-strand breaks by homologous recombination between directly repeated sequences in the plant genome. Plant Cell. 2002;14(5):1121–1131. doi: 10.1105/tpc.001727.

45. Petolino JF, Worden A, Curlee K, Connell J, Strange Moynahan TL, Larsen C, et al. Zinc finger nuclease-mediated transgene deletion. Plant Mol Biol. 2010;73(6):617–628. doi: 10.1007/s11103-010-9641-4.

46. Antunes MS, Smith JJ, Jantz D, Medford JI. Targeted DNA excision in Arabidopsis by a re-engineered homing endonuclease. BMC Biotechnol. 2012;12:86. doi: 10.1186/1472-6750-12-86.

47. Qi Y, Li X, Zhang Y, Starker CG, Baltes NJ, Zhang F, et al. Targeted deletion and inversion of tandemly arrayed genes in Arabidopsis thaliana using zinc finger nucleases. G3 (Bethesda) 2013;3(10):1707–1715. doi: 10.1534/g3.113.006270.

48. Zhou H, Liu B, Weeks DP, Spalding MH, Yang B. Large chromosomal deletions and heritable small genetic changes induced by CRISPR/Cas9 in rice. Nucleic Acids Res. 2014;42(17):10903–10914. doi: 10.1093/nar/gku806.

49. Rong YS, Golic KG. Gene targeting by homologous recombination in Drosophila. Science.2000;288(5473):2013–2018. doi: 10.1126/science.288.5473.2013.

50. Rong YS, Golic KG. A targeted gene knockout in Drosophila. Genetics. 2001;157(3):1307–1312.

51. Maggert KA, Gong WJ, Golic KG. Methods for homologous recombination in Drosophila. Methods Mol Biol. 2008;420:155–174. doi: 10.1007/978-1-59745-583-1_9.

52. Bi X, Rong YS. Genome manipulation by homologous recombination in Drosophila. Brief Funct Genomic Proteomic. 2003;2(2):142–146. doi: 10.1093/bfgp/2.2.142.

53. Kumar S, Fladung M. Controlling transgene integration in plants. Trends Plant Sci. 2001;6(4):155–159. doi: 10.1016/S1360-1385(01)01890-8.

54. Fauser F, Roth N, Pacher M, Ilg G, Sanchez-Fernandez R, Biesgen C, et al. In planta gene targeting.Proc Natl Acad Sci U S A. 2012;109(19):7535–7540. doi: 10.1073/pnas.1202191109.

55. Schiml S, Fauser F, Puchta H. The CRISPR/Cas system can be used as nuclease for in planta gene targeting and as paired nickases for directed mutagenesis in Arabidopsis resulting in heritable progeny. Plant J. 2014;80(6):1139–1150. doi: 10.1111/tpj.12704.

56. Ayar A, Wehrkamp-Richter S, Laffaire JB, Le Goff S, Levy J, Chaignon S, et al. Gene targeting in maize by somatic ectopic recombination. Plant Biotechnol J. 2013;11(3):305–314. doi: 10.1111/pbi.12014.

57. Even-Faitelson L, Samach A, Melamed-Bessudo C, Avivi-Ragolsky N, Levy AA. Localized egg-cell expression of effector proteins for targeted modification of the Arabidopsis genome. Plant J. 2011;68(5):929–937. doi: 10.1111/j.1365-313X.2011.04741.x.

58. Wang ZP, Xing HL, Dong L, Zhang HY, Han CY, Wang XC, et al. Egg cell-specific promoter-controlled CRISPR/Cas9 efficiently generates homozygous mutants for multiple target genes in Arabidopsis in a single generation. Genome Biol. 2015;16:144. doi: 10.1186/s13059-015-0715-0.

59. Mao Y, Zhang Z, Feng Z, Wei P, Zhang H, Botella JR, et al. Development of germ-line-specific CRISPR-Cas9 systems to improve the production of heritable gene modifications in Arabidopsis. Plant Biotechnol J. 2015

CHAPTER 7

IMPROVED NUTRITIVE QUALITY AND SALT RESISTANCE IN TRANSGENIC MAIZE BY SIMULTANEOUSLY OVEREXPRESSION OF A NATURAL LYSINE-RICH PROTEIN GENE, SBGLR, AND AN ERF TRANSCRIPTION FACTOR GENE, TSRF1

Meizhen Wang[1, 2], Chen Liu[1], Shixue Li[1], Dengyun Zhu[1], Qian Zhao[1] and Jingjuan Yu[1]

[1] State Key Laboratory of Agrobiotechnology, College of Biological Sciences, China Agricultural University, No. 2 Yuanmingyuan West Road, Haidian District, Beijing 100193, China

[2] Institute of Medicinal Plant Development, Chinese Academy of Medical Sciences & Peking Union Medical College, No. 151, Malianwa North Road, Haidian District, Beijing 100193, China

ABSTRACT

Maize (Zea mays L.), as one of the most important crops in the world, is deficient in lysine and tryptophan. Environmental conditions greatly impact plant growth, development and productivity. In this study, we used particle bombardment mediated co-transformation to obtain marker-free transgenic maize inbred X178 lines harboring a lysine-rich protein gene SBgLR from potato and an ethylene responsive factor (ERF) transcription factor gene, TSRF1, from tomato. Both of the target genes were successfully expressed and showed various expression levels in different transgenic lines. Analysis showed that the protein and lysine content in T_1 transgenic maize seeds increased significantly. Compared to non-transformed maize, the protein and lysine content increased by 7.7% to 24.38% and 8.70% to 30.43%, respectively. Moreover, transgenic maize exhibited more tolerance to salt stress. When treated with 200 mM NaCl for 48 h, both non-transformed and transgenic plant leaves displayed wilting and losing green symptoms and dramatic increase of the free proline contents. However, the degree of control seedlings was much more serious than that of transgenic lines and much more increases of the free proline contents in the transgenic lines than that in the control seedlings were observed. Meanwhile, lower extent decreases of the chlorophyll contents were detected in the transgenic seedlings. Quantitative RT-PCR was performed to analyze the expression of ten stress-related genes, including stress responsive transcription factor genes, ZmMYB59 and ZmMYC1, proline synthesis related genes, ZmP5CS1 and ZmP5CS2, photosynthesis-related genes, ZmELIP, ZmPSI-N,ZmOEE, Zmrbcs and ZmPLAS, and one ABA biosynthesis related gene, ZmSDR. The results showed that with the exception of ZmP5CS1 and ZmP5CS2 in line 9–10 and 19–11, ZmMYC1 in line 19–11 and ZmSDRin line 19–11, the expression of other stress-related genes were inhibited in transgenic lines under normal conditions. After salt treatment, the expressions of the ten stress-related genes were significantly induced in both wild-type (WT) and transgenic lines. However, compared to WT, the increases of ZmP5CS1 in all these three transgenic lines and ZmP5CS2 in line 9–10 were less than WT plants. This study provides an effective approach of maize genetic engineering for improved nutritive quality and salt tolerance.

Citation: Meizhen Wang, Chen Liu, Shixue Li, Dengyun Zhu, Qian Zhao and Jingjuan Yu, Improved Nutritive Quality and Salt Resistance in Transgenic Maize by Simultaneously Overexpression of a Natural Lysine-Rich Protein Gene, SBgLR, and an ERF Transcription Factor Gene, TSRF1, doi:10.3390/ijms14059459.

INTRODUCTION

Maize (*Zea mays* L.), also known as corn, is one of the most widely cultivated crops in the world. It is mainly used as human food, livestock feed and industrial raw material. However, malnutrition is common in the countries, where corn is the sole or primary food source, due to the deficiency of essential amino acids, like lysine and tryptophan [1]. Actually, major efforts have been made to identify high-lysine corn varieties by genetic approaches since the mid-twentieth century. *Opaque 2* (*o2*) is a natural high-lysine mutant, which has twice of the lysine content than in the normal maize [2]. However, several drawbacks were observed in this mutant, such as soft chalky endosperm [2]. Even though quality protein maize (QPM) has been produced by backcrossing of *o2* modifier genes into *o2* mutant [3], it is a lengthy progress (~30 years) that limits the spread of QPM. Genetic engineering technologies are also utilized aiming to increase lysine content by zein reduction [4–6]. Unfortunately, all these transgenic plants exhibit opaque kernel phenotypes, which were very similar to that of *o2* mutant [4–6]. Moreover, increase of lysine content in corn grain has been achieved by manipulation of both lysine biosynthesis and metabolic pathways [7,8], including the commercialized high-lysine maize variety LY038 [9]. However, only free lysine content is significantly increased in these transgenic plants [8].

Recently, seed-specific expression of natural proteins with high-lysine concentration has been proved to be an effective approach to increase the lysine content of corn grain. Among them, milk proteins are attractive choice due to their balanced amino acid composition and good digestibility. As reported, lysine content is enhanced obviously in maize endosperm when expressing the milk protein, α-lactalbumin [10,11], whereas total protein content is not significantly different from negative kernels [11]. Liu *et al.* (1997) cloned a pollen-specific cDNA encoding a lysine-rich protein SB401 from potato [12]. In our previous studies, seed-specific expression of *SB401* increases lysine content from 16.1% to 54.8%, and meanwhile, the total protein content is increased from 11.6% to 39.0% in T_1 transgenic maize seeds, compared with the non-transgenic lines [13]. Thereafter, the natural lysine-rich protein gene *SBgLR*, homologous to *SB401*, was isolated from potato and transformed into maize, resulting in greater lysine and protein contents increase in transgenic maize [14]. It provides an effective way for maize quality improvement.

Environmental conditions, such as drought, salinity, extreme temperatures and pathogen attack, greatly affect plant growth, development and productivity. To survive under these conditions, plants have developed a series of defense responsive pathways [15]. In this signaling network, transcription factors play important roles through regulating expression of stress-related genes [16,17]. Ethylene responsive factor (ERF) proteins are a class of plant specific transcription factors, which have been reported to modulate plant development and multiple stress responses [18–22]. It was found that ERF proteins could function not only in plant pathogen resistance by interacting with GCC-box (AGCCGCC) [23–25], but also in abiotic resistance through interacting with the DRE element (CCGAC) [18,21,26–28]. TSRF1, an ERF protein from tomato, can activate the expression of PR genes by binding to the GCC-box and positively regulates pathogen resistance in transgenic tomato and tobacco [29,30]. ABA enhances the binding of TSRF1 to GCC-box and modifies the pathogen resistance of transgenic tobacco [31]. In addition, TSRF1 negatively regulates the osmotic response in transgenic tobacco [30], but improves the osmotic and drought tolerance in transgenic rice, indicating different mechanisms of TSRF1 in dicots and monocots [32]. All these results suggest that genetic engineering of ERF transcription factor genes is an effective way to improve plant stress tolerance [33].

Genetically modified (GM) crops are the fast-adopted technology in agriculture [34]. It has increased to 29 countries growing GM crops in recent years [35]. However, intense debates have been addressed regarding the safety of GM crops to the environment and customers. Selectable marker genes, like herbicide or antibiotic resistance genes, are supposed to be dangerous, although there is no strong scientific evidence [36]. By far, various strategies to eliminate the selectable marker genes have been developed and successfully used, such as co-transformation, site-specific recombination, multi-autotransformation vector, transposition system and homologous recombination (see more in review [35]). For *Agrobacterium*-mediated co-transformation, there are three methods: (i) two different vectors carried by different *Agrobacterium* strains [37]; (ii) two different vectors in the same *Agrobacterium* strain [38] and (iii) one binary vector with twin T-DNAs [39]. For biolistic bombardment mediated co-transformation, two different plasmids were introduced into the same tissue [40,41]. Among them, co-transformation is widely used due to its simplicity. It can be carried out either by *Agrobacterium*-mediated transformation or by biolistic bombardment. Selectable marker genes and target genes are usually integrated into the separate loci and segregate independently in the subsequent generations [42–44].

In this study, two plasmids containing the target genes, a lysine-rich protein gene, *SBgLR*, and an ERF transcription factor gene, *TSRF1*, and the selectable marker gene, *Hpt*, respectively, were introduced into maize elite inbred X178 embryogenic calli by particle bombardment. Marker-free transgenic maize lines were obtained. Further analysis revealed that both of the target genes were successfully expressed. Further, the lysine and protein content of transgenic maize seeds were detected and the stress tolerance was analyzed by measuring the free proline and chlorophyll contents. The results showed that transgenic lines had not only increased lysine and protein content, but also more tolerance to salt stress. Moreover, ten stress-related gene expression changes were analyzed using quantitative RT-PCR. Marker-free transgenic maize lines with high quality and stress resistance obtained in this research will be good materials for corn breeding.

RESULTS AND DISCUSSION

TSRF1, but Not SBgLR Protein Has Homologs in Maize

To obtain marker-free transgenic maize inbred lines, immature embryogenic calli were transformed by particle bombardment-mediated co-transformation. In this system, two constructs pTSSB and pHpt were used, pTSSB contains a potato lysine-rich protein gene, *SBgLR*, under the control of seed-specific promoter, 19Z, and a tomato ERF transcription factor gene, *TSRF1*, under the control of 35S constitutive promoter. The vector pHpt contains a selectable marker gene, *Hpt*, for hygromycin selection. By searching SBgLR protein sequence against the maize reference genome using the tBLASTn program [45,46], no homologs of SBgLR gene were identified (data not shown). However, two homologs for TSRF1 gene were found by similar method (Figure 1). The TSRF1 contains four motifs: an AP2/ERF domain, a CMIX-1 motif, a CMIX-3 motif and a CMIX-4 motif [32]. No ERF proteins containing all the four motifs of TSRF1 have been identified in the japonica rice genome by bioinformatic analysis [47]. To compare the structure of TSRF1 with its homologs in maize, we performed a multiple alignment using Clustal W (http://www.ebi.ac.uk/Tools/clustalw2/index.html) [48]. In contrast to rice, the homologs of maize contain all the four conserved motifs (Figure 1). Therefore, we designed the specific primer pairs in the non-conserved region of *TSRF1* for subsequent PCR and RT-PCR detection of transgenic lines.

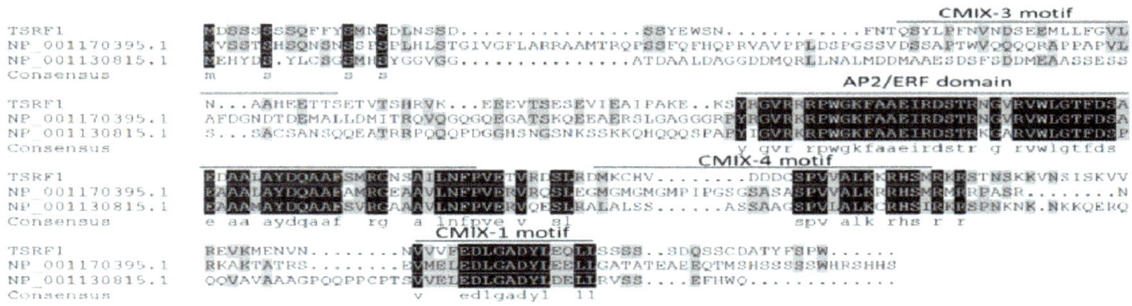

Figure 1: Comparison of the putative motifs of *TSRF1* and its homologs in maize. Motifs of TSRF1 were marked by black lines on the top of the sequences.

Generation of Transgenic Maize Inbred Lines

For co-transformation, two constructs pTSSB and pHpt mixture at a mole ratio of 1.5:1 were co-bombarded into maize embryogenic calli. A total of 114 fertile plants were obtained and self-pollinated for seeds set. In order to confirm the integration of transgenes, we conducted PCR analysis for *SBgLR*, *TSRF1* and *Hpt* genes. Twenty-six transgenic lines positive for all three genes were identified. Transformation efficiency of 1.08% in this study was obtained (Table 1). Both target genes tend to insert into the same loci of one transgenic line. A similar result was reported in maize that an entire 10-member kafirin gene cluster was transformed into maize genome by particle bombardment method without gene silencing [49].

Table 1: Efficiency of particle bombardment-mediated maize co-transformation

Ratio of pTSSB/pHpt	No. of transformed calli	No. of hygromycin-resistant events recovered	No. of regenerated plants	PCR analysis *SBgLR* (+) *TSRF1* (+) *Hpt* (+)	Co-transformation efficiency (%)
1.5:1	2411	224	114	26	1.08 [a]

[a] Co-transformation efficiency = (No. of all three genes PCR positive plants/No. of calli transformed) × 100%.

Overexpression of SBgLR and TSRF1 in Transgenic Maize

To test whether *SBgLR* and *TSRF1* expressed in transgenic maize, we conducted semi-quantitative RT-PCR using cDNAs from T_1 transgenic maize immature seeds at 22 days after pollination (DAP) and leaves as templates, respectively (Figure 2A,B). The results revealed that both *SBgLR* and *TSRF1* were expressed in the progeny from 16 transgenic lines, and they showed various expression levels among different transgenic lines.

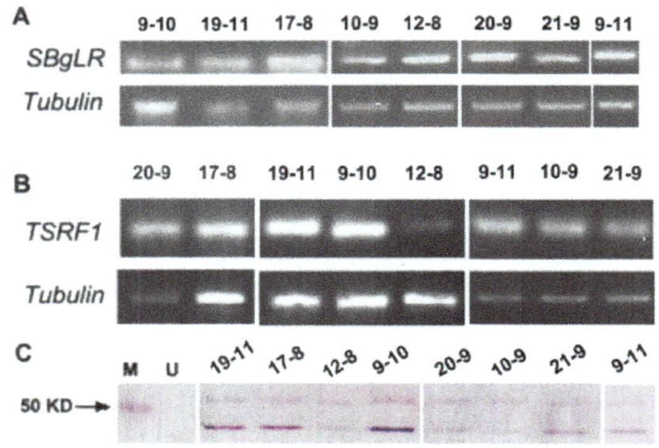

Figure 2: Expression of transgenes in T1 maize (partial results are shown). (A) Semi-quantification RT-PCR analysis of SBgLR in transgenic maize immature seeds (22 DAP); (B) semi-quantification RT-PCR analysis of TSRF1 in transgenic maize leaves; (C) Western blot analysis of SBgLR in transgenic maize immature seeds (20 DAP). M, protein marker; U, non-transformed maize plants. Numbers on the top of each figure represent the transgenic line number.

Water-soluble proteins were extracted from the immature seeds (20 DAP) of T_1 generation of *SBgLR* and*TSRF1* RT-PCR positive lines. Western blot was further carried out to confirm SBgLR protein accumulation in transgenic maize seeds (Figure 2C). Specific rabbit polyclonal antiserum against SBgLR at 1:400 was used. The predicted molecular mass of the SBgLR was 23 kD, whereas it was much larger (about 50 kD) after SDS/PAGE separation. This discrepancy was also observed in previous Western blot analysis of SBgLR and SB401 in transgenic maize [13,14]. Studies have shown that SB401 is a microtubule-associated protein and has a reduced mobility on SDS/PAGE [13,50], which is a common feature of microtubule-associated proteins [51,52]. In addition, it should be noted that two specific bands were observed in Western blot analysis of SBgLR (Figure 2C). The glycosylation site existed in the *N*-terminal of SBgLR may result in the reduction of mobility. However, further confirmation is needed. The transgenic lines with relative high*SBgLR* and *TSRF1* expression levels were selected for subsequent nutritive quality and stress tolerance analysis.

Marker-Free Transgenic Lines Obtained by Segregation in T_2 Generation

In the particle bombardment-mediated co-transformation system, marker gene elimination is dependent on the segregation of the target gene and selectable marker gene in the progeny. To examine whether the selectable marker gene *Hpt* was removed successfully, we performed genomic DNA dot blot to analyze T_2 transgenic lines (Figure 3). Purified PCR fragments of *SBgLR* and *Hpt* were used as probes, respectively. The results indicated that segregation of target genes and *Hpt* occurred in T_2 generation. Two transgenic lines 17 and 15 in R_0 generation contained both target genes and selectable marker gene, but their corresponding T_2generation transgenic lines 17-8-3 and 15-9-1 harbored only target genes as highlighted by arrows. These results indicated that it is a feasible approach to obtain marker-free transgenic lines via particle bombardment mediated co-transformation.

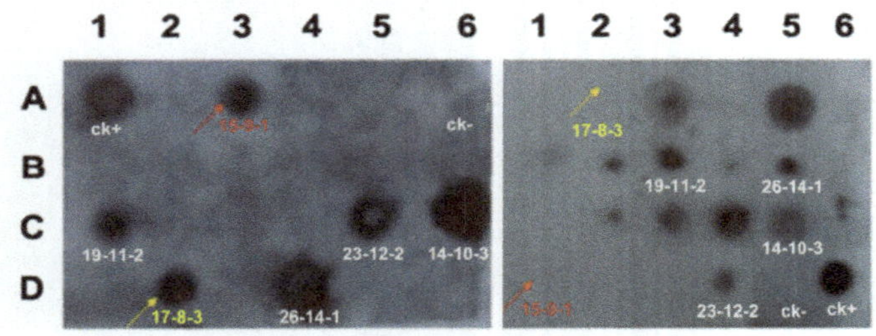

Figure 3: Segregation of target genes and selectable marker gene in T_2 transgenic plants by DNA dot blot analysis. Two segregated transgenic lines are highlighted in red for 15-9-1 and yellow for 17-8-3, respectively. Transgenic lines that have only Hpt are not indicated. For SBgLR (left), A1 represents positive control and A6 represents negative control. For Hpt (right), D6 represents positive control and D5 represents negative control.

SBgLR Enhanced Crude Protein and Lysine Contents in Transgenic Maize

Twenty kernels from each of the self-pollinated transgenic and non-transformed T_1 plants were randomly selected and the lysine and protein contents were analyzed. The results showed that the lysine and protein contents in the seeds were increased at various levels in different transgenic maize lines. Compared to non-transformed maize, the increases of protein and lysine content ranged from 7.7% to 24.38% (Figure 4A) and from 8.70% to 30.43% (Figure 4B), respectively. Line 19–11 showed high SBgLR accumulation level (Figure 2C) and total protein content, but its lysine content was not analyzed, due to limited amount of seeds. Subsequently, we detected the lysine content of T_2 transgenic seeds. The result showed that the lysine contents in all these transgenic lines were increased. In addition, it should be noted that the lysine contents in T_2 transgenic lines were increased to a much greater extent than in T_1 generation, indicating much more protein and lysine accumulation, because the *SBgLR* gene became homozygous. However, more evidence is needed. These transgenics will be good materials in quality maize breeding programs.

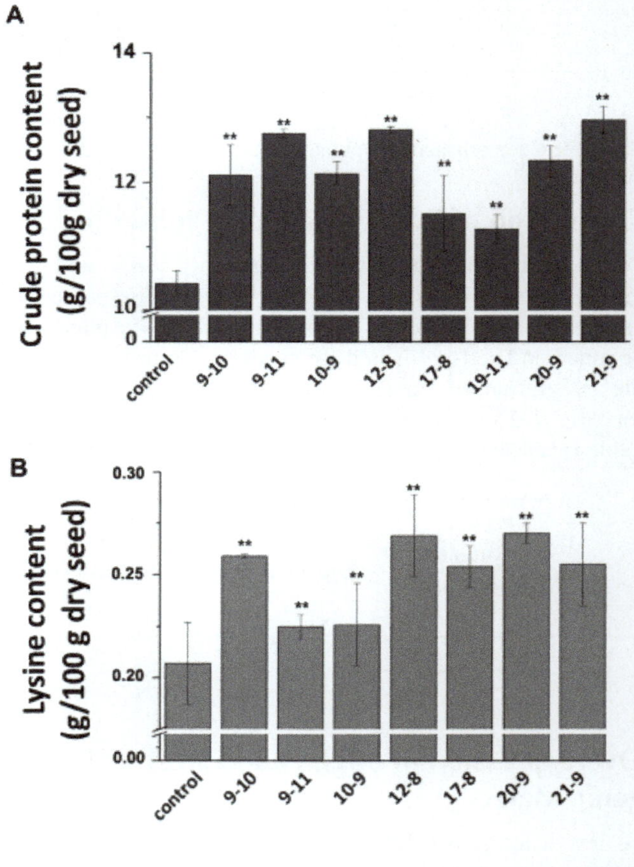

Figure 4: Protein and lysine levels were simultaneously increased in T_1 transgenic seeds. (A) Protein content; (B) lysine content. The crude protein and lysine contents are expressed as g/100 g dry seed. Data are averages of triplicates ± standard deviations. Asterisks, **, denote transgenic lines statistically different from Control by Student t-test at $p < 0.01$. The lysine content of line 19–11 was not measured due to limited amount of seeds.

TSRF1 Increased Salt Tolerance in Transgenic Maize

The overexpression of *TSRF1* increases pathogen tolerance in tobacco and tomato [29] and drought resistance in rice [32]. To investigate whether TSRF1 have salt tolerance in transgenic maize, four-leaf stage seedlings of three transgenic lines were used for salt treatment. The result was shown in Figure 5A.

Before treatment, these transgenic lines displayed similar growth situation as non-transformed control, suggesting that the expression and insertion of *TSRF1* into maize genome did not influence the growth of transgenic lines. After treated in Hoagland solution containing 200 mM NaCl for 48 h, both non-transformed and transgenic plant leaves displayed wilting and losing green symptoms. However, the degree of control seedlings was much more serious than that of transgenic lines. All the leaves of control seedlings were wilting and rolling. However, the first and second new leaves were still flat and green in transgenic seedlings.

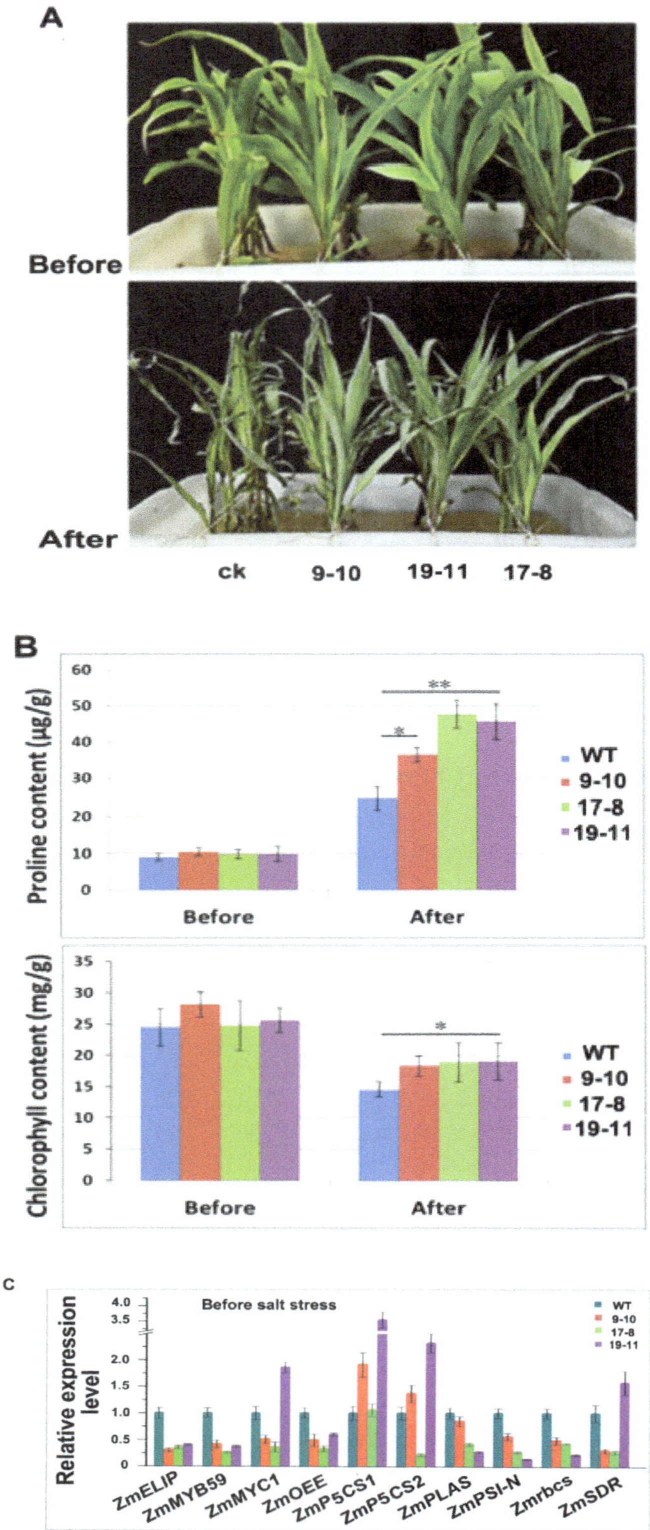

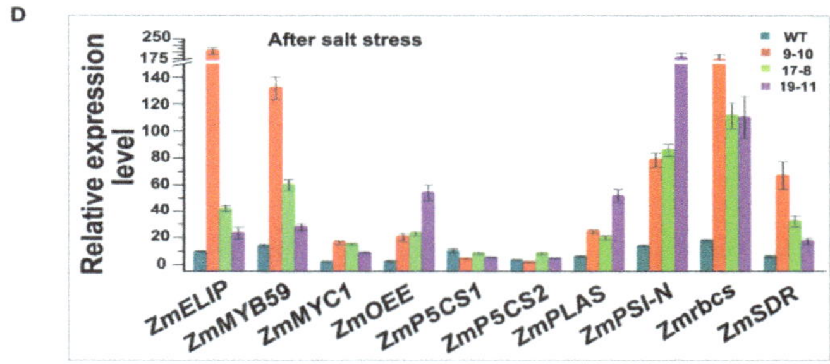

Figure 5: TSRF1 improved salt tolerance in maize. (A) Salt tolerance of TSRF1 transgenic maize. Before, control experiment without drought treatment; after, maize seedlings at four-leaf stage were subjected to 200 mM NaCl stress for 48 h; (B) proline and chlorophyll content in the control and transgenic maize. Data are averages of triplicate ± standard deviations. Asterisks, * and **, denote statistically different from wild-type by Student t-test at $p < 0.05$ and $p < 0.01$, respectively; (C) relative expression of stress-related genes in wild-type and TSRF1 transgenic maize seedlings before salt treatment; (D) relative expression of stress-related genes in wild-type and TSRF1 transgenic maize seedlings after salt treatment. Total RNA extracted from leaves of maize seedlings before and after 200 mM NaCl treatment for 48 h was used, respectively. Tubulin was used as a reference gene. Data are averages of triplicate ± standard deviations.

Proline was considered to be an osmoprotectant to protect plants against the damage resulting from abiotic stress [53]. Plants accumulate high level of proline under stress conditions to prevent the detrimental change. In addition, there is strong evidence that salt stress affects photosynthetic enzymes, chlorophyll and carotenoids in plants [54]. Therefore, we measured the free proline and chlorophyll content change in control and transgenic lines after salt treatment. As it is shown in Figure 5B, the free proline contents in control and transgenic seedlings were at similar level before treatment and increased dramatically after salt treatment. However, the proline content in the three transgenic lines increased at a much larger extent. Similarly, the chlorophyll content in transgenic seedlings decreased at a lower extent after salt treatment, indicating the more tolerance of *TSRF1* overexpressing transgenic maize.

It was previously shown that the constitutive expression of *TSRF1* enhanced rice salt and drought resistance by activating the expression of MYB, MYC, proline synthesis and photosynthesis-related genes [32]. Therefore, we further investigated whether TSRF1 affects the expression of these genes in maize. Ten genes were selected to be analyzed, and the results were shown in Figure 5C,D. The expressions of these genes were found to be different between transgenic lines. This might be due to the different transgenic lines with different target gene insertion sites and the expression levels of *TSRF1*. Under normal conditions, the expression level of *ZmMYC1* in line 19–11, *ZmP5CS1* (Δ^1-pyrroline-5-carboxylate synthetase 1) and*ZmP5CS2* (Δ^1-pyrroline-5-carboxylate synthetase 2) in line 9–10 and 19–11 and *ZmSDR* (short-chain dehydrogenase/reductase) in line 19–11 were increased, while the expression of these four genes in other lines were inhibited. The other six genes, including stress responsive transcription factor genes—*ZmMYB59*—photosynthesis-related genes—*ZmELIP* (low molecular mass early light-inducible protein HV90), *ZmPSI-N* (photosystem I reaction center subunit N), *ZmOEE* (photosystem II oxygen-evolving complex protein),*Zmrbcs* (encoding ribulose 1, 5-bisphosphate carboxylase small subunit) and *ZmPLAS* (encoding plastocyanin)—were inhibited in all these three transgenic lines, which was different from that in rice [32]. This indicated that TSRF1 was probably involved in different regulatory pathways in maize from that in rice. Furthermore, we detected the expression changes of these genes in wild-type and transgenic lines after salt stress. As is shown in Figure 5C, the expression of all these ten genes were increased by 2–16-fold in wild-type after salt treatment, indicating the endogenous regulation of salt responsive genes. Intriguingly, much more fold accumulations of these genes were induced after salt stress in *TSRF1* overexpressing lines. For example, the expression of *Zmrbcs* had increased as high as 171-, 111- and 110-fold in transgenic lines 9–10, 17–8 and 19–11, respectively, compared with wild-type. The sharp induction of photosynthesis-related genes (*ZmELIP*, *Zmrbcs* and *ZmPSI-N*) was consistent with the result of chlorophyll content change in Figure 5B. TSRF1 increases ABA sensitivity in tobacco [31,55] and rice [32]. *SDR*, an ABA biosynthesis related gene, was increased in *TSRF1* overexpressing rice [32]. On the contrary, it was slightly suppressed in line 9–10 and 17–8 under normal condition. After salt treatment, surprisingly, TSRF1 triggered significant *ZmSDR*induction. All these results suggested the complicated interaction of TSRF1 and ABA in plants. The accumulation of transcription factor gene *ZmMYB59* increased much more than *ZmMYC1* in all the transgenic lines, demonstrating the different roles of stress responsive transcription factors in salt stress. They may induce different other downstream salt related genes. These results indicated the complexity mechanism of *TSRF1* that involved in maize salt stress regulatory network.

Moreover, we noted that the expression levels of *ZmP5CS1* and *ZmP5CS2* were not or slightly increased in transgenic lines under salt condition, compared to wild-type, even though high level of proline accumulation was observed (Figure 5B,C). It is reported that the activity of P5CS, a key enzyme representing a rate-limiting step in proline biosynthesis, is regulated through feedback inhibition by proline [56–59]. It is possible that a large amount of proline is synthesized at the early stage of salt stress and maintained this high level throughout the salty condition. Therefore, high activity of P5CS or high level of *P5CS* transcript might be not necessary for high proline content maintaining under salt condition.

EXPERIMENTAL SECTION

Plasmid Construction

The plasmid pTSRF1 containing the*TSRF1* expression cassette was kindly provided by Prof. Rongfeng Huang, Biotechnology Research Institute, China Academy of Agricultural Sciences (CAAS). The TSRF1 expression cassette was obtained by digesting pTSRF1 with *Bam*HI and *Hin*dIII, blunt-ended with Klenow enzyme (Promega, Madison, WI, USA) and then inserted into the blunt-ended *Hin*dIII site of p19SB, harboring the *SBgLR* expression cassette. The plasmid pHpt was constructed by Wang *et al.* 2006 [44].

Plant Transformation and Detection

Maize inbred line X178, the male parent of the leading commercial cultivar Nongda108 in China, was grown in the test field and was self-pollinated. Embryogenic calli were induced from immature embryos at 10–12 DAP and transformed by particle bombardment, as described by Yu *et al.* 2004 [13].

For transgene detection, genomic DNA was extracted from leaf of transformed plants using SDS method [60], and PCR amplification was performed as follows: 95 °C for 10 min, 30 cycles of 95 °C for 30 s, 54 °C for 30 s and 72 °C for 30 s.

The expression of *SBgLR* and *TSRF1* in transgenic plants was further determined by RT-PCR. Total RNA was extracted from immature seeds (22 DAP) and young leaves, respectively, using TRIzol reagent (Invitrogen, Carlsbad, CA, USA). The first strand of cDNA was synthesized according to the manufacturer's protocol of Reverse Transcription System (Promega, Madison, WI, USA). *Tubulin* was used as the endogenous reference. PCR reactions were carried out for 5 min at 95 °C, followed by 25 cycles of 40 s at 94 °C, 40 s at 54 °C (*SBgLR* and *TSRF1*) or 52 °C (*Tubulin*), 40 s at 72 °C and a final extension of 10 min at 72 °C.

Genomic DNA dot blot analysis was used to detect marker-free progeny of T_2 plants. Total genomic DNA was isolated from fresh leaves using the cetyl trimethyl ammonium bromide (CTAB) procedure [61]. A 10 µg aliquot of DNA was spotted onto nylon membrane (Amersham Pharmacia, Arlington Heights, IL, USA). Filters were hybridized using specific purified PCR fragments of *SBgLR* and *Hpt*, respectively, as described for Southern blot by Yu *et al.* 2004 [13]. The primer pair for Hpt gene was the same as reported [14]. The primer pairs for *SBgLR*, *TSRF1* and *Tubulin* were listed.

The lysine-rich protein SBgLR accumulation in transgenic plant seeds was examined by Western blot, as described by Lang *et al.* 2004 [14]. Specific rabbit polyclonal SBgLR antiserum was prepared by the Institute of Genetics and Development Biology, Chinese Academy of Sciences (CAS).

Salt Stress Treatment

We determined the salt tolerance assay of transgenic maize at T_3 generation. Seeds were sowed in nurse soil and grown to four-leaf stage at the condition of 28 °C and 16 h light/8 h dark. Before salt treatment, leaves of one half wide type and transgenic lines were collected and immediately stored in liquid nitrogen, respectively. The other half was used for subsequent free proline and chlorophyll content analysis. For salt tolerance assay, after washing the soil off the roots, the seedlings of wild-type and transgenic lines were treated in Hoagland solution with 200 mM NaCl for 48 h, respectively. As described above, leaves of one-half salt treated wild-type and transgenic seedlings were collected and immediately stored in liquid nitrogen, respectively. The other half was used for subsequent free proline and chlorophyll content analysis.

Detection of Free Proline and Chlorophyll Content

Free proline was extracted from 0.5 g maize leaves (dry weight) using 3% salicylsulfonic acid and then reacted with ninhydrin, followed by determination based on colorimetric method [62].

For chlorophyll a and b content detection, total chlorophyll was extracted from 0.5 g leaves (dry weight), grinded into homogenate in 95% ethanol, and then absorbance at the wavelength 649 and 665 nm was measured under a spectrophotometer (UV2300, Shanghai, China), respectively. The chlorophyll content of each sample was calculated using the equation as follows: $C_a = 13.7A_{665} - 5.76A_{649}$; $C_b = 25.8A_{649} - 7.6A_{665}$; $C_{a+b} = C_a + C_b$.

Gene Expression Analysis by Quantitative Real-Time PCR

Total RNA of maize seedling leaves were extracted using TRIzol reagent (Invitrogen, Carlsbad, CA, USA). After digested with DNase I, 5 μg total RNA was used for cDNA first strand synthesis according to the manufacturer's protocol of Reverse Transcription System (Promega A3500, Madison, WI, USA). For quantitative real-time PCR amplification, 100 ng cDNA was used as template in a 20 μL reaction system, containing 10 μL 2× SYBR Premix Ex Taq II (TaKaRa, Shiga, Japan), 0.5 μM each specific forward and reverse primer. Amplification was carried out on a Bio-Rad Real-Time System CFX96 C1000 Thermal Cycler (Bio-Rad, Hercules, CA, USA) using the following conditions: 95 °C for 30 s, 35 cycles of 95 °C for 10 s, 60 °C for 10 sec and 72 °C for 10 s. Four reference genes, including *Tubulin*, *EF1*-α, *GAPDH* and*Ubiquitin*, were chosen. Expression levels of these four genes were determined as the Ct values [63]. The Ct values were then converted into relative quantities and imported into the geNorm v3,5 software [64] for stability analysis. According to the result, *Tubulin* was selected to be used as the endogenous reference.

Crude Protein and Lysine Content Analysis

Samples were ground meal of 20 mature kernels from individual self-fertilized ears. Due to limited seeds, only 8 T_1 mature kernels for line 19–11 were used. For crude protein content, total nitrogen was first measured based on the principle of Kjeldahl determination under national standard GB2905-82, Beijing Academy of Agriculture and Forestry Science. Then, the nitrogen content was converted to protein content by multiplying a conversion factor of 6.25.

Lysine content was analyzed based on the principle of ninhydrin reaction. Briefly, about 10 mg of defatted powder, 1 mL double distilled water and 2 mL ninhydrin reaction reagent were added into a 30 mL tube, and then the mixture was thoroughly vortexed (approximately 5 s). Subsequently, the tube was incubated in boiling water for 20 min, and thereafter, 3 mL 50% ethanol was added to cooled sample, followed by a centrifugation at 13,400× *g* for 10 min. The supernatant was collected, and the absorbance at 570 nm was recorded using a spectrophotometer (Techcomp UV2300, Shanghai, China). The standard curve was conducted by using 50 μg/mL leucine solutions in parallel. Lysine content of each sample was calculated using the equation as follows: Lys content (g/100 g dry seed weight) = (measured lysine content × hydrolysis volume)/sample weight.

CONCLUSIONS

In summary, we obtained marker-free transgenic maize inbred X178 lines harboring a potato lysine-rich protein gene, *SBgLR*, and a tomato ERF transcription factor gene, *TSRF1*, via particle bombardment mediated co-transformation. Molecular analysis revealed that both of these two target genes were successfully expressed and showed various expression levels in different transgenic lines. Quantification of protein and lysine content in T_1 maize seeds showed that transgenic lines had both increased protein and lysine content. The protein and lysine content increased by 7.7% to 24.38% and 8.70% to 30.43%, compared to non-transformed maize, respectively. Moreover, transgenic maize exhibited more tolerance to salt stress. This study provides an effective way of corn molecular breeding and materials for improved nutritive quality and salt tolerance.

ACKNOWLEDGMENTS

We are grateful to Xiaojuan Wang and Hanwen Liang for assistance with the embryogenic calli preparation. This work was sponsored by the National Transgenic Major Program of China (Grant No. 2011ZX003-002) and the National Basic Research Program of China (2012CB215301).

REFERENCES

1. Osborne T.B., Mendel L.B. Nutritional properties of proteins of the maize kernel. J. Biol. Chem.1914;18:1–16.
2. Mertz E.T., Bates L.S., Nelson O.E. A mutant gene that changes the protein composition and increases the lysine content of maize endosperm. Science. 1964;145:279–280.
3. Vasal S.K., Villegas E., Bjarnason M., Gelaw B., Goertz P. Genetic Modifiers and Breeding Strategies in Developing Hard Endosperm opaque-2 Materials. In: Pollmer W.G., Phipps R.H., editors. Improvement of Quality Traits of Maize for Grain and Silage Use. 2nd ed. Martinus Nijhoff; London, UK: 1980. pp. 37–73.
4. Segal G., Song R., Messing J. A new opaque variant of maize by a single dominant RNA interference-inducing transgene. Genetics. 2003;165:387–397.
5. Huang S., Adams W.R., Zhou Q., Malloy K.P., Voyles D.A., Anthony J., Kriz A.L., Luethy M.H. Improving nutritional quality of maize proteins by expressing sense and antisense zein genes. J. Agric. Food Chem. 2004;52:1958–1964.
6. Huang S., Frizzi A., Florida C.A., Kruger D.E., Luethy M.H. High lysine and high tryptophan transgenic maize resulting from the reduction of both 19- and 22-kD-zeins. Plant Mol. Biol. 2006;61:525–535.
7. Huang S., Kruger D.E., Frizzi A., D'Ordine R.L., Florida C.A., Adams W.R., Brown W.E., Luethy M.H. High-lysine corn produced by the combination of enhanced lysine biosynthesis and reduced zein accumulation. Plant Biotechnol. J. 2005;3:555–569.
8. Houmard N.M., Mainville J.L., Bonin C.P., Huang S., Luethy M.H., Malvar T.M. High lysine corn generated by endosperm specific suppression of lysine catabolism using RNAi. Plant Biotechnol. J.2007;5:605–614.
9. Lucas D.M., Glenn K.C., Bu J.Y. Petition for determination of nonregulated status for lysine maize LY038. [(accessed on 13 August 2004)]. Available online:http://www.aphis.usda.gov/brs/aphisdocs/0422901p.pdf.
10. Yang S.H., Moran D.L., Jia H.W., Bicar E.H., Lee M., Scott M.P. Expression of a synthetic porcine-lactalbumin

gene in the kernels of transgenic maize. Transgenic Res. 2002;11:11–20.

11. Bicar E.H., Woodman-Clikeman W., Sangtong V., Peterson J.M., Yang S.S., Lee M., Scott M.P. Transgenic maize endosperm containing a milk protein has improved amino acid balance. Transgenic Res.2008;17:59–71.

12. Liu J.Q., Seul U., Thompson R. Cloning and characterization of pollen-specific cDNA encoding a glutamic acid-rich protein (GARP) from potato *Solanum berthautii*. Plant Mol. Biol. 1997;33:291–300.

13. Yu J., Peng P., Zhang X., Zhao Q., Zhu D., Sun X., Liu J., Ao G. Seed-specific expression of a lysine rich protein sb401 gene significantly increases both lysine and total protein content in maize seeds. Mol. Breed. 2004;14:1–7.

14. Lang Z., Zhao Q., Yu J., Zhu D., Ao G. Cloning of potato SBgLR gene and its intron splicing in transgenic maize. Plant Sci. 2004;166:1227–1233.

15. Xiong L., Schumaker K.S., Zhu J. Cell signaling during cold, drought and salt stress. Plant Cell.2002;14:S165–S183.

16. Xiong L., Zhu J.K. Regulation of abscisic acid biosynthesis. Plant Physiol. 2003;133:29–36.

17. Guo H., Ecker J.R. The ethylene signaling pathway: New insights. Curr. Opin. Plant Biol. 2004;7:40–49.

18. Stockinger E.J., Gilmour A.J., Thomashow M.F. *Arabidopsis thaliana* CBF1 encodes an AP2 domain-containing transcriptional activator that binds to the C-repeat/DRE, a cis-acting DNA regulatory element that stimulates transcription in response to low temperature and water deficit. Proc. Natl. Acad. Sci. USA.1997;94:1035–1040.

19. Chakravarthy S., Tuori R.P., D'Ascenzo M.D., Fobert P.R., Despres C., Martin G.B. The tomato transcription factor Pti4 regulates defense-related gene expression via GCC box and non-GCC box cis elements. Plant Cell. 2003;15:3033–3050.

20. Guo Z.J., Chen X.J., Wu X.L., Ling J.Q., Xu P. Overexpression of the AP2/EREBP transcription factor OPBP1 enhances disease resistance and salt tolerance in tobacco. Plant Mol. Biol. 2004;55:607–618.

21. Yamaguchi-Shinozaki K., Shinozaki K. Transcriptional regulatory networks in cellular responses and tolerance to dehydration and cold stresses. Annu. Rev. Plant Biol. 2006;57:781–803.

22. Mantiri F.R., Kurdyukov S., Lohar D.P., Sharopova N., Saeed N.A., Wang X.D., Vandenbosch K.A., Rose R.J. The transcription factor MtSERF1 of the ERF subfamily identiWed by transcriptional prowling is required for somatic embryogenesis induced by auxin plus cytokinin in *Medicago truncatula*. Plant Physiol.2008;146:1622–1636.

23. Ohme-Takagi M., Shinshi H. Ethylene-inducible DNA binding proteins that interact with an ethylene-responsive element. Plant Cell. 1995;7:173–182.

24. Fujimoto S.Y., Ohta M., Usui A., Shinshi H., Ohme-Takagi M. Arabidopsis ethylene-responsive element binding factors act as transcriptional activators or repressors of GCC box-mediated gene expression. Plant Cell. 2000;12:393–404.

25. Gu Y.Q., Wildermuth M.C., Chakravarthy S., Loh Y.T., Yang C., He X., Han Y., Martin G.B. Tomato transcription factors pti4, pti5, and pti6 activate defense responses when expressed in Arabidopsis. Plant Cell.2002;14:817–831.

26. Liu Q., Kasuga M., Sakuma Y., Abe H., Miura S., Yamaguchi-Shinozaki K., Shinozaki K. Two transcription factors, DREB1 and DREB2, with an EREBP/AP2 DNA binding domain separate two cellular signal transduction pathways in drought- and low-temperature-responsive gene expression, respectively, in Arabidopsis. Plant Cell. 1998;10:1391–1406.

27. Hong J.P., Kim W.T. Isolation and functional characterization of the Ca-DREBLP1 gene encoding a dehydration-responsive element binding-factor-like protein 1 in hot pepper (*Capsicum annuum* L. cv. Pukang) Planta. 2005;220:875–888.

28. Liu N., Zhong N.Q., Wang G.L., Li L.J., Liu X.L., He Y.K., Xia G.X. Cloning and functional characterization of PpDBF1 gene encoding a DRE-binding transcription factor from *Physcomitrella patens*.Planta. 2007;226:827–838.

29. Zhang H., Zhang D., Chen J., Yang Y., Huang Z., Huang D., Wang X., Huang R. Tomato stress-responsive factor TSRF1 interacts with ethylene responsive element GCC box and regulates pathogen resistance to *Ralstonia solanacearum*. Plant Mol. Biol. 2004;55:825–834.

30. Zhang H., Li W., Chen J., Yang Y., Zhang Z., Wang X.C., Huang R. Transcriptional activator TSRF1 reversely regulates pathogen resistance and osmotic stress tolerance in tobacco. Plant Mol. Biol. 2007;63:63–71.

31. Zhou J., Zhang H., Yang Y., Zhang Z., Zhang H., Hu X., Chen J., Wang X.C., Huang R. Abscisic acid regulates TSRF1-mediated resistance to *Ralstonia solanacearum* by modifying the expression of GCC box-containing genes in tobacco. J. Exp. Bot. 2008;59:645–652.

32. Quan R., Hu S., Zhang Z., Zhang H., Zhang Z., Huang R. Overexpression of an ERF transcription factor TSRF1 improves rice drought tolerance. Plant Biotech. J. 2010;8:476–488.

33. Bhatnagar-Mathur P., Vadez V., Sharma K. Transgenic approaches for abiotic stress tolerance in plants: Retrospect and prospects. Plant Cell Rep. 2008;27:411–424.

34. James C. Global Status of Commercialized Biotech/GM Crops: 2010. ISAAA Brief No. 42. ISAAA; Ithaca, NY, USA: 2010.

35. Tuteja N., Verma S., Sahoo R.K., Raveendar S., Reddy B.L. Recent advances in development of marker-free transgenic plants: Regulation and biosafety concern. J. Biosci. 2012;37:167–197.

36. Holger P. Removing selectable marker genes: Taking shortcut. Trends Plant Sci. 2000;7:273–274.

37. De Neve M., de Buck S., Jacobs A., van Montagu M., Depicker A. T-DNA integration patterns in co-

transformed plant cells suggest that T-DNA repeats originate from co-integration of separate T-DNAs. Plant J. 1997;11:15–29.

38. Sripriya R., Raghupathy V., Veluthambi K. Generation of selectable marker-free sheath blight resistant transgenic rice plants by efficient co-transformation of a cointegrate vector T-DNA and a binary vector T-DNA in one *Agrobacterium tumefaciens* strain. Plant Cell Rep. 2008;27:1635–1644.

39. Miller M., Tagliani L., Wang N., Berka B., Bidney D. High efficiency transgene segregation in co-transformed maize plants using an *Agrobacterium tumefaciens* 2 T-DNA binary system. Transgenic Res.2002;11:381–396.

40. Shiva P.N., Bhojaraja R., Shivbachan S.K., Hari P.G.G., Nagraj T.K., Prasad V., Srikanth B.V., Jayaprakash T.L., Dasgupta S., Spencer T.M., et al. Marker-free transgenic corn plant production through co-bombardment. Plant Cell Rep. 2009;28:1655–1668.

41. Kumar S., Arul L., Talwar D. Generation of marker-free Bt transgenic indica rice and evaluation of its yellow stem borer resistance. J. Appl. Genet. 2010;51:243–257.

42. Qi Y., Ye S., Lu Y., Jin Q., Zhang X. Development of marker-free transgenic Cry1Ab rice with Lepidopteran pest resistance by *Agrobacterium* mixture-mediated co-transformation. Rice Sci. 2009;16:181–186.

43. Ramana R.M.V., Parameswari C., Sripriya R., Veluthambi K. Transgene stacking and marker elimination in transgenic rice by sequential *Agrobacterium*-mediated co-transformation with the same selectable marker gene. Plant Cell Rep. 2011;30:1241–1252.

44. Wang D., Zhao Q., Zhu D., Ao G., Yu J. Particle-bombardment-mediated co-transformation of maize with a lysine rich protein gene (sb401) from potato. Euphytica. 2006;150:75–85.

45. Schnable P.S., Ware D., Fulton R.S., Stein J.C., Wei F., Pasternak S., Liang C., Zhang J., Fulton L., Graves T.A., et al. The B73 maize genome: Complexity, diversity, and dynamics. Science. 2009;20:1112–1115.

46. Altschul S.F., Madden T.L., Schaffer A.A., Zhang J., Zhang Z., Miller W., Lipman D.J. Gapped BLAST and PSI-BLAST: A new generation of protein database search programs. Nucleic Acids Res. 1997;25:3389–3402.

47. Nakano T., Suzuki K., Fujimura T., Shinshi H. Genome-wide analysis of the ERF gene family in*Arabidopsis* and rice. Plant Physiol. 2006;140:411–432.

48. Clustal W. [(accessed on 10 September 2007)]. Available online:http://www.ebi.ac.uk/Tools/clustalw2/index.html.

49. Song R., Segal G., Messing J. Expression of the sorghum 10-member kafirin gene cluster in maize endosperm. Nucleic Acids Res. 2004;32:e189.

50. Huang S., Jin L., Du J., Li H., Zhao Q., Ou G., Ao G., Yuan M. SB401, a pollen-specific protein from*Solanum berthaultii*, binds to and bundles microtubules and F-actin. Plant J. 2007;51:406–418.

51. Noble M., Lewis S., Cowan N.J. The microtubule binding domain of microtubule-associated protein MAP1B contains a repeated sequence motif unrelated to that of MAP2 and Tau. J. Cell Biol.1989;109:3367–3376.

52. Wiche G., Oberkanins C., Himmler A. Molecular structure and function of microtubule-associated proteins. Int. Rev. Cytol. 1991;124:217–273.

53. Kishor P.B.K., Sangam S., Amrutha R.N., Laxmi P.S., Naidu K.R., Rao K.R.S.S., Rao S., Reddy K.J., Theriappan P. Regulation of proline biosynthesis, degradation, uptake and transport in higher plants: Its implications in plant growth and abiotic stress tolerance. Curr. Sci. 2005;88:424–438.

54. Stepien P., Klobus G. Water relations and photosynthesi in *Cucumis sativus* L. leaves under salt stress.Biologia Plantarum. 2006;50:610–616.

55. Zhang H., Yang Y., Zhang Z., Chen J., Wang X.C., Huang R. Expression of the ethylene response factor gene TSRF1 enhances abscisic acid responses during seedling development in tobacco. Planta.2008;228:777–787.

56. Savoure A., Jaoua S., Hua X.J., Ardiles W., van Montagu M., Verbruggen N. Isolation, characterization, and chromosomal location of a gene encoding the DELTA-1-pyrroline-5- carboxylate synthetase in*Arabidopsis thaliana*. FEBS Lett. 1995;372:13–19.

57. Yoshiba Y., Kiyosue T., Katagiri T., Ueda H., Mizoguchi T., Yamaguchi-Shinozaki K., Wada K., Harada Y., Shinozaki K. Correlation between the induction of a gene for delta 1-pyrroline-5-carboxylate synthase and the accumulation of proline in *Arabidopsis thaliana* under osmotic stress. Plant J. 1995;7:751–760.

58. Zhang C.S., Lu Q., Verma D.P.S. Removal of feedback inhibition of delta 1-pyrroline-5- carboxylate synthase, a bifunctional enzyme catalyzing the first two steps of proline biosynthesis in plants. J. Biol. Chem.1995;270:20491–20496.

59. Strizhov N., Abraham E., Okresz L., Blickling S., Zilberstein A., Schell J., Koncz C., Szabados L. Differential expression of two P5CS genes controlling proline accumulation during salt-stress requires ABA and is regulated by ABA1, ABI1 and AXR2 in Arabidopsis. Plant J. 1997;12:557–569.

60. Sambrook J., Fritsch E.F., Maniatis T. Molecular Cloning: A Laboratory Manual. 2nd ed. Cold Spring Harbor Press; New York, NY, USA: 1989.

61. Saghai-Maroof M.A., Soliman K.M., Jorgensen R.A., Allard R.W. Ribosomal DNA spacer-length polymorphisms in barley: Mendelian inheritance, chromosomal location, and population dynamics. Proc. Natl. Acad. Sci. USA. 1984;81:8014–8018.

62. Bates L.S., Waldren R.P., Teare I.D. Rapid determination of free proline for water-stress studies. Plant Soil. 1973;39:205–207.

63. Walker N.J. A technique whose time has come. Science. 2002;296:557.

64. geNorm v3.5 software. [(accessed on 7 May 2002)]. Available online:http://medgen.ugent.be/~jvdesomp/genorm/

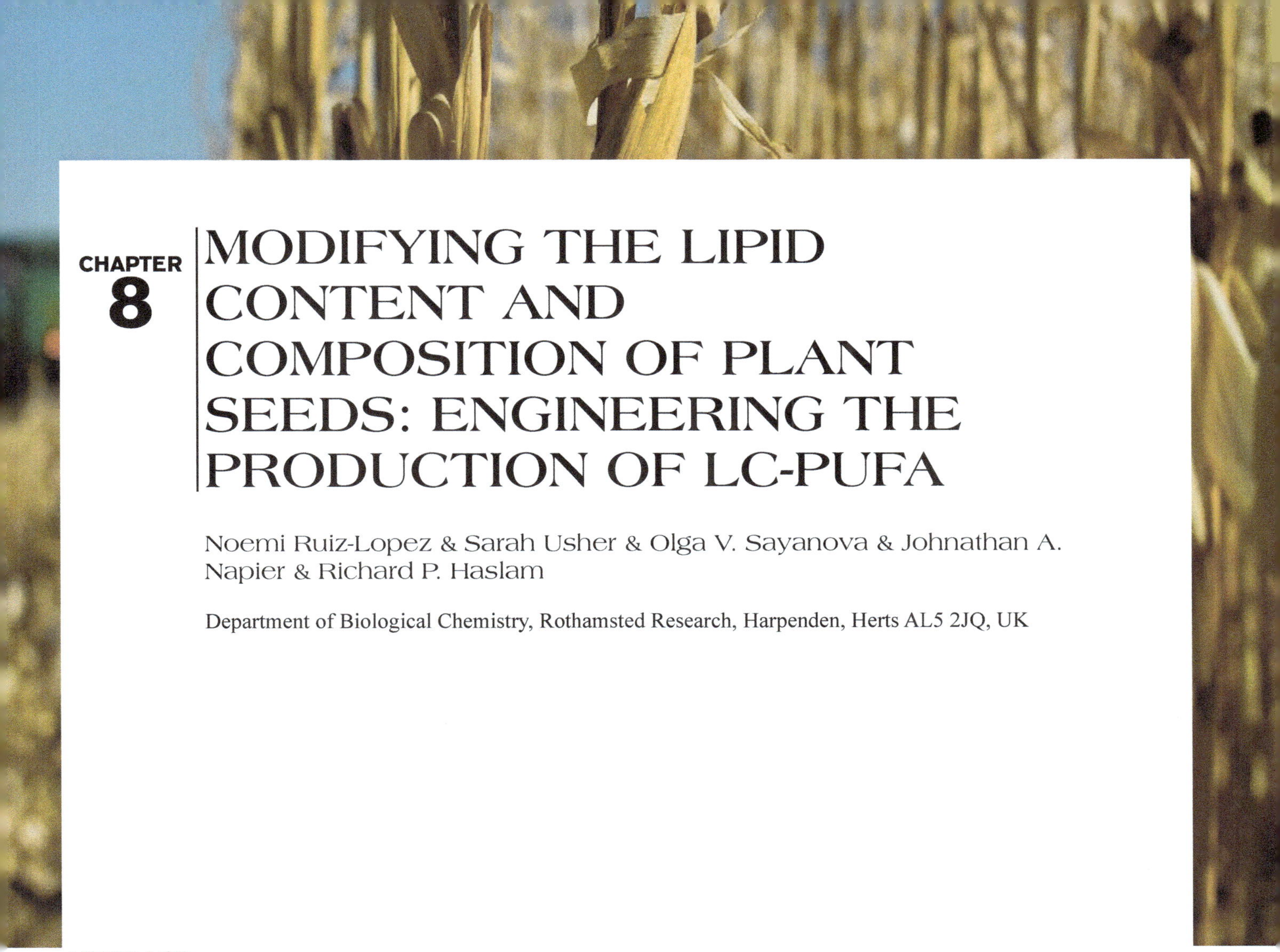

CHAPTER 8

MODIFYING THE LIPID CONTENT AND COMPOSITION OF PLANT SEEDS: ENGINEERING THE PRODUCTION OF LC-PUFA

Noemi Ruiz-Lopez & Sarah Usher & Olga V. Sayanova & Johnathan A. Napier & Richard P. Haslam

Department of Biological Chemistry, Rothamsted Research, Harpenden, Herts AL5 2JQ, UK

ABSTRACT

Omega-3 fatty acids are characterized by a double bond at the third carbon atom from the end of the carbon chain. Latterly, long chain polyunsaturated omega-3 fatty acids such as eicosapentaenoic acid (EPA; 20:5Δ5,8,11,14,17) and docosahexanoic acid (DHA; 22:6 Δ4,7,10,13,16,19), which typically only enter the human diet via the consumption of oily fish, have attracted much attention. The health benefits of the omega-3 LC-PUFAs EPA and DHA are now well established. Given the desire for a sustainable supply of omega-LC-PUFA, efforts have focused on enhancing the composition of vegetable oils to include these important fatty acids. Specifically, EPA and DHA have been the focus of much study, with the ultimate goal of producing a terrestrial plant-based source of these so-called fish oils. Over the last decade, many genes encoding the primary LC-PUFA biosynthetic activities have been identified and characterized. This has allowed the reconstitution of the LC-PUFA biosynthetic pathway in oilseed crops, producing transgenic plants engineered to accumulate omega-3 LC-PUFA to levels similar to that found in fish oil. In this review, we will describe the most recent developments in this field and the challenges of overwriting endogenous seed lipid metabolism to maximize the accumulation of these important fatty acids.

INTRODUCTION

Lipids (oils and fatty acids) are indispensable for the growth and survival of all organisms. They are important structural components of membranes and they also play a crucial role in energy storage and signaling. Furthermore, several polyunsaturated fatty acids (FA) can act as metabolic precursors for eicosanoids. Natural sources of lipids include plants, animals, and microorganisms. Oils from plants (e.g., olive oil, sunflower oil) and animals (e.g., butter, lard) have always played a vital role in human life, providing multiple sources of food and fuel; recognized mostly for their applications in nutrition, but also their role as raw material in industrial processes. Currently, plant oils account for the majority of the natural oils and fats in the world, about 150 million metric tons of vegetable oils are consumed annually and this number is expected to increase in the future (OECD_FAO Agricultural Outlook 2012–2021). Plant oils are relatively inexpensive and are commonly considered to be healthier than animal fats, as they contain relatively high amounts of unsaturated fatty acids. However, over recent years, there has been increased interest in some specific animal oils, namely the long-chain polyunsaturated fatty acids (LC-PUFAs), found in fish oil. LC-PUFAs are vital constituents of human metabolism and have

Citation: Noemi Ruiz-Lopez & Sarah Usher & Olga V. Sayanova & Johnathan A. Napier & Richard P. Haslam, Modifying the lipid content and composition of plant seeds: engineering the production of LC-PUFA, DOI 10.1007/s00253-014-6217-2.

a recognized role in human health. Thus, the nutritional value of omega-3 LC-PUFAs is now widely appreciated. Plant oils are rich in C18 FA, including the essential FA linoleic acid (18:2Δ9,12 *n*-6; LA) and α-linolenic acid (18:3Δ9,12,15 *n*-3; ALA), but are devoid of LC-PUFAs, such as arachidonic acid (20:4Δ5,8,11,14, *n*-6; ARA), eicosapentaenoic acid (20:5Δ5,8,11,14,17 *n*-3; EPA) and docosahexaenoic acid (22:6Δ4,7,10,13,16,19 *n*-3; DHA), which typically only enter the human diet as oily fish. Although humans can synthesis both EPA and DHA (approximately 8 % of dietary ALA is converted to EPA and 0–4 % is converted to DHA), the supply does not meet nutritional demands. Equally, the decline in global fish stocks is also well established and compounded by the use of fish oils in aquaculture. For several years, we and others have suggested that transgenic oilseeds metabolically engineered with the heterologous capacity to synthesize omega-3 LC-PUFAs represent a novel (terrestrial) production system, which could substitute for some or all of the applications currently using fish oils.

GENERAL PRINCIPLES OF OILSEED ENGINEERING

Success in agriculture has always depended on innovation; inventive approaches drive crop improvement, the selection of plants with desirable traits, and their successful propagation. In recent years, a better understanding of lipid assembly and storage, improved molecular genetics, and plant biotechnology have set the stage for a real breakthrough in the manipulation of oil crops for sustainable development. Specifically, genetic engineering based on recombinant DNA technology has now enabled the production of seed oils with a predictable FA composition; the incorporation of characteristics that are impossible to achieve by traditional breeding techniques.

Metabolic engineering of seed oil metabolism can be achieved via a wide range of methods. The most commonly used approaches are infection with *Agrobacterium tumefaciens* and biolistic gene transfer. *A. tumefaciens* is a widespread naturally occurring Gram-negative soil bacterium that causes crown gall, and has the ability to transfer a portion of DNA (transferred DNA or T-DNA) from its tumor-inducing (Ti) plasmid to plant cells (Gelvin 2003). The T-DNA contains the genes for inducing tumor formation and opine biosynthesis. In the laboratory, *Agrobacterium* bacterial T-DNA genes have been replaced by genes of interest. *Agrobacterium* is capable of transforming a range of dicotyledonous plant species; however, monocots are less responsive towards this method. The principle advantage of *Agrobacterium*-mediated gene transfer is the high occurrence of single-copy T-DNA integration with a relatively stable high level of transgene expression. In methods based on biolistics (or microprojectile bombardment), transformation is achieved by coating gold or tungsten microparticles with the desired DNA, and accelerating them using high pressure gas such that they are able to penetrate the cell (wall and membrane) and enter the plant nucleus. This method is routinely used for the transformation of monocots, cereals, legumes, and microalgae; species that are typically recalcitrant to *Agrobacterium*-mediated transformation (Sparks and Jones 2004). Bombardment offers a relatively high efficiency, the ability of delivering DNA without vector backbone, and also the opportunity to deliver DNA directly to organelle genomes. However, there can be substantial variation between experiments (due to differences in bombardment conditions), integrations at multiple loci, frequent truncations of the transgene and substantial damage to bombarded tissue, lowering regeneration capacity.

Several methods have been developed to overcome the limitations of *Agrobacterium*-mediated and particle bombardment gene delivery. One of them is a method of transformation using cell-penetrating peptides (CPPs). Chugh et al. (2009) demonstrated that the Tat monomer (Tat) and its dimer (Tat2) belong to the CPP family and are able to deliver DNA into plant cells. So far, this method has been used to deliver the transgene into microspores, but somatic cell targeting is still under development. This method has also been used in combination with an artificially produced T-DNA complex, allowing the production of transgenic triticale plants with great efficiency (Ziemienowicz et al. 2012). Latterly, efforts to improve crop performance have benefited from the arrival of a number of exciting techniques including mutagenesis and genome editing using Zinc finger nucleases (ZFN-1, ZFN-2, and ZFN-3); transcription activator-like effector nucleases (TALENs); oligonucleotide-directed mutagenesis (ODM); and gene inactivation using the bacterial CRISPR/Cas system (Wang et al. 2013; Balter et al. 2014; Lozano-Juste and Cutler 2014). Many of these approaches remain to be tested for their efficacy in manipulating seed oil assembly.

Oilseed crops and their different cultivars can have a highly variable fatty acid content and composition in their seeds with no significant differences in plant physiology. Therefore, the manipulation of oil content or/and composition should not dramatically alter plant physiology. Over the last two decades, there has been a substantial research effort to decipher the genes encoding the most important enzymes involved in FA and triacylglycerol (TAG) biosynthesis (for reviews, see Ruiz-Lopez et al. 2012a; Haslam et al. 2013; Li-Beisson et al. 2013; Napier et al. 2014), but further successful improvement of oil crops requires a good understanding of the endogenous biochemical processes that underpin seed oil assembly (Bates et al. 2013). With steady progress in genomics, proteomics, metabolomics, lipidomics, and genetic engineering as well as an increased number of genomes sequenced and annotated, the modification of oilseed crops becomes a more straightforward practice.

LONG CHAIN POLYUNSATURATED FATTY ACIDS

LC-PUFAs are composed of a long hydrocarbon chain (consisting of 20 or more carbon atoms) and a terminal carboxylate group containing more than two double bonds in their backbone. They are classified according to the position of the first double bond, as counted from the methyl terminus, which is denoted in the form (*n*-*x*), where *n* is the chain length of the fatty acid and *x* is the number of carbon atoms from the last double bond. Thus linoleate (LA) and α-linolenate (ALA) are 18:2(*n*-6) and 18:3(*n*-3), respectively (18:2ω6 and 18:3ω3 in the older literature). Double bonds in PUFAs may also be counted from the carboxylate group and are then represented by the symbol Δ. In Table 1, examples of several *n*-3 PUFAs are listed. Mammals lack the ability to introduce double bonds in FAs beyond carbon 9 and 10. Linoleic acid and α-linolenic acid found in plant oils are therefore an essential human dietary requirement (Hansen and Burr 1946) due to the lack of Δ-12- and Δ-15 desaturase activity. LA and ALA are precursors for the synthesis of LC-PUFAs. Today omega-3 LC-PUFAs such as EPA and DHA are particularly lacking in modern diets, leading to sub-optimal organ function and an increased risk of disease. A situation that, together with the fact that humans synthesize a limited amount of these PUFAs, has led many experts to recommend increased omega-3 LC-PUFA intake and make observable health benefits (FAO 2010).

Table 1: Lists of most common omega-3 and omega-6 fatty acids and their accumulation in microalgae and oilseed crops

Common name	Systematic name	Synonyms	Algae	Oilseeds
OMEGA3				
α-Linolenic acid (ALA)	9Z,12Z,15Z-octadecatrienoic acid	18:3*n*-3; 18:3 $^{\Delta9,12,15}$	√	√
Stearidonic acid (SDA)	6Z,9Z,12Z,15Z-octadecatetraenoic acid	18:4*n*-3; 18:4 $^{\Delta6,9,12,15}$	√	√[a]
Eicosatrienoic acid (ETE)	11Z, 14Z, 17Z-eicosatrienoic acid	20:3*n*-3; 20:3 $^{\Delta11,14,17}$		
Eicosatetraenoic acid (ETA)	8Z,11Z,14Z,17Z-eicosatetraenoic acid	20:4*n*-3; 20:4 $^{\Delta8,11,14,17}$	√	
Eicosapentaenoic acid (EPA)	5Z,8Z,11Z,14Z,17Z-eicosapentaenoic acid	20:5*n*-3; 20:5 $^{\Delta5,8,11,14,17}$	√	
		22:5*n*-3;	√	
Docosapentaenoic acid (DPA)	7Z,10Z,13Z,16Z,19Z-docosapentaenoic acid	22:5 $^{\Delta7,10,13,16,19}$		
Docosahexaenoic acid (DHA)	4Z,7Z,10Z,13Z,16Z,19Z-docosahexaenoic acid	22:6*n*-3; 22:6$^{\Delta4,7,10,13,16,19}$	√	
OMEGA6				
Linoleic acid (LA)	9Z,12Z-octadecadienoic acid	18:2*n*-6; 18:2 $^{\Delta9,12}$	√	√
γ-Linolenic acid (GLA)	6Z,9Z,12Z-octadecatrienoic acid	18:3*n*-6; 18:3 $^{\Delta6,9,12}$	√	√[a]
Dihomo-γ-linolenic acid (DGLA)	8Z,11Z,14Z-eicosatrienoic acid	20:3*n*-6; 20:3 $^{\Delta8,11,14}$	√	
Arachidonic acid (ARA)	5Z,8Z,11Z,14Z-eicosatetraenoic acid	20:4*n*-6; 20:4 $^{\Delta5,8,11,14}$	√	
Adrenic acid (DTA)	7Z,10Z,13Z,16Z-docosatetraenoic acid	22:4*n*-6; 20:4 $^{\Delta7,10,13,16}$		
Docosapentaenoic acid (DPA*n*-6)	4Z,7Z,10Z,13Z,16Z-docosapentaenoic acid	22:5*n*-6; 20:4 $^{\Delta4,7,10,13,16}$		

[a] Some Boraginaceae such as *Echium plantagineum* contain SDA, although this species is not generally recognized as an oilseed crop species. GLA is found in a few seed oils, and those of evening primrose, borage, and blackcurrant have some commercial importance

HEALTH ASPECTS OF *N*-3 LC-PUFAS

The position of the double bond in a fatty acid strongly affects the properties of its derivatives. For instance, eicosanoids derived from the *n*-6 LC-PUFA arachidonic acid have strong inflammatory properties, whereas those produced from *n*-3 LC-PUFA, e.g., eicosapentaenoic acid, are anti-inflammatory (Gill and Valivety1997; Calder 2014). For reference, the chemical structures of DHA and EPA are shown in Fig. 1. The biological functions of omega-3 LC-PUFA (especially EPA and DHA) are now well established (Schmitz and Ecker 2007; Saravanan et al. 2010). DHA is an essential component of the lipids in cell membranes where it clearly exerts a structural and functional role, e.g., modifying membrane phospholipid composition, for instance, it accounts for over 60 % of the total FAs in the rod outer segment in the retina (Giusto et al.2000). DHA also modulates properties of the hydrophobic core of the membrane bilayer, directly interacting with membrane proteins and is involved in lipid raft formation (Innis 2007). Furthermore, DHA is regarded to be essential for the proper visual and neurological development of infants (Crawford et al. 1997; Das and Fams 2003). Both EPA and DHA give rise to anti-inflammatory and inflammation resolving mediators called resolvins, protectins, and maresins; promoting the resolution of the inflammatory response back to a noninflamed state.

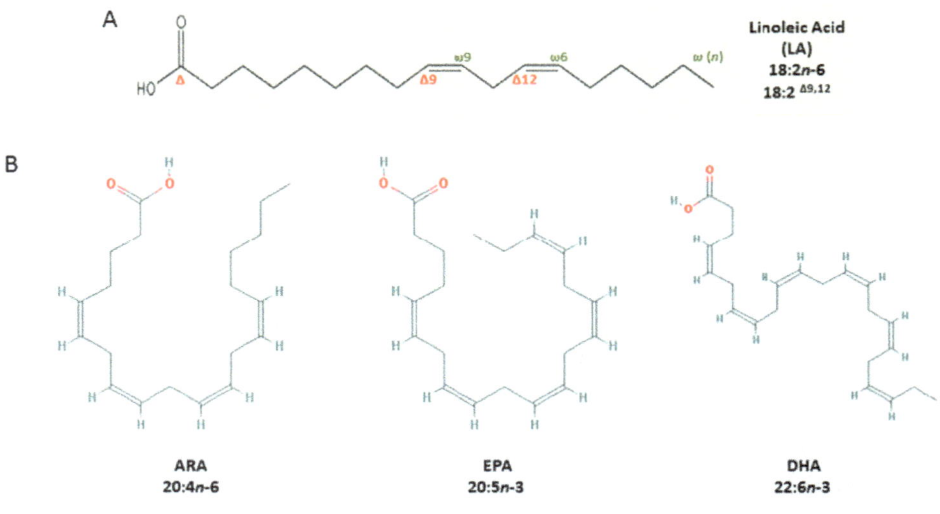

Figure 1: a Chemical structure of linoleic acid (LA, 18:2Δ9,12), consisting of a chain of 18 carbons with two double bonds on carbons Δ9,12, also named n-6,9. This nomenclature is taken from the location of the first double bond, counted from the carboxylic (Δ-) carbon or counted from the methyl (n) end. b Schematic representation of the 2D structures of arachidonic acid (ARA, 20:4n-6), eicosapentaenoic acid (EPA, 20:5n-3) and docosahexaenoic acid (DHA, 22:6n-3).

The protective effect of EPA and DHA in cardiovascular disease has been well documented in humans (Kromann and Green 1980; Kang and Leaf 1996; Nordøy et al. 2001; Balk et al. 2007). One of the main effects of dietary *n*-3 PUFA is to reduce elevated plasma TAG (Yokoyama et al. 2007). The hypotriglyceridemic effect of *n*-3 PUFA is believed to be due to their potent enhancement of lipolysis through activation of peroxisome proliferator-activated receptors (Nagao and Yanagita 2008). Additionally, long-chain PUFA are viewed as potent modulators of inflammation. Most of the mediators formed from EPA and DHA (leukotrienes, resolvins, and neuroprotectin/protectin D1) are anti-inflammatory, whereas those formed from the *n*-6 PUFA, arachidonic acid are mostly pro-inflammatory (Schmitz and Ecker 2007). Oxidative stress results from an imbalance between formation and degradation of pro-oxidants or decreased cellular antioxidant protection mechanisms, and may result in increased cell damage and apoptosis. DHA administration exerts antioxidant activity as shown by increasing glutathione reductase activity and decreasing accumulation of lipid peroxide and reactive oxygen species in the cortex and hippocampus of AD model rats (Guillot et al. 2008; Hashimoto et al. 2002, 2008). However, the few human studies examining the impact of DHA supplementation on oxidative stress have yielded inconsistent results. Finally, it has also been established that omega-3 LC-PUFA have some positive effects on diseases such as hypertension, arthritis, arteriosclerosis, and thrombosis (Horrocks and Yeo 1999), and are involved in the protection, might possibly even enhance, the effect of medical treatment diseases such as Alzheimer's, multiple sclerosis, and cancer. At present, most FAs entering the human diet originate from plant oils and belong to the *n*-6 group. It is probable that humans evolved on a diet with a ratio of omega-6 to omega-3 essential fatty acids (EFA) of approximately 1 whereas in Western diets the ratio is now 15/1–16.7/1. In order to restore the fatty acid balance, generally seen as optimal for human health, an increase in *n*-3 (LC)-PUFA consumption and a reduction in *n*-6 PUFAs is needed. Collectively, there is a demand for a sustainable solution to EPA and DHA supply.

CURRENT SOURCES OF DHA

Currently marine fish (oily fish species, such as herring, mackerel, sardine, and salmon) and seafood are the primary dietary source of these beneficial omega-3 LC-PUFAs (Gunstone 1996), including DHA and EPA. Fish concentrate them by ingesting marine microalgae that are able to synthesize PUFAs de novo. Besides microalgae, some lower plants, fungi, bacteria, and moss can also synthesize LC-PUFA; however, crucially no higher plants can produce LC-PUFAs. The quality of fish oil is variable and depends on fish species, season, and location of catching sites. Fish oils cannot meet the ever increasing global demand for *n*-3 LC-PUFAs because worldwide fish stocks are declining. Additionally, the contamination of the fish oil by environmental pollution of marine ecosystems has become pervasive and global problem (Domingo et al.2007; Tocher 2009). Furthermore, as marine fish oil is a complex mixture of FAs with varying lengths and degrees of unsaturation, expensive DHA purification may be required. Notwithstanding these issues, fish oil is sometimes perceived to have an unpleasant smell and taste. Krill oil has also been suggested as an alternative to fish and microalgal oils to fulfill the dietary demand for EPA and DHA. The EPA and DHA in fish are mostly in the form of TAG, whereas krill oil has TAG, but also phospholipids and nonesterified fatty acid forms of these FAs. Some studies have suggested that the EPA and DHA in krill oil is more bioavailable, however this may in fact be due to study design limitations (Salem and Kuratko 2014). The demands for omega-3 LC-PUFA continues to rise, due to a rapid increase in aquaculture and applications in food and pharmacy. It is therefore expected that the production of LC-PUFAs from current sources will become inadequate for supplying the expanding market in the near future. Collectively, these circumstances have led to an increasing interest in the search for alternative sustainable sources of omega-3 LC-PUFAs. Presently, there are two potential alternatives to fish oils: EPA and DHA from metabolically engineered plant oilseeds and microbial single cell oils (see Adarme-Vega et al. 2014).

LC-PUFA BIOSYNTHETIC PATHWAYS

The primary producers of LC-PUFA mostly consist of marine bacteria, fungi, protists, and microalgae. Within these organisms, two separate biochemical biosynthetic pathways have been identified: (1) the aerobic desaturase/elongase pathway (Sayanova and Napier 2004); and (2) the anaerobic polyketide synthase (PKS) pathway (Metz et al. 2001). A large amount of research effort has been committed to characterizing these two biosynthetic pathways.

Aerobic pathway

EPA, DHA, and other LC-PUFAs are traditionally considered to be products of alternating desaturation and elongation steps acting on long-chain (C18) polyunsaturated substrates; therefore, two distinct types of primary biosynthetic activities (desaturases and elongases) are required. To date, several sets of genes encoding these enzymes have been isolated and identified from a wide range of prokaryotic and eukaryotic species (Pereira et al. 2003; Hashimoto et al. 2008). The first step is a desaturation which is catalyzed by a front-end desaturase, which introduces a double bond between the pre-existing double bond and the carboxyl end of the fatty acid substrate, as opposed to methyl-end desaturases that insert a double bond between the pre-existing double bond and the methyl end (Meesapyodsuk and Qiu 2012). Desaturases involved in LC-PUFA synthesis normally include three highly conserved histidine-rich sequences (His boxes) with the general motifs H-X[3,4]H, H-X[2,3]H-H, and H/Q-X[2,3]H-H (Shanklin and Cahoon 1998; Sperling et al.2003). Front-end desaturases from eukaryotes (Mitchell and Martin 1995; Sayanova et al. 2001; Napier et al.1999, 2003) have a number of unique characteristics; namely, they contain an N-terminal cytochrome b5 fusion domain (Sayanova et al. 1997) that presumably acts as electron receptor for the desaturation reaction and they have a histidine/glutamine substitution in the third His box. To date, these front-end desaturases have been mostly identified in animals and microorganisms. The majority of the microsomal desaturases from lower eukaryotes use glycerolipid-linked substrates, in particular FAs esterified to the *sn*-2 position of glycerolipids. This is in contrast to animals and microalgae, in which the substrates for these enzyme activities are thought to be acyl-CoAs (Stymne and

Appelqvist 1978; Griffiths et al. 1988; Jackson et al.1998; Domergue et al. 2005). Methyl-end desaturases lacking the cytochrome b5 domain, such as membrane-bound *n*-3, Δ12-, and Δ15-desaturases, also commonly occur in plants, algae, and some fungi.

The second step in LC-PUFA biosynthesis is microsomal FA elongation by two carbons, catalyzed by an elongation complex consisting of four discrete subunits: a β-ketoacyl-CoA synthase (KCS, the condensing enzyme), a ketoacyl-CoA reductase, a hydroxyl acyl-CoA dehydratase, and an enoyl-CoA reductase (Fehling et al. 1992). However, the heterologous expression of just the initial condensing enzyme is capable of reconstituting the heterologous elongating activity (Millar and Kunst 1997; Paul et al. 2006) and for this reason KCS are often referred to as "elongases". It is widely accepted that the substrate specificity of the elongation complex (in terms of chain length and degree of unsaturation) is primarily determined by the KCS, rather than the other three components (Millar and Kunst 1997). Furthermore, the turnover of KCS activity is considered to be the rate limiting step for the elongase complex. In terms of classification, there are two groups of KCS which encompasses all those identified from plants, animals, and eukaryotic microorganisms: the FAE1-like plant-specific KCS activities involved in the biosynthesis of saturated and monounsaturated FAs with C18 to C22+ chain length (James et al. 1995; Leonard et al. 2004) and the ELO-like enzymes (some of which are involved in LC-PUFA biosynthesis). Several ELO-like sequences have been identified and heterologous studies have demonstrated that they can elongate PUFAs with chain lengths of 18-carbons or more (Dittrich et al. 1998; Parker-Barnes et al. 2000; Meyer et al. 2004; Pereira et al. 2004). In contrast to microsomal desaturation, all the microsomal elongation described so far, both ELO-type and FAE1-like, exclusively uses acyl-CoAs as substrates (Jakobsson et al. 2006).

The synthesis of LC-PUFAs in plants begins with the production of FAs by the multi-subunit fatty acid synthase (FAS) complex in the plastid (Harwood 1988; Somerville and Browse 1991). Stearic acid (18:0), the final product of FAS, is typically desaturated to oleic acid (OA, 18:1Δ9) by a stearoyl-acyl carrier protein (ACP) Δ9-desaturase. Almost all higher plants, some cyanobacteria, and many *n*-3 LC-PUFA producing microorganisms have Δ12-, and Δ15-/*n*-3 desaturases to convert OA into linoleic acid and α-linolenic acid while animals, including humans, are deficient in them. Additional conversion of LA and ALA to *n*-3 LC-PUFAs needs the introduction of appropriate front-end desaturases, and C18/C20-PUFA-specific elongases (Fig. 2). There are two separate and converging pathways in LC-PUFA synthesizing organisms: the conventional pathway (in most *n*-3 LC-PUFA synthesizing eukaryotic organisms) and the alternative pathway found in the protists such as *Euglena gracilis*, *Tetrahymena pyroformis*, *Acanthamoeba* spp., and some species of microalgae (Lees and Korn 1966; Ulsamer et al. 1969; Wallis and Browse 1999). In the conventional pathway, biosynthesis of LC-PUFAs starts with the Δ6-desaturation of LA (*n*-6) and ALA (*n*-3), generating γ-linolenic acid (GLA, 18:3 Δ6,9,12) and stearidonic acid (SDA, 18:4 Δ6,9,12,15), respectively. These Δ6-desaturated FAs are then elongated to yield dihomo-γ-linolenic acid (DGLA, 20:3 Δ8,11,14) and eicosatetraenoic acid (ETA, 20:4 Δ8,11,14,17), respectively. Finally, a Δ5-desaturase carries out one more desaturation to produce ARA (*n*-6) and EPA (*n*-3), respectively (Michaelson et al. 1998). In the alternative pathway, biosynthesis of LC-PUFAs is initiated by the Δ9-elongation of LA and ALA generating eicosadienoic acid (EDA, 20:2 Δ11,14) and eicosatrienoic acid (ERA, 20:3Δ 11,14,17), respectively (Wallis and Browse 1999; Qi et al. 2002; Li et al. 2011). These Δ9-elongated FAs are then desaturated by a Δ8-desaturase to yield DGLA and ETA, respectively. Subsequently, these FAs can then be Δ5-desaturated to form ARA and EPA, respectively. The two *n*-6 and *n*-3 pathways can be interconnected by omega-3 desaturases, which convert *n*-6 FAs into their *n*-3 counterparts. From this point, the biosynthesis of DHA may follow two different routes: either the "linear" or Sprecher pathway. In DHA-producing microbes, DHA is synthesized by the "linear" pathway in which a specific Δ5-elongation of EPA by a Δ5/ C20-elongase yields docosapentaenoic acid (DPA, 22:5 Δ7,10,13,16,19), followed by a Δ 4-desaturation to form DHA. Several genes encoding Δ5-elongating or Δ4-desaturation activities have been isolated and functionally characterized (Pereira et al. 2004; Qiu et al. 2001; Meyer et al. 2003). Conversely, mammals are thought to synthesize DHA by the Sprecher pathway in which EPA is twice elongated, yielding 24:5 *n*-3, desaturated to 24:6 *n*-3, and finally shortened to DHA via beta oxidation, a route independent of Δ4-desaturation (Sprecher et al. 1999; Burdge 2006).

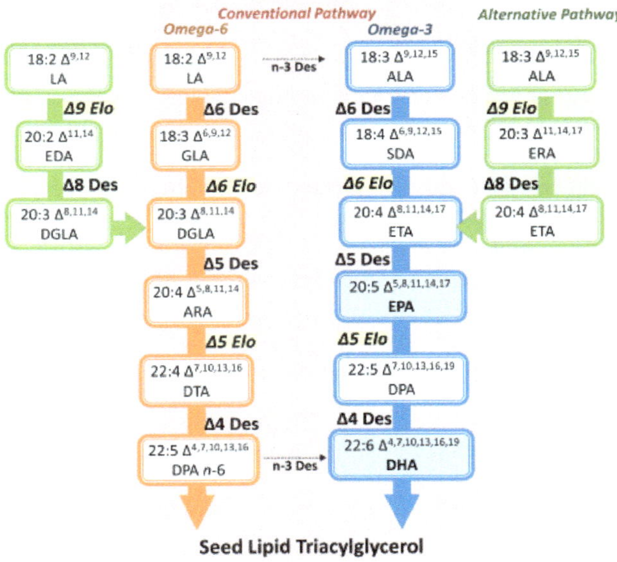

Figure 2: A schematic representation of the conventional LC-PUFA Δ6- biosynthetic pathway from linoleic and α-linolenic acid precursors. The alternative Δ8-pathway utilizes a Δ9-elongase and Δ8-desaturase to produce DGLA and ETA, respectively.

Anaerobic pathway

The anaerobic or polyketide synthase (PKS) pathway represents an alternative route to EPA and DHA; identified in both prokaryotic and eukaryotic microorganisms (Metz et al. 2001). First described in the marine bacteria *Shewanella pneumatophori* strain SCRC-2378 (Yazawa 1996), the complex consists of eight different PKS protein domains, including malonyl-CoA:ACP acyltransferase (MAT), acyl carrier protein (ACP), 3-ketoacyl synthase (KS), 3-ketoacyl-ACP reductase (KR), acyltransferase (AT), chain length factor (CLF), enoyl reductase (ER), and a 3-hydroxyacyl-ACP dehydratase/isomerase (DH/IS). Although the precise mechanism by which some marine bacteria and low eukaryotes synthesize large amounts of EPA or DHA within the cell remains uncertain, a hypothetical pathway has been proposed (Wallis et al. 2002; Napier2002). Duplicating the aerobic pathway, biosynthesis via PKS also involves several rounds of reduction, dehydration, reduction, and condensation; in each cycle the fatty acyl chain is extended by a two-carbon unit while a double bond is introduced every three carbons along the acyl chain (Kaulmann and Hertweck 2002; Cao et al. 2012). Several rounds of sequential reactions are repeated and result in the formation of LC-PUFA with 20 or 22 carbons and several double bonds that are methylene-interrupted, i.e., EPA or DHA. Unlike the aerobic pathway, the anaerobic PKS pathway inserts double bonds most likely through the action of a FabA-like dehydratase/isomerase module during the iterative extension of the fatty acyl chain and utilizes acetyl-CoA as the putative primer molecule to synthesize EPA or DHA. Interestingly, these two distinct LC-PUFA biosynthetic pathways (aerobic and anaerobic) have been found to coexist in the same microbial species such as *Schizochytrium* and *Thraustochytrium aureum* (Metz et al. 2001; Qiu et al. 2001).

ENGINEERING LC-PUFA IN HIGHER PLANTS

Although it has been some years since the successful reconstitution of the omega-3 LC-PUFA biosynthetic pathway in plants using multiple desaturases and elongases, only recently have seed oils been engineered to produce EPA and/or DHA at levels matching fish oils (Ruiz-Lopez et al. 2014). Today, the stable transformation of different oilseeds crops with multiple genes is not the technical barrier; rather issues around endogenous acyl-exchange have caused bottlenecks for engineering plants to produce high levels of *n*-3 LC-PUFAs. To successfully convert native plant FAs such as LA and ALA to LC-PUFAs such as EPA and DHA in seeds require the coordinated expression of multiple genes, as a minimum of three sequential nonnative enzymatic reactions are required (e.g., two desaturations and an acyl-CoA elongation; see Fig. 1).

Substrate limitations on LC-PUFA synthesis

The first bottleneck (substrate dichotomy) was identified as researchers endeavored to reconstruct either partially or entirely the LC-PUFA biosynthetic pathways in the yeast *Saccharomyces cerevisiae*. In this in vivo study, six Δ6-desaturases, five Δ5-desaturases, and two Δ12-desaturases genes from different organisms were expressed in *S. cerevisiae* either separately or in combination. Domergue et al. (2003) demonstrated that unlike mammalian desaturases that use acyl-CoAs as substrates, the front-end desaturases used in the study showed preferential specificity for acyl groups esterified at the *sn*-2 position of phospholipids, e.g., phosphatidylcholine (PC).

In plants, the seed-specific expression in transgenic tobacco (*Nicotiana tabacum*) and linseed (*Linum usitatissimum*) of cDNAs encoding fatty acyl-desaturases and elongases, resulted in the very high accumulation of Δ6-desaturated C18 FAs and a low accumulation of ARA and EPA (Table 2; Abbadi et al.2004). Moreover, detailed lipid analyses of developing seeds revealed that after desaturation on PC, Δ6-desaturated products were immediately channeled to the triacylglycerols and effectively bypassed the acyl-CoA pool. Therefore, the lack of Δ6-desaturated acyl-CoA substrates in the acyl-CoA pool limited the synthesis of elongated C20 FAs, and rendered the production of LC-PUFA in transgenic oilseeds unsuccessful. This bottleneck was described as a "substrate dichotomy" (Napier 2007), and it is the consequence of desaturase and elongase activities requiring different substrates, namely phospholipid-linked substrates for desaturases and acyl-CoA for elongases. Qi et al. (2004) reconstituted the alternative Δ8-desaturase pathway in *Arabidopsis* plants with genes expressing a Δ9-elongase, a Δ8-desaturase, and Δ5-desaturase under the control of the constitutive 35S promoter. The accumulation of 6.6 % ARA and 3.0 % EPA in total lipids of leaf tissues represented a "proof-of-concept" demonstration for the plant synthesis of LC-PUFAs and for the functionality of the alternative pathway in plants. However, further detailed analyses of leaf lipids indicated an inefficient transfer of these nonnative FAs from the acyl-CoA pool into extraplastidial phospholipids (Fraser et al. 2004; Sayanova et al. 2006).

Table 2: Comparison of published transgenic lines producing LC-PUFA and biosynthetic intermediates

References	Plant species	Tissue	GLA	SDA	DGLA	ARA	ETA	EPA	DPA	DHA
Conventional pathway										
Abbadi et al. 2004	*N. tabacum*	Seed	29.3	–	1.8	1.5	–	–	–	–
	L. usitatissimum	Seed	16.8	11.4	1.2	1.0	0.9	0.8	–	–
Kinney et al. 2004	*G. max*	Embryo	22.7	3.1	4.0	0.4	3.3	13.3	0.9	–
	G. max	Embryo	2.7	3.6	3.1	2.5	2.1	5.2	1.0	3.3
	G. max	Seed	11.7	1.1	10.1	2.2	2.4	19.6	0.8	–
Wu et al. 2005	*B. juncea*	Seed	27.3	2.2	1.9	4.0	1.1	8.1	0.1	0.2
Ruiz-López et al. 2009	*L. usitatissimum*	Seed	>0.5	11.8	–	–	–	–	–	–
Cheng et al. 2010	*B. carinata*	Seed	26.9	5.4	2.2	5.7	2.5	20.4	4.0	–
Ruiz-Lopez et al. 2012a	*A. thaliana*	Seed	17.7	8.1	0.3	2.1	nd	4.1	–	–
Alternative pathway										
Qi et al. 2004	*A. thaliana*	Leaf	–	–	1.3	6.6	1.2	3.0	–	–
Using Acyl-CoA desaturases										
Robert et al. 2005	*A. thaliana*	Seed	0.6	1.8	1.9	1.6	0.4	3.2	0.1	–
	A. thaliana	Seed	0.4	1.5	1.5	1.0	0.8	2.4	0.1	0.5
Hoffmann et al. 2008	*A. thaliana*	S seed	>0.5	>0.1	0.8	0.1	0.9	0.05	–	–
Petrie et al. 2010a, b	*N. benthamiana*	Leaves	2.1	1.5	–	0.6	0.6	10.7	0.3	–
Ruiz-Lopez et al. 2012b	*A. thaliana*	Seeds	1.7	0.8	0.2	6.2	nd	4.0	–	–
Petrie et al. 2012a	*A. thaliana*	T4 seeds	0.4	4.8	–	–	0.8	1.5	1.1	13.3
Ruiz-Lopez et al. 2013	*A. thaliana*	T3 seeds	1.9	1.1	1.5	3.2	0.9	13.2	–	–
	A. thaliana	T2 seeds	2.8	1.6	0.6	1.0	nd	3.4	1.1	2.5
Petrie et al. 2014	*C. sativa*	S seed	1.2	8.9	–	–	0.4	3.3	0.8	12.4
Mansour et al. 2014	*C. sativa*	Line (seeds)	–	3.3	–	–	0.4	0.2	1.1	4.2
Ruiz-Lopez et al. 2014	*C. sativa*	S seed	1.6	1.4	0.4	1.2	2.7	12.6	np	13.7
	C. sativa	Line (seeds)	2.7	2.2	0.9	2.2	3.5	11.3	np	7.7
PKS system										
Metz et al. 2006	*A. thaliana*	Seed	nd	nd	nd	nd	nd	nd	1.8	2.4

Resolving the substrate dichotomy problem led researchers to identify at least three acyl-CoA-dependent Δ6-desaturases from microalgae. Heterologous expression of the acyl-CoA Δ6-desaturase from *Ostreococcus tauri* in *S. cerevisiae* revealed very high desaturation activity with Δ6-regioselectivity, desaturating ALA to SDA with a conversion rate of 71–73 %, much higher than that of other lipid-linked Δ6-desaturases (Domergue et al. 2005). Moreover, short-time kinetic experiments showed that the desaturase product was detected in the acyl-CoA pool. In addition, using LA or ALA as exogenous substrate, both the ARA and EPA biosynthetic pathways were efficiently reconstituted in yeast by co-expressing OtD6 together with a moss Δ6-elongase and a microalgal Δ5-desaturase. Using this combination, the synthesis of relatively high yields of LC-PUFAs was achieved and they were associated with very high elongation efficiency of OtD6 desaturated intermediates (95 %), which bypasses the endogenous acyl-exchange between PC and the acyl-CoA pool. Sayanova et al. (2012) evaluated this activity in transgenic yeast, *Arabidopsis thaliana* and*Camelina sativa*, demonstrating that the use of acyl-CoA-dependent Δ6-desaturases almost completely abolished the accumulation of unwanted biosynthetic intermediates such as γ-linolenic acid in total seed lipids and that the expression of acyl-CoA Δ6-desaturases resulted in increased distribution of long-chain polyunsaturated FAs in the polar lipids of transgenic plants, reflecting a larger substrate pool available for acylation by enzymes of the Kennedy pathway. By contrast, MsD6, another acyl-CoA Δ6-desaturase isolated from the microalga *Mantoniella squamata*, only converted ALA into SDA (Hoffmann et al. 2008). The EPA biosynthetic pathway was also reconstituted by co-expressing the moss Δ6-elongase (PSE1; Zank et al.2002) together with MsD6 and acyl-CoA Δ5-desaturase (MsD5). Transgenic seeds accumulated low levels (<0.5 %) of EPA, but lacked the accumulation of Δ6-desaturation products previously observed. A third acyl-CoA Δ6-desaturase was recently identified in the marine microalga *Micromonas pusilla*, and its heterologous expression in yeast also led to Δ6-desaturation of 71 % of exogenous ALA into SDA with no enrichment in the PC (Petrie et al. 2010a), indicating a strong preference for *n*-3 substrates.

Petrie et al. (2010b) went on to demonstrate how up to 26 % EPA could be produced in *N. benthamiana* leaf TAGs (10.7 % in total FAs) using this *M. pusilla* putative acyl-CoA-dependent Δ6-desaturase. These experiments make use of a very useful and exciting tool reported by Wood et al. (2009); a leaf-based transient expression was used to identify a set of omega-3 LC-PUFA biosynthetic genes with high enzyme activities and the desired substrate (acyl-CoA) preference. This new approach could certainly enhance our capacity to identify some optimal combinations of FA biosynthetic activities for the production of LC-PUFAs in plants, provided that any differences between leaf cells and developing cotyledon cells in seeds are accounted for. The utility of this approach has allowed for the rapid validation of seed-specific constructs, which would otherwise be dependent on stable transformation (Petrie et al. 2010b). With this knowledge, the authors were then able to assemble a large T-DNA construct for stable seed-specific transformation of *Arabidopsis*, and reported a high level of DHA (but not EPA) in seed oil (Petrie et al. 2012a) and separately yielded lines accumulating significant ARA (Petrie et al. 2012b).

In an attempt to identify the optimal combination of biosynthetic enzymes (desaturases and elongases) for LC-PUFA production in seeds, a systematic study was carried out evaluating 12 different constructs (combinations of three to seven transgenes) in *Arabidopsis* (Ruiz-Lopez et al. 2013). The authors evaluated the contribution of the different transgene enzyme activities, as well as the contribution of endogenous fatty acid metabolism. Successive iterations were then informed by lipidomic analysis; an approach that enabled a significant improvement on levels previously reported for the accumulation of EPA in *Arabidopsis* seeds and also facilitated the successful engineering of the high-value polyunsaturated fatty acid DHA to ten-fold higher levels. Collectively, these studies demonstrated that the accumulation of significant (meaning similar to that found in fish oils) levels of EPA and/or DHA is achievable in plants. More recently, Ruiz-Lopez et al. (2014) described two different iterations carrying a set of heterologous genes capable of efficiently directing synthesis of these FAs in the seed oil of the crop *C. sativa*, while simultaneously avoiding accumulation of undesirable intermediate FAs, and reporting the highest levels of C20+ omega-3 LC-PUFAs in a common oilseed crop. In the first iteration, seeds contained EPA levels of up to 31 % (mean 24 %) and in the second one, seeds accumulate up to 12 % EPA and 14 % DHA (mean 11 % EPA and 8 % DHA). These omega-3 LC-PUFA levels are equivalent to those in fish oils, and represent a sustainable, terrestrial source of these FAs. The engineering of oilseed crops has led to the accumulation of omega-3 LC-PUFAs at fish oil levels, demonstrating the efficacy of acyl-CoA desaturases over previously used lipid-linked desaturases.

Triacylglycerol assembly and endogenous acyl-exchange

The de novo assembly of TAG from glycerol-3-phosphate (G3P) (also known as the Kennedy pathway) involves four enzymatic steps: first, two acylations of G3P by *sn*-1 glycerol-3-phosphate acyltransferase (GPAT) and lyso-phosphatidic acid acyltransferase (LPAAT), followed by phosphatidic acid phosphatase (PAP), and a third acylation by diacylglycerol acyltransferase (DGAT). However, TAG biosynthesis in seeds is not a simple linear pathway, but a complex network of reactions involving multiple subcellular compartments, parallel paths, cycles, and crossover with membrane lipid synthesis. The relative flux of FAs through different parts of the network affects the availability of the FAs for modification, e.g., desaturation and elongation (Fig. 3). FAs are synthesized with up to 18 carbons and one double bond in the plastid; subsequent Δ12(*n*-6)- and Δ15(*n*-3)-desaturations occur in the ER after incorporation of oleoyl groups as ester components into PC. In seeds engineered with new gene activities, additional desaturation and elongation continues in the acyl-CoA pool to produce LC-PUFA. The assembly of LC-PUFA in TAG then requires the coordinated flux of FA between the acyl-CoA pool and PC (Bates et al. 2013). Plants contain two main mechanisms for FA flux into and out of PC for desaturation: (1) Acyl editing, a PC deacylation and lyso-PC reacylation cycle that exchanges FAs with the acyl-CoA pool; (2) utilization of a PC-derived diacylglycerol (DAG) pool for TAG synthesis, rather than DAG synthesized de novo from acyl-CoA (Fig. 3). The exchange of FA from the PC pool to DAG producing TAG via phospholipid:diacylglycerol acyltransferase (PDAT) activity is unclear. PDATs from *Arabidopsis* and other plants have been shown to have high activity with PC containing unusual FA suggesting that PDAT may be involved in the removal of damaged or unusual FA from the membrane and sequestering them in TAG, a role which is potentially beneficial to the production of LC-PUFA in seed oil. The level of complexity is compounded in seeds by three other routes to DAG from PC; (1) phosphatidylcholine:diacylgly cerol cholinephosphotransferase (PDCT), which exchanges phosphocholine between PC and DAG; (2) the reverse reaction of CDP-choline:diacylglycerol cholinephosphotransferase (CPT) and (3) lipase-based mechanism utilizing phospholipase C, or phospholipase D. Finally, rather than a circuitous route via PC and DAG, acyl-CoA can directly be incorporated into the *sn*-3 position of TAG via diacylglycerol acyltransferase (DGAT) activity. Plants are known to have multiple DGAT enzymes, which appear to have roles in the assembly of seed oil. Individually or collectively, each mechanism differently affects the availability of LC-PUFAs as substrates for TAG synthesis, and thus the composition of the seed oil. Moreover, the flux through the respective pathways of TAG synthesis appears to differ between plant species ranging from just a simple Kennedy pathway to a pathway where >90 % of the FAs within the seed fluxes through PC before incorporation into TAG (Bates and Browse 2012), therefore the suitability of different oilseed crops for metabolic engineering is species dependent and must be accommodated in any seed oil modification strategy.

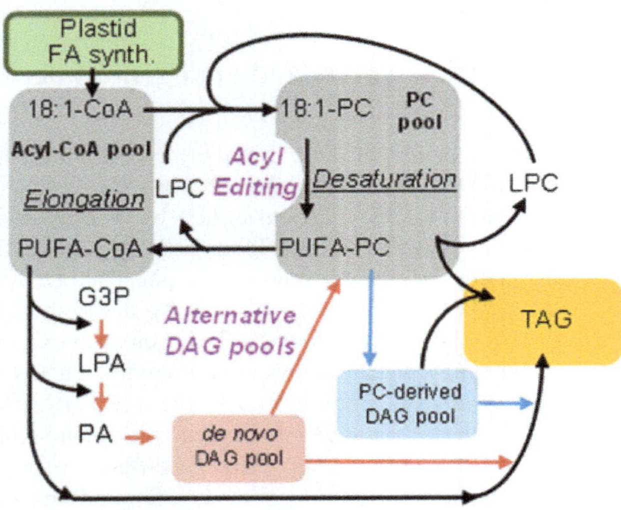

Figure 3: Overview of the major reactions involved in fatty acid and triacylglycerol synthesis (adapted with permission from Bates et al.2013).

CONCLUSIONS AND FUTURE OUTLOOK

After the initial successful introduction into plants of the primary biosynthetic enzymes for the production of LC-PUFA was demonstrated, the recent engineering of economically viable levels of EPA and DHA in the oilseed crop *Camelina* now represents a tangible success. Equally, the numerous engineering iterations that have arisen during this process have highlighted our incomplete appreciation of the pathways of fatty acid synthesis and lipid remodeling. The benefits of producing LC-PUFA in plants are clear: a sustainable and noncontaminated source of important fatty acids essential to human nutrition. Once a route for a LC-PUFA trait through performance field trials and regulatory approval is secured, the entry of transgenic LC-PUFA into the human food chain is possible. This might occur directly through products formulated with transgene-derived LC-PUFA, e.g., yogurt, margarine or indirectly, in animal feeds containing transgene-derived LC-PUFA, e.g., most obviously in farmed fish fed on a diet of modified terrestrial oils. Solving bottlenecks that limit the synthesis and accumulation of novel fatty acids has demonstrated the need for a more in-depth understanding of fatty acid metabolic pathways in seeds. Successful seed oil engineering necessitates a detailed understanding of the relative contributions of different enzymes specialized for TAG assembly in the native species and to possibly downregulate nonproductive or competing pathways. The desire to produce LC-PUFA has focused attention on fundamental metabolic processes and the application of new technology to characterize their interaction; further advances in our knowledge of plant lipid biochemistry will undoubtedly occur. It is clear that the future success of metabolic engineering of specialty oil traits will require greater predictability of genetic modifications on fatty acid and oil metabolism in seeds by use of techniques, such as mass spectrometry-based lipidomics. Ultimately, the challenging task of modifying seed oil remains the integration of a few transgene-derived activities into a greater number of endogenous metabolic processes.

ACKNOWLEDGMENTS

Research on the nutritional enhancement of oilseeds is funded by an Institute Strategic Program grant "Designing Seeds" from the Biotechnology and Biological Sciences Research Council (BBSRC, U.K).

REFERENCES

1. Abbadi A, Domergue F, Bauer J, Napier JA, Welti R, Zahringer U, Cirpus P, Heinz E. Biosynthesis of very-long-chain polyunsaturated fatty acids in transgenic oilseeds: constraints on their accumulation.Plant Cell. 2004;16:2734–2748. doi: 10.1105/tpc.104.026070.
2. Adarme-Vega TC, Thomas-Hall SR, Schenk PM. Towards sustainable sources for omega-3 fatty acid production. Curr Opin Biotechnol. 2014;26:14–18. doi: 10.1016/j.copbio.2013.08.003.
3. Balk EM, Horsley TA, Newberry SJ, Lichtenstein AH, Yetley EA, Schachter HM, Moher D, MacLean CH, Lau J. A collaborative effort to apply the evidence-based review process to the field of nutrition: challenges, benefits, and lessons learned. Am J Clin Nutr. 2007;85:1448–1456.
4. Balter NJ, Gil-Humanes J, Cermak T, Atkins PA, Voytas DF. DNA replicons for plant genome engineering. Plant Cell. 2014;26(1):151–163. doi: 10.1105/tpc.113.119792.
5. Bates PD, Browse J. The significance of different diaclglycerol synthesis on plant oil composition and bioengineering. Front Plant Sci. 2012;3:147–152. doi: 10.3389/fpls.2012.00147.

6. Bates PD, Stymne S, Ohlrogge J. Biochemical pathways in seed oil synthesis. Curr Opin Plant Biol.2013;16:358–364. doi: 10.1016/j.pbi.2013.02.015.

7. Burdge GC. Metabolism of alpha-linolenic acid in humans. Prostaglandins Leukot Essent Fat Acids.2006;75:161–168. doi: 10.1016/j.plefa.2006.05.013.

8. Calder PC. Marine omega-3 fatty acids and inflammatory processes: effects, mechanism and clinical relevance. Biochim Biophys Mol Cell Biol. 2014

9. Cao Y, Cao Y, Zhao M. Biotechnological production of eicosapentaenoic acid: from a metabolic engineering point of view. Proc Biochem. 2012;47:1320–1326. doi: 10.1016/j.procbio.2012.05.011.

10. Cheng BF, Wu GH, Vrinten P, Falk K, Bauer J, Qiu X. Towards the production of high levels of eicosapentaenoic acid in transgenic plants: the effects of different host species, genes and promoters.Transgenic Res. 2010;19:221–229. doi: 10.1007/s11248-009-9302-z.

11. Chugh A, Amundsen E, Eudes F. Translocation of cell-penetrating peptides and delivery of their cargoes in triticale microspores. Plant Cell Rep. 2009;28(5):801–810. doi: 10.1007/s00299-009-0692-4.

12. Crawford MA, Costeloe K, Ghebremeskel K, Phlactos A, Skirvin L, Stacey F. Are deficits of arachidonic and docosahexaenoic acids responsible for the neural and vascular complications of pre-term babies? Am J Clin Nutr. 1997;66:1032S–1041S.

13. Das UN, Fams MD. Long-chain polyunsaturated fatty acids in the growth and development of the brain and memory. Nutrition. 2003;19:62–65. doi: 10.1016/S0899-9007(02)00852-3.

14. Dittrich F, Zajonc D, Hühne K, Hoja U, Ekici A, Greiner E, Klein H, Hofmann J, Bessoule JJ, Sperling P, Schweizer E. Fatty acid elongation in yeast-biochemical characteristics of the enzyme system and isolation of elongation-defective mutants. Eur J Biochem. 1998;252:477–485. doi: 10.1046/j.1432-1327.1998.2520477.x.

15. Domergue F, Abbadi A, Ott C, Zank TK, Zähringer U, Heinz E. Acyl carrier used as substrates by the desaturases and elongases involved in very long-chain polyunsaturated fatty acids biosynthesis reconstituted in yeast. J Biol Chem. 2003;278:35115–35126. doi: 10.1074/jbc.M305990200.

16. Domergue F, Abbadi A, Zähringer U, Moreau H, Heinz E. *In vivo* characterization of the first acyl-CoA Δ6-desaturase from a member of the plant kingdom, the microalga *Ostreococcus tauri*. Biochem J. 2005;389:483–490. doi: 10.1042/BJ20050111.

17. Domingo JL, Bocio A, Falcó G, Llobet JM. Benefits and risks of fish consumption Part I. A quantitative analysis of the intake of n-3 fatty acids and chemical contaminants. Toxicology.2007;230:219–226. doi: 10.1016/j.tox.2006.11.054.

18. Fehling E, Lessire R, Cassagne C, Mukherjee KD. Solubilization and partial purification of constituents of acyl-CoA elongase from *Lunaria annua*. Biochim Biophys Acta. 1992;1126:88–94. doi: 10.1016/0005-2760(92)90221-G.

19. Food and Agriculture Organization of the United Nations (2010) Fats and fatty acids in human nutrition: report of an expert consultation. FAO Food and Nutrition Paper 91. Rome

20. Fraser TCM, Qi B, Elhussein S, Chatrattanakunchai S, Stobart AK, Lazarus CM (2004) Expression of the *Isochrysis* C18-Δ9 polyunsaturated fatty acid specific elongase component alters arabidopsis glycerolipid profiles. Plant Physiol 135:859–866

21. Gelvin BS. *Agrobacterium* mediated plant transformation: the biology behind the "Gene-Jockeying" tool. Microbiol Mol Biol Rev. 2003;67(1):16–37. doi: 10.1128/MMBR.67.1.16-37.2003.

22. Gill I, Valivety R. Polyunsaturated fatty acids, part 1: occurrence, biological activities and applications.Trends Biotechnol. 1997;15:401–409. doi: 10.1016/S0167-7799(97)01076-7.

23. Giusto NM, Pasquare SJ, Salvador PI, Roque MG. Lipid metabolism in vertebrate retinal rod outer segments. Prog Lipid Res. 2000;39:315–391. doi: 10.1016/S0163-7827(00)00009-6.

24. Griffiths G, Stobart AK, Stymne S. Delta 6- and delta 12-desaturase activities and phosphatidic acid formation in microsomal preparations from the developing cotyledons of common Borage (*Borago officinalis*) Biochem J. 1988;252:641–647.

25. Guillot N, Debard C, Calzada C, Vidal H, Lagarde M, Vericel E. Effects of docosahexaenoic acid on some megakaryocytic cell gene expression of some enzymes controlling prostanoid synthesis.Biochem Biophys Res Commun. 2008;372:924–928. doi: 10.1016/j.bbrc.2008.05.155.

26. Gunstone FD. Fatty acid and lipid chemistry. London: Blackie; 1996.

27. Hansen AE, Burr GO. Essential fatty acids and human nutrition. JAMA. 1946;132:855–859. doi: 10.1001/jama.1946.02870490033008.

28. Harwood JL. Fatty acid metabolism. Annu Rev Plant Physiol Plant Mol Biol. 1988;39:101–188. doi: 10.1146/annurev.pp.39.060188.000533.

29. Hashimoto M, Hossain S, Shimada T, Sugioka K, Yamasaki H, Fujii Y, Ishibashi Y, Oka J, Shido O. Docosahexaenoic acid provides protection from impairment of learning ability in Alzheimer's disease model rats. J Neurochem. 2002;81:1084–1091. doi: 10.1046/j.1471-4159.2002.00905.x.

30. Hashimoto K, Yoshizawa AC, Okuda S, Kuma K, Goto S, Kanehisa M. The repertoire of desaturases and elongases reveals fatty acid variations in 56 eukaryotic genomics. J Lipid Res. 2008;49:183–191. doi: 10.1194/jlr.M700377-JLR200.

31. Haslam RP, Ruiz-Lopez N, Eastmond P, Moloney M, Sayanova O, Napier JA. The modification of plant oil composition via metabolic engineering—better nutrition by design. Plant Biotech J.2013;11:157–168. doi: 10.1111/pbi.12012.

32. Hoffmann M, Wagner M, Abbadi A, Fulda M, Feussner I. Metabolic engineering of -3 very long chain polyunsaturated fatty acid production by an exclusively acyl-CoA-dependent pathway. J Biol Chem. 2008;283:22352–22362. doi: 10.1074/jbc.M802377200.

33. Horrocks LA, Yeo YK. Health benefits of docosahexaenoic acid (DHA) Pharmacol Res.1999;40(3):211–225. doi: 10.1006/phrs.1999.0495.

34. Innis SM. Dietary (*n*-3) fatty acids and brain development. J Nutr. 2007;137(4):855–859.

35. Jackson FM, Fraser TC, Smith MA, Lazarus C, Stobart AK, Griffiths G. Biosynthesis of C_{18}polyunsaturated fatty acids in microsomal membrane preparations from the filamentous fungus *Mucor circinelloides*. Eur J Biochem. 1998;252:513–519. doi: 10.1046/j.1432-1327.1998.2520513.x.

36. Jakobsson A, Westerberg R, Jacobsson A. Fatty acid elongases in mammals: their regulation and roles in metabolism. Prog Lipid Res. 2006;45:237–249. doi: 10.1016/j.plipres.2006.01.004.

37. James DW, Lim E, Keller J, Plooy I, Ralston E, Dooner HK. Directed tagging of the Arabidopsis FATTY ACID ELONGATION1 (FAE1) gene with the maize transposon activator. Plant Cell.1995;7:309–319. doi: 10.1105/tpc.7.3.309.

38. Kang JX, Leaf A. The cardiac antiarrhythmic effects of polyunsaturated fatty acid. Lipids.1996;31:S41–S44. doi: 10.1007/BF02637049.

39. Kaulmann U, Hertweck C. Biosynthesis of polyunsaturated fatty acids by polyketide synthases. Angew Chem Int Ed. 2002;41:1866–1869. doi: 10.1002/1521-3773(20020603)41:11<1866::AID-ANIE1866>3.0.CO;2-3.

40. Kinney AJ, Cahoon EB, Damude HG, Hitz WD, Kolar CW, Liu ZB (2004) Production of very long chain polyunsaturated fatty acids in oilseed plants. International Patent Application WO 2004/071467 A2

41. Kromann N, Green A. Epidemiological studies in the Upernavik district, Greenland. Incidence of some chronic diseases 1950–1974. Acta Med Scand. 1980;208:401–406. doi: 10.1111/j.0954-6820.1980.tb01221.x.

42. Lees AM, Korn ED. Metabolism of unsaturated fatty acids in protozoa. Biochemistry. 1966;5:1457–14581.

43. Leonard AE, Pereira SL, Sprecher H, Huang Y. Elongation of long chain fatty acids. Prog Lipid Res.2004;43:36–54. doi: 10.1016/S0163-7827(03)00040-7.

44. Li M, Ou XY, Yang XD, Guo DQ, Qian XY, Xing LJ, Mingchun L. Isolation of a novel C18-Δ9 polyunsaturated fatty acid specific elongase gene from DHA-producing *Isochrysis galbana* H29 and its use for the reconstitution of the alternative Δ8 pathway in *Saccharomyces cerevisiae*. Biotechnol Lett. 2011;33:1823–1830. doi: 10.1007/s10529-011-0626-4.

45. Li-Beisson Y, Shorrosh B, Beisson F, Andersson MX, Arondel V, Bates PD, Baud S, Bird D, DeBono A, Durrett TP, Franke RB, Graham IA, Katayama K, Kelly AA, Larson TR, Markham JE, Miquel M, Molina I, Nishida I, Rowland O, Samuels L, Schmid KM, Wada H, Welti R, Xu C, Zallot R, Ohlrogge J. Acyl-lipid metabolism. Arabidopsis Book. 2013;11:e0161. doi: 10.1199/tab.0161.

46. Lozano-Juste J, Cutler SR. Plant genome editing in full bloom. Trends Plant Sci. 2014;19(5):284–287. doi: 10.1016/j.tplants.2014.02.014.

47. Mansour MP, Shrestha P, Belide S, Petrie JR, Nichols PD, Singh SP. Characterization of oilseed lipids from “DHA-producing *Camelina sativa*”: a new transformed land plant containing long-chain omega-3 oils. Nutrients. 2014;6(2):776–789. doi: 10.3390/nu6020776.

48. Meesapyodsuk D, Qiu X. The front-end desaturases: structure, function, evolution and biotechnological use. Lipids. 2012;47:227–237. doi: 10.1007/s11745-011-3617-2.

49. Metz JG, Roessler P, Facciotti D, Levering C, Dittrich F, Lassner M, Valentine R, Lardizabal K, Domergue F, Yamada A, Yazawa K, Knauf V, Browse J. Production of polyunsaturated fatty acids by polyketide synthase in both prokaryotes and eukaryotes. Science. 2001;293:290–293. doi: 10.1126/science.1059593.

50. Metz JG, Flatt JH, Kuner JM (2006) The genes for the enzymes of the polyunsaturated fatty acid polyketide synthase of Schizochytrium and their use in the manufacture of polyunsaturated fatty acids. International Patent Publication WO 2006135866

51. Meyer A, Cirpus P, Ott C, Schlecker R, Zahringer U, Heinz E. Biosynthesis of DHA in *Euglena gracillis*: biochemical and molecular evidence for the involvement of a Δ4-fatty acyl group desaturase. Biochemistry. 2003;19:9779–9788. doi: 10.1021/bi034731y.

52. Meyer A, Kirsch H, Domergue F, Abbadi A, Sperling P, Bauer J, Cirpus P, Zank TK, Moreau H, Roscoe TJ, Zähringer U, Heinz E. Novel fatty acid elongases and their use for the reconstitution of docosahexaenoic acid biosynthesis. J Lipid Res. 2004;45:1899–1909. doi: 10.1194/jlr.M400181-JLR200.

53. Michaelson LV, Lazarus CM, Griffiths G, Napier JA, Stobart AK. Isolation of a Δ5-fatty acid desaturase gene from *Mortierella alpina*. J Biol Chem. 1998;273:19055–19059. doi: 10.1074/jbc.273.30.19055.

54. Millar AA, Kunst L. Very-long-chain fatty acid biosynthesis is controlled through the expression and specificity of the condensing enzyme. Plant J. 1997;12:121–131. doi: 10.1046/j.1365-313X.1997.12010121.x.

55. Mitchell AG, Martin CE. A novel cytochrome b5-like domain is linked to the carboxyl terminus of the*Saccharomyces cerevisiae* Δ9 fatty acid desaturase. J Biol Chem. 1995;270:29766–29772. doi: 10.1074/jbc.270.50.29766.

56. Nagao K, Yanagita T. Bioactive lipids in metabolic syndrome. Prog Lipid Res. 2008;47:127–135. doi: 10.1016/j.plipres.2007.12.002.

57. Napier JA. Plumbing the depths of PUFA biosynthesis: a novel polyketide synthase-like pathway from marine

organisms. Trends Plant Sci. 2002;7:51–54. doi: 10.1016/S1360-1385(01)02191-4.

58. Napier JA. The production of unusual fatty acids in transgenic plants. Ann Rev Plant Biol.2007;58:295–319. doi: 10.1146/annurev.arplant.58.032806.103811.

59. Napier JA, Sayanova O, Sperling P, Heinz E. A growing family of cytochrome b5 fusion desaturases.Trends Plant Sci. 1999;4:2–5. doi: 10.1016/S1360-1385(98)01357-0.

60. Napier JA, Michaelson LV, Sayanova O. The role of cytochrome b5 fusion desaturases in the synthesis of polyunsaturated fatty acids. Prostaglandins Leukot Essent Fat Acids. 2003;68:135–143. doi: 10.1016/S0952-3278(02)00263-6.

61. Napier JA, Haslam RP, Beaudoin F, Cahoon EB. Understanding and manipulating lipid composition: metabolic engineering leads the way. Curr Opin Plant Biol. 2014;19:68–75. doi: 10.1016/j.pbi.2014.04.001.

62. Nordøy A, Marchioli R, Arnesen H, Videbaek J. n-3 polyunsaturated fatty acids and cardiovascular diseases. Lipids. 2001;36:S127–S129. doi: 10.1007/s11745-001-0695-7.

63. Parker-Barnes JM, Das T, Bobik E, Leonard AE, Thurmond JM, Chaung LT, Huand YS, Mukerji P. Identification and characterization of an enzyme involved in the elongation of *n*-6 and *n*-3 polyunsaturated fatty acids. PNAS. 2000;97:8284–8289. doi: 10.1073/pnas.97.15.8284.

64. Paul S, Gable K, Beaudoin F, Cahoon E, Jaworski J, Napier JA, Dunn TM. Members of the Arabidopsis FAE1-like 3-ketoacyl- CoA synthase gene family substitute for the Elop proteins of*Saccharomyces cerevisiae*. J Biol Chem. 2006;281:9018–9029. doi: 10.1074/jbc.M507723200.

65. Pereira SL, Leonard AE, Mukerji P. Recent advances in the study of fatty acid desaturases from animals and lower eukaryotes. Prostaglandins Leukot Essent Fat Acids. 2003;68:97–106. doi: 10.1016/S0952-3278(02)00259-4.

66. Pereira SL, Leonard AE, Huang YS, Chuang LT, Mukerji P. Identification of two novel microalgal enzymes involved in the conversion of the omega3-fatty acid, eicosapentaenoic acid, into docosahexaenoic acid. Biochem J. 2004;384:357–366. doi: 10.1042/BJ20040970.

67. Petrie JR, Shrestha P, Mansour MP, Nichols PD, Liu Q, Singh SP. Metabolic engineering of omega-3 long-chain polyunsaturated fatty acids in plants using an acyl-CoA Δ6-desaturase with ω3-preference from the marine microalga *Micromonas pusilla*. Metab Eng. 2010;12:233–240. doi: 10.1016/j.ymben.2009.12.001.

68. Petrie JR, Shrestha P, Liu Q, Mansour MP, Wood CC, Zhou XR, Nichols PD, Green AG, Singh SP. Rapid expression of transgenes driven by seed-specific constructs in leaf tissue DHA production. Plant Methods. 2010;6:8–12. doi: 10.1186/1746-4811-6-8.

69. Petrie JR, Shrestha P, Zhou XR, Mansour MP, Liu Q, Belide S, Nichols PD, Singh SP. Metabolic engineering plant seeds with fish oil-like levels of DHA. PLoS ONE. 2012;7:e49165. doi: 10.1371/journal.pone.0049165.

70. Petrie JR, Shrestha P, Belide S, Mansour MP, Liu Q, Horne J, Nichols PD, Singh SP. Transgenic production of arachidonic acid in oilseeds. Transgenic Res. 2012;21:139–147. doi: 10.1007/s11248-011-9517-7.

71. Petrie JR, Shrestha P, Belide S, Kennedy Y, Lester G, Liu Q, Divi UK, Mulder RJ, Mansour MP, Nichols PD, Singh SP. Metabolic engineering of *Camelina sativa* with fish-oil levels of DHA. PLoS One. 2014;9(1):e85061. doi: 10.1371/journal.pone.0085061.

72. Qi B, Beaudoin F, Fraser T, Stobart AK, Napier JA, Lazarus CM. Identification of a cDNA encoding a novel C18-Δ9 polyunsaturated fatty acid-specific elongation activity from the docosahexaenoic acid (DHA)-producing microalga, *Isochrysis galbana*. FEBS Lett. 2002;510:159–165. doi: 10.1016/S0014-5793(01)03247-1.

73. Qi B, Fraser T, Mugford S, Dobson G, Sayanova O, Butler J, Napier JA, Stobart AK, Lazarus CM. Production of very long chain polyunsaturated *n*-3 and *n*-6 fatty acids in plants. Nat Biotechnol.2004;22:739–745. doi: 10.1038/nbt972.

74. Qiu X, Hong H, MacKenzie SL. Identification of a Δ4 fatty acid desaturase from *Thraustochytrium*sp. involved in the biosynthesis of docosahexaenoic acid by heterologous expression in*Saccharomyces cerevisiae* and *Brassica juncea*. J Biol Chem. 2001;276:31561–31566. doi: 10.1074/jbc.M102971200.

75. Robert S, Singh SP, Zhou XR, Petrie JR, Blackburn SI, Mansour PM, Nichols PD, Liu Q, Green A. Metabolic engineering of Arabidopsis to produce nutritionally important DHA in seed oil. Funct Plant Biol. 2005;32:473–479. doi: 10.1071/FP05084.

76. Ruiz-López N, Haslam RP, Venegas-Calerón M, Larson TR, Graham IA, Napier JA, Sayanova O. The synthesis and accumulation of stearidonic acid in transgenic plants: a novel source of 'heart-healthy' ω-3 fatty acids. Plant Biotech J. 2009;7:704–716. doi: 10.1111/j.1467-7652.2009.00436.x.

77. Ruiz-Lopez N, Sayanova O, Napier JA, Haslam RP. Metabolic engineering of the omega-3 long chain polyunsaturated fatty acid biosynthetic pathway into transgenic plants. J Exp Bot. 2012;63:2397–2410. doi: 10.1093/jxb/err454.

78. Ruiz-Lopez N, Haslam RP, Venegas-Caleron M, Li T, Bauer J, Napier JA, Sayanova O. Enhancing the accumulation of omega-3 long chain polyunsaturated fatty acids in transgenic *Arabidopsis thaliana*via iterative metabolic engineering and genetic crossing. Transgenic Res. 2012;21(6):1233–1243. doi: 10.1007/s11248-012-9596-0.

79. Ruiz-Lopez N, Haslam RP, Usher SL, Napier JA, Sayanova O. Reconstitution of EPA and DHA biosynthesis in Arabidopsis: iterative metabolic engineering for the synthesis of n-3 LCPUFAs in transgenic plants. Metab Eng. 2013;17:30–41. doi: 10.1016/j.ymben.2013.03.001.

80. Ruiz-Lopez N, Haslam RP, Napier JA, Sayanova O. Successful high-level accumulation of fish oil omega-3 long-chain polyunsaturated fatty acids in a transgenic oilseed crop. Plant J. 2014;77(2):198–208. doi: 10.1111/tpj.12378.

81. Salem N, Kuratko CN. A reexamination of krill oil bioavailability studies. Lipids Health Dis.2014;13:137. doi: 10.1186/1476-511X-13-137.

82. Saravanan P, Davidson NC, Schmidt EB, Calder PC. Cardiovascular effects of marine omega-3 fatty acids. Lancet. 2010;376:540–550. doi: 10.1016/S0140-6736(10)60445-X.

83. Sayanova OV, Napier JA. Eicosapentaenoic acid: biosynthetic route and the potential for synthesis in transgenic plants. Phytochemistry. 2004;65:147–158. doi: 10.1016/j.phytochem.2003.10.017.

84. Sayanova O, Smith MA, Lapinskas P, Stobart AK, Dobson G, Christie WW, Shewry PR, Napier JA. Expression of a borage desaturase cDNA containing an N-terminal cytochrome b5 domain results in the accumulation of high levels of Δ6-desaturated fatty acids in transgenic tobacco. PNAS.1997;94:4211–4216. doi: 10.1073/pnas.94.8.4211.

85. Sayanova O, Beaudoin F, Libisch B, Castel A, Shewry PR, Napier JA. Mutagenesis and heterologous expression in yeast of a plant Δ6-fatty acid desaturase. J Exp Bot. 2001;360:1581–1585. doi: 10.1093/jexbot/52.360.1581.

86. Sayanova O, Haslam RP, Qi B, Lazarus CM, Napier JA. The alternative pathway C20 Δ8-desaturase from the nonphotosynthetic organism *Acanthamoeba castellanii* is an atypical cytochrome b5-fusion desaturase. FEBS Lett. 2006;580:1946–1952. doi: 10.1016/j.febslet.2006.02.062.

87. Sayanova O, Ruiz-Lopez N, Haslam RP, Napier JA. The role ofΔ6-desaturase acyl-carrier specificity in the efficient synthesis of long-chain polyunsaturated fatty acids in transgenic plants. Plant Biotechnol J. 2012;10:195–206. doi: 10.1111/j.1467-7652.2011.00653.x.

88. Schmitz G, Ecker J. The opposing effects of *n*-3 and *n*-6 fatty acids. Prog Lipid Res. 2007;47(2):147–155. doi: 10.1016/j.plipres.2007.12.004.

89. Shanklin J, Cahoon EB. Desaturation and related modifications of fatty acids. Annu Rev Plant Physiol Plant Mol Biol. 1998;49:611–642. doi: 10.1146/annurev.arplant.49.1.611.

90. Somerville C, Browse J. Plant lipids: metabolism, mutants and membranes. Science. 1991;252:80–87. doi: 10.1126/science.252.5002.80.

91. Sparks CA, Jones HD (2004) Transformation of wheat by biolistics. Transgenic crops of the world: essential protocols 19–34

92. Sperling P, Ternes P, Zank TK, Heinz E. The evolution of desaturases. Prostaglandins Leukot Essent Fat Acids. 2003;68:73–95. doi: 10.1016/S0952-3278(02)00258-2.

93. Sprecher H, Chen Q, Yin FQ. Regulation of the biosynthesis of 22:5 *n*-6 and 22:6 *n*-3: a complex intracellular process. Lipids. 1999;34:153–156. doi: 10.1007/BF02562271.

94. Stymne S, Appelqvist LA. The biosynthesis of linoleate from oleoyl-CoA via oleoyl-phosphatidylcholine in microsomes of developing safflower seeds. Eur J Biochem. 1978;90:223–229. doi: 10.1111/j.1432-1033.1978.tb12594.x.

95. Tocher D. Issues surrounding fish as a source of omega-3 LC-PUFA. Lipid Technol. 2009;21:13–16. doi: 10.1002/lite.200800079.

96. Ulsamer AG, Smith FR, Korn ED. Composition and effects of phagocytosis on incorporation of radioactive precursors. J Cell Biol. 1969;43:105–114. doi: 10.1083/jcb.43.1.105.

97. Wallis JG, Browse J. The Δ8-desaturase of *Euglena gracilis*: an alternate pathway for synthesis of 20-carbon polyunsaturated fatty acids. Arch Biochem Biophys. 1999;365:307–316. doi: 10.1006/abbi.1999.1167.

98. Wallis JG, Watts JL, Browse J. Polyunsaturated fatty acid synthesis: what will they think of next?Trends Biochem Sci. 2002;27:467–473. doi: 10.1016/S0968-0004(02)02168-0.

99. Wang H, Yang H, Shivalila CS, Dawlaty MM, Cheng AW, Zhang F, Jaenisch R. One-step generation of mice carrying mutations in multiple genes by CRISPR /Cas-mediated genome engineering. Cell.2013;153(4):910–918. doi: 10.1016/j.cell.2013.04.025.

100. Wood CC, Petrie JR, Shrestha P, Mansour MP, Nichols PD, Green AG, Singh SP. A leaf-based assay using interchangeable design principles to rapidly assemble multistep recombinant pathways. Plant Biotech J. 2009;7:914–924. doi: 10.1111/j.1467-7652.2009.00453.x.

101. Wu G, Truksa M, Datla N, Vrinten P, Bauer J, Zank T, Cirpus P, Heinz E, Qiu X. Stepwise engineering to produce high yields of very long-chain polyunsaturated fatty acids in plants. Nat Biotechnol. 2005;23:1013–1017. doi: 10.1038/nbt1107.

102. Yazawa K. Production of eicosapentaenoic acid from marine bacteria. Lipids. 1996;31(Suppl):S297–S300. doi: 10.1007/BF02637095.

103. Yokoyama M, Origasa H, Matsuzaki M, Matsuzawa Y, Saito Y, Ishikawa Y, Oikawa S, Sasaki J, Hishida H, Itakura H, Kita T, Kitabatake A, Nakaya N, Sakata T, Shimada K, Shirato K, Japan EPA lipid intervention study (JELIS) Investigators Effects of eicosapentaenoic acid on major coronary events in hypercholesterolaemic patients (JELIS): a randomised open-label, blinded endpoint analysis.Lancet. 2007;369:1090–1098. doi: 10.1016/S0140-6736(07)60527-3.

104. Zank TK, Zahringer U, Beckmann C, Pohnert G, Boland W, Holtorf H, Reski R, Lerchl J, Heinz E. Cloning and functional characterisation of an enzyme involved in the elongation of Δ6- polyunsaturated fatty acids from the moss *Physcomitrella patens*. Plant J. 2002;31:255–268. doi: 10.1046/j.1365-313X.2002.01354.x.

105. Ziemienowicz A, Shim YS, Matsuoka A, Eudes F, Kovalchuk I. A novel method of transgene delivery into triticale plants using the *Agrobacterium* T-DNA-derived nano- complex. Plant Physiol.2012;158(4):1503–1513. doi: 10.1104/pp.111.192856.

CHAPTER 9

MODULATION OF PHYTOALEXIN BIOSYNTHESIS IN ENGINEERED PLANTS FOR DISEASE RESISTANCE

Philippe Jeandet, Christophe Clément, Eric Courot and Sylvain Cordelier

Laboratory of Stress, Defenses and Plant Reproduction, Research Unit "Vines and Wines of Champagne", UPRES EA 4707, Faculty of Sciences, University of Reims, Reims 51687, France

ABSTRACT

Phytoalexins are antimicrobial substances of low molecular weight produced by plants in response to infection or stress, which form part of their active defense mechanisms. Starting in the 1950's, research on phytoalexins has begun with biochemistry and bio-organic chemistry, resulting in the determination of their structure, their biological activity as well as mechanisms of their synthesis and their catabolism by microorganisms. Elucidation of the biosynthesis of numerous phytoalexins has permitted the use of molecular biology tools for the exploration of the genes encoding enzymes of their synthesis pathways and their regulators. Genetic manipulation of phytoalexins has been investigated to increase the disease resistance of plants. The first example of a disease resistance resulting from foreign phytoalexin expression in a novel plant has concerned a phytoalexin from grapevine which was transferred to tobacco. Transformations were then operated to investigate the potential of other phytoalexin biosynthetic genes to confer resistance to pathogens. Unexpectedly, engineering phytoalexins for disease resistance in plants seem to have been limited to exploiting only a few phytoalexin biosynthetic genes, especially those encoding stilbenes and some isoflavonoids. Research has rather focused on indirect approaches which allow modulation of the accumulation of phytoalexin employing transcriptional regulators or components of upstream regulatory pathways. Genetic approaches using gain- or less-of functions in phytoalexin engineering together with modulation of phytoalexin accumulation through molecular engineering of plant hormones and defense-related marker and elicitor genes have been reviewed.

THE CONCEPT OF PHYTOALEXINS

The concept of phytoalexins, that is antimicrobial substances of low molecular weight produced by plants in response to infection or stress, was proposed in 1940 by Müller and Börger [1] in the study of the interaction between potato and *Phytophthora infestans*. Briefly, the inhibition of the development of a race of *P. infestans* which normally develops profusely on potato tubers following infection of the potato tubers with another race of *P. infestans* able to initiate hypersensitive reaction (HR), was caused by a "principle" produced by the hypersensitively-reacting plant cells. The principle was named a phytoalexin. The term phytoalexin was derived from Greek to mean "warding-off agents in plants" and the proposal of this elegant concept was made after deliberating two important phenomena in plant pathology. The first one is the active response of the plant cell to attempted infection; the second one is the acquisition of resistance by plants after exposure to microorganisms [2].

Citation: Philippe Jeandet, Christophe Clément, Eric Courot and Sylvain Cordelier, Modulation of Phytoalexin Biosynthesis in Engineered Plants for Disease Resistance, doi:10.3390/ijms140714136.

A fundamental feature of this definition is that it restricts phytoalexins to compounds which are produced from remote precursors, through *de novo* synthesis of enzymes. At the time they have been discovered, phytoalexins were, therefore, clearly distinguished from other antimicrobial substances which are pre-formed or arise directly from inactive precursors [3]. As a corollary, the fact that these compounds constitute a defense response of plants to biotic and abiotic stresses, makes the regulatory networks involved in their induction extremely complex. For instance, studies on the signaling pathways controlling the induction of camalexin, the major phytoalexin from *Arabidopsis*, particularly underline the complexity of these regulatory mechanisms, which in turn depend on the infecting pathogen [4]. From reports evaluating the response of*Arabidopsis*-jasmonic acid (JA) mutants to *Alternaria brassicicola* it ensued that synthesis of this phytoalexin was not JA-dependent [5,6]. Conversely, JA has been found to form part of the regulatory pathways for camalexin biosynthesis in the response of that plant to *Botrytis cinerea* [7]. In fact, absence of functional jasmonate signaling in *Arabidopsis* mutants resulted in dramatic differences in phytoalexin accumulation upon infection by *B. cinerea*. Other results showed that the regulation of camalexin production can be controlled by salicylic acid (SA)-independent [8,9] or SA-dependent [10] signaling pathways. Camalexin production was also found to be decreased in response to bacterial or fungal infection in *Arabidopsis* mutants impaired in ethylene signaling [5,11]. Suppression of auxin signaling in *Arabidopsis* re-directed secondary metabolism towards glucosinolates and away from camalexin [12] (see section 4). Recent works have reported a regulation of camalexin biosynthesis involving two Mitogen-Activated Protein Kinases (MAPKs), inducing in turn the responses further downstream, especally camalexin accumulation [13] (see section 5). The two transcriptions factors WRKY40 and WRKY18 were also found to be involved in the regulation of some enzymes of the camalexin pathway, namely regulation of the cytochrome P450 monoxoygenase 71A13 catalyzing transformation of indole acetaldoxime to indole-3-acetonitrile on the route to camalexin [14]. Based on this example, it thus appears that regulatory pathways for phytoalexin induction are complex. Modulation of phytoalexin biosynthesis in engineered plants for disease resistance typically reflects this complexity.

Phytoalexins form part of the defense mechanisms of plants against phytopathogenic microorganisms. These are antibacterial, antifungal and antiviral compounds showing phytotoxic activities as well. It is now clear that phytoalexins exhibit toxicity across much of the biological spectrum, prokaryotic and eukaryotic, and that their activities are by no means confined to phytopathogenic fungi though considerable work has been done on plant/fungus interactions [15].

Are phytoalexins active? Relatively ancient studies reporting on the direct comparison of the protectant activity of isoflavonoid phytoalexins and the fungicides benomyl and mancozeb, have shown that phytoalexins are considerably less toxic than chemical fungicides [16]. Although differences in activity do occur between and within phytoalexin families, effective doses of phytoalexins generally fall within order of magnitude 10^{-5} to 10^{-4} M [17–19]. Beside their fungistatic action, phytoalexins can exert some fungitoxic activities as their action results in cytological abnormalities in fungal cells. Resveratrol and pterostilbene, for instance, two phytoalexins from grapevine, induce the formation of curved germ tubes, cessation of growth of some germ tubes with protoplasmic retraction in the dead hyphal tip cell, cytoplasmic granulation of the cellular content, disorganization of mitochondria and disruption of the plasma membrane in *B. cinerea*conidia [17–19]. Phytoalexins may also act as uncoupling agents of electron transport and photophosphorylation, leading to a rapid and complete cessation of respiration in *B. cinerea* conidia [17] and are able to induce fungal apoptotic-like programmed cell death as a result of the action of camalexin on *B. cinerea* [20].

The question as to whether phytoalexins are produced by all plants is remaining as there are few reports of thorough examinations concluding that these compounds are not synthesized by a particular plant species [3]. Even today too few families have been studied to permit the assertion that phytoalexin synthesis is a universal defense mechanism of plants [2]. Knowledge on phytoalexins from 1960 to 1980 was based essentially on extensive works in two plant families, Leguminosae or Fabaceae [21] and Solanaceae [22], and investigations of one or of a few species in each of a number of other plant families (Vitaceae, Poaceae, Malvaceae, Euphorbiaceae, Moraceae, Orchidaceae, Ginkgoaceae...) [4,23].

Phytoalexins display an enormous chemical diversity. The most studied members of the phytoalexins from Leguminosae (Fabaceae) are the simple isoflavones, daidzein [24], formononetin [25], and genistein [26]; the complex isoflavone, wighteone [27]; the simple isoflavanones, vestitone [28] and kievitone [29]; the pterocarpans, medicarpin [30], maackiain [31], pisatin [31], phaseolin [32] and glyceollin [33]; the simple isoflavan, vestitol [34]; the simple coumestan, coumestrol [35]; the furanoacetylenes, wyerone acid [36] and wyerone [37] and stilbenes [38] (Figure 1). Four main classes of phytolexins have been reported in Solanaceae: the phenylpropanoid-related phytoalexins caffeic acid and chlorogenic acid [39]; the steroid glycoalkaloids, α-solanine and α-chaconine [40]; the norsesquiterpenoids/sesquiterpenoids, rishitin [41,42] lubimin [43], capsidiol [44] and the coumarin, scopoletin [45] (Figures 1 and and2).2). Maize and rice are Poaceae crop plants. The main phytoalexins in rice are the two 9-β-pimaradiene diterpenes, momilactones A and B [46], the flavanone, sakuranetin [47] and the diterpenes, phytocassanes A–E [48]. Two new phytoalexin families have recently been discovered in maize: the *ent*-kaurane-related diterpenoids, kauralexins [49] and the acidic sesquiterpenoids, zealexins [50] (Figures 1 and and2).2). The 3-deoxyanthocyanidins including luteolinidin (Figure 1) and apigeninidin, form an unusual group of flavonoid phytoalexins produced by members of the Poaceae, for instance, as a response of sorghum mesocotyls to infection by pathogenic and nonpathogenic fungi [51–53]. Gossypol and its congeners, that is, naphthaldehyde compounds, constitute the main phytoalexins from cotton plants (*Gossypium* spp.) of the Malvaceae family [54]. We will end this list, which is far from being exhaustive, with camalexin, the major phytolexin from the Brassicaceae (also known as Cruciferae) [55] and resveratrol, the major phytoalexin from the Vitaceae [56] (Figures 1 and and22).

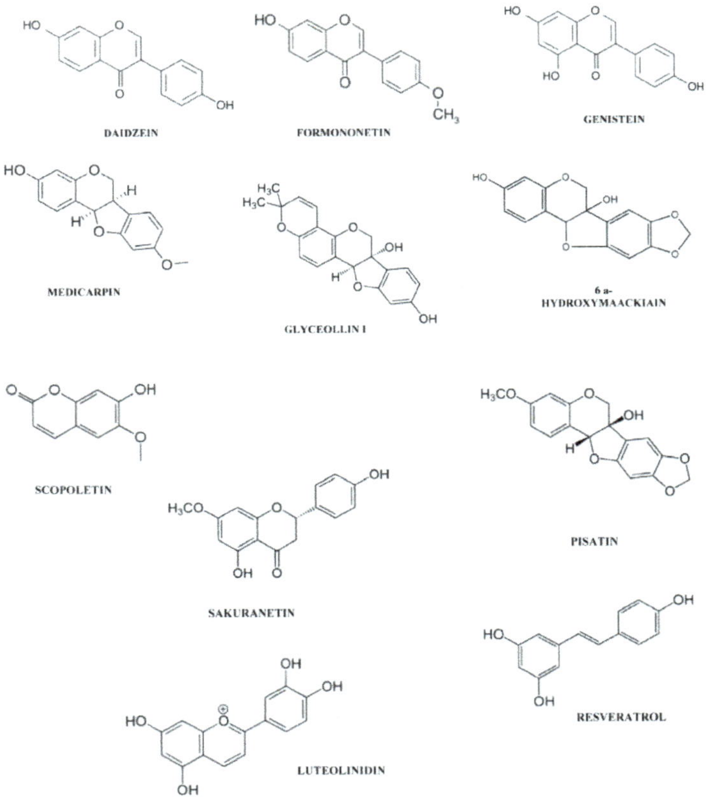

Figure 1: Chemical structures of some phytoalexins cited in this work: daidzein, formononetin, sakuranetin and genistein (simple isoflavones); medicarpin, maackiain, pisatin and glyceollin I (pterocarpans); resveratrol (stilbenes); scopoletin (coumarins); luteolinidin (3-desoxyanthocyanidins).

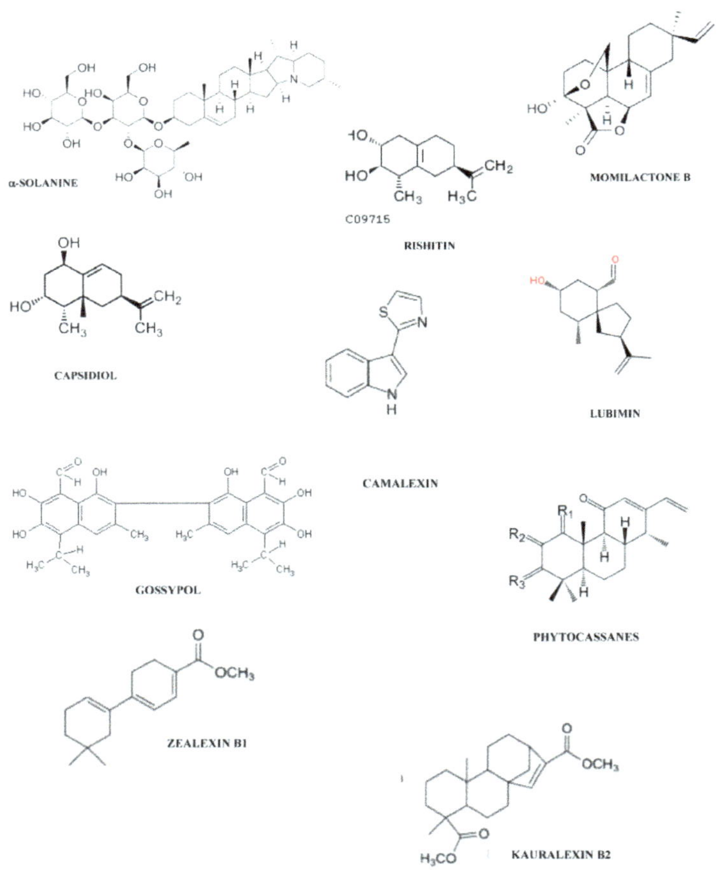

Figure 2: Chemical structures of some phytoalexins cited in this work: α-solanine (steroid glycoalkaloids); rishitin, lubimin, capsidiol, gossypol, kauralexin B2; zealexin B1 and momilactone B (norsesquiterpenoids/sesquiterpenoids); and phytocassanes.

How can our knowledge of phytoalexins be used to develop new approaches to plant disease control? Phytoalexins have long been recognized to as being important in the defense mechanisms of plants against phytopathogenic microorganisms. Their potential biological properties have stimulated a ferment of activity concerning the biosynthesis and the metabolism of these compounds in the plant. Research has gone forward because these compounds are thought to help agriculturally and economically important crop plants withstand colonization by pathogens. In many instances, a close correlation has been found between phytoalexin production and resistance to diseases.

Use of modern molecular biology tools for elucidating the control mechanisms of phytoalexin synthesis and for engineering disease resistant plants is based on the expression of stress- or disease-related genes [57]. Genetic engineering of phytoalexins for disease resistance requires manipulation of a single or a few genes directly involved in their biosynthetic pathways or involved in their signaling/regulatory pathways [58]. We will describe in this review, efforts that have been done for engineering phytoalexin production in plants as well as provide information on how phytoalexin accumulation can be modulated through engineering of plant hormones, defense-related markers or elicitors.

GENETIC MANIPULATION OF PHYTOALEXIN PRODUCTION AND DISEASE RESISTANCE ADDRESSING GAIN- OR LOSS-OF-FUNCTION GENETIC APPROACHES

As surprising as it may seem, there have been not so much reports of attempts to manipulate phytoalexins by genetically modifying expression of biosynthetic pathway genes for disease resistance in plants. Works in this area address both gain- or loss-of-functions genetic approaches, with the most abundant literature concerning resveratrol, the major phytoalexin from Vitaceae (for reviews see [59–63]).

Resveratrol is obtained through the universal phenylpropanoic acid pathway beginning with phenylalanine (via the phenylalanine ammonia lyase, PAL) or with tyrosine (via the tyrosine ammonia lyase, TAL) (Figure 3). The obtained cinnamic acid is activated by ligation to coenzyme A by 4-coumaroyl:CoA ligase (C4L). Resveratrol is then synthesized by the stilbene synthase (STS) which acts to condense three successive units of malonyl-CoA with *p*-coumaroyl-CoA. STS, the key enzyme of resveratrol synthesis, thus uses as substrates precursor molecules that are present throughout the plant kingdom. Therefore, the introduction of a single gene is sufficient to synthesize resveratrol in heterologous plant species. Transformations have been then operated to investigate the potential of stilbene biosynthetic genes to confer resistance to pathogens. Interest in resveratrol engineering has increased due to its antimicrobial properties acting as an allelochemical or a phytoalexin, but also due to its antioxidant and antitumor capabilities (for reviews see [59–64]). Transfer experiments began with two grapevine STS genes (*Vst1* and *Vst2*) introduced into tobacco under the control of the stress responsive promoter pVst1, conferring that plant a higher resistance to *B. cinerea* infection [65]. This study constituted the first report of a disease resistance resulting from foreign phytoalexin expression in a novel plant. In the same way, Hipskind and Paiva [66] have transformed alfalfa (*Medicago sativa*) with the *Arachis hypogea* STS *AhRS* gene transcriptionally regulated by an enhanced cauliflower mosaic virus (CaMV) 35S promoter, conferring increased resistance to *Phoma medicaginis*. The grapevine rootstock 41-B overexpressing the grapevine *Vst1* stilbene synthase gene under the control of the fungus inducible promoter *ms* PR 10.1 produced high stilbene levels and exhibited *in vitro*resistance to *B. cinerea* [67]. Similar transformations have improved the resistance of rice to *Piricularia orizae* (grapevine STS gene *Vst1* with the stress responsive promoter pVst1) [68], of barley and wheat to *B. cinerea* and to powdery mildew (grapevine STS gene *Vst1* with the stress responsive promoter pVst1 + CaMV 35S) [69,70], of tomato to *P. infestans* (two grapevine STS genes *Vst1* and *Vst2* under the control of the stress responsive promoter pVst1) [71], of wheat to *Puccinia recondita* and *Septoria nodorum* (two grapevine STS genes *Vst1* and *Vst2* with the stress responsive promoter pVst1 + CaMV 35S) [72], of papaya to *Phytphthora palmivora* (grapevine STS gene *Vst1* with the stress responsive promoter pVst1) [73] and of*Rehmannia glutinosa* to *Fusarium oxysporum* with the *Arachis hypogea* STS *AhRS3* gene transcriptionally regulated by the constitutive CaMV 35S promoter [74]. More recently, overexpression of a resveratrol gene*PcRS* from *Polygonum cuspidatum* under the control of the constitutive CaMV 35S promoter in transgenic*Arabidopsis* was shown to result in the accumulation of *trans*-piceid (the 3-*O*-β-d-resveratrol glucoside) and an increased resistance to the fungal pathogen, *Colletotrichm higginsianum* [75].

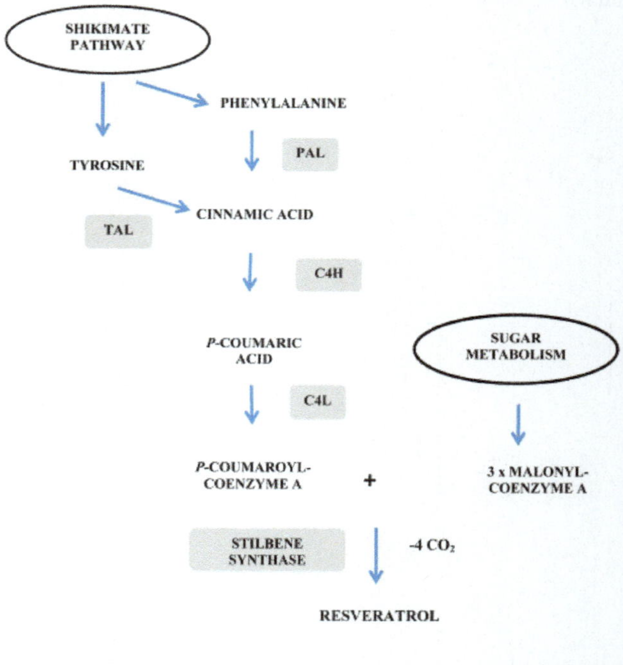

Figure 3: In vivo biosynthesis of resveratrol from phenylalanine or tyrosine via the phenylpropanoid/polymalonate pathway. PAL/TAL: phenylalanine/tyrosine ammonia lyase; C4H: cinnamate-4-hydroxylase; C4L: coumarate:coenzyme A ligase; STS: stilbene (resveratrol) synthase.

All these results show that resveratrol is a determinant factor in the expression of resistance of plants to phytopathogens and underline the above-mentioned interest for engineering this phytoalexin in plants. However, there were some cases for which no resistance was observed after transforming the plants with *STS*genes. For example, transformation of white poplar (*Populus alba*) with the grapevine *StSy* or with a pinosylvin-synthase-encoding gene from *Pinus sylvestris* did not confer any increased resistance to*Melaspora pulcherrina* (rust disease) [78,79]. Similarly, no increased resistance against *B. cinerea* has been observed in STS transgenic kiwi plants expressing the grapevine STS *pSV25* under the control of pCAMV 35S [80]. These contradictory results suggest that pathogen control by transgenic STS plants can be significant but may be considered as empiric and not predictable. Generally, STS expression does not trigger any detrimental effect on plant development, growth, morphology or fertility. However, in few cases, STS overexpression can lead to morphological alterations in transgenic plants such as color modification and male sterility as observed in tobacco and petunia [81] or in conifers [82]. Such events may be explained by the possible transgene insertion site, such as, for instance, insertions within a gene involved in male fertility for sterile lines.

Isoflavonoids constitute the second major phytoalexin family that has attracted considerable attention for plant engineering though this gain of interest was due mainly to the antioxidant and anticancer activities of these compounds. 4′-*O*-methylation of isoflavonoid-type compounds is a prerequisite for further substitutions of the isoflavonoid core, leading to the pterocarpan phytoalexin medicarpin. Genetic manipulation of the isoflavone 7-*O*-methyltransferase has been reported in alfalfa (*Medicago sativa*) in relation to increased resistance of that plant to *Phoma medicaginis*, the causal agent for spring black stem and leaf spot disease [83]. Gene constructs harboring the full-length isoflavone-*O*-methyltransferase from alfalfa (IOMT8) cDNA in the sense and antisense orientations under control of the constitutive promoter 35S were obtained. Unexpectedly, overexpression of IOMT led to 7-*O*-methylisoflavonoid derivatives *in vitro* as well as *in vivo*unchallenged leaves of transgenic alfalfa when ectopically expressed but mainly produced 4′-*O*-methylderivatives, such as formononetin (4′-*O*-methyldaidzein) and medicarpin in plants elicited with metallic cations or infected with *P. medicaginis*. Such apparently different regiospecificities of IOMT expressed *in vivo* or *in vitro* will not be discussed further but underlined potential metabolic channeling at the entry point into the isoflavonoid-type phytoalexin pathway (Figure 4).

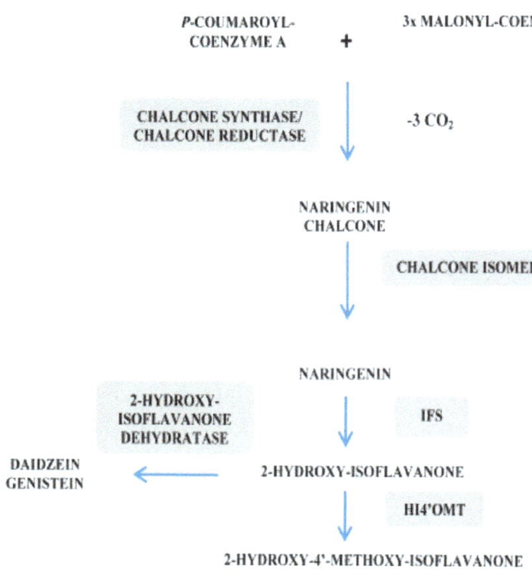

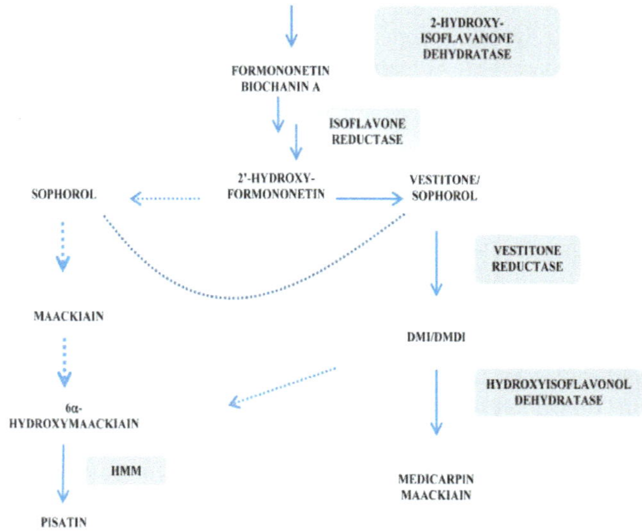

Figure 4: Scheme of biosynthetic pathways to the main phytoalexins from Leguminosae. IFS: 2-hydroxy isoflavanone synthase; DMI: 7,2'-dihydroxy-4'-methoxy-isoflavanol; DMDI: 7,2'-dihydroxy-4',5'-methylenedioxy-isoflavanol; HMM: 6α-hydroxymaackiain 3-O-methyltransferase; HI4'OMT: SAM: 2,7,4'-trihydroxy-isoflavanone 4'-O-methyltransferase (adapted from [76,77]). The dashed arrows represent hypothetic steps and the solid arrows denote reactions for which the catalyzing enzymes have been cloned.

In antisense IOMT transgenic lines, root tissues from those plants exhibited markedly reduced phytoalexin concentrations as a consequence of a potential downregulation of the isoflavonoid content. Ectopically overexpression of IOMT in unchallenged transgenic leaves produced 7-*O*-methylisoflavonoids but no 4′-*O*-methylisoflavonoids (see above). Conversely, abiotic elicitation using spores of *P. medicaginis* led to the accumulation of the phytoalexins formononetin glucoside and medicarpin both in empty-vector control plants and IOMT over-expressing plants 2 days post infection (dpi), being the content of medicarpin 2.5 fold higher in IOMT overexpressing lines (67 µg/g fresh WT *vs.* 30 µg/g fresh WT). Interestingly, high medicarpin accumulation rates in IOMT-overexpressing plants were associated with a marked reduction in the size of the brown *P. medicaginis* lesions on leaves measured 5 dpi (around 60% decrease as compared to empty-vector control plants), being this reduction of the disease symptoms well correlated with postinoculation increases in medicarpin concentrations, confirming the role of this compound in disease resistance in alfalfa. In contrast, authors failed to demonstrate any increased susceptibility in antisense IOMT lines with decreased amounts of medicarpin, as this may be the result of an incomplete inhibition of accumulation of the phytoalexin in those lines. Such a situation was not observed in *Arabidopsis* mutants deficient in camalexin where susceptibility to the necrotrophic fungus *Alternaria brassicicola* is highly correlated with the absence of this phytoalexin [5]. These different behaviors underline limitations that can be encountered by using loss-of-function genetic approaches for the understanding of plant/ pathogen interactions.

In a similar approach, biosynthesis of the simple isoflavone genistein glucoside was engineered in leaves of alfalfa by constitutive expression of isoflavone synthase from *Medicago truncatula* [76] (Figure 4). The Mt*IFS1* cDNA from *M. truncatula* was subcloned into the pRTL2 vector and the expression cassette containing Mt*IFS1* flanked by CaMV 35S promoter and terminator sequences was cloned into the *Hin*dIII site of the binary vector pCAMBIA 2300. Mt*IFS1* was highly expressed in all organs, while genistein accumulation was limited to leaves. Other simple isoflavone-type phytoalexins were also found such as daidzein, biochanin A (4′-*O*-methylgenistein) and 3′-hydroxybiochanin A (pratensein) in transgenic lines with constitutive expression of Mt*IFS1*. Additionally, together with daidzein, other 5-deoxyflavone phytoalexins accumulated, for instance formonotenin, in transgenic lines irradiated with UV-B for 6h or as a result of plant infection with *P. medicaginis*. The pterocarpan phytoalexin, medicarpin, was also found to be accumulated, likely in relation to its antifungal activity [84]. MtIFS1-expressing plants contained much higher concentrations of formononetin and medicarpin than the vector control lines, being the levels of formononetin between 46 and 77 µg/g fresh WT in leaves 24 h to 72 h after infection, that is, concentrations able to inhibit the pathogen development.

Loss-of-function approaches were applied to study the role of the phytoalexin pisatin in the disease resistance of hairy roots of pea (*Pisum sativum* L.) to the fungal pathogen *Nectria haematococca* through the genetic manipulation of enzymes involved in both the biosynthesis and the catabolism of this compound [85] (Figure 4). Six genetic constructs were thus designed: sense and antisense-oriented cDNAs of isoflavone reductase (IFR), (+) 6α-hydroxymaackiain 3-*O*-methyltransferase (HMM) fused to the 35S CaMV promoter, being these two enzymes involved in pisatin biosynthesis, and a pisatin demethylating activity (PDA) from*N. haematococca*. Reduced amounts of pisatin were observed in hairy root cultures transformed with antisense-*Hmm* which led to fewer *Hmm* transcripts, resulting in less HMM protein and enzyme activity in response to elicitation with $CuCl_2$ or infection with the pea-pathogenic fungus, *N. haematococca*. As a consequence of the transformation with antisense *Hmm*, roots became more susceptible to *N. haematococca*. As expected, cultures with the lowest levels of pisatin were those containing the pisatin demethylating activity. At the time this paper was published, it appeared to be the first case of producing transgenic plant tissues with reduced capability to accumulate a given phytoalexin and demonstrating that such a tissue was less resistant to fungal infection [85]. Additionally, this loss-of-function genetic approach afforded direct evidence of the role of pterocarpan phytoalexins in plant defense mechanisms [77].

One of the most highly characterized systems for the delineation of the roles of secondary metabolites in plant disease resistance is the model pathosystem soybean (*Glycine max*)-*Phytophthora sojae*. An RNAi silencing approach was used to illustrate the role played by 5-deoxyisoflavonoid phytoalexins in the resistance and hypersensitive cell death of soybean to *P. sojae* [86]. Silencing either isoflavone synthase (IFS) or chalcone reductase (CHR) led to suppression (90%) of 5-deoxyisoflavones (daidzein and its derivatives) and glyceollin as well as disease resistance to *P. sojae* upon treatment with a cell wall glucan elicitor (WGE) from this pathogen. Loss of resistance was accompanied by suppression of HR cell death. Thus, 5-deoxyisoflavonoid-type phytoalexins from soybean were shown to play a critical role in the establishment of HR and race-specific resistance. Interestingly a third gene silencing RNAi approach targeting the gene *PR-2* which encodes the elicitor-releasing endoglucanase, led to loss of HR, loss of disease resistance and most importantly to suppression of the phytoalexin response upon treatment with WGE. This underlined links existing between a general resistance elicitor (pathogen-associated molecular pattern), HR cell death, race-specific resistance and accumulation of phytoalexin compounds in a given plant—pathogen system.

Loss-of-functions approaches were also used for the study of the role of 3-deoxyanthocyanidin phytoalexins in the resistance of sorghum (*Sorghum bicolor*) to anthracnose leaf blight caused by the phytopathogenic fungus *Colletotrichum sublineolum* as well as for the delineation of the biosynthesis pathway of this particular group of phytoalexins [87]. In the sorghum, it was shown that the corresponding *yellow seed1* gene (*y1*) encoded a R2R3 type of MYB domain protein which regulates the coordinate expression of *chalcone synthase (chs1)*, *chalcone isomerase*, *dihydroflavonol reductase* and *flavonoid 3′-hydroxylase* [88,89] genes required for the biosynthesis of 3-deoxyflavonoids including 3-deoxyanthocyanidin phytoalexins [87,90,91]. Interestingly, loss-of-function alleles of *y1* were found to be deficient in the synthesis of 3-deoxyanthocyanidins, these plants exhibiting severe symptoms characteristic of the anthracnose disease [87]. This directly supported the involvement of a functional *y1* gene in the increased resistance of sorghum to *C. sublineolum* [87]. The upper stream of the biosynthesis pathway leading to 3-deoxyflavonoids, especially formation of luteoferol from apiforol, that is, the precursor of the phytoalexin luteoninidin, was also shown to be under the control of the *red aleurone 1* (*pr1*) gene which encodes a CYP450-dependent flavonoid 3′-hydroxylase in maize, by using a loss-of-function approach [89].

Finally, loss-of-function mutants have been extensively used for the understanding of the role of the phytoalexin camalexin in the model plant *Arabidopsis* upon challenge with various bacterial or fungal pathogens. Five non-allelic phytoalexin-deficient mutants of *Arabidopsis* producing reduced amounts of camalexin during inoculation with bacteria or with biotrophic fungi (*pad1-1*, *pad2-1*, *pad3-1*, *pad4-1* and*pad5-1*) have been identified [92–94]. Some studies focused on the phytoalexin *Arabidopsis* mutant pad3-1, showing that the effect of the *PAD* mutation on the level of camalexin production was dependent on the pathogen [5]. This mutant was indeed found not to display altered susceptibility with *Pseudomonas syringae*nor with the biotrophic fungi *Perenospora parasitica* and *Erysiphe orontii* [92–94]. In another work, the same mutant was reported to be markedly more susceptible than its wild-type parental line upon infection with *Alternaria brassicicola* but this was not the case in transgenic lines upon inoculation with *B. cinerea*. In the latter, camalexin was undetectable whereas a strong camalexin response was reported in wild-type plants [5]. In the non phytoalexin producer-mutants *pad3-1* which exhibited resistance to *B. cinerea*, it was suggested that effectors other than camalexin which are not disabled in the transgenes, especially JA-or ethylene-dependent pathogenesis-related genes such as *PR3* and *PR4*, were good determinants for disease resistance [5]. Despite the observed discrepancies within these loss-of-function genetic approaches, the use of*PAD* mutants offers interesting models for studying the role of camalexin in the defense response of*Arabidopsis* against pathogens.

GENETIC MANIPULATION OF PHYTOALEXIN GLUCOSYLATION AND DISEASE RESISTANCE

Glucosylation of scopoletin, a coumarin phytoalexin from tobacco (*Nicotiana tabacum* L.), is a particular example illustrating the crucial role of plant UDP-Glc:phenylpropanoid glucosyltransferases (UGTs) in plant defense responses and their relevance to the more general context of plant/pathogen interactions [95,96]. Glucosylation indeed intervenes in the accumulation, storage, transport and water solubility of a wide range of compounds [97].

Based on two different and consecutive studies including an antisense strategy, on one hand [98], and a gain-of-function approach, on the other hand [99], Saindrenan and colleagues have underlined the role of glucosyltransferases in scopoletin accumulation in tobacco plants following inoculation with Tobacco Mosaic Virus (TMV). In the first approach, downregulation of the expression of two tobacco salicylic acid- and pathogen-inducible UDP-glucosyltransferase genes *Togt 1* and *Togt 2* (acting on scopoletin glucosylation) was obtained by introducing the complete coding sequence of Togt 1 in the antisense orientation downstream from the constitutive CaMV 35S promoter in tobacco [98]. Expectedly, TOGT downregulation resulted in a decrease in scopoletin UGT activity together with a decline in the contents of both the aglycone and the glucoside of scopoletin named scopolin. Scopolin accumulation was reduced by 70% to 75% in tobacco reacting hypersensitively to TMV.

The TOGT inhibition-mediated decrease in both the scopolin and scopoletin contents could have been the result of a feedback inhibition of upstream enzymes of the scopoletin biosynthesis. Antisensing *Togt 1* was found to have no effects on other enzymes involved in plant defense mechanisms such as *O*-methyltransferases of class II (COMT II) or acidic β-1,3-glucanases [98]. Interestingly, transgenic lines downregulated for TOGT showed a 63% increase in TMV lesion surfaces *vs.* control plants relative to changes in scopolin and the TMV-replicate inhibiting-phytoalexin, scopoletin. As the antisense strategy used previously left open the possibility of a non-specific downregulation of a similar glucosyltransferase or of enzymes involved in the biosynthetic route to scopoletin, a gain-of-function approach was needed to further confirm the role of this pathogen-inducible glucosyltransferase during the hypersensitive response of *N. tabacum* to TMV.

Clonage of the *togt 1* full-length cDNA was carried out in the sense orientation downstream the 35S CaMV promoter into the pFB8 vector [99]. Basal scopoletin UGT activity in lines harboring the sense orientation construct was about 3-fold higher than in vector control plants and the resulting accumulation of scopoletin and scopolin was found to be 2-fold higher. Acceleration of the development of TMV lesions in transgenic TOGT overexpressing lines, despite increases in the phytoalexin content, was thought to result in a better restriction of TMV, thus preventing virus spread. These results thus clearly demonstrated the involvement of TOGT in scopoletin glucosylation *in planta*, controlling the accumulation of this antiviral compound during the hypersensitive reaction of tobacco to TMV; they also suggested the overall relevance of metabolic engineering of glucosyltransferases in the outcome of plant/pathogen interactions. Conversely, TOGT activity for scopoletin glucosylation was reported to be moderately expressed upon cytokinine-mediated resistance of tobacco to *Pseudomonas Syringae* pv *tabaci* [100].

MODULATION OF PHYTOALEXIN ACCUMULATION THROUGH ENGINEERING OF PLANT HORMONES

The essential role played by phytohormones in the induction of plant defense responses has been reviewed elsewhere [101–109]. Defense responses against biotrophs and necrotrophs are classically based on two antagonistic pathways, SA signaling controlling defense against biotrophs, JA/ethylene signaling controlling defense against necrotrophs. Some aspects of the involvement of SA, JA/ethylene in phytoalexin induction have been discussed in the particular case of camalexin (see section 1). There are also some examples of phytoalexin accumulation being up- or downregulated by overexpression of other signaling molecules such as auxins, abscisic acid, cytokinins and gibberellins.

Auxin homeostasis is considered as one of the components participating in the regulation of plant defense responses [110,111]. In many instances, auxins appear to negatively regulate plant defense mechanisms. In the case of plant/virus interactions, there are some examples relating auxin depletion and pathogen resistance [112–114]. Overexpression of *OsGH3.8*, a member of the *GH3* gene family which encodes an indole-3-acetic acid-amido synthase in rice, acts as an activator of disease resistance by transient and cell-specific suppression of auxin signaling [115]. Constitutive overexpression of *OsGH3.1* was shown to reduce auxin content and to enhance defense responses and resistance to *Magnaporthe grisae* in rice [111]. Thirty four defense genes were up- and down-regulated in the transgenic rice plants overexpressing *OsGH3.1* though no information were provided concerning phytoalexin gene induction in this case. Jones and colleagues have shown recently that the increased susceptibility to biotrophic pathogens and re-direction of phytoalexin metabolism observed in *Arabidopsis* plants overexpressing the *AFB1* gene (which encodes an auxin receptor and whose mRNA is partially resistant to a targeting auxin receptor- microRNA), were probably due to the negative effect of auxin on SA signaling [12]. Conversely, targeting of the auxin receptor mRNAs by overexpression of the microRNA miR393 resulted in direct suppression of auxin signaling and stabilization of inactive AUX/IAA-ARF, which negatively regulate auxin signaling through binding and inactivation of auxin response factors. Suppression of auxin signaling allowed the plant to mount a full SA response and to re-direct secondary metabolism towards the more fungitoxic phytoanticipins, glucosinolates [116] and away from the less fungitoxic phytoalexin camalexin.

The plant hormone abscisic acid (ABA) regulates the adaptive response of plants to environmental stresses such as drought, salinity, and chilling as well as biotic stresses *via* diverse physiological and developmental processes. As for auxins, it seems that ABA depressed fungal, bacterial and viral resistances of plants by negatively regulating genes involved in plant defense mechanisms though there are some reports of stress-induced elevated ABA concentrations being associated with decreased susceptibility to fungal pathogens, for instance in tomato challenged with *Phytophthora parasitica* [117]. Loss-of-function mutations in the ABA pathway led to ABA deficiencies and increase in disease resistance in tomato to, respectively, the necrotrophic fungus *B. cinerea* [118] and the pathogenic bacterium *Erwinia chrysanthemi* [119]. Exogenous applications of ABA made tobacco plants more susceptible to Tobacco Mosaic Virus (TMV) [120] and led to the down-regulation of genes encoding PR protein β-1,3-glucanase isoforms in tobacco cell cultures [121]. ABA was also involved in the regulation of phytoalexins. The synthesis of the bean (*Phaseolus vulgaris*) phytoalexin kievitone was downregulated by ABA [122]. ABA reduced glyceollin production in soybean as a response to *Phytophthora megasperma* [123] or *P. sojae* [124], as well. Rishitin and lubimin production, two phytoalexins from Solanaceae, was also found to be diminished by ABA in potato tubers infected by *P. infestans* [125].

To understand the mechanisms of the regulation of phytoalexin expression under the control of ABA, capsidiol biosynthesis was assessed in two tobacco (*N. plumbaginifolia*) nonallelic mutants, *Npaba2* and*Npaba1* deficient in ABA *vs.* wild-type plants as a response to abiotic (induction with cellulose or arachidonic acid) or biotic stresses [126]. The mutant *Npaba2* was impaired in zeaxanthin epoxidase (ZEP) and *Npaba1* was deficient in abscisic aldehyde oxidase. Total capsidiol phytoalexin accumulation in the two mutants doubled the amount of capsidiol in wild-type plants, even in those inoculated with *B. cinerea* which is known for its ability to produce ABA. This was the consequence of an enhanced expression of two genes acting on the capsidiol biosynthesis pathway, *5-epi-aristolochene synthase* (*EAS*) and *5-epi-aristolochene hydroxylase* (*EAH*). At the same time, the transcription activity of ZEP and 9-*cis*-epoxy-carotenoid dioxygenase which encode enzymes of the upstream pathway of ABA synthesis, decreased. ABA thus negatively regulates elicitor-induced phytoalexin production in tobacco. Induction of plant hormone catabolic pathways was also observed during capsidiol accumulation, being this catabolism not limited to ABA, but also affecting the gibberellins. In fact, both the expression of the gene encoding *ABA-8'-hydroxylase* (*ABAH*) and of that encoding *gibberellin-2-oxidase* (*GA-2Ox*) [127] were induced in parallel to the expression of *EAS*and *EAH* as well as the 2-fold increase in capsidiol. The expression pattern observed for *ABAH* and *GA-2Ox*implied a regulatory role of abscisic acid and likely gibberellins catabolic pathways on capsidiol accumulation in tobacco [126]. This clearly illustrates plant hormone cross talk for phytoalexin regulation.

There is little evidence of phytoalexin expression being regulated by gibberellins with the exception of the possible implication of gibberellin catabolism in the accumulation of capsidiol in tobacco (see above) [126]. Altered disease development was reported in rice plants overexpressing Elongated uppermost internode (*Eui*) gene encoding a gibberellin deactivating enzyme [128]. Conversely, knockout of this gene was found to compromise resistance to disease, indicating that gibberellins negatively regulate rice basal disease resistance. Although the induction of a PR1a pathogenesis-related protein was reported during pathogen resistance, no mention of a possible modulation of phytoalexins by gibberellins was made.

Cytokinins (CKs) are phytohormones involved in various regulatory processes during plant development [129]. Reports of the role of CKs in plant disease resistance are still scarce. Elevated CK levels were associated to resistance against viruses [130,131] and suppression of the HR in tobacco [132]. Plant-derived CKs were also shown to promote resistance of *Arabidopsis* to *Pseudomonas syringae* pv. *tomato DC3000(Pst)* via the salicylic acid response factor TAG3/ Non-expressor of pathogenesis-related genes-1 (*NPR1*) [133]. Modulating CK levels or signaling activity in cytokinine-oxydase- or isopentenyl transferase-overexpressing plants in *ahk2 ahk3* mutants (double knockout of the CK receptor) correlated with altered resistance. Induction of phytoalexin synthesis upon resistance of *Arabidopsis* to *P. syringae* was not reported in this study.

More recently, CK overexpression was shown to enhance resistance against *P. syringae* pv *tabaci* in tobacco (*Nicotiana tabacum*) [100]. The CK-mediated resistance in tobacco strongly correlated with up-regulated synthesis of two major phytoalexins from that plant, the diterpenoid capsidiol and the coumarin scopoletin. Since CKs are aminopurine-derived compounds containing an isopentenyl moiety, the role of this phytohormone in the pathosystem between tobacco and *P. syringae* pv *tabaci* was addressed through the overexpression of the *isopentenyl transferase* (*ipt*) gene from *Agrobacterium tumefaciens* driven by the synthetic pathogen-inducible *4xJERE* promoter in the *4xJERE:ipt* plasmid. Isopentenyl transferase indeed catalyzes the rate-limiting reaction in the route to CKs.

In *ipt*-overexpressing tissues challenged with *P. syringae*, CK production was enhanced and this correlated well with increased resistance to this pathogen. As the protocol used for pathogen inoculation included pre-infiltration of leaves with *A. tumefaciens* which can interfere with symptom development, experiments were completed by two additional genetic approaches leading to ectopic production of CKs independent of prior infiltration with *A. tumefaciens*. The *ipt* gene was successively expressed under the control of a senescence-induced promoter (SAG12:ipt) and under the control of a chemically inducible tetracycline-dependent promoter (TET:ipt). In both transformed tobacco lines, there was a strong CK accumulation well correlating with disease symptom reduction.

Changes in endogenous CK concentrations enhanced the production of capsidiol and scopoletin by respectively 4.6 and 6.4-fold compared with the controls. Ectopic phytoalexin synthesis resulting from overexpression of the *4xJERE:ipt* construct was associated with the CK-mediated resistance against *P. syringae*. Interestingly, there was an induced expression of 5-*epi*-aristolochene synthase, catalyzing a rate-limiting step in the route to capsidiol, together with an induction of the cinnamate-4-hydroxylase which is implied in the phenylpropanoid pathway leading, namely, to the coumarin scopoletin. Conversely, expression of other enzymes related to phytoalexin biosynthesis such as the glucosyltransferase TOGT involved in the sorage of scopoletin (see section 3), phenylalanine ammonia lyase, the key enzyme of the phenylpropanoid pathway and flavonol synthase 1, were respectively moderate, not significantly affected or even repressed upon CK-mediated resistance of tobacco to *P. syringae*. Importantly, this CK-mediated resistance was also shown to be SA- and JA- independent.

PHYTOALEXIN ACCUMULATION IN THE UPREGULATION OF DEFENSE-RELATED MARKER AND ELICITOR GENES

In the new concept of innate immunity for plants, defense responses are triggered after recognition by the plant cell of conserved Pathogen/Microbe-Associated Molecular Patterns (PAMPs/MAMPs) or by endogenous molecules released by pathogen ingress and called Danger-Associated Molecular Patterns (DAMPs) [134–136]. Early responses after elicitor perception are characterized by signaling processes including ion fluxes, Mitogen-Activated Protein Kinase (MAPK) cascade [137] and production of reactive oxygen species [138].

In this context, MAPKs of *Arabidopsis*, respectively, MPK3 and MPK6, were shown to be activated upon MAMP treatment or following infection by a pathogen with correlated induction of the responses further downstream such as phytoalexin accumulation by using both gain-of-function and loss-of-function transgenic systems. [13]. MPK3 and MPK6 take part in a pathogen-responsive MAPK cascade, MAPKKKa/MEKK1-MKK4/MKK5-MPK3/MPK6, involved in the up-regulation of numerous enzymes of the biosynthetic route from chorismate to camalexin via tryptophan (Trp) [139,140] (Figure 5). Constitutively active mutants of the upstream MAPKKs, that is, tobacco NtMEK2DD and *Arabidopsis* MKK4DD or MKK5DD, were overexpressed under the control of a steroid-inducible promoter. Upon treatment with the steroid dexamethasone (DEX), induction of the active MAPKs, MPK3/MPK6 activated the endogenous MAPKs, which in turn induced the responses further downstream such as accumulation of the phytoalexin camalexin. Camalexin production was associated with the induction of numerous genes encoding enzymes acting on the biosynthesis of this phytoalexin, including the route from chorismate to Trp, on one hand, and the pathway from Trp to camalexin (Figure 5). Specifically, genes encoding anthranilate synthase a and b (ASA and ASB), phosphoribosylanthranilate transferase (PAT), indole-3-glycerolphosphate synthase (IGPS) and Trp synthase b (TSB), were all highly induced with levels of induction ranging from 12-fold (PAT) to more than 35-fold (TSB) 6h after DEX treatment. Genes encoding the P450 enzymes operating in the biosynthetic pathway from Trp to camalexin, namely the one catalyzing conversion of Trp to indole-3-acetaldoxime (*CYP79B2* gene) showed a 50-fold increase. Expression of P450 enzyme genes acting at the end of the pathway, for instance the *CYP71B15* (*PHYTOALEXIN DEFICIENT 3*, *PAD3*) gene encoding the multifunctional enzyme converting cysteine-3-indolacetonitrile to camalexin, increased by 400-fold (Figure 5).

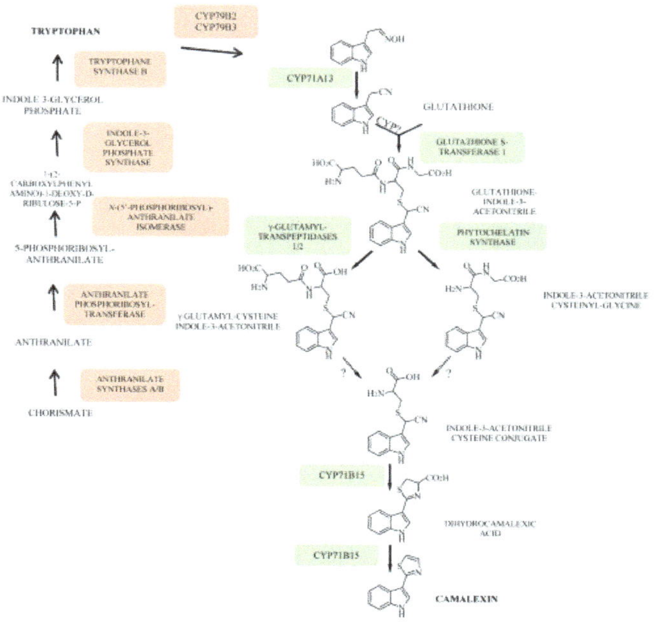

Figure 5: Biosynthetic pathways from chorismate to tryptophan and from tryptophan to camalexin (adapted from [139,140]).

Mutations in MPK3 and MPK36 impaired *B. cinerea*-induced camalexin production and disease resistance, providing loss-of-function evidence supporting the role of this MAPK cascade in phytoalexin biosynthesis upon pathogen infection. Camalexin accumulation was reduced by 50% in *mpk3* mutant, delayed in *mpk6*mutant and completely abolished in *mpk3/mpk6* double mutant, resulting in increased susceptibility to *B. cinerea*. All together, these experiments clearly showed that MPK3 and MPK6, two MAP kinases of*Arabidopsis* play key roles in camalexin biosynthesis and resistance to fungal pathogens. Induction of camalexin after MAPK activation was considered to be independent of ethylene [13].

Phytoalexin accumulation was reported in the upregulation of defense-related marker genes in rice plants (*Oryza sativa*) exhibiting spontaneous hypersensitive response upon overexpression of the *Oryza sativa*Accelerated Cell Death and Resistance 1 (*OsACDR1*) gene which encodes a putative Raf-like Mitogen-Activated Protein Kinase Kinase Kinase (MAPKKK) [141]. Two compounds, sakuranetin and momilactone A, respectively, flavanone and diterpenoid phytoalexins from rice had their concentrations increased by 4-fold and 32-fold following infection by the rice blast fungal pathogen *Magnaporthe grisea*, in *OsACDR1*overexpressing lines while their accumulation in vector control plants were nil or negligible. The fact that*OsACDR1* overexpressing plants exhibited enhanced disease resistance to *M. grisea* through a reduction in the ability of fungal appressoria to penetrate into plant cells could be indirectly related with phytoalexin accumulation in transgenic lines. Conversely, *Osacdr1* knock-out and OsACDR1-RNAi rice plants showed increased susceptibilities to the pathogen though no information regarding a possible relationship between the loss-of-function of *OsACDR1*, decreased-disease resistance and phytoalexin content were provided in this study.

Systemic Acquired Resistance (SAR) is a long lasting defense response that is induced by localized infection providing subsequent protection against a broad spectrum of pathogens [142]. Resistance to several fungal and bacterial diseases has been obtained by overexpressing the Non-expressor of Pathogenesis-Related genes-1 (*NPR1*) in many plant species. NPRs1 play a critical role in SAR [143] and validation of their involvement in various defense networks arose from a number of overexpression studies. For instance, overexpression of *NPR1* was shown to confer resistance to bacteria, viruses and fungi in rice, [144], tomato [145], wheat [146] and carrot [147]. Recently, resistance against various fungal pathogens and a reniform nematode was obtained in transgenic cotton overexpressing *Arabidopsis NPR1* (*AtNPR1*) [148]. The*AtNPR1*-expressing lines which displayed namely high level of resistance to *Verticillium dahliae* had corresponding high levels of gossypol, a phytoalexin from cotton and derivated terpenoids, forming part of the overall defense response that protects the *AtNPR1*-transformants from disease.

The Rac family also named Rop family belongs to the Ras superfamily of GTPases. This family constitutes one of the most important regulators of signal transduction in plants, participating to their adaptation to various environmental situations [149]. Rac proteins are also involved in innate immunity namely in rice and barley [149–151]. Importantly, the *Oryza sativa* Rac family OsRac1 forms a complex termed defensome with downstream proteins which regulate reactive oxygen intermediate (ROI) production, hypersensitive-like responses, lignin biosynthesis and phytoalexin accumulation as well as the transcription of pathogenesis-related genes [150,152]. Transgenic rice plants expressing the constitutively active OsRac1 exhibited HR-like responses and increased resistance both to a virulent race of the rice blast fungus and a virulent race of bacterial blight together with an enhancement of the production of the rice phytoalexin momilactone A (19- to 180-fold higher than the levels of untransformed rice plants) [150]. Phytoalexin content in the transgenic lines expressing the dominant-negative OsRac1 was similar to control plants.

Overexpression of an homologue of mamalian selenium-binding proteins, the rice *O. sativa* selenium-binding protein homologue (*OsSBP*) led to increased resistance to both rice blast fungus and rice bacterial blight *via*activation of plant defense mechanisms including accumulation of the rice phytoalexin, momilactone A [153]. Although there are some reports of the occurence of selenium binding proteins in plants, their functions remain unclear [154,155]. A full length *OsSBP* cDNA fragment was inserted in sense- or antisense-orientation under the control of the CaMV 35S promoter. *OsSBP* overexpressing rice lines exhibited delayed disease symptoms upon rice blast fungus while downregulated expression of *OsSBP*resulted in symptoms slightly more severe than in controls. Correspondingly, overexpression of *OsSBP* led to a 20–25-fold increase in momilactone A, being phytoalexin levels in *OsSBP* downregulated lines similar as those in the wild-type parental rice. Besides, there was also an induction in the PR gene expression as well as enhancement of H_2O_2 accumulation. The fact that overexpressing *OsSBP* rice plants also exhibited disease resistance to bacterial blight was consistent with the overall relevance of this protein in modulating the defense mechanisms of that plant.

Calcium regulates various cellular processes in plants as an ubiquitous internal second messenger, transferring signals received at the cell surface to the inside of the cell *via* spatial and temporal changes in its concentration. The information encoded within the Ca^{2+} transients is decoded and transmitted by a toolkit of Ca^{2+}-binding proteins that regulate transcription *via* Ca^{2+}-responsive promoter elements, regulating protein phosphorylation [156]. Several families of calcium sensors have been identified so far of which the best known are calmodulins (CaMs) and CaM-related proteins [157]. The second major class is the Ca^{2+}-dependent protein kinases [158]. The third family consists of calcineurin B-like proteins (CBLs) acting as Ca^{2+} sensors to activate specific protein kinases, CBL interacting protein kinases (CIPKs) [159]. Since Ca^{2+}mobilization as a reponse to pathogen-associated molecular patterns plays a pivotal role in innate immunity in plants, the molecular links between this cation and downstream defense responses, namely phytoalexin production, were investigated using gain-of-function approaches and an RNAi strategy to address regulation of PAMP-induced defenses by two genes *OsCIPK14* and *OSCIPK15* encoding CIPKs in rice (*O. sativa*) cultured cells [160]. These two genes were rapidly induced in rice transgenic lines by microbe-associated molecular patterns, including chitooligosaccarides and a *Trichoderma viride*/ethylene-inducing xylanase (TvX/EIX). *OsCIPK14/15* overexpressing plants showed an increase in the two major phytoalexins of rice, momilactones and phytocassanes, as evidenced by LC-MS-MS analyses. This was the result of a slow and prolonged expression of genes encoding cyclase enzymes involved in the biosynthesis of these phytoalexins, namely, *ent*-copalyl diphosphate synthase 4 and *ent*-kaurene synthase-like 4 for momilactones and *ent*-copalyl diphosphate synthase 2 and *ent*-kaurene synthase -like 7 for phytocassanes, respectively. The TvX/EIX-induced phytoalexin biosynthesis was significantly suppressed in the *OsCIPK14/15-RNAi* lines.

Momilactones and phytocassanes are also regulated in rice by the *Oryza sativa* TGA factor for phytoalexin production 1 named *OsTGAP1* [161]. Transformed rice plants constitutively expressing *OsTGAP1* indeed showed enhanced accumulation of both phytoalexins. OstGAP1 regulates momilactone phytoalexin biosynthesis through positive control of the *OsKSL4* gene expression, responsible for the first committed step in the route to momilactone. Specifically, clustered genes for momilactone biosynthesis were shown to be coordinately regulated by OsTGAP1, leading to a hyper-accumulation of this phytoalexin as a response to elicitor treatment.

Camalexin biosynthesis in *Arabidopsis* was also shown to be regulated by microbial virulence factors of the Nep1-like protein family (NLP) [162]. Beside their role as toxin-like virulence factors, Nep1-like proteins (NLPs) act as signals triggering the plant innate immune responses involving a signaling cascade which includes reactive oxygen species and ethylene production, and ending in callose deposition, programme cell death and phytoalexin production [163,164]. Overexpression of a Nep1-like protein from *Pythium aphanidermatum* in *Arabidopsis* under the control of an ethanol-inducible promoter resulted in the strong transcriptional activation of genes working on the camalexin biosynthetic pathway (genes *ASA1* and *TSA*) (see above).

Activation-tagging is a powerful system for creating gain-of-function mutants in plants. Upon random introduction into the genome of a T-DNA containing enhancer sequences, there is an increased expression of neighboring genes on either side of it, resulting in a gain-of-function phenotype. A T-DNA tag was used to identify a neighbor gene overexpressed in a *Spotted leaf 18* mutant (*Spl18*) displaying strong resistance to blast disease and obtained by an activation tagging system in rice [165]. An ORF was located about 500 bp downstream of the inserted T-DNA and the deduced protein named OsAT1 presented a sequence similarity to an acyltransferase whose expression was induced in tobacco reacting hypersensitively. Combination of*OsAT1* genomic DNA downstream of a modified 35S promoter was introduced to rice. Overexpression of*OsAT1* induced several defense responses, including transcriptional activation of PR protein genes, accumulation of two phytoalexins from rice, sakuranetin and momilactone A, resulting in the plant resistance to blast disease and bacterial blight. Since the only one report of an acyltransferase being involved in the plant disease resistance mechanisms was *Arabidopsis* [166,167], it was unexpected that the overexpression of this acyltransferase could lead to upregulation of phytoalexins in rice.

We have seen in this section that, in addition to hormone signaling (section 4), phosphorylation relays and cascades, defense-related markers genes, calcium sensors and elicitors are also potentially important regulators for the modulation of phytoalexin production and pathogen resistance.

PHYTOALEXIN ENGINEERING: WHERE PLANT AND HUMAN INTERESTS DO NOT MEET

As some phytoalexins, especially cottonseed and potato phytoalexins, display a certain level of toxicity for humans, there was a crucial interest in engineering those plants for abolishing accumulation of these undesirable compounds.

Genetic manipulation of plants for technological applications can affect phytoalexin expression. This is the case with potatoes which produce a number of antinutritional phytoalexins such as the steroid glycoalkaloids α-solanine and α-chaconine [168] (Figure 2). Potatoes genetically modified with an antisense potato invertase gene, exhibited a modified phytoalexin profile regarding glycoalkaloids. Specifically, transgenic lines were obtained by introducing a potato acid invertase gene sequence in antisense orientation driven by a patatin promoter and subtended by the *Agrobacterium*-derived nopaline synthase terminator. This was done to genetically manipulate through a reduction in the amount of invertase, the accumulation of reducing sugars responsible for subsequent browning problems during cooking at high temperatures. Such transformation namely decreased glycoalkaloid accumulation in potatoes *vs.* controls. It was speculated that diminutions in the carbohydrate pool in antisense invertase lines may reduce the UDP-glucose:solanidine glucosyltransferase and the UPD-galactose:solanidine galactosyltransferase-mediated conversion of solanidine to γ-chaconine and γ-solanidine, respectively, decreasing the amounts of the toxic glycoalkaloids α-chaconine and α-solanidine. Modifications of these toxic secondary metabolites in potatoes were related to a direct effect of the transgene rather than to interactions of the plant with its environment.

Gossypol and its congeners are dimeric sesquiterpene phytoalexins of the cadinane family from cotton (*Gossypium hirsutum*) deriving from the plastidic methylerythritol phosphate (GAP-pyruvate) pathway [161]. The latter leads to two C_5 units, respectively isopentenyl-pyrophosphate (IPP) and its isomer dimethylallyl-pyrophosphate (DMAPP). Condensation of these two C_5 units and larger IPP- and DMAPP-derived building blocks such as the C_{10} unit, geranyl pyrophosphate (GPP), the C_{15} unit, farnesyl pyrophosphate (FPP) and the C_{20} unit, geranylgeranylpyrophosphate (GGPP), yields monoterpenes, sesquiterpenes (such as gossypol), triterpenes (steroids), tetraterpenes (carotenoids) and diterpenes. The starter of the biosynthetic route to gossypol is farnesyl diphosphate (FPP) (Figure 6). In cotton, (+)-δ-cadinene synthase (CDNS) catalyzes the cyclization of *E,Z*-FPP to (+)-δ-cadinene, involving isomerization of FPP to a nerolidyl intermediate, the cyclization of which leads to a *cis*-germacradienyl cation. This is followed by 1,3-hydride shift, cyclization to a cadinanyl cation and deprotonation to form (+)-δ-cadinene [169,170]. CDNS catalyzes the first committed step in the route of the cadinane sesquiterpenoids from FPP, making this enzyme a crucial control component for gossypol engineering (Figure 6). According to the works of Chen and colleagues [171,172], four cDNAs of *cdn* synthase have been isolated from *Gossypium arboreum*, *i.e.*, *cdn*1-C1, *cdn*1-C14, *cdn*1-A and *cdn*1-C2.

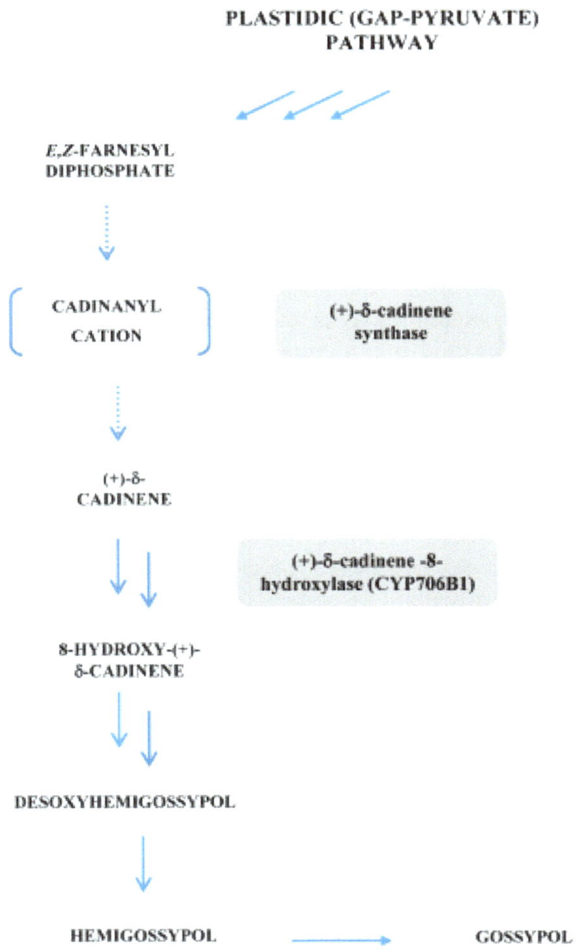

Figure 6: Scheme of biosynthetic pathway from farnesyl diphosphate to gossypol.

Gossypol plays a role as a phytoalexin produced in response to the fungal pathogen *Verticillium dahlia* Kleb. and to bacterial blight, caused by *Xanthomonas campestris* pv. *malvacearum* (Smith) Dye. Beside its presence in the epidermis and hairs of developing roots together with its storage within the very typical lysigenous gossypol glands in the foliage, gossypol is the main cadinane sesquiterpenoid in developing and mature seeds as well as in the cottonseed glands. Gossypol is known to be toxic and to have antifertility effects on animals and humans ([173] and references therein) and must be removed from cottonseed oil prior to human consumption. To address this problem and for generating the trait of gossypol-free seeds in cultivated cotton species, genetic manipulation of the pathway leading to this compound through overexpression of antisense (+)-δ-cadinene synthase was carried out.

Two genetic approaches using *a priori* silencing strategy with the introduction of a (+)-δ-cadinene synthase gene, respectively, *cdn*-1-C1 from *G. arboretum* [171], and *cdn*-1-C4 from *G. Hirsutum* [174], in the antisense orientation were employed. The *cdn*-1-C1 from *G. arboretum* was found to be highly homologous to *cdn*-1-C4 from *G. Hirsutum* [174]. In the first set of experiments [171], cotton plants were harbored with a*cdn*-1C1 cDNA from *G. arboretum* in antisense transformation construct driven by the CaMV 35S promoter. Here, the expression of the antisense construct was sufficient to negate the transcripts from the expression of multiple constitutive *cdn* synthase genes, resulting in reduced CDN synthase activity and decreasing the cottonseed gossypol content by up to 70%. A new cotton variant with a markedly lowered gossypol content, but with no reduction in the number of lysigenous seed glands, was isolated by the same group from the progeny of hemizygous cotton plants (*G. hirsutum* cv. Coker 312) transformed with antisense *cdn*1-C1 cDNA [175]. Townsend and colleagues [174] made diametrically opposite observations in that the *cdn*synthase gene they used never affected gossypol levels in the seeds of transformants with either a constitutive 35S promoter or a seed-specific soybean lectin promoter driven-antisense construct, or in stems of the constitutive antisense lines challenged with *V. dahliae*. The converse situation was observed when cotyledons were subjected to infection with bacterial blight, *X. campestris*. From these observations it ensued that gossypol production in cotton implicates a complex system of differential regulation of the CDNS. Given the high DNA sequence conservation among the different *cdn* synthase gene families, the ability of the antisense*cdn*1-C4 constructs to silence only blight-induced CDNS genes and neither the cotyledons nor the stems infected by *V. dahlia*e was somewhat intriguing.

Due to equivocal and apparently contradictory data recovered from the use of the antisense δ-cadinene synthase gene strategy aiming at reducing gossypol amounts in cotton, the feasibility of a targeted RNAi-based approach for suppression of the δ-cadinene synthase gene in that plant with resultant, but tissue-specific disruption of gossypol accumulation, was clearly demonstrated [176]. A 604-bp sequence from a δ-cadinene synthase (*Cad*1-A gene) cDNA clone from a *G. hirsutum* staged-embryo library was chosen as the trigger sequence. Interestingly, the selected portion of the clone had 80.9% to 99.8% homology to other gene sequences of other δ-cadinene synthase genes from the diploid, *G. arboreum*, and the tetraploid, *G. hirsutum*, cotton plants, aiming at targeting all members of this gene family. This sequence was used to make an intron-containing hairpin (ihp) construct and transcription of the ihpRNA sequence was under the control of a highly seed-specific α-globulin B gene promoter from cotton [177]. *Agrobacterium*-mediated transformation of *G. hirsutum* cv. Coker 312 with the final hairpin vector allowed the gossypol concentration in seeds to be decreased by 99%. Fortunately, RNAi-mediated silencing remained confined to the tissues expressing the hairpin RNA-encoding transgene in cotton, and it was observed that the presence of the transgene did not reduce the phytoalexin content in foliage, floral organs and roots. Here, the RNAi approach coupled to a highly tissue-specific promoter made feasible a selective and significant diminution in the amounts of a given phytoalexin which is deleterious for human health, without affecting its levels and the defense responses in other organs of the plant.

A patent describing a method for reducing levels of gossypol in cottonseed was derived from these results [178]. This included selective inducing RNA gene silencing in the seeds of a transgenic cotton plant, to interfere with expression of the δ-cadinene synthase gene or the δ-cadinene-8-hydroxylase gene (encoding formation of 8-hydroxycadinene on the route to gossypol) (Figure 6) in cottonseed without substantially affecting expression of that gene in the foliage, floral parts, and roots of the plant.

CONCLUSIONS

Starting in the 1950's, research on phytoalexins has begun with biochemistry and bio-organic chemistry, resulting in the determination of their structure, their biological activity as well as mechanisms of their synthesis and their catabolism by microorganisms. Elucidation of the biosynthesis of numerous phytoalexins has permitted the use of molecular biology tools for the exploration of the genes encoding enzymes of their synthesis pathways and their regulators. Genetic manipulation of phytoalexins requires *a priori* a sound knowledge of the genes involved in their biosynthesis and how accumulation of a given phytoalexin can be modulated. Success of the transformation will also depend on the inability of the pathogen to counteract the phytoalexin action. It is well known that phytopathogenic fungi, particularly, are able to metabolize the phytoalexins to which they are exposed. Engineering of fungal genes responsible for detoxification of phytoalexins in plants has pointed out their role in the interactions between plants and pathogens. For example, overexpression in hairy roots of pea (*Pisum sativum* L.) of a pisatin demethylating activity (PDA) from the pea-pathogen fungus *Nectria haematococca* reduced the amounts of pisatin. As a result, transgenic plant tissues with reduced capability to accumulate pisatin were found to be less resistant to fungal infection [85]. Also, it appears clearly evident that, in phytopathogenetic fungi, ATP-binding cassette (ABC) transporters, which may extrude plant defense products as well as fungicides, act as virulence factors, providing protection against defense compounds produced by the host. Many factors thus interplay which could affect the outcome of the interaction between plants and pathogens.

Unexpectedly, engineering phytoalexins for disease resistance in plants seems to have been limited to exploiting only a few phytoalexin biosynthetic genes, especially those encoding stilbenes and some isoflavonoids. The first example of a disease resistance resulting from foreign phytoalexin expression in a novel plant was published only in 1993 [65]. One can imagine that such a success would have opened the way for a ferment of activity in this area, but, if transformations were then operated to investigate the potential of stilbene biosynthetic genes to confer resistance to pathogens [59,60,62], strategy in engineering phytoalexins from other plant species did not receive as much applications as might be expected. As previously stated in this review, interest in secondary metabolite engineering deals with their implications in human health and disease. The extraordinary success obtained with resveratrol, the phytoalexin of Vitaceae, is linked to the fact that in this case, the engineered system requires a relatively simple genetic construct. Resveratrol is indeed obtained in one single step from *p*-coumaroyl-CoA and three malonyl-CoA units catalyzed by stilbene synthase. As these substrate precursor molecules are present throughout the plant kingdom, the introduction of a single gene is therefore sufficient to synthesize resveratrol in heterologous plant species. Except the case of resveratrol biosynthesis which appears to be very simple, other phytoalexins (isoflavonoids, terpenoids) are formed through very complex biosynthetic pathways. Engineering the entire pathway is not feasible and the problem is to choose the right enzyme catalyzing the limitant reaction of the given pathway. Moreover, modalities of the expression of genes encoding a given enzyme can be unpredictable. In fact, overexpression of 7-*O*-methyltransferase in alfalfa led to different regiospecificities, producing mainly 4′-*O*-methyl derivatives instead of the expected 7-*O*-methyl derivatives [83].

New techniques for metabolic engineering have thus to be exploited in the next years to come. Namely, methodologies for generating high-quality libraries of enzyme variants and novel high-throughput screening (HTS) technologies will open the way for the engineering of enzymes for the biosynthesis of various phytoalexins with potent biological activities. Specifically, HTS technologies can rapidly lead to the identification of genes which modulate a particular biosynthesis pathway. Gathering all genes encoding for a biomolecular pathway will allow the assembly of genetic constructs for the synthesis of a given phytoalexin.

Limitations in the strategy of engineering phytoalexins also arose from the fact that the obtained resistance is sometimes too weak since it is well known that phytoalexins are less phytotoxic than chemical fungicides [16]. This thus justifies indirect approaches which allow modulation of the accumulation of phytoalexin employing transcriptional regulators or components of upstream regulatory pathways (see above sections 4 and 5). For example, overexpression in *Arabidopsis* of the active MAPKs, MPK3/MPK6 activated the endogenous MAPKs, which in turn induced genes encoding the P450 enzymes operating in the biosynthetic pathway from tryptophan to camalexin, showing their activities a 50-fold increase or in some cases a 400-fold increase [13]. As a consequence, the data presented in this review show undoubtedly that indirect modulation of phytoalexin levels through transgenic approaches paves the way for the creation of novel plants with improved pathogen resistance traits.

REFERENCES

1. Müller K.O., Börger H. Experimentelle untersuchungen über die *Phytophthora* resistenz der *Kartoffel*.Arbeit. Biol. Reichsant Land Forstwirtsch. 1940;23:189–231.
2. Deverall B.J. Introduction. In: Bailey J.A., Mansfield J.W., editors. Phytoalexins. Blackie; Glasgow and London, UK: 1982. pp. 1–20.
3. Mansfield J.W., Bailey J.A. Phytoalexins: Current Problems and Future Prospects. In: Bailey J.A., Mansfield J.W., editors. Phytoalexins. Blackie; Glasgow and London, UK: 1982. pp. 319–323.
4. Ahuja I., Kissen R., Bones A.M. Phytoalexins in defense against pathogens. Trends Plant Sci.2012;17:73–90.
5. Thomma B.P.H.J., Nelissen I., Eggermont K., Broekaert W.F. Deficiency in phytoalexin production causes enhanced susceptibilty of *Arabidopsis thaliana* to the fungus *Alternaria brassicola*. Plant J.1999;19:163–171.
6. VanWees S.C., Chang H.S., Zhu T., Glazebrook J. Characterization of the early response of Arabidopsis to *Alternaria brassicicola* infection using expression profiling. Plant Physiol. 2003;132:606–617.
7. Rowe H.C., Walley J.W., Corwin J., Chan E.K.F., Dehesh K., Kliebenstein D.J. Deficiencies in jasmonate-mediated plant defense reveal quantitative variation in *Botrytis cinerea* pathogenesis. PLoS Pathog. 2010;6:e1000861.
8. Nawrath C., Métraux J.P. Salicylic acid induction-deficient mutants of *Arabidopsis* express *PR-2* and *PR-5*and accumulate high levels of camalexin after pathogen inoculation. Plant Cell. 1999;11:1393–1404.
9. Roetschi A., Si-Ammour A., Belbahri L., Mauch F., Mauch-Mani B. Characterization of an *Arabidopsis-Phytophthora* pathosystem: Resistance requires a functional *PAD2* gene and is independent of salicylic acid, ethylene and jasmonic acid signalling. Plant J. 2001;28:293–305.
10. Denby K.J., Jason L.J.M., Murray S.L., Last R.L. *ups1*, an *Arabidopsis thaliana* camalexin accumulation mutant defective in multiple defence signalling pathways. Plant J. 2005;41:673–684.
11. Heck S., Grau T., Buchala A., Métraux J.P., Nawrath C. Genetic evidence that expression of NahG modifies defence pathways independent of salicylic acid biosynthesis in the *Arabidopsis-Pseudomonas syringae* pv. tomato interaction. Plant J. 2003;36:342–352.
12. Robert-Seilaniantz A., MacLean D., Jikumaru Y., Hill L., Yamaguchi S., Kamiya Y., Jones J.D.G. The microRNA miR393 re-directs secondary metabolite biosynthesis away from camalexin and towards glucosinates. Plant J. 2011;67:218–231.
13. Ren D., Liu Y., Yang K.Y., Han L., Mao G., Glazebrook J., Zhang S. A fungal-responsive MAPK cascade regulates phytoalexin biosynthesis in *Arabidopsis*. Proc. Natl Acad. Sci. USA. 2008;105:5638–5643.
14. Pandey S.P., Rocarro M., Schön M., Logemann E., Somssich I.E. Transcriptional reprogramming regulated by WRKY18 and WRKY40 facilitates powdery mildew infection of *Arabidopsis*. Plant J.2010;64:912–923.
15. Smith D.A. Toxicity of Phytoalexins. In: Bailey J.A., Mansfield J.W., editors. Phytoalexins. Blackie; Glasgow and London, UK: 1982. pp. 218–252.

16. Rathmell W.G., Smith D.A. Lack of activity of selected isoflavonoid phytoalexins as protectant fungicides. Pestic. Sci. 1980;11:568–572.

17. Pezet R., Pont V. Ultrastructural observations of pterostilbene fungitoxicity in dormant conidia of *Botrytis cinerea* Pers. J. Phytopathol. 1990;129:29–30.

18. Adrian M., Jeandet P., Veneau J., Weston L.A., Bessis R. Biological activity of resveratrol, a stilbenic compound from grapevines, against *Botrytis cinerea*, the causal agent for gray mold. J. Chem. Ecol.1997;23:1689–1702.

19. Adrian M., Jeandet P. Effects of resveratrol on the ultrastructure of *Botrytis cinerea* conidia and biological significance in plant/pathogen interactions. Fitoterapia. 2012;83:1345–1350.

20. Shlezinger N., Minz A., Gur Y., Hatam I., Dagdas Y.F., Talbot N.J., Sharon A. Anti-apoptotic machinery protects the necrotrophic fungus *Botrytis cinerea* from host-induced apoptotic-like cell death during plant infection. PLoS Pathog. 2011;7:e1002185.

21. Ingham J.L. Phytoalexins from the Leguminosae. In: Bailey J.A., Mansfield J.W., editors. Phytoalexins. Blackie; Glasgow and London, UK: 1982. pp. 21–80.

22. Kuc J. Phytoalexins from the Solanaceae. In: Bailey J.A., Mansfield J.W., editors. Phytoalexins. Blackie; Glasgow and London, UK: 1982. pp. 81–105.

23. Coxon D.T. Phytoalexins from other Plant Families. In: Bailey J.A., Mansfield J.W., editors. Phytoalexins. Blackie; Glasgow and London, UK: 1982. pp. 106–132.

24. Keen N.T., Zaki A., Sims J.J. Biosynthesis of hydoxyphaseollin and related isoflavanoids in disease-resistant soybean hypocotyls. Phytochemistry. 1972;11:1031–1039.

25. Markham K.R., Ingham J.L. Tectorigenin, a phytoalexin from *Centrosema haitiense* and other*Centrosema* species. Z. Naturforsch. C. 1980;35:919–922.

26. Ingham J.L. Induced isoflavanoids from fungus-infected stems of pigeon pea (*Cajanus cajan*) Z. Naturforsch. C. 1976;31:504–508.

27. Ingham J.L., Keen N.T., Hymowitz T. A new isoflavone phytoalexin from fungus-inoculated stems of*Glycine wightii*. Phytochemistry. 1977;16:1943–1946.

28. Ingham J.L. Flavonoid and isoflavonoid compounds from leaves of sainfoin (*Onobrychis viciifolia*) Z. Naturforsch. C. 1978;33:146–148.

29. Ingham J.L. Phytoalexins from hyacinth bean (*Lablab niger*) Z. Naturforsch. C. 1977;32:1018–1020.

30. Hargreaves J.A., Mansfield J.W., Coxon D.T. Identification of medicarpin as a phytoalexin in the broad bean plant (*Vicia faba* L.) Nature. 1976;262:318–319.

31. Ingham J.L. Phytolexin production by species of the genus *Caragana*. Z. Naturforsch. C. 1979;34:293–295.

32. Gnanamanickam S.S. Isolation of isoflavonoid phytoalexins from seeds of *Phaseolus vulgaris*. Experientia. 1979;35:323.

33. Burden R.S., Bailey J.A. Structure of the phytoalexin from soybean. Phytochemistry. 1975;14:1389–1390.

34. Bonde M.R., Millar R.L., Ingham J.L. Induction and identification of sativan and vestitol as two phytoalexins from *Lotus cornicolatus*. Phytochemistry. 1973;12:2957–2959.

35. Lyon F.M., Wood R.K.S. Production of phaseollin, coumestrol and related compounds in bean leaves inoculated with *Pseudomonas* spp. Physiol. Plant Pathol. 1975;6:117–124.

36. Mansfield J.W., Porter A.E.A., Smallman R.V. Dihydrowyerone derivatives as components of the furanoacetylene phytoalexin response of tissues of *Vicia faba*. Phytochemistry. 1980;19:1057–1061.

37. Hargreaves J.A., Mansfield J.W., Coxon D.T. Conversion of wyerone to wyerol by *Botrytis cinerea* and*B. fabae in vitro*. Phytochemistry. 1976;15:651–653.

38. Ingham J.L. 3,5,4′-trihydroxystilbene as a phytoalexin from groundnuts (*Arachis hypogea*) Phytochemistry. 1976;15:1791–1793.

39. Kuc J. A biochemical study of the resistance of potato tuber to attack by various fungi. Phytopathology.1957;47:676–680.

40. Locci R., Kuc J. Steroid glycoalkaloids as compounds produced by potato tuber slices under stress. Phytopathology. 1967;57:1272–1273.

41. Tomiyama K., Sakuma T., Ishizaka N., Sato N., Takasugi M., Katsui T. A new antifungal substance isolated from potato tuber tissue infected by pathogens. Phytopathology. 1968;58:115–116.

42. Katsui N., Murai A., Takasugi M., Imaizumi K., Masamune T. The structure of rishitin, a new anfifungal compound from diseased potato tubers. J. Chem. Soc. Chem. Comm. 1968:43–44.

43. Katsui N., Matsunaga A., Masamune T. The structure of lubimin, oxylubimin, antifungal metabolites from diseased potato tubers. Tetrahedron Lett. 1974;15:4483–4486.

44. Birnbaum G.I., Stoessl A., Grover S.H., Stothers J.B. The complete stereostructure of capsidiol, X-ray analysis and ^{13}C nuclear magnetic resonance of eremophilane derivatives having trans-vicinal methyl groups.Can. J. Chem. 1974;52:993–1005.

45. Reuveni M., Cohen Y. Growth retardation and changes in phenolic compounds, with special reference to scopoletin, in mildewed and ethylene-treated tobacco plants. Physiol. Plant Pathol. 1978;12:179–189.

46. Cartwright D.W., Langcake P., Pryce R.J., Leworthy D.P., Ride J.P. Isolation and characterization of two phytoalexins from rice as momilactones A and B. Phytochemistry. 1981;20:535–537.

47. Kodama O., Miyakawa J., Akatsuka T., Kiyosawa S. Sakuranetin, a flavanone phytoalexin from ultraviolet-irradiated rice leaves. Phytochemistry. 1992;31:3807–3809.

48. Umemura K., Ogawa N., Shimura M., Koga J., Usami H., Kono T. Possible role of phytocassane, rice phytoalexin, in disease resistance of rice against the

blast fungus *Magnaporthe grisae*. Biosci. Biotechnol. Biochem. 2003;67:899–902.

49. Schmelz E.A., Kaplan F., Huffaker A., Dafoe N.J., Vaughan M.M., Ni X., Rocca J.R., Alborn H.T., Teal P.E.A. Identity, regulation, and activity of inducible diterpenoid phytoalexins in maize. Proc. Natl. Acad. Sci. USA. 2011;108:5455–5460.
50. Huffaker A., Kaplan F., Vaughan M.M., Dafoe N.J., Ni X., Rocca J.R., Alborn H.T., Teal P.E.A., Schmelz E.A. Novel acidic sesquiterpenoids constitute a dominant class of pathogen-induced phytoalexins in maize. Plant Physiol. 2011;156:2082–2097.
51. Nicholson F., Kollipara S.S., Vincent J.R., Lyons P.C., Cadena-Gomez G. Phytoalexin synthesis by the sorghum mesocotyl in response to infection by pathogenic and nonpathogenic fungi. Proc. Natl. Acad. Sci. USA. 1987;84:5520–5524.
52. Lo S.C., de Verdier K., Nicholson R. Accumulation of 3-deoxyanthocyanidin phytoalexins and resistance to *Colletotrichum sublineolum* in sorghum. Physiol. Mol. Plant Pathol. 1999;55:263–273.
53. Ejike C.E.C.C., Gong M., Udenigwe C.C. Phytoalexins from the Poaceae: Biosynthesis, function and prospects in food preservation. Food Res. Int. 2013;52:167–177.
54. Bell A.A. Formation of gossypol in infected or chemically irritated tissues of *Gossypium* species. Phytopathology. 1967;57:759–764.
55. Browne L.M., Conn K.L., Ayert W.A., Tewari J.P. The camalexins: New phytoalexins produced in the leaves of *Camelia sativa* (Cruciferae) Tetrahedron. 1991;47:3909–3914.
56. Langcake P., Pryce R.J. The production of resveratrol by *Vitis vinifera* and other members of the *Vitaceae*as a response to infection or injury. Physiol. Plant Pathol. 1976;9:77–86.
57. Pérez-Clemente R.M., Vives V., Zandalinas S.I., Lopez-Climent M.F., Munoz V., Gomez-Cadenas A. Biotechnological approaches to study plant responses to stress. BioMed. Res. Int. 2013 doi: 10.1155/2013/654120.
58. Grosskinsky D.K., van der Graaff E., Roitsch T. Phytoalexin transgenics in crop protection-Fairy tale with a happy end? Plant Sci. 2012;195:54–70.
59. Halls C., Yu O. Potential for metabolic engineering of resveratrol biosynthesis. Trends Biotechnol.2008;26:77–81.
60. Delaunois B., Cordelier S., Conreux A., Clément C., Jeandet P. Molecular engineering of resveratrol in plants. Plant Biotechnol. J. 2009;7:2–12.
61. Donnez D., Jeandet P., Clément C., Courot E. Bioproduction of resveratrol and stilbene derivatives by plant cells and microorganisms. Trends Biotechnol. 2009;27:706–713.
62. Jeandet P., Delaunois B., Conreux A., Donnez D., Nuzzo V., Cordelier S., Clément C., Courot E. Biosynthesis, metabolism, molecular engineering and biological functions of stilbene phytoalexins in plants. BioFactors. 2010;36:331–341.
63. Jeandet P., Delaunois B., Aziz A., Donnez D., Vasserot Y., Cordelier S., Courot E. Metabolic engineering of yeast and plants for the production of the biologically active hydroxystilbene, resveratrol. J. Biomed. Biotechnol. 2012 doi: 10.1155/2012/579089.
64. Jeandet P., Vasserot Y., Chastang T., Courot E. Engineering microbial cells for the biosynthesis of natural compounds of pharmaceutical significance. BioMed. Res. Int. 2013 doi: 10.1155/2013/780145.
65. Hain R., Reif H.J., Krause E., Langebartels R., Kindl H., Vornam B., Wiese W., Schmelzer E., Schreier P., Stöcker R., Stenzel K. Disease resistance results from foreign phytoalexin expression in a novel plant. Nature. 1993;361:153–156.
66. Hipskind J.D., Paiva N.L. Constitutive accumulation of a resveratrol glucoside in transgenic alfalfa increases resistance to *Phoma medicaginis*. Mol. Plant-Microbe Interact. 2000;13:551–562.
67. Coutos-Thévenot P., Poinssot B., Bonomelli A., Yean H., Breda C., Buffard D., Esnault R., Hain R., Boulay M. *In vitro* tolerance to *Botrytis cinerea* of grapevine 41B rootstock in transgenic plants expressing the stilbene synthase *Vst 1* gene under the control of a pathogen-inducible PR 10 promoter. J. Exp. Bot.2001;52:901–910.
68. Stark-Lorenzen P., Nelke B., Hänbler G., Mühlbach H.P., Thomzik J.E. Transfer of a grapevine stilbene synthase gene to rice (*Oryza sativa* L.) Plant Cell Rep. 1997;16:668–673.
69. Leckband G., Lörz H. Transformation and expression of a stilbene synthase gene of *Vitis vinifera* L. in barley and wheat for increased fungal resistance. Theor. Appl. Genet. 1998;96:1004–1012.
70. Liang H., Zheng J., Shuange J.I.A., Wang D., Ouyang J., Li J., Li L., Tian W., Jia X., Duan X., Sheng B., Hain R. A transgenic wheat with a stilbene synthase gene resistant to powdery mildew obtained by biolistic method. Chin. Sci. Bull. 2000;45:634–638.
71. Thomzik J.E., Stenzel K., Stöcker R., Schreier P.H., Hain R., Stahl D.J. Synthesis of a grapevine phytoalexin in transgenic tomatoes (*Lycopersicon esculentum* Mill.) conditions resistance against*Phytophthora infestans*. Physiol. Mol. Plant Pathol. 1997;51:265–278.
72. Serazetdinova L., Oldach K., Lörz H. Expression of transgenic stilbene synthases in wheat causes the accumulation of unknown stilbene derivatives with antifungal activity. J. Plant Physiol. 2005;162:985–1002.
73. Zhu Y.J., Agbayani R., Jazckson M.C., Tang C.S., Moore P.M. Expression of the grapevine stilbene synthase gene *VST1* in papaya provides increased resistance against diseaes caused by *Phytophthora palmivora*. Planta. 2004;12:807–812.
74. Lim J.D., Yun S.J., Chung I.M., Yu C.Y. Resveratrol synthase transgene expression and accumulation of resveratrol glycoside in *Rehmannia glutinosa*. Mol. Breed. 2005;16:219–233.
75. Liu Z.Y., Zhuang C.X., Sheng S.J. Overexpression of a resveratrol synthase gene (*PcRs*) from*Polygonum cuspidatum* in transgenic *Arabidopsis* causes

the accumulation of *trans*-piceid with antifungal activity. Plant Cell Rep. 2011;30:2027–2036.

76. Deavours B.E., Dixon R.A. Metabolic engineering of isoflavonoid biosynthesis in alfalfa. Plant Physiol.2005;138:2245–2259.

77. Kaimoyo E., VanEtten H.D. Inactivation of pea genes by RNAi supports the involvement of two similar*O*-methyltransferases in the biosynthesis of (+)-pisatin and of chiral intermediates with a configuration opposite that found in (+)-pisatin. Phytochemistry. 2008;69:76–87.

78. Giorcelli A., Sparvoli F., Mattivi F., Tava A., Balestrazzi A., Vrhovsek U., Calligari P., Bollini R., Confalonieri M. Expression of the stilbene synthase (StSy) gene from grapevine in transgenic white poplar results in high accumulation of the antioxidant resveratrol glucosides. Transgenic Res. 2004;13:203–214.

79. Seppänen S.K., Syrjälä L., von Weissenberg K., Teeri T.H., Paajanen L., Pappinen A. Antifungal activity of stilbenes *in vitro* bioassays and in transgenic *Populus* expressing a gene encoding pinosylvin synthase. Plant Cell Rep. 2004;22:584–593.

80. Kobayashi S., Ding C.K., Nakamura Y., Nakajima I., Matsumoto R. Kiwifruits (*Actinidia deliciosa*) transformed with a *Vitis* stilbene synthase gene produce piceid (resveratrol-glucoside) Plant Cell Rep.2000;19:904–910.

81. Fischer R., Budde I., Hain R. Stilbene synthase gene expression causes changes in flower colour and male sterility in tobacco. Plant J. 1997;11:489–498.

82. Höfig K.P., Möller R., Donaldson L., Putterill J., Walter C. Towards male sterility in *Pinus radiata*—A stilbene synthase approach to genetically engineer nuclear male sterility. Plant Biotechnol. J. 2006;4:333–343.

83. He X.Z., Dixon R.A. Genetic manipulation of isoflavone 7-*O*-methyltransferase enhances biosynthesis of 4′-*O*-methylated isoflavonoid phytoalexins and disease resistance in alfalfa. Plant Cell. 2000;12:1689–1702.

84. Blount J.W., Dixon R.A., Paiva N.L. Stress responses in alfalfa (*Medicago sativa* L.) XVI. Antifungal activity of medicarpin and its biosynthetic precursors; implications for the genetic manipulation of stress metabolites. Physiol. Mol. Plant Pathol. 1993;41:333–349.

85. Wu Q., VanEtten H.D. Introduction of plant and fungal genes into pea (*Pisum sativum* L.) hairy roots reduces their ability to produce pisatin and affects their response to a fungal pathogen. Mol. Plant-Microbe Interact. 2004;17:798–804.

86. Graham T.L., Graham M.Y., Subramanian S., Yu O. RNAi silencing of genes for elicitation or biosynthesis of 5-deoxyisoflavonoids suppresses race-specific resistance and hypersensitive cell death in*Phytophthora sojae* infected tissues. Plant Physiol. 2007;144:728–740.

87. Ibraheem F., Gaffoor I., Chopra S. Flavonoid phytoalexin-dependent resistance to anthracnose leaf blight requires a functional *yellow seed1* in *Sorghum bicolor*. Genetics. 2010;184:915–926.

88. Shih C.H., Chu I.K., Yip W.K., Lo C.S. Differential expression of two flavonoid 3’-hydroxylase cDNAs involved in biosynthesis of anthocyanin pigments and 3-deoxyanthocyanidin phytoalexins in sorghum. Plant Cell Physiol. 2006;47:1412–1419.

89. Sharma M., Chai C., Morohashi K., Grotewold E., Snook M.E., Chopra S. Expression of flavonoid 3′-hydroxylase is controlled by P1, the regulator of 3-deoxyflavonoid biosynthesis in maize. BMC Plant Biol.2012;12:196.

90. Boddu J., Svabek C., Ibraheem F., Jones A.D., Chopra S. Characterization of a deletion allele of a sorghum Myb gene, *yellow seed1* showing loss of 3-deoxyflavonoids. Plant Sci. 2005;169:542–552.

91. Boddu J., Jiang C., Sangar V., Olson T., Peterson T., Chopra S. Comparative structural and functional characterization of sorghum and maize duplications containing orthologous myb transcription regulators of 3-deoxyflavonoid biosynthesis. Plant Mol. Biol. 2006;60:185–199.

92. Glazebrook J., Ausubel F.M. Isolation of phytoalexin-deficient mutants of *Arabidopsis thaliana* and characterization of their interactions with bacterial pathogens. Proc. Natl. Acad. Sci. USA. 1994;91:8955–8959.

93. Glazebrook J., Zook M., Mert F., Kagan I., Rogers E.E., Crute I., Holub E., Hammerschmidt R., Ausubel F.M. Phytoalexin-deficient mutants of *Arabidopsis* reveal that *PAD4* encodes a regulatory factor and that four PAD genes contribute to downy mildew resistance. Genetics. 1997;146:381–392.

94. Reuber T.L., Plotnikova J.M., Dewdnev J., Rogers E.E., Wood W., Ausubel F.M. Correlation of defense gene induction defects with powdery mildew susceptibility in *Arabidopsis* enhanced disease susceptibility mutants. Plant J. 1998;16:473–485.

95. Ahl-Goy P., Signer H., Reist R., Aichholz R., Blum B., Schmidt E., Kessmann H. Accumulation of scopoletin is associated with the high disease resistance of the hybrid *Nicotiana tabacum* × *Nicotiana debneyi*. Planta. 1993;191:200–206.

96. Gnonlonfin G.J.B., Sanni A., Brimer L. Scopoletin—A coumarin phytoalexin with medicinal properties.Crit. Rev. Plant Sci. 2012;31:47–56.

97. Li Y., Baldauf S., Lim E.K., Bowles J.D. Phylogenetic analysis of the UDP-glycosyltransferase multigene family of *Arabidopsis thaliana*. J. Biol. Chem. 2001;276:4338–4343.

98. Chong J., Baltz R., Schmitt C., Beffa R., Fritig B., Saindrenan P. Downregulation of a pathogen-responsive tobacco UDP-Glc:phenylpropanoid glucosyltransferase reduces scopoletin glucoside accumulation, enhances oxidative stress, and weakens virus resistance. Plant Cell. 2002;14:1093–1107.

99. Gachon C., Baltz R., Saindrenan P. Over-expression of a scopoletin glucosyltransferase in *Nicotiana tabacum* leads to precocious lesion formation during the hypersensitive response to tobacco mosaic virus but does not affect virus resistance. Plant Mol. Biol. 2004;54:137–146.

100. Grosskinsky D.K., Naseem M., Abdelmoshem U.A., Plickert N., Engelke T., Griebel T., Zeier J., Novak O., Strand M., Pfeifhofer H., et al. Cytokinins mediate resistance against *Pseudomonas syringae* in tobacco through increased antimicrobial phytoalexin synthesis independent of salicylic acid signaling. Plant Physiol. 2011;157:815–830.

101. Mauch-Mani B., Mauch F. The role of abscissic acid in plant-pathogen interactions. Curr. Opin. Plant Biol. 2005;8:409–414.

102. Jones J.D., Dangles J.L. The plant immune system. Nature. 2006;444:323–329.

103. Robert-Seilaniantz A., Navarro L., Bari R., Jones J.D. Pathological hormone imbalances. Curr. Opin. Plant Biol. 2007;10:372–379.

104. Asselbergh B., de Vleesschauwer D., Höfte M. Global switches and fine-tuning-ABA modulates plant pathogen defense. Mol. Plant-Microbe Interact. 2008;21:709–719.

105. Lopez M.A., Bannenberg G., Castresana C. Controlling hormone signaling is a plant and pathogen challenge for growth and survival. Curr. Opin. Plant Biol. 2008;11:420–427.

106. Spoel S.H., Dong X. Making sense of hormone crosstalk during plant immune responses. Cell Host Microbe. 2008;3:348–351.

107. Bari R., Jones J.D. Role of plant hormones in plant defence responses. Plant Mol. Biol. 2009;69:473–488.

108. Verhage A., van Wees S.C.M., Pieterse C.M.J. Plant immunity: It's the hormones talking, but what do they say? Plant Physiol. 2010;154:536–540.

109. Erb M., Meldau S., Howe G.A. Role of phytohormones in insect-specific plant reactions. Trends Plant Sci. 2012;17:250–259.

110. Navarro L., Dunoyer P., Jay F., Arnold B., Dharmasiri N., Estelle M., Voinnet O., Jones J.D.G. A plant miRNA contributes to antibacterial resistance by repressing auxin signaling. Science. 2006;312:436–439.

111. Domingo C., Andrès F., Tharreau D., Iglesias D.J., Talon M. Constitutive expression of *OsGH3.1*reduces auxin content and enhances defense responses and resistance to a fungal pathogen in rice. Mol. Plant-Microbe Interact. 2009;22:201–210.

112. Mayda E., Marqués M.C., Conejero V., Vera P. Expression of a pathogen-induced gene can be mimicked by auxin insensitivity. Mol. Plant-Microbe Interact. 2000;13:23–31.

113. Padmanabhan M.S., Goregaoker S.P., Golem S., Shiferaw H., Culver J.N. Interaction of the tobacco mosaic virus replicase protein with the Aux/IAA protein PAP1/IAA26 is associated with disease development. J. Virol. 2005;79:2549–2558.

114. Padmanabhan M.S., Shiferaw H., Culver J.N. The tobacco mosaic virus replicase protein disrupts the localization and function of interacting Aux/IAA proteins. Mol. Plant-Microbe Interact. 2006;19:864–873.

115. Ding X., Cao Y., Huang L., Zhao J., Xu C., Li X., Wang S. Activation of the indole-3-acetic acid-amido synthetase GH3-8 suppresses expansin expression and promotes salicylate- and jasmonate- independent basal immunity in rice. Plant Cell. 2008;20:228–240.

116. Iriti M., Faro F. Chemical diversity and defence metabolism: How plant cope with pathogens and ozone pollution. Int. J. Mol. Sci. 2009;10:3371–3399.

117. Ristaino J.B., Duniway J.M. Effect of preinoculation and postinoculation water-stress on the severity of*Phytophthora* root-rot in processing tomatoes. Plant Dis. 1989;73:349–352.

118. Audenaert K., de Meyer G.B., Höfte M.M. Abscisic acid determines basal susceptibility of tomato to*Botrytis cinerea* and suppresses salicylic acid-dependent signaling mechanisms. Plant Physiol.2002;128:491–501.

119. Asselbergh B., Achuo A.E., Höfte M., van Breusegem F. Abscisic acid deficiency leads to rapid activation of tomato defence responses upon infection with *Erwinia chrysanthemi*. Mol. Plant Pathol.2008;9:11–24.

120. Balazs E., Gaborjanyi R., Kiraly Z. Leaf senescence and increased virus susceptibility in tobacco: The effect of abscisic acid. Physiol. Mol. Plant Pathol. 1973;3:341–346.

121. Rezzonico E., Flury N., Meins F., Jr, Beffa R. transcriptional down-regulation by abscisic acid of pathogenesis-related beta-1,3-glucanase genes in tobacco cell cultures. Plant Physiol. 1998;117:585–592.

122. Goosens J.F.V., Vendrig J.C. effects of abscissic acid, cytokinins, and light on isoflavonoid phytoalexin accumulation in *Phaseolus vulgaris*. Planta. 1982;154:441–446.

123. Ward E.W., Cahill D.M., Bhattacharyya M.K. Abscisic acid suppression of phenylalanine ammonia-lyase activity and mRNA, and resistance of soybeans to *Phytophthora megasperma* f.s.p. glycinea. Plant Physiol. 1989;91:23–27.

124. Mohr P., Cahill D.M. Relative roles of glyceollin, lignin and the hypersensitive response and the influence of ABA in compatible and incompatible interactions of soybeans with *Phytophthora sojae*.Physiol. Mol. Plant Pathol. 2001;58:31–41.

125. Henfling J.W.D.M., Bostock R.M., Kuc J. Effect of abscisic acid on rishitin and lubimin accumulation and resistance to *Phytophthora infestans* and *Cladosporium cucumerinum* in potato tuber tissue slices. Phytopathology. 1980;70:1074–1078.

126. Mialoundama A.S., Heintz D., Debayle D., Rahier A., Camara B., Bouvier F. Abscisic acid negatively regulates elicitor-induced synthesis of capsidiol in wild tobacco. Plant Physiol. 2009;150:1556–1566.

127. Thomas S.G., Phillips A.L., Hedden P. Molecular cloning and functional expression of gibberellin 2-oxidases, multifunctional enzymes involved in gibberellin deactivation. Proc. Natl. Acad. Sci. USA.1999;96:4698–4703.

128. Yang D.L., Li Q., Deng Y.W., Lou Y.G., Wang M.Y., Zhou G.X., Zhang Y.Y., He Z.H. Altered disease development in the *eui* mutants and *Eui* overexpressors indicates that gibberellins negatively regulate rice basal disease resistance. Mol. Plant. 2008;1:528–537.

129. Sakakibara H. Cytokinins: Activity, biosynthesis, and translocation. Annu. Rev. Plant Biol.2006;57:431–449.

130. Sano H., Seo S., Koizumi N., Niki T., Iwamura H., Ohashi Y. Regulation by cytokinins of endogenous levels of jasmonic and salicylic acids in mechanically wounded tobacco plants. Plant Cell Physiol.1996;37:762–769.

131. Pogany M., Koehl J., Heiser I., Elstner E., Barna B. Juvenility of tobacco induced by cytokinin gene introduction decreases susceptibility to *Tobacco necrosis* virus and confers tolerance to oxidative stress. Physiol. Mol. Plant Pathol. 2004;65:39–47.

132. Barna B., Smigocki A.C., Baker J.C. Transgenic production of cytokinin suppresses bacterially induced hypersensitive response symptoms and increases antioxidative enzyme levels in *Nicotiana* spp. Phytopathology. 2008;98:1242–1247.

133. Choi J., Huh S.U., Kojima M., Sakakibara H., Paek K.H., Hwang I. The cytokinin-activated transcription factor A RR2 promotes plant immunity via TGA3/NPR1-dependent salicylic acid signaling in*Arabidopsis*. Dev. Cell. 2010;19:284–295.

134. Boller T., Felix G. A renaissance of elicitors: Perception of microbe-associated molecular patterns and danger signals by pattern-recognition receptors. Annu. Rev. Plant Biol. 2009;60:379–406.

135. Dodds P.N., Rathjen J.P. Plant mmunity: Towards an integrated view of plant-pathogen interactions.Nat. Rev. Genet. 2010;11:539–548.

136. Newmann M.A., Sundelin T., Nielsen J.T., Erbs G. MAMP (microbe-associated molecular pattern) triggered immunity in plants. Front. Plant Sci. 2013;4:139.

137. Pitzschke A., Schikora A., Hirt H. MAPK cascade signaling networks in plant defence. Curr. Opin. Plant Biol. 2009;12:421–426.

138. Garcia-Brugger A., Lamotte O., Vandelle E., Bourque S., Lecourieux D., Poinssot B., Wendehenne D., Pugin A. Early signaling events induced by elicitors of plant defenses. Mol. Plant-Microbe Interact.2006;19:711–724.

139. Zhao J., Last R.L. Coordinate regulation of the tryptophan biosynthetic pathway and indolic phytoalexin accumulation in *Arabidopsis*. Plant Cell. 1996;8:2235–2244.

140. Su T., Xu J., Li Y., Lei L., Zhao L., Yang H., Feng J., Liu G., Ren D. Glutathione-indole-3-acetonitrile is required for camalexin biosynthesis in *Arabidopsis thaliana*. Plant Cell. 2011;23:364–380.

141. Kim J.A., Cho K., Singh R., Jung Y.H., Jeong S.H., Kim S.H., Lee J.E., Cho Y.S., Agrawal G.K., Rakwal R., et al. Rice *OsACDR1* (*Oryza sativa* accelerated cell death and resistance 1) is a potential positive regulator of fungal disease resistance. Mol. Cells. 2009;28:431–439.

142. Grant M., Lamb C. Systemic immunity. Curr. Opin. Plant Biol. 2006;9:414–420.

143. Cao H., Li X., Dong X. Generation of broad-spectrum disease resistance by overexpression of an essential regulatory gene in systemic acquired resistance. Proc. Natl. Acad. Sci. USA. 1998;95:6531–6536.

144. Quilis J., Penas G., Messeguer J., Brugidou C., San Segundo B. The Arabidopsis AtNPR1 inversely modulates defense responses against fungal, bacterial, or viral pathogens while conferring hypersensitivity to abiotic stresses in transgenic rice. Mol. Plant-Microbe Interact. 2008;21:1215–1231.

145. Lin W.C., Lu C.F., Wu J.W., Cheng M.L., Lin Y.M., Yang N.S., Black L., Green S.K., Wang J.F., Cheng C.P. Transgenic tomato plants expressing the *Arabidopsis* NPR1 gene display enhanced resistance to a spectrum of fungal and bacterial diseases. Transgenic Res. 2004;13:567–581.

146. Makandar R., Essig J.S., Schapaugh M.A., Trick H.N., Shah J. Genetically engineered resistance to Fusarium head blight in wheat by expression of *Arabidopsis* NPR1. Mol. Plant-Microbe Interact.2006;19:123–129.

147. Wally O., Jayaraj J., Punja Z.K. Broad-spectrum disease resistance to necrotrophic and biotrophic pathogens in trnsgnic carrots (*Daucus carota* L.) expressing an *Arabidopsis NPR1* gene. Planta.2009;231:131–141.

148. Parkhi V., Kumar V., Campbell L.M., Bell A.A., Shah J., Rathore K.S. Resistance against various fungal pathogens and reniform nematode in transgenic cotton plants expressing *Arabidopsis NRP1*.Transgenic Res. 2010;19:959–975.

149. Berken A. ROPs in the spotlight of plant signal transduction. Cell. Mol. Life Sci. 2006;63:2446–2459.

150. Ono E., Wong H.L., Kawasaki T., Hasegawa M., Kodama O., Shimamoto K. Essential role of the small GTPase Rac in disease resistance of rice. Proc. Natl. Acad. Sci. USA. 2001;98:759–764.

151. Kawano Y., Akamatsu A., Hayashi K., Housen Y., Okuda J., Yao A., Nakashima A., Takahashi H., Yoshida H., Wong H.L., et al. Activation of a Rac GTPase by the NLR family disease resistance protein pit plays a critical role in rice innate immunity. Cell Host Microbe. 2010;7:362–375.

152. Chen L., Shiotani K., Togashi T., Miki D., Aoyama M., Wong H.L., Kawasaki T., Shimamoto K. Analysis of the Rac/Rop small GTPase family in rice: Expression, subcellular localization and role in disease resistance. Plant Cell Physiol. 2010;51:85–95.

153. Sadawa K., Hasegawa M., Tokuda L., Kameyama J., Kodama O., Kohchi T., Yoshida K., Shinmyo A. Enhanced resistance to blast fungus and bacterial blight in transgenic rice constitutively expressing *OsSBP*, a rice homologue of mamalian selenium-binding proteins. Biosci. Biotechnol. Biochem. 2004;68:873–880.

154. Flemetakis E., Agalou A., Kavroulakis N., Dimou M., Martsikovskaya A., Slater A., Spaink H.P., Roussis A., Katinakis P. *Lotus japonicus* gene *Ljsbp* is highly conserved among plants and animals and encodes a homologue to the mamalian selenium-binding proteins. Mol. Plant-Microbe Interact. 2002;15:313–322.

155. Sawada K., Tokuda L., Shinmyo A. Characterization of the rice blast fungal elicitor-responsive gene*OsSBP* encoding a homolog to the mamalian selenium-binding proteins. Plant Biotechnol. 2003;20:179–183.

156. Dodd A.N., Kudla J., Sanders D. The language of calcium signalling. Annu. Rev. Plant Biol.2010;61:593–620.

157. Reddy A.S., Ali G.S., Celesnik H., Day I.S. Coping with stresses: Roles of calcium- and calcium/calmodulin-regulated gene expression. Plant Cell. 2011;23:2010–2032.

158. Romeis T., Ludwig A., Martin R., Jones J.D.G. Calcium-dependent protein kinases play an essential role in a plant defence response. EMBO J. 2001;20:5556–5567.

159. Luan S. The CBL-CIPK network in plant calcium signaling. Trends Plant Sci. 2009;14:37–42.

160. Kurusu T., Hamada J., Nokajima H., Kitagawa Y., Kiyoduka M., Takahashi A., Hanamata S., Ohno R., Hayashi T., Okada K., et al. Regulation of microbe-associated molecular pattern-induced hypersensitive cell death, phytoalexin production, and defense gene expression by calcineurin B-like protein-interacting protein kinases, OsCIPK14/15, in rice cultured cells. Plant Physiol. 2010;153:678–692.

161. Okada K. The biosynthesis of isoprenoids and the mechanisms regulating it in plants. Biosci. Biotechnol. Biochem. 2011;75:1219–1225.

162. Rauhut T., Luberacki B., Seitz H.U., Glawischnig E. Inducible expression of a Nep1-like protein serves as a model trigger system of camalexin biosynthesis. Phytochemistry. 2009;70:185–189.

163. Pemberton C.L., Salmond G.P.C. The Nep1-like proteins—A growing family of microbial elicitors of plant necrosis. Mol. Plant Pathol. 2004;5:353–359.

164. Qutob D., Kemmerling B., Brunner F., Küfner L., Engelhardt S., Gust A.A., Luberacki B., Seitz H.U., Stahl D., Rauhut T., et al. Phytotoxicity and innate immune responses induced by Nep1-like proteins. Plant Cell. 2006;18:3721–3744.

165. Mori M., Tomita C., Sugimoto K., Hasegawa M., Hayashi N., Dubouzet J.G., Ochiai H., Sekimoto H., Hirochika H., Kikuchi S. Isolation and molecular characterization of a *Spotted leaf 18* mutant by modified activation-tagging in rice. Plant Mol. Biol. 2007;63:847–860.

166. Kachroo A., Lapchyk L., Fukushige H., Hildebrand D., Klessig D., Kachroo P. Plastidial fatty acid signaling modulates salicylic acid-and jasmonic acid-mediated defense pathways in the *Arabidopsis* ssi2 mutant. Plant Cell. 2003;15:2952–2965.

167. Kachroo A., Venugopal S.C., Lapchyk L., Falcone D., Hildebrand D., Kachroo P. Oleic acid levels regulated by glycerolipid metabolism modulate defense gene expression in *Arabidopsis*. Proc. Natl. Acad. Sci. USA. 2004;101:5152–5157.

168. Matthews D., Jones H., Gans P., Coates S., Smith L.M.J. Toxic secondary metabolite production in genetically modified potatoes in response to stress. J. Agric. Food Chem. 2005;53:7766–7776.

169. Benedict C.R., Lu J.L., Pettigrew D.W., Liu J., Stipanovic R.D., Williams H.J. The cyclization of farnesyl diphosphate and nerolidyl diphosphate by a purified recombinant δ-cadinene synthase. Plant Physiol. 2001;125:1754–1765.

170. Martin G.S., Liu J., Benedict C.R., Stipanovic R.D., Magill W.M. Reduced levels of cadinane sesquiterpenoids in cotton plants expressing antisense (+)-δ-cadinene synthase. Phytochemistry. 2003;62:31–38.

171. Chen X.Y., Chen Y., Heinstein P., Davisson V.J. Cloning, expression, and characterization of (+)-δ-cadinene synthase: A catalyst for cotton phytoalexin biosynthesis. Arch. Biochem. Biophys. 1995;324:255–266.

172. Chen X.Y., Wang M., Chen Y., Davisson V.J., Heinstein P. Cloning and heterologous expression, and characterization of a second (+)-δ-cadinene synthase from *Gossypium arboreum*. J. Nat. Prod. 1996;59:944–951.

173. Kalani R., Gamboa D.A., Calhoun M.C., Haq A.U., Bailey C.A. Relative toxicity of cottonseed gossypol enantiomers in broilers. Open Toxicol. J. 2010;4:26–31.

174. Townsend B.J., Poole A., Blake C.J., Llewellyn D.J. Antisense suppression of a (+)-δ-cadinene synthase gene in cotton prevents the induction of this defense response gene during blight infection but not its constitutive expression. Plant Physiol. 2005;138:516–528.

175. Benedict C.R., Martin G.S., Liu J., Puckhaber L., Magill C.W. Terpenoid aldehyde formation and lysigenous gland storage sites in cotton: Variant with mature glands but suppressed levels of terpenoid aldehydes. Phytochemistry. 2004;65:1351–1359.

176. Sunilkumar G., Campbell L.M., Pukhaber L., Stipanovic R.D., Rathore K.S. Engineering cottonseed for use in human nutrition by tissue-specific reduction of toxic gossypol. Proc. Natl Acad. Sci. USA.2006;103:18054–18059.

177. Sunilkumar G., Connell J.P., Smith C.W., Reddy A.S., Rathore K.S. Cotton α-globulin promoter: Isolation and functional characterization in transgenic cotton, *Arabidospsis*, and tobacco. Transgenic Res.2002;11:347–359.

178. Rathore K.S., Sunilkumar G., Campbell L.M. Cotton plant with seed specific reduction in gossypol. US Patent N° 20070199098. 2007 Feb 16;

CHAPTER 10

N-GLYCOSYLATION MODIFICATION OF PLANT-DERIVED VIRUS-LIKE PARTICLES: AN APPLICATION IN VACCINES

Hyun-Soon Kim[1], Jae-Heung Jeon[1], Kyung Jin Lee[2] and Kisung Ko[2]

[1] Plant Systems Engineering Research Center, KRIBB, 125 Gwahangno, Yuseong-gu, Daejeon 305-806, Republic of Korea

[2] Department of Medicine, College of Medicine, Chung-Ang University, Seoul 156-756, Republic of Korea

ABSTRACT

Plants have been developed as an alternative system to mammalian cells for production of recombinant prophylactic or therapeutic proteins for human and animal use. Effective plant expression systems for recombinant proteins have been established with the optimal combination of gene expression regulatory elements and control of posttranslational processing of recombinant glycoproteins. In plant, virus-like particles (VLPs), viral "empty shells" which maintain the same structural characteristics of virions but are genome-free, are considered extremely promising as vaccine platforms and therapeutic delivery systems. Unlike microbial fermentation, plants are capable of carrying out N-glycosylation as a posttranslational modification of glycoproteins. Recent advances in the glycoengineering in plant allow human-like glycomodification and optimization of desired glycan structures for enhancing safety and functionality of recombinant pharmaceutical glycoproteins. In this review, the current plant-derived VLP approaches are focused, and N-glycosylation and its in planta modifications are discussed.

PLANT-DERIVED VIRUS-LIKE PARTICLE (VLP)

Viruses are able to form the quaternary structure of viral capsids through molecular self-assembly of repetitive building blocks [1, 2]. Plant viruses can be easily multiplied, which are structurally uniform, robust, and biodegradable with a size particularly suitable for nanoscale applications. Virus-like particles (VLPs) are multimeric self-assembled protein complexes mimicking the organization and conformation of native viruses but lack the viral genome making them replication-deficient and noninfectious [3]. VLPs consist of protein shells (termed as capsids), and the capsids are typically composed of identical coat protein subunits. Peptide-based vaccines are in general poorly immunogenic and for this reason they require multiple injections and adjuvants in order to increase their effectiveness. VLPs offer a promising approach to the production of vaccines against many diseases, because their repetitive, high density display of epitopes is potentially highly effective in eliciting strong immune responses [4]. VLPs lacking viral nucleic acid are noninfectious. Nevertheless, they are self-assembled protein structures mimicking infectious viruses and thus constitute a safe and effective approach for the induction of neutralizing antibodies to surface proteins, where soluble forms of their protein subunits have failed. It has been also reported that viral structures are regarded as a vaccine platform to display foreign epitopes [5].

Citation: Hyun-Soon Kim, Jae-Heung Jeon, Kyung Jin Lee, and Kisung Ko, N-Glycosylation Modification of Plant-Derived Virus-Like Particles: An Application in Vaccines, http://dx.doi.org/10.1155/2014/249519.

In general, bacteria, yeast, insect, and animal cells have been applied as cell-based systems to produce VLPs. The bacterial cell cultures have been explored as a VLP production platform with advantages in terms of scalability and production cost [6]. However, bacteria are prokaryotes which lack glycosylation process essential for proper immunogenicity and antigen stability when VLPs are applied as vaccines. In contrast to bacteria, yeast cells have glycosylation apparatus [7]. However, their glycoforms are mainly high mannose type, which is not desirable for the most therapeutic glycoproteins [8]. The matured glycoforms in baculovirus-insect cell system also are mainly high mannose type [9]. The glycosylation of envelope proteins affects their folding and thus is essential for formation and immunogenicity of VLPs [10, 11]. In glycosylation process, bacteria, yeast, and insect cells have fundamental limitations. The mammalian cells have proper glycosylation apparatus and ability to fold the envelope proteins of virus, which facilitate functional VLPs production. However, the mammalian cell-based systems require manufacturing facilities including fermentation bioreactors for large-scale upstream processing, which is too expensive to establish. This high production cost is a major disadvantage of the mammalian cell-based system. Plants do not need such expensive facilities to produce biomass. Thus, plants are considered as a potential bioreactor system for VLPs with advantages such as low cost of upstream biomass process, flexible production scalability, and the lack of human pathogen contaminants [12, 13]. Nevertheless, plants for VLP production platform are not perfectly acceptable due to relatively lower VLP production level than animal systems and plant-specific *N*-glycosylation of glycoproteins [14, 15]. Development of new plant expression system and advanced *N*-glycosylation engineering overcome such hurdles.

VIRUS-LIKE PARTICLES IN PLANT EXPRESSION SYSTEMS

VLPs can be generated through different types of viral vectors and expression strategies in plants [16, 17]. The plant-derived viral vectors used for VLP expression can be classified into full virus vectors such as the potato virus X (PVX) [18, 19] and the cowpea mosaic virus (CPMV) [20] and the deconstructed vectors such as bean yellow dwarf virus (BeYDV) [21, 22] and MagniCON based on tobacco mosaic virus (TMV) [23, 24]. The earlier plant VLPs were Hepatitis B core antigen (HBcAg) VLPs [25] and Hepatitis B surface antigen (HBsAg) VLPs fused to soybean vegetable storage protein vspA (VSPαS) in transgenic tobacco leaves obtained by *Agrobacterium*-mediated transformation [25, 26]. PVX and CPMV based viral vectors were applied to generate HBcAg VLPs [27]. Transgenic plants using *Agrobacterium*-mediated DNA transfer have been used for the stable gene expression system for VLPs; however the VLP expression level is low [10~24 μg/g fresh leaf weight (FLW)] [27, 28]. The human papilloma virus (HPV) L-1 based VLP vaccines were also successfully expressed and assembled in stable transgenic potato and tobacco [29–31]. Unlike stable transformation, the transient expression with deconstructed geminiviral vectors showed 80 times higher accumulation of HBcAg VLPs in *N. benthamiana* compared to both PVX and CPMV vectors [32]. Another transient expression system using deconstructed TMV-based MagnICON vector agroinfiltration has allowed the production of HBcAg VLPs with a yield of 2.38 mg/g (FLW), almost 3 times higher compared to the geminiviral vector within a short period time [17, 32–35]. The intraperitoneally injected HBcAg VLPs obtained from the MagnICON system efficiently induced immune responses generating HBcAg specific IgGs in mice. These results indicate that, among currently available VLP expression systems, the production of large quantities of VLPs for vaccine applications is more feasible using MagniCON systems. In many cases genetic manipulation of plant-derived VLPs has been performed to modify the external surface of the particle. To this end, the heterologous polypeptide has been fused at the N- or C-terminus of the CP. VLPs can also be exploited as "platforms" for the presentation of foreign epitopes and/or targeting molecules on chimeric VLPs (cVLPs) [1, 2, 6]. Indeed, the VLPs can display multicomponent vaccine candidate epitopes as a fusion form between two different proteins [5]. For instance, the green fluorescent protein (GFP) and the HB surface antigen (HBsAg) S-protein were transiently expressed and heterodimerized with the native HBsAg sequentially forming chimeric VLPs (cVLPs) in *N. benthamiana* [36]. The HBsAg fusion with GFP was showed to be more stable and immunogenic than native HBsAg in *in vivo* mice experiment, indicating that cVLPs can be applied to display heterologous antigens to generate more immunogenic vaccines [5]. The fusion proteins between domain III (DIII) of West Nile virus (WNV) and HBcAg were expressed and displayed as cVLPs with geminiviral transient expression vectors in *N. benthamiana* [37]. In addition, the influenza virus M2 epitope [38] or HPV16 epitopes [27, 39, 40] individually were fused to HBcAg induced strong immune responses generating specific antibodies. The cVLPs displaying both HPV16 E6 and E7 proteins triggered their specific antibodies, respectively [39]. In general, vaccines are administered through intramuscular, subcutaneous, and intravenous injections. In addition, vaccines can be orally or nasally applied to induce mucosal immune responses [17]. Indeed, various results indicate that VLPs can be applied safely as oral vaccines carrying multiple epitopes without needle injection. For example, oral delivery of purified Norwalk virus CP (NVCP) VLPs produced in tobacco and tomato stimulated mucosal and serum immune responses to produce IgA and IgG [41] and oral administration with HBsAg displaying HIV-1 ENV and GAG epitopes provoked strong serum and mucosal antibody responses in mice [42]. These results indicate that VLPs can be applied safely as oral vaccines carrying multiple epitopes without needle injection.

GLYCOSYLATION OF VLP VACCINES

Even though virus-like particles- (VLPs-) based vaccines have shown promising results, commercial production systems are currently limited to eukaryotic cells such as yeast, insect, and mammalian [14]. For instance, Lassa virus (LASV) VLPs cannot be easily produced in bacterial cell systems, because bacteria are incapable of performing glycosylation and other posttranslational protein modifications which are a key feature in most VLP-based proteins [14]. The glycosylation pattern of GP1 and GP2 glycoproteins of Lassa virus (LASV) has been shown to play a critical structural and functional role in preserving protein stability and allowing binding and fusion to host cells [43]. The glycosylation of VLP proteins has major impact on their structure and function, and thus it is important to determine the choice of platforms for their production. As the viral glycoproteins localize, guide, and potentiate the process of enveloped virus assembly, it becomes important to study their individual and combined behavior upon expression in both animal and plant cells, in order to identify domains within the glycoproteins responsible for the critical differences between the intracellular targeting in either cell system. The large structural protein of lettuce necrotic yellow virus was glycosylated with complex oligosaccharides containing *N*-acetylglucosamine *N*-linked to asparagine residues [44]. The potato virus X CP and PPV CP were also glycosylated [45]. Glycosylated CP of beet western yellows virus plays a role in the virus/aphis interaction and promotes the aphid transmission of the virus [46]. Reviewed earlier, plants offer an attractive alternative system for VLP vaccine production with cost-effective, scalable, versatile, appropriate glycosylation, efficient assembly of VLP, and safety from adventitious human pathogens [12].

Although it yielded encouraging results, expression of VLPs expressed in plants suffers from plant-specific glycosylation of glycoproteins [14, 15]. Most proteins in eukaryotic multicellular organisms including plants are synthesized as glycoproteins with *N*- and *O*-glycosylation, which are important posttranslational protein modifications [47]. *N*-glycans attached to proteins are crucial for protein folding, assembly, and their stability but also involved in cell to cell adhesion, protein targeting, and immune responses as biological activity [48,49]. During the glycoprotein transportation through the secretory pathway, the oligosaccharide *N*-linked to the asparagine residue (Asn) undergoes several maturation steps involving the removal of glucose and mannose residues by different exoglycosidase to generate high mannose type *N*-glycan in the endoplasmic reticulum (ER) and the Golgi apparatus and, eventually, it is characterized by the addition of new oligosaccharide residues in the Golgi apparatus to form the matured complex type *N*-glycan (Figure 1). In the ER, the first step of *N*-glycosylation of plant proteins is the transfer of the oligosaccharide precursor $Glc_3Man_9GlcNAc_2$ from a dolichol lipid to specific Asn residues on the nascent polypeptide chain [50]. Processing of this oligosaccharide into high mannose, complex, hybrid, or paucimannosidic type *N*-glycan occurs during the secretory pathway. Particularly in plant, the β-mannose is substituted by a bisecting β1,2-xylose that is not found in mammalian *N*-glycans, and the proximal *N*-acetylglucosamine of the core is substituted by an α1,3-fucose, instead of an α1,6-fucose in mammalians. In addition, β1,3-galactose and fucose that are α1,4-linked to the terminal *N*-acetylglucosamine of plant *N*-glycans form Lewis a (Le^a) glycosylation (Figure 1) [51]. These modifications are not present in mammalian. Many mammalian complex*N*-glycans have an α1,6-fucose on the first core *N*-acetylglucosamine of *N*-glycan and are characterized by terminal β1,4-galactose and sialic acid which are not observed in plants (Figure 1) [52–54]. Most plant-derived therapeutic proteins are complex glycoproteins requiring posttranslational modifications. The β1,2-xylose, core α1,3-fucose, and Le^a containing epitopes have been considered as immunogenic glycan epitopes found in plant-specific *N*-glycans. Such glycan residues are not present in humans, and thus proteins could cause immune rejection inducing plant-glycan specific antibodies causing protein clearance in blood stream as well as potential allergenic effects [54–56]. These hurdles can be overcome by recent progress in plant glycoengineering. The plant expression with glycoengineering will allow the novel application of plant-made VLPs, including vessels for the delivery of small therapeutics, DNA fragments, and adjuvants (Figure 2).

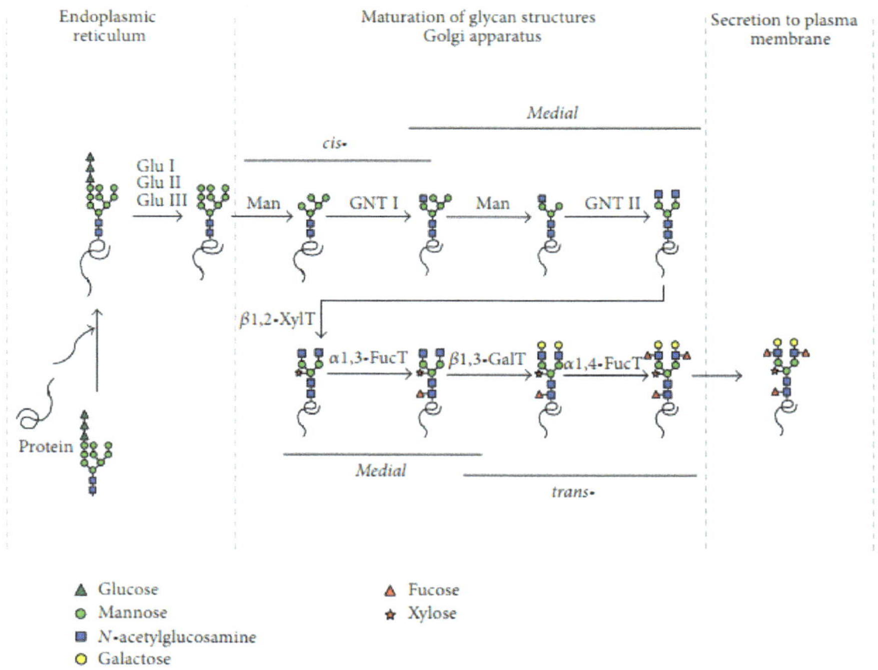

Figure 1: N-glycosylation pathway in plant. The primary glycosylation pathways with consequent series of steps occur in different subcellular compartments, ER, Golgi complex within the plant cell. During the pathway, glycosidase digestion and additional glycosyltransferase result in additional different branches and terminal glycan residues. GuI: glucosidase I, GuII: glucosidase II, GuIII: glucosidase III, Man: mannosidase, GNT I: N-acetylglucosaminyltransferase I, GNT II: N-acetylglucosaminyltransferase II, *β*1,2-XylT: *β*1,2-xylose transferase, *α*1,3-FucT: *α*1,3-fucose transferase, *β*1,3-GalT: *β*1,3-galactosidase, *α*1,4-FucT: *α*1,4-fucose transferase.

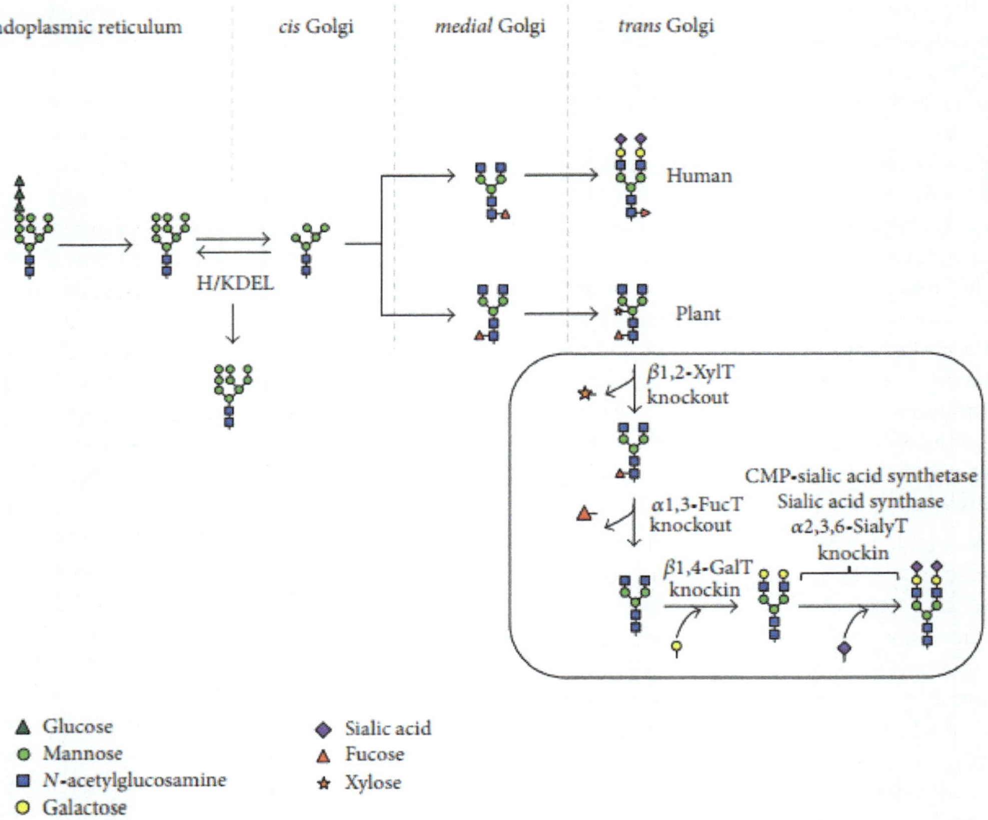

Figure 2: Schematic diagram of humanization of the glycosylation pathway in plant. In ER, protein is glycosylated and three glucoses are removed from the attached glycan. The glycoproteins then are transferred to the Golgi complex where mannoses are trimmed, and the glyco-residues are sequentially attached. When the ER retention signal KDEL sequence is attached to the C-terminal of glycoproteins, glycoproteins are retained and accumulated in the ER. Plant glycans carry β1,2-xylose and α1,3-fucose residues attached to the Nacetylglucosamine whereas human glycans contain α1,6-fucose, β1,4-galactose, and α2,3,6-sialic acid. In humanization glycoengineering process the β1,2-XylT and α1,3-FucT should be knocked out to remove xylose and fucose, respectively. The β1,4-GalT should be knocked in to add β1,4-galactose. Furthermore, finally CMP-sialic acid synthetase, sialic acid synthase, and α2,3,6-sialic transferase should be knocked in to attach α2,3,6-sialic acid to the terminal galactose. KDEL: ER retention motif (Lys-Asp-Glu-Leu), β1,2-XylT: β1,2-xylose transferase, α1,3-FucT: α1,3-fucose transferase, β1,4-GalT: β1,4-galactosidase, α2,3,6-SialyT: α2,3,6-sialic transferase.

N-GLYCOMODIFICATION IN PLANTS

Targeted Expression to the ER

N-glycan structures influence biofunctionality and stability of therapeutic proteins and even directly affect immunogenicity of glycosylated subunit vaccines displayed on VLP surfaces. In plants, thus, *N*-glycosylation pathway has been modified in order to humanize the glycan structures of glycoproteins [57, 58].

A commonly used approach to express recombinant glycoproteins in plants is their accumulation in ER by addition of C-terminal signal H/KDEL ER retention motif [59]. The ER-retained proteins contain high mannose type *N*-glycans structurally similar between plant and mammalian cells [58, 60, 61]. The high mannose type *N*-glycans are oligosaccharide structures that mammals and plants have in common and thus are probably not immunogenic [62]. This strategy is largely devoid of plant-specific, immunogenic β1,2-xylose and core α1,3-fucose. Additionally, some studies have reported enhanced accumulation of KDEL-tagged proteins in the ER. Such ER retention of proteins usually increases the production level compared to that without KDEL in plant [63, 64]. Plant-derived monoclonal antibody (mAb) with high mannose *N*-glycan structure has shorter half-life than that of the mammalian-derived mAb with mammalian specific glycan structures [65]. However, mAb with high mannose glycans had relatively similar biological activities compared to the mammalian-derived mAb overcoming concerns about plant-specific glycoepitopes expressed by others [58]. In addition, the high mannose type glycan structure would be expected to cause an enhanced immune response through the mannose receptor (MR) on macrophages and dendritic cells recognizing the oligomannose of glycoproteins [66], which is an advantage for vaccine development. According to a previous study [67], the high mannose glycans on antigenic protein can render the protein more immunogenic, producing IgG against the high mannose glycosylated protein.

Knockout of Plant-Specific Glycosyltransferases

Gene inactivation or silencing may be used to reduce or eliminate the activity of plant- specific glycosyltransferases. In a plant cell, the specific enzymes are β1,2-xylosyltransferase and core α1,3-fucosyltransferase, which are responsible for transfer of the plant-specific xylose and fucose onto the attached*N*-glycan. Such glycan residues are not present in humans and are thus unwanted on proteins intended for therapeutic use. The knockout of the genes that are responsible for the synthesis of these glycan epitopesβ1,2-xylosyltransferase and core α1,3-fucosyltransferase provides an easy strategy to solve this problem. The feasibility of this strategy has been proven by the generation of knockout *Arabidopsis thaliana* plant lacking xylosyltransferase and fucosyltransferase [68, 69]. In addition, biological activity assays of such glycoengineered mAbs showed that their antigen binding activity was not altered but significantly enhanced antibody-dependent cell-mediated cytotoxicity (ADCC) effect [70, 71]. Therapeutic antibodies without fucosylation have higher binding affinity for FcγRIIIa than for fucosylated human serum IgG, which is desirable to overcome the interference by human plasma IgG. Thus, the therapeutic antibodies without fucosylation can avoid the inhibitory effect of human plasma IgG on ADCC through their high FcγRIIIa binding affinity.

Humanization of Plant *N*-Glycosylation

The immunogenic and allergenic reactions of the β1,2-xylose and core α1,3-fucose *N*-glycan epitopes on plant-derived glycoproteins have been a problem for application of therapeutic proteins produced from plant expression system [72]. Glycoengineering strategies using transgenic plants and the availability of mutant plants lacking xylosyltransferase and fucosyltransferase genes for humanization of *N*-glycosylation allow producing recombinant proteins with more mammalian-like *N*-glycan structures in plant expression system. Most proteins used for therapy of human diseases are glycosylated, and the glycan structures have been shown to affect safety and efficacy of therapeutic glycoproteins [73]. Particularly, nonsialylation significantly causes shorter *in vivo* half-life of circulating glycoproteins, because exposed galactose glycan residues are recognized and captured by asialoglycoprotein receptors resulting in internalization of the glycoproteins in hepatocytes [74]. Terminal acid residues in *N*-linked glycans of most therapeutic glycoproteins affect important roles in *in vivo* physical stability, immunogenicity, and enzymatic activity [68, 70, 71, 74, 75]. Previous studies have demonstrated the importance of fully sialylated *N*-linked glycans and of consistency of homogeneous *N*-linked glycan structures on therapeutic glycoproteins in heterologous expression systems [75–77]. For instance, the sialylated recombinant erythropoietin (EPO) had longer plasma half-life (5-6 h) compared to that (2 min) of desialylated EPO [78]. This beneficial effect of sialic acid on protein stability likely explains why knockin strategies for plant glycoengineering in glycosylation have mainly focused on the addition of terminal β1,4-galactose and sialic acid residues to humanize *N*-glycan in mutant plants lacking plant-specific *N*-glycan residues [75]. It was claimed that plant virus-based transient expression systems can be applied as the knockin strategy of β1,4-galactose and sialic acid transferring genes in the mutant plants, allowing the generation of abundant amount of therapeutic proteins within 1 week after virus infection, provide a feasible advantage over existing glycoprotein expression systems [75].

CONCLUSIONS

Taken together, plant-derived VLPs are considered safe because plants do not bear human pathogens and promising in terms of cost-effective scalability and speed of production. In fact, as far as upstream and downstream processing are concerned, plant-derived VLPs can take advantage of what has been done so far in the broader field of plant-made pharmaceuticals. Also compared to prokaryotes host cells, plants host guarantees the appropriate posttranslational modifications, such as glycosylation, often needed for proper protein function.

In plants, glycoengineering has been improved to create plants able to perform the ideal glycosylation enhancing efficacy and potency of VLPs-based therapeutics. As described in this review, several strategies, focused on the inactivation and/or addition of key enzymes, can be adopted to decorate tailor-made glycoforms of VLPs in plants. Thus, plant expression systems will be further improved for production of VLPs-based vaccines with respect to their proper glycomodification and the rapid and cost-effective expression.

ACKNOWLEDGMENTS

This research was supported by Evaluation for Technology of Food, Agriculture, Forestry, and Fisheries, Korea (iPET-Code nos. 312037-5 and 111096-03-1-SB010) and a Grant (Code no. PJ009419022014) from the Korean Rural Development Administration, Korea institute of Planning.

REFERENCES

1. Lico C, Schoubben A, Baschieri S, Blasi P, Santi L. Nanoparticles in biomedicine: new insights from plant viruses. *Current Medicinal Chemistry*. 2013;20(28):3471–3487.
2. Grasso S, Santi L. Viral nanoparticles as macromolecular devices for new therapeutic and pharmaceutical approaches. *International Journal of Physiology, Pathophysiology and Pharmacology*. 2010;2(2):161–178.
3. Pokorski JK, Steinmetz NF. The art of engineering viral nanoparticles. *Molecular Pharmaceutics*.2011;8(1):29–43.
4. Bachmann AS, Corpuz G, Hareld WP, Wang G, Coller BA. A simple method for the rapid purification of copia virus-like particles from Drosophila Schneider 2 cells. *Journal of Virological Methods*.2004;115(2):159–165.
5. Chen Q, Lai H. Plant-derived virus-like particles as vaccines. *Human Vaccines and Immunotherapeutics*.2013;9(1):26–49.
6. Kushnir N, Streatfield SJ, Yusibov V. Virus-like particles as a highly efficient vaccine platform: diversity of targets and production systems and advances in clinical development. *Vaccine*. 2012;31(1):58–83.
7. Herscovics A, Orlean P. Glycoprotein biosynthesis in yeast. *The FASEB Journal*. 1993;7(6):540–550.
8. Wildt S, Gerngross TU. The humanization of *N*-glycosylation pathways in yeast. *Nature Reviews Microbiology*. 2005;3(2):119–128.
9. Harrison RL, Jarvis DL. Protein *N*-glycosylation in the baculovirus-insect cell expression system and engineering of insect cells to produce "mammalianized" recombinant glycoproteins. *Advances in Virus Research*. 2006;68:159–191.
10. Schwarzer J, Rapp E, Reichl U. *N*-glycan analysis by CGE-LIF: profiling influenza A virus

hemagglutinin *N*-glycosylation during vaccine production. *Electrophoresis*. 2008;29(20):4203–4214.

11. Hamby SE, Hirst JD. Prediction of glycosylation sites using random forests. *BMC Bioinformatics*.2008;9, article 500
12. Faye L, Gomord V. Success stories in molecular farming-a brief overview. *Plant Biotechnology Journal*.2010;8(5):525–528.
13. Giddings G, Allison G, Brooks D, Carter A. Transgenic plants as factories for biopharmaceuticals.*Nature Biotechnology*. 2000;18(11):1151–1155.
14. Roldao A, Mellado MC, Castilho LR, Carrondo MJ, Alves PM. Virus-like particles in vaccine development. *Expert Review of Vaccines*. 2010;9(10):1149–1176.
15. Lai H, Chen Q. Bioprocessing of plant-derived virus-like particles of Norwalk virus capsid protein under current Good Manufacture Practice regulations. *Plant Cell Reports*. 2012;31(3):573–584.
16. Gleba Y, Klimyuk V, Marillonnet S. Viral vectors for the expression of proteins in plants. *Current Opinion in Biotechnology*. 2007;18(2):134–141.
17. Santi L, Batchelor L, Huang Z, et al. An efficient plant viral expression system generating orally immunogenic Norwalk virus-like particles. *Vaccine*. 2008;26(15):1846–1854.
18. Chapman S, Kavanagh T, Baulcombe D. Potato virus X as a vector for gene expression in plants. *The Plant Journal*. 1992;2(4):549–557.
19. Cerovska N, Hoffmeisterova H, Moravec T, et al. Transient expression of *Human papillomavirus* type 16 L2 epitope fused to N- and C-terminus of coat protein of *Potato virus* X in plants. *Journal of Biosciences*.2012;37(1):125–133.
20. Chatterji A, Ochoa WF, Paine M, Ratna BR, Johnson JE, Lin T. New addresses on an addressable virus nanoblock: uniquely reactive Lys residues on cowpea mosaic virus. *Chemistry and Biology*. 2004;11(6):855–863.
21. Zhang X, Mason H. Bean Yellow Dwarf Virus replicons for high-level transgene expression in transgenic plants and cell cultures. *Biotechnology and Bioengineering*. 2006;93(2):271–279.
22. Liu L, Davies JW, Stanley J. Mutational analysis of bean yellow dwarf virus, a geminivirus of the genus Mastrevirus that is adapted to dicotyledonous plants. *Journal of General Virology*. 1998;79:2265–2274.
23. Smith ML, Lindbo JA, Dillard-Telm S, et al. Modified tobacco mosaic virus particles as scaffolds for display of protein antigens for vaccine applications. *Virology*. 2006;348(2):475–488.
24. Conti G, Rodriguez MC, Manacorda CA, Asurmendi S. Transgenic expression of *Tobacco mosaic virus*capsid and movement proteins modulate plant basal defense and biotic stress responses in *Nicotiana tabacum*. *Molecular Plant-Microbe Interactions*. 2012;25(10):1370–1384.
25. Tsuda S, Yoshioka K, Tanaka T, et al. Application of the human hepatitis B virus core antigen from transgenic tobacco plants for serological diagnosis. *Vox Sanguinis*. 1998;74(3):148–155.
26. Sojikul P, Buehner N, Mason HS. A plant signal peptide-hepatitis B surface antigen fusion protein with enhanced stability and immunogenicity expressed in plant cells. *Proceedings of the National Academy of Sciences of the United States of America*. 2003;100(5):2209–2214.
27. Mechtcheriakova IA, Eldarov MA, Nicholson L, Shanks M, Skryabin KG, Lomonossoff GP. The use of viral vectors to produce hepatitis B virus core particles in plants. *Journal of Virological Methods*.2006;131(1):10–15.
28. Huang Z, Mason HS. Conformational analysis of hepatitis B surface antigen fusions in an*Agrobacterium*-mediated transient expression system. *Plant Biotechnology Journal*. 2004;2(3):241–249.
29. WarzechaH,MasonHS,LaneC,etal.Oralimmunogenicity of human papillomavirus-like particles expressed in potato. *Journal of Virology*. 2003;77(16):8702–8711.
30. Varsani A, Williamson AL, Stewart D, Rybicki EP. Transient expression of *Human papillomavirus* type 16 L1 protein in *Nicotiana benthamiana* using an infectious tobamovirus vector. *Virus Research*.2006;120(1-2):91–96.
31. Varsani A, Williamson A-L, Rose RC, Jaffer M, Rybicki EP. Expression of *Human papillomavirus* type 16 major capsid protein in transgenic *Nicotiana tabacum* cv. Xanthi. *Archives of Virology*.2003;148(9):1771–1786.
32. Huang Z, Chen Q, Hjelm B, Arntzen C, Mason H. A DNA replicon system for rapid high-level production of virus-like particles in plants. *Biotechnology and Bioengineering*. 2009;103(4):706–714.
33. He J, Lai H, Brock C, Chen Q. A novel system for rapid and cost-effective production of detection and diagnostic reagents of West Nile virus in plants. *Journal of Biomedicine and Biotechnology*. 2012;2012:10 pages.106783
34. Hefferon KL. Broadly neutralizing antibodies and the promise of universal vaccines: where are we now?*Immunotherapy*. 2014;6(1):51–57.
35. Huang Z, Santi L, LePore K, Kilbourne J, Arntzen CJ, Mason HS. Rapid, high-level production of hepatitis B core antigen in plant leaf and its immunogenicity in mice. *Vaccine*. 2006;24(14):2506–2513.
36. Lambert C, Thome N, Kluck CJ, Prange R. Functional incorporation of green fluorescent protein into hepatitis B virus envelope particles. *Virology*. 2004;330(1):158–167.
37. Chen Q, He J, Phoolcharoen W, Mason HS. Geminiviral vectors based on bean yellow dwarf virus for production of vaccine antigens and monoclonal antibodies in plants. *Human Vaccines*. 2011;7(3):331–338.
38. Ravin NV, Kotlyarov RY, Mardanova ES, et al. Plant-produced recombinant influenza vaccine based on virus-like HBc particles carrying an extracellular domain of M2 protein. *Biochemistry*. 2012;77(1):33–40.

39. de la Rosa GP, Monroy-Garcia A, de Lourdes Mora-García M, et al. An HPV 16 L1-based chimeric human papilloma virus-like particles containing a string of epitopes produced in plants is able to elicit humoral and cytotoxic T-cell activity in mice. *Virology Journal*. 2009;6, article 2

40. Santi L, Huang Z, Mason H. Virus-like particles production in green plants. *Methods*. 2006;40(1):66–76.

41. Huang Z, Elkin G, Maloney BJ, et al. Virus-like particle expression and assembly in plants: hepatitis B and Norwalk viruses. *Vaccine*. 2005;23(15):1851–1858.

42. Shchelkunov SN, Salyaev RK, Pozdnyakov SG, et al. Immunogenicity of a novel, bivalent, plant-based oral vaccine against hepatitis B and human immunodeficiency viruses. *Biotechnology Letters*.2006;28(13):959–967.

43. Branco LM, Grove JN, Geske FJ, et al. Lassa virus-like particles displaying all major immunological determinants as a vaccine candidate for Lassa hemorrhagic fever. *Virology Journal*. 2010;7, article 279

44. Dietzgen RG, Francki RI. Analysis of lettuce necrotic yellows virus structural proteins with monoclonal antibodies and concanavalin A. *Virology*. 1988;166(2):486–494.

45. Baratova LA, Fedorova NV, Dobrov EN, et al. *N*-terminal segment of potato virus X coat protein subunits is glycosylated and mediates formation of a bound water shell on the virion surface. *European Journal of Biochemistry*. 2004;271(15):3136–3145.

46. Seddas P, Boissinot S. Glycosylation of beet Western yellows virus proteins is implicated in the aphid transmission of the virus. *Archives of Virology*. 2006;151(5):967–984.

47. Helenius A, Aebi M. Intracellular functions of *N*-linked glycans. *Science*. 2001;291(5512):2364–2369.

48. Lerouge P, Bardor M, Pagny S, Gomord V, Faye L. *N*-glycosylation of recombinant pharmaceutical glycoproteins produced in transgenic plants: towards an humanisation of plant *N*-glycans. *Current Pharmaceutical Biotechnology*. 2000;1(4):347–354.

49. Varki A. Biological roles of oligosaccharides: all of the theories are correct. *Glycobiology*. 1993;3(2):97–130.

50. Gomord V, Fitchette A, Menu-Bouaouiche L, et al. Plant-specific glycosylation patterns in the context of therapeutic protein production. *Plant Biotechnology Journal*. 2010;8(5):564–587.

51. Saint-Jore-Dupas C, Faye L, Gomord V. From planta to pharma with glycosylation in the toolbox.*Trends in Biotechnology*. 2007;25(7):317–323.

52. Bencurova M, Hemmer W, Focke-Tejkl M, Wilson IB, Altmann F. Specificity of IgG and IgE antibodies against plant and insect glycoprotein glycans determined with artificial glycoforms of human transferrin. *Glycobiology*. 2004;14(5):457–466.

53. Fotisch K, Vieths S. N- and O-linked oligosaccharides of allergenic glycoproteins. *Glycoconjugate Journal*. 2001;18(5):373–390.

54. van Ree R, Cabanes-Macheteau M, Akkerdaas J, et al. β(1,2)-xylose and α(1,3)-fucose residues have a strong contribution in IgE binding to plant glycoallergens. *The Journal of Biological Chemistry*.2000;275(15):11451–11458.

55. Altmann F. The role of protein glycosylation in allergy. *International Archives of Allergy and Immunology*. 2007;142(2):99–115.

56. Bosch D, Castilho A, Loos A, Schots A, Steinkellner H. *N*-glycosylation of plant-produced recombinant proteins. *Current Pharmaceutical Design*. 2013;19(31):5503–5512.

57. Strasser R, Stadlmann J, Svoboda B, Altmann F, Glossl J, Mach L. Molecular basis of *N*-acetylglucosaminyltransferase I deficiency in *Arabidopsis thaliana* plants lacking complex *N*-glycans. *Biochemical Journal*. 2005;387:385–391.

58. Ko K, Tekoah Y, Rudd PM, et al. Function and glycosylation of plant-derived antiviral monoclonal antibody. *Proceedings of the National Academy of Sciences of the United States of America*.2003;100(13):8013–8018.

59. Gomord V, Denmat LA, Fitchette-Laine AC, Satiat-Jeunemaitre B, Hawes C, Faye L. The C-terminal HDEL sequence is sufficient for retention of secretory proteins in the endoplasmic reticulum (ER) but promotes vacuolar targeting of proteins that escape the ER. *The Plant Journal*. 1997;11(2):313–325.

60. Floss DM, Sack M, Stadlmann J, et al. Biochemical and functional characterization of anti-HIV antibody-ELP fusion proteins from transgenic plants. *Plant Biotechnology Journal*. 2008;6(4):379–391.

61. Petruccelli S, Otegui MS, Lareu F, et al. A KDEL-tagged monoclonal antibody is efficiently retained in the endoplasmic reticulum in leaves, but is both partially secreted and sorted to protein storage vacuoles in seeds. *Plant Biotechnology Journal*. 2006;4(5):511–527.

62. Gomord V, Sourrouille C, Fitchette AC, et al. Production and glycosylation of plant-made pharmaceuticals: the antibodies as a challenge. *Plant Biotechnology Journal*. 2004;2(2):83–100.

63. Conrad U, Fiedler U. Compartment-specific accumulation of recombinant immunoglobulins in plant cells: an essential tool for antibody production and immunomodulation of physiological functions and pathogen activity. *Plant Molecular Biology*. 1998;38(1-2):101–109.

64. Sharp JM, Doran PM. Strategies for enhancing monoclonal antibody accumulation in plant cell and organ cultures. *Biotechnology Progress*. 2001;17(6):979–992.

65. Lee JH, Park DY, Lee KJ, et al. Intracellular reprogramming of expression, glycosylation, and function of a plant-derived antiviral therapeutic monoclonal antibody. *PLoS ONE*. 2013;8(8)e68772

66. Apostolopoulos V, McKenzie IF. Role of the mannose receptor in the immune response. *Current Molecular Medicine*. 2001;1(4):469–474.

67. Lu Z, Lee KJ, Shao Y, et al. Expression of GA733-Fc fusion protein as a vaccine candidate for colorectal cancer in transgenic plants. *Journal of Biomedicine and*

Biotechnology. 2012;2012:11 pages.364240

68. Schahs M, Strasser R, Stadlmann J, Kunert R, Rademacher T, Steinkellner H. Production of a monoclonal antibody in plants with a humanized *N*-glycosylation pattern. *Plant Biotechnology Journal*.2007;5(5):657–663.

69. Strasser R, Altmann F, Mach L, Glossl J, Steinkellner H. Generation of *Arabidopsis thaliana* plants with complex *N*-glycans lacking β1,2-linked xylose and core α1,3-linked fucose. *FEBS Letters*. 2004;561(1–3):132–136.

70. Cox KM, Sterling JD, Regan JT, et al. Glycan optimization of a human monoclonal antibody in the aquatic plant Lemna minor. *Nature Biotechnology*. 2006;24(12):1591–1597.

71. Forthal DN, Gach JS, Landucci G, et al. Fc-glycosylation influences Fcγ receptor binding and cell-mediated anti-hivactivity of monoclonal antibody 2G12. *The Journal of Immunology*. 2010;185(11):6876–6882.

72. Cabanes-Macheteau M, Fitchette-Laine A, Loutelier-Bourhis C, et al. *N*-glycosylation of a mouse IgG expressed in transgenic tobacco plants. *Glycobiology*. 1999;9(4):365–372.

73. Apweiler R, Hermjakob H, Sharon N. On the frequency of protein glycosylation, as deduced from analysis of the SWISS-PROT database. *Biochimica et Biophysica Acta*. 1999;1473(1):4–8.

74. Fukuda MN, Sasaki H, Lopez L, Fukuda M. Survival of recombinant erythropoietin in the circulation: the role of carbohydrates. *Blood*. 1989;73(1):84–89.

75. Castilho A, Strasser R, Stadlmann J, et al. In planta protein sialylation through overexpression of the respective mammalian pathway. *The Journal of Biological Chemistry*. 2010;285(21):15923–15930.

76. Egrie JC, Browne JK. Development and characterization of novel erythropoiesis stimulating protein (NESP) *Nephrology Dialysis Transplantation*. 2001;16(supplement 3):3–13.

77. Ngantung FA, Miller PG, Brushett FR, Tang GL, Wang DI. RNA interference of sialidase improves glycoprotein sialic acid content consistency. *Biotechnology and Bioengineering*. 2006;95(1):106–119.

78. Erbayraktar S, Grasso G, Sfacteria A, et al. Asialoerythropoietin is a nonerythropoietic cytokine with broad neuroprotective activity *in vivo*. *Proceedings of the National Academy of Sciences of the United States of America*. 2003;100(11):6741–6746.

CHAPTER 11

PLANT VIRUS EXPRESSION VECTOR DEVELOPMENT: NEW PERSPECTIVES

Kathleen Hefferon

Cornell University, Ithaca, NY 14850, USA

ABSTRACT

Plant made biologics have elicited much attention over recent years for their potential in assisting those in developing countries who have poor access to modern medicine. Additional applications such as the stockpiling of vaccines against pandemic infectious diseases or potential biological warfare agents are also under investigation. Plant virus expression vectors represent a technology that enables high levels of pharmaceutical proteins to be produced in a very short period of time. Recent advances in research and development have brought about the generation of superior virus expression systems which can be readily delivered to the host plant in a manner that is both efficient and cost effective. This review presents recent innovations in plant virus expression systems and their uses for producing biologics from plants.

INTRODUCTION

It takes just a quick glance at the number and variety of novel biologics that are currently emerging from the commercial sector to realize that the use of plants as production platforms for vaccines and other therapeutic proteins has over time moved from mere theory and into actual practice. Plant-derived biologics are safe, efficacious, and easy to mass produce. They can offer new solutions to the challenge of providing inexpensive medicines that lack cold chain requirements or an established medical infrastructure for the world's rural poor. While assisting developing countries has always been a fundamental driving force for the development of plant-made biologics, other uses have also been presented and range from facilitating the stockpiling of vaccines against pandemic infectious diseases to the applications of plant-made proteins in the field of personalized medicine [1, 2].

Originally, plant-made biologics were engineered to be expressed from stably transformed plants. As this research area progressed to the point of determining plant-based pharmaceutical performance in actual clinical trials, it became increasingly apparent that lengthy tissue culture and plant regeneration procedures and concerns regarding public perception over the use of GM technology were all major stumbling blocks with regard to the use of transgenic plants that needed to be addressed. The design and implementation of novel transient expression systems through the use of plant virus expression vectors has been one technology that circumvents many of these hurdles. Not only are plant viruses able to produce extremely high levels of foreign protein at low cost and in a matter of a few days postinoculation, but also their use as a technology is more appealing to the public because it lacks the negative connotations that are often associated with GM plants. Indeed, removal of the genes encoding virus

Citation: Kathleen Hefferon, Plant Virus Expression Vector Development: New Perspectives, doi: 10.1155/2014/785382.

movement and/or coat proteins prevents plant viruses from moving from plant to plant and as a result reduces concerns about transmission and cross-contamination of pharmaceutical proteins to weedy relatives.

As research and development of plant virus expression vectors progress, the means of introducing them to the host plant has also advanced. Rather than merely infecting plants with the appropriate virus vector, plant leaves can now be inoculated by agroinfiltration, that is, by incorporating the virus vector into Agrobacterium tumefaciens and infiltrating the leaf using a syringe or a vacuum. Both methods have their own advantages, and both are rapid and scalable and lack the requirement of sophisticated equipment. The universality of these technologies makes them amenable for both commercialization and classroom alike [3].

Over the past decade, plant viruses have been developed both as vectors for biopharmaceutical protein production and also as research tools for plant functional genomics studies, by incorporating virus induced gene silencing (VIGS) as a means to target and downregulate specific host transcripts. The latter technology has great implications for the development of crops with improved characteristics, while the former has the potential to launch a new era for the production of biologics. This review will discuss several of the plant virus expression systems that have been developed to date and will provide selected examples of their application in the field of plant-made biologics.

EXPRESSION SYSTEMS BASED ON POSITIVE SENSE RNA VIRUSES

Tombusviruses

Tobacco mosaic virus (TMV) is one of the most well characterized plant viruses and was the first plant virus examined for vector development. While original TMV-based viral vectors depended on usage of the entire viral genome, second generation vectors were "deconstructed" and consisted of the only sections of genomes that were important for replication; these were subcloned into an assortment of plasmid constructs. The first of these new series of vectors, the MagnICON vector system of ICON Genetics, is a core technology of Dr. Gleba and coworkers [4]. A technique for transfecting plants with these recombinant virus vector modules, known as "magnifection," involved the infiltration of a suspension of *A. tumefaciens* using a vacuum into all mature leaves of a tobacco plant, thus infecting the whole plant in its entirety [5]. These constructs are part of the T-DNA of a binary vector, and researchers need to merely mix different Agrobacterium strains which harbour these constructs prior to agroinfecting plants. As an example, yields of recombinant human growth hormone protein (hGH) reaching up to 10% of total soluble protein or 1 mg/g of fresh weight leaf biomass have been achieved using this system of expression [5]. Further modifications, including alteration of transcript splicing sites, modification of codon usage patterns, and the introduction of introns into TMV coding sequences, have improved protein expression further.

The technology has been built upon so that it could be incorporated into the Gateway cloning system, a series of plasmids that are used for a wide variety of expression systems [6]. In this case, the need for traditional cloning has been circumvented by a system that involves site-specific recombination. In this way, foreign genes and other DNA fragments alike can be transferred between plasmids to enable the desired recombination event to take place. Protein expression can also be enhanced 10- to 25-fold by the coexpression of the RNA silencing suppressor gene of Tomato bushy stunt virus known as P19. For example, Lindbo coinoculated plants with a TMV-based vector and a viral suppressor of RNA silencing; this culminated in the production of tremendous levels of recombinant protein (between 600 and 1200 micrograms of GFP per gram of infiltrated tissue) after one week postinfection [7].

Other researchers have found that gene expression can be increased several times by placing the foreign gene open reading frame (ORF) closer to the 3′ end of the TMV RNA. This TRBO (TMV RNA-based overexpression) vector lacks the coat protein coding sequence and can produce 100-fold more recombinant protein in plants than the P19-enhanced transient expression system. Lindbo›s group found that this vector could generate 100 times more of a variety of recombinant proteins than their P19 silencer system [8].

The number of examples of the use of TMV-based vector technology as a platform for biopharmaceutical production in plants is rapidly increasing. Since this review cannot adequately describe all in the space constructions allowed, only a few are provided in the following section.

The first example illustrates the use of TMV to generate the broadly neutralizing antibody (bnMAb) known as VRCO1 against HIV-1 [9]. VRC01 binds to the CD4-binding site of gp120 and was isolated recently from a slowly progressing HIV-1-infected donor. Broadly neutralizing antibodies are known for their ability to block infection of a wide number of different strains of a particular virus and thus are considered to be used as part of a topical microbicide to block HIV-1 transmission. Monoclonal antibodies are expensive to manufacture on a large scale, and plants offer an attractive alternative for generating biologically active, inexpensive versions of these much needed pharmaceuticals. *Nicotiana benthamiana* plants were used as the host for a TMV vector which contained a full-length version of the immunoglobulin IgG1 from a single polypeptide. This VRCO1 MAb was produced at approximately 150 mg/kg fresh leaf material within days 5–7 postagroinoculation and could be purified on a protein A affinity column. These plant-made antibodies not only were shown to be biologically active in a neutralization assay but also worked synergistically with other microbicides, demonstrating their potential as part of a topically applied microbicide cocktail to act as a prophylactic against infection by HIV-1 [9].

The use of TMV to produce a vaccine for pandemic influenza virus represents the second example. Seasonal influenza virus has become a serious health threat across the globe. New vaccines must be provided every new year to protect against emerging new virus subtypes that circulate every new flu season. The requirement to supply the demand for flu vaccines worldwide is enormous and cannot currently be met. Recently, Petukhova et al. have used a TMV vector to express three different versions of the epitope M2e of influenza virus [10]. The authors used the CP of TMV as a carrier molecule and expressed the epitope as part of a fusion protein that is exposed on the surface of TMV as a nanoparticle. Antibodies raised in mice against the nanoparticle were specific to M2e and mice immunized with the M2e:TMV recombinant virus were resistant to inoculation with lethal doses of influenza H1N1 virus [10].

More extensive investigation into the utilization of TMV as a feasible vector for influenza vaccine production has been pursued by Yusibov et al.›s research group at Fraunhofer USA. The Fraunhofer group has generated a subunit vaccine based on recombinant hemagglutinin from the 2009 pandemic A/California/04/2009 (H1N1) strain of influenza virus [11]. This TMV-based vaccine was demonstrated to be efficacious and free of adverse effects in human clinical trials in the presence or absence of Alhydrogel as an adjuvant, in a manner that is highly comparable with the currently used approved vaccine for H1N1 [12,13]. This represents the first study that demonstrates a plant produced subunit influenza H1N1 vaccine in healthy adults. In a further study, Neuhaus et al., tested the ability of this tobacco produced vaccine in conjunction with a silica nanoparticle-based drug system to induce an antigen-specific recall response at the site of virus entry in human precision-cut lung slices (PCLS) [14]. The authors demonstrated that the plant produced vaccine was capable of reactivating an established antigen-specific T cell response at the site of virus entry [15].

As a third example, Li et al. have used the TMV-based TRBO vector to express the allergen R8 from dust mites in tobacco plants [16]. The authors used murine asthmatic models to investigate the possibility of using these plant-derived antigens for immunotherapy. The plant-derived antigen behaved the same way as the native antigen, offering the possibility that it might be used in the future for the diagnosis of asthma or the production of a candidate vaccine for allergen-specific immunotherapy of asthma [16].

As a final example, MAGNICON vectors were used to compare expression of the highly unstable recombinant protein, human complement factor 5a (C5a) in tobacco plants [17]. Transient expression of C5a that was subcellular targeted to the ER or vacuole using the MagnICON vector was increased from 0.0003 and 0.001% total soluble protein (TSP) to 0.2 and 0.7% of TSP, respectively, demonstrating the utility of this system to produce biopharmaceutical proteins [17].

Potexviruses

Potato virus X (PVX) and related potexviruses have also been constructed into expression vectors for vaccine production. For example, the L1 protein of canine oral papillomavirus has been expressed in transgenic tobacco chloroplasts using a Potato virus X vector [18]. Similarly, the Human Papillomavirus-16 L2 minor capsid protein has been expressed in plants as part of a fusion protein with the PVX CP [19]. Other potexviruses, including white clover mosaic virus, foxtail mosaic virus, and alternanthera mosaic virus, have also been developed for foreign gene expression. Recently, an expression vector based on plantago asiatica mosaic virus (PlAMV) has been constructed [20]. PlAMV is unique among potexviruses due to an overlap at the third gene of its triple gene block and its CP. As a result, the generation of a recombinant PlAMV expression system resulted in a duplicated CP promoter sequence that is much longer than that generated for other potexviruses. Due to this unusual feature, the PlAMV-based vector was more stable and had the ability to exert a stronger suppression of gene silencing activity than its PVX-based counterpart.

Cucumovirus

The RNA virus Cucumber mosaic virus (CMV) has also been developed extensively for use in the production of plant-made biologics. Although it has a highly diverse host range, CMV's trimeric RNAs are each required for virus infection to be successful and are packaged into icosahedral capsids, thus enforcing a size limitation to the foreign gene that can be inserted. Recently, Hwang et al. constructed the complete tripartite genome on a binary plasmid and replaced the coat protein gene with one for α-1-antitrypsin [AAT] [21]. The authors placed one of the protein components of the viral replicase (1a) under the control of the XVE chemically inducible promoter in such a way that recombinant viral amplicons could be induced upon addition of the inducer (β-estradiol). This enables all of the components that are required for virus replication to be introduced simultaneously into the same cell, thus improving the efficiency of transgene expression. The authors tried to further improve expression by generating multiple plasmids of reduced size which, when combined, incorporate all of the essential components of the genome that are required for replication. By using both GFP and the *Acidothermus cellulolyticus* endo-1, 4-β-glucanase (E1), a cellulose degrading enzyme, the authors demonstrated that this CMV advanced replication system can effectively achieve recombinant proteins at levels comparable to transgenic plants, but without the same time and effort required [21].

CMV has also been utilized as an antigen presentation system to express epitopes of porcine circovirus type 2 (PCV2) capsid protein [22]. Chimeric CMV:PCV2 particles were injected into mice and pigs and analyzed for their ability to induce a PCV specific antibody response. Furthermore, pigs challenged with the virus were able to demonstrate partial protection against infection, suggesting that generation of a plant virus-based vaccine may be a feasible and affordable approach to combat this animal pathogen [22].

OTHER PLANT VIRUSES INVOLVED IN OPTIMIZING EXPRESSION

Cowpea Mosaic Virus

The icosahedral, positive sense RNA virus Cowpea mosaic virus (CPMV) has also been designed for the generation of vaccine and other therapeutic proteins in plants. Initially designed as an epitope presentation system, CPMV chimeric particles have been shown to elicit a robust immune response for a variety of diseases [23]. CPMV has also been used to express entire proteins as fusion products with the capsid protein or movement protein of the virus, which can undergo proteolytic cleavage to release the therapeutic protein.

Medicago, Inc., has also developed their own CPMV vector-based technology which produces virus-like particles (VLPs) carrying influenza virus antigens [24]. These VLPs express a lipid-anchored recombinant HA and are fully protected against lethal viral challenge in both mice and ferrets. This vaccine was placed under further study in a Phase

1 clinical trial of 48 healthy volunteers, who showed that the vaccine was both safe and well tolerated. A Phase 2 clinical trial with over 250 volunteers is now underway. Medicago's production system offers the advantage of inexpensively generating a vaccine within 3 weeks of the release of influenza strain sequence information, with easily adaptable upscaling capacity [24].

Nonreplicating CPMV expression systems have also been developed. These involve the positioning of the foreign gene between the 5′ leader sequence and 3′ untranslated region (UTR) of RNA-2 [25]. Deletion of an in-frame initiation codon found upstream of the main translation initiation site of RNA-2 brought about a substantial increase in foreign protein production. This new series of vectors based on CPMV, known as the pEAQ vectors, have been explored as a means to mass produce large amounts of pharmaceutical proteins in plants without the need for virus replication. Using a high translational efficiency, the Cowpea Mosaic Virus hypertranslational "CPMV-HT" expression system generates extremely high yields of recombinant protein [23]. Many pharmaceutical proteins have been successfully produced in plants using this system, including anti-HIV-1 monoclonal antibodies and influenza A virus vaccines [25]. Recently, Thuenemann et al. used the nonreplicating CPMV system known as CPMV-HT (Hyper-Trans) to generate the ruminant Bluetongue virus-(BTV-) like particles in benthamiana plants. This required the synchronous expression of four distinct capsid proteins in each cell and enabling them to self-assemble into virus-like particles. These plant-derived VLPs were demonstrated to protect sheep against live virus infection [26].

Plant DNA Viruses

Plant DNA viruses, including geminivirus and Cauliflower mosaic virus (CaMV), have also been designed as pharmaceutical protein production platforms. Geminiviruses are small single stranded DNA viruses and are named for their twinned capsid morphology; they infect a broad range of plants and can replicate to extremely high copy numbers. Two geminiviruses have been developed for biopharmaceutical production, Bean yellow dwarf virus (BeYDV), a mastrevirus, and Beet curly top virus, (BCTV), a curtovirus [27, 28]. Both viruses have been constructed in such a way that the replication initiator protein (Rep) is expressed independently from the viral genome. A variety of biologics have been produced using this approach, ranging from a vaccine against hepatitis A virus to a monoclonal antibody against Ebola virus [27, 28].

ADDITIONAL APPLICATIONS FOR PLANT VIRUS VECTORS IN MEDICINE

Besides offering a novel production platform for plant-made biopharmaceuticals, plant viruses have been engineered to provide other medical applications. For example, virus-like particles based on *Cowpea mosaic virus* (CPMV) have been designed which can incorporate fluorescent dyes, polyethylene glycol (PEG) polymers, and various targeting moieties on their surfaces for the purpose of creating novel tumour-targeted molecular imaging agents [29, 30]. These CPMV VLPs exhibit high selectivity for molecular targets that are cancer-specific and as a result are effective for in vivo imaging of tumors. CPMV represents an icosahedral nanoparticle and the exterior of its capsid displays 300 accessible lysine residues; each of these can be conjugated to various chemical moieties [30]. Examples of the use of this technology include the construction of CPMV nanoparticles displaying gastrin releasing peptide receptor (GRPR) or vascular endothelial growth factor receptor 1 (VEGFR1) [31].

As another example, Cowpea Chlorotic Mottle Virus (CCMV) has been shown to stably assemble in vitro and package the RNA derived from a mammalian virus, Sindbis virus. The hybrid VLPs were able to deliver and release their RNA contents within the cytoplasm of mammalian cells. The CCMV-based VLP was shown to protect against RNA degradation by cellular nucleases. By conjugating subcellular targeting moieties, these hybrid VLPs could be directed toward distinct sites within the cell [32].

Plant viruses have also been engineered to act as adjuvants to elicit an immune response that is more potent and effective. Recently, nanoparticles that are constructed from the coat protein of the rod-shaped Papaya mosaic virus (PapMV) have been shown to be highly immunogenic and are taken up by dendritic cells [33]. These nanoparticles have been engineered to express an influenza epitope on their surface, and mice and ferrets immunized with these recombinant nanoparticles exhibit an increase in robustness of humoral response to influenza virus infection [34]. TMV has also been demonstrated to stimulate cellular immunity. Incubation of TMV CP fused to T cell tumor-specific epitopes elicited immune responses and protected against tumor challenge in mice, indicating that the virus could act as an antigen carrier and induce an adequate immune response [35].

CONCLUSIONS

The development of plant virus vectors continues to evolve with respect to their ease and breadth of use in the field of plant-made biologics. Improved expression vectors based on deconstructed viruses will most likely remain a trend in the near future. Efforts will be made to address other challenges including enhanced expression levels, the generation of proteins with more complex quaternary structures, and issues concerning host specificity. With the advancement of techniques to facilitate their introduction and transient expression in plants, plant virus expression vectors and components derived from them offer strategic advantages for the rapid and cost-effective production of biologics for the world's poor. The fact that this technology lends itself to other applications as well, including the development of vaccines to combat global pandemics and even cancer targeting therapeutics, offers much needed vigor to address a broadening spectrum of needs for medical researchers for many years to come.

REFERENCES

1. Hamorsky KT, Kouokam JC, Bennett LJ, et al. Rapid and scalable plant-based production of a cholera toxin B subunit variant to aid in mass vaccination against cholera outbreaks. *Public Library of Sciences Neglected Tropical Diseases*. 2013;7(3, article e2046)
2. McCormick AA, Reddy S, Reinl SJ, et al. Plant-produced idiotype vaccines for the treatment of non-Hodgkin's lymphoma: safety and immunogenicity

in a phase I clinical study. *Proceedings of the National Academy of Sciences of the United States of America*. 2008;105(29):10131–10136.

3. Leuzinger K, Dent M, Hurtado J, et al. Efficient agroinfiltration of plants for high-level transient expression of recombinant proteins. *Journal of Visual Experimentation*. 2013;(77)
4. Gleba Y, Marillonnet S, Klimyuk V. Engineering viral expression vectors for plants: the "full virus" and the "deconstructed virus" strategies. *Current Opinion in Plant Biology*. 2004;7(2):182–188.
5. Klimyuk V, Pogue G, Herz S, Butler J, Haydon H. Production of recombinant antigens and antibodies in*Nicotiana benthamiana* using "Magnifection" technology: GMP-compliant facilities for small- and large-scale manufacturing. *Current Topics in Microbiology and Immunology*. 2014;375:127–154.
6. Kagale S, Uzuhashi S, Wigness M, et al. TMV-Gate vectors: gateway compatible tobacco mosaic virus based expression vectors for functional analysis of proteins. *Science Reports*. 2012;2, article 874
7. Lindbo JA. High-efficiency protein expression in plants from agroinfection-compatible Tobacco mosaic virus expression vectors. *BMC Biotechnology*. 2007;7, article 52
8. Lindbo JA. TRBO: a high-efficiency tobacco mosaic virus RNA-based overexpression vector. *Plant Physiology*. 2007;145(4):1232–1240.
9. Hamorsky KT, Grooms-Williams TW, Husk AS, Bennett LJ, Palmer KE, Nobuyuki Matoba N. Efficient single tobamoviral vector-based bioproduction of broadly neutralizing anti-HIV-1 monoclonal antibody VRC01 in *Nicotiana benthamiana* plants and utility of VRC01 in combination microbicides. *Antimicrobial Agents and Chemotherapy*. 2013;57(5):2076–2086.
10. Petukhova NV, Gasanova TV, Stepanova LA, et al. Immunogenicity and protective efficacy of candidate universal influenza A nanovaccines produced in plants by *Tobacco mosaic* virus-based vectors. *Current Pharmaceutical Design*. 2013;19(31):5587–5600.
11. Yusibov V, Streatfield SJ, Kushnir N, Roy G, Padmanaban A. Hybrid viral vectors for vaccine and antibody production in plants. *Current Pharmaceutical Design*. 2013;19(31):5574–5586.
12. Shoji Y, Jones RM, Mett V, et al. A plant-produced H1N1 trimeric hemagglutinin protects mice from a lethal influenza virus challenge. *Human Vaccine Immunotherapy*. 2013;9(3)
13. Cummings JF, Guerrero ML, Moon JE, et al. Safety and immunogenicity of a plant-produced recombinantmonomer hemagglutinin-based influenza vaccine derived frominfluenza A (H1N1)pdm09 virus: a Phase 1 dose-escalation study in healthy adults. *Vaccine Journal*. 2013
14. Neuhaus V, Schwarz K, Klee A, et al. Functional testing of an inhalable nanoparticle based influenza vaccine using a human precision cut lung slice technique. *PLoS ONE*. 2013;8(8)e71728
15. Vezina L-P, D'Aoust MA, Landry N, et al. Plants as an innovative and accelerated vaccine-manufacturing solution. *BioPharmaceutical International Supplementary*. 2011;24(5):s27–s30.
16. Li C, Jiang Y, Guo W, Liu Z. Production of a chimeric allergen derived from the major allergen group 1 of house dust mite species in *Nicotiana benthamiana*. *Human Immunology*. 2013;74(5):531–537.
17. Nausch H, Mikschofsky H, Koslowski R, Meyer U, Broer I, Huckauf J. Expression and subcellular targeting of human complement factor C5a in *Nicotiana species*. *PLoS ONE*. 2012;7(12)
18. Azhakanandam K, Weissinger SM, Nicholson JS, Qu R, Weissinger AK. Amplicon-plus targeting technology (APTT) for rapid production of a highly unstable vaccine protein in tobacco plants. *Plant Molecular Biology*. 2007;63(3):393–404.
19. Cerovska N, Hoffmeisterova H, Moravec T, et al. Transient expression of Human papillomavirus type 16 L2 epitope fused to N- and C-terminus of coat protein of Potato virus X in plants. *Journal of Biosciences*.2012;37(1):125–133.
20. Minato N, Komatsu K, Okano Y, et al. Efficient foreign gene expression in planta using a plantago asiatica mosaic virus-based vector achieved by the strong RNA-silencing suppressor activity of TGBp1.*Archives of Virology*. In press.
21. Hwang MS, Lindenmuth BE, McDonald KA, Bryce W, Falk B. Bipartite and tripartite *Cucumber mosaic* virus-based vectors for producing the *Acidothermus cellulolyticus* endo-1,4- -glucanase and other proteins in non-transgenic plants. *BMC Biotechnology*. 2012;12, article 66
22. Gellért A, Salánki K, Tombácz K, Tuboly T, Balázs E. A cucumber mosaic virus based expression system for the production of porcine circovirus specific vaccines. *PLoS ONE*. 2012;7(12)e52688
23. Peyret H, Lomonossoff GP. The pEAQ vector series: the easy and quick way to produce recombinant proteins in plants. *Plant Molecular Biology*. 2013;(1-2):51–58.
24. D'Aoust M-A, Lavoie P-O, Couture MM-J, et al. Influenza virus-like particles produced by transient expression in *Nicotiana benthamiana* induce a protective immune response against a lethal viral challenge in mice. *Plant Biotechnology Journal*. 2008;6(9):930–940.
25. Sainsbury F, Sack M, Stadlmann J, Quendler H, Fischer R, Lomonossoff GP. Rapid transient production in plants by replicating and non-replicating vectors yields high quality functional anti-HIV antibody. *PLoS ONE*. 2010;5(11)e13976
26. Thuenemann EC, Meyers AE, Verwey J, Rybicki EP, Lomonossoff GP. A method for rapid production of heteromultimeric protein complexes in plants: assembly of protective bluetongue virus-like particles. *Plant Biotechnology Journal*. 2013;11(7):839–846.
27. Chen Q, He J, Phoolcharoen W, Mason HS. Geminiviral vectors based on bean yellow dwarf virus for production of vaccine antigens and monoclonal antibodies in plants. *Human Vaccines*. 2011;7(3):331–338.

28. Chung HY, Lee HH, Kim KI, et al. Expression of a recombinant chimeric protein of hepatitis A virus VP1-Fc using a replicating vector based on *Beet curly* top virus in tobacco leaves and its immunogenicity in mice. *Plant Cell Reports*. 2011;30(8):1513–1521.

29. Cho C-F, Shukla S, Simpson EJ, et al. *Virus Hybrids as Nanomaterials: Methods and Protocols, Methods in Molecular Biology*. Vol. 1108. New York, NY, USA: Springer Science, Business Media; 2014. Molecular targeted viral nanoparticles as tools for imaging cancer.

30. Steinmetz NF, Cho C-F, Ablack A, Lewis JD, Manchester M. Cowpea mosaic virus nanoparticles target surface vimentin on cancer cells. *Nanomedicine*. 2011;6(2):351–364.

31. Steinmetz NF, Ablack AL, Hickey JL, et al. Intravital imaging of human prostate cancer using viral nanoparticles targeted to gastrin-releasing peptide receptors. *Small*. 2011;7(12):1664–1672.

32. Azizgolshani O, Garmann RF, Cadena-Nava R, Knobler CM, Gelbart WM. Reconstituted plant viral capsids can release genes to mammalian cells. *Virology*. 2013;441(1):12–17.

33. Lacasse P, Denis J, Lapointe R, Leclerc D, Lamarre A. Novel plant virus-based vaccine induces protective cytotoxic T-lymphocyte-mediated antiviral immunity through dendritic cell maturation. *Journal of Virology*. 2008;82(2):785–794.

34. Savard C, Guérin A, Drouin K, et al. Improvement of the trivalent inactivated flu vaccine using papmv nanoparticles. *PLoS ONE*. 2011;6(6)e21522

35. McCormick AA, Corbo TA, Wykoff-Clary S, et al. TMV-peptide fusion vaccines induce cell-mediated immune responses and tumor protection in two murine models. *Vaccine*. 2006;24(40-41):6414–6423.

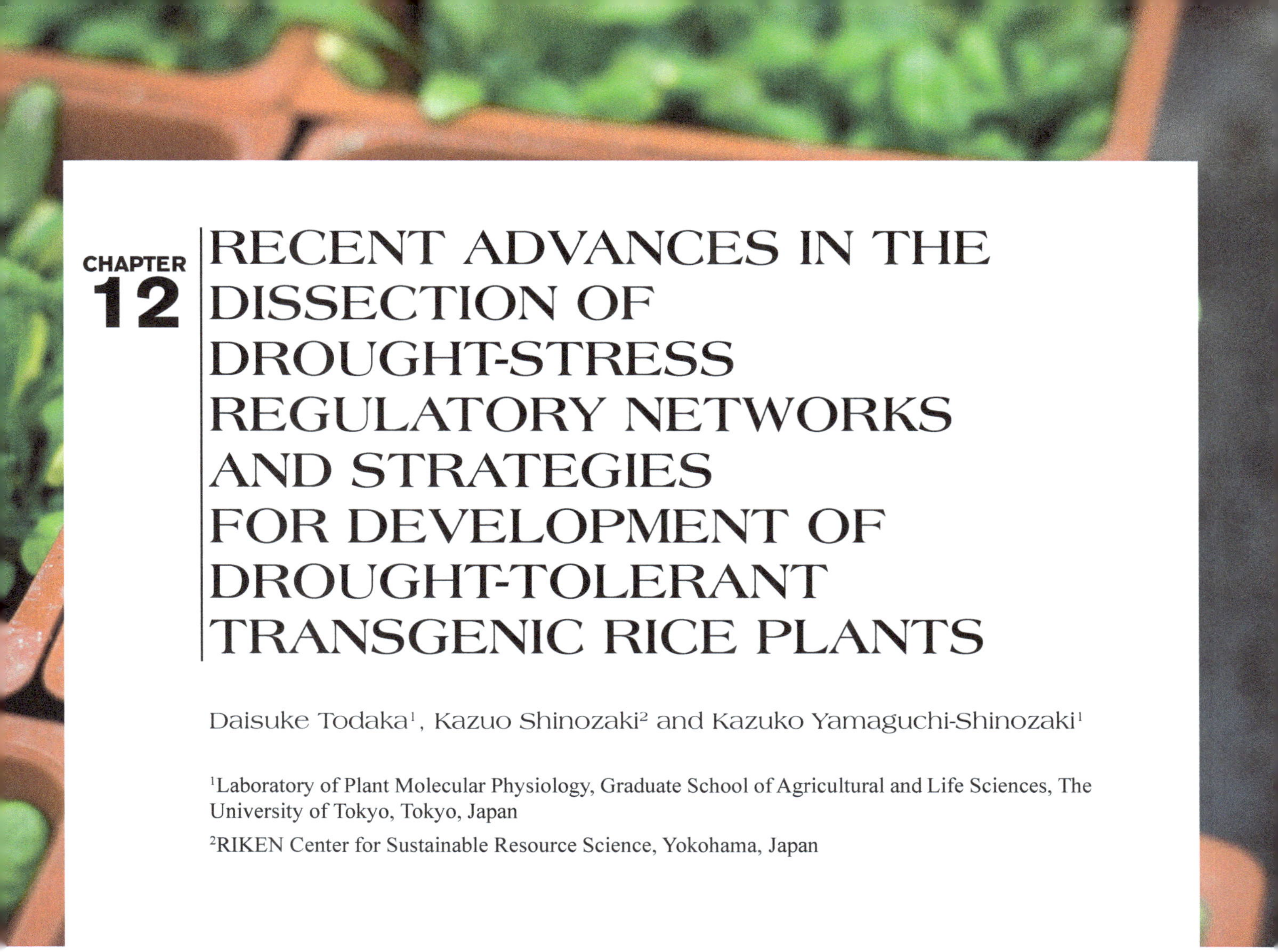

CHAPTER 12

RECENT ADVANCES IN THE DISSECTION OF DROUGHT-STRESS REGULATORY NETWORKS AND STRATEGIES FOR DEVELOPMENT OF DROUGHT-TOLERANT TRANSGENIC RICE PLANTS

Daisuke Todaka[1], Kazuo Shinozaki[2] and Kazuko Yamaguchi-Shinozaki[1]

[1]Laboratory of Plant Molecular Physiology, Graduate School of Agricultural and Life Sciences, The University of Tokyo, Tokyo, Japan

[2]RIKEN Center for Sustainable Resource Science, Yokohama, Japan

ABSTRACT

Advances have been made in the development of drought-tolerant transgenic plants, including cereals. Rice, one of the most important cereals, is considered to be a critical target for improving drought tolerance, as present-day rice cultivation requires large quantities of water and as drought-tolerant rice plants should be able to grow in small amounts of water. Numerous transgenic rice plants showing enhanced drought tolerance have been developed to date. Such genetically engineered plants have generally been developed using genes encoding proteins that control drought regulatory networks. These proteins include transcription factors, protein kinases, receptor-like kinases, enzymes related to osmoprotectant or plant hormone synthesis, and other regulatory or functional proteins. Of the drought-tolerant transgenic rice plants described in this review, approximately one-third show decreased plant height under non-stressed conditions or in response to abscisic acid treatment. In cereal crops, plant height is a very important agronomic trait directly affecting yield, although the improvement of lodging resistance should also be taken into consideration. Understanding the regulatory mechanisms of plant growth reduction under drought stress conditions holds promise for developing transgenic plants that produce high yields under drought stress conditions. Plant growth rates are reduced more rapidly than photosynthetic activity under drought conditions, implying that plants actively reduce growth in response to drought stress. In this review, we summarize studies on molecular regulatory networks involved in response to drought stress. In a separate section, we highlight progress in the development of transgenic drought-tolerant rice plants, with special attention paid to field trial investigations.

INTRODUCTION

Drought is inevitable. For example, the U.S. suffered an agricultural drought in 2012, in which a 12% decrease in corn production was recorded compared with the previous year (USDA, 2014). Because such decreases in crop production cause enormous economic disruption, demand for the development of drought-tolerant crops is increasing.

Rice (*Oryza sativa* L.) is one of the world's most important cereals, with production comparable to that of wheat. In 2013, rice and wheat were cultivated in 124 and 126 countries, respectively, with corresponding worldwide production of 745 and 713 million tons (FAOSTAT). Compared with other cereal crops such as maize and wheat, rice is sensitive to decreases in soil water content because rice cultivars have been historically

Citation: Daisuke Todaka, Kazuo Shinozaki and Kazuko Yamaguchi-Shinozaki, Recent advances in the dissection of drought-stress regulatory networks and strategies for development of drought-tolerant transgenic rice plants, doi: 10.3389/fpls.2015.00084.

grown under flood irrigation conditions where the soil matric potential is zero. As a consequence, large amounts of water are required for production of rice compared with other crops. Production of 1 kg of rice seed requires 3,000 to 5,000 L of water, with less than half that amount needed for 1-kg seed production in other crops such as maize or wheat (Singh et al., 2002). Improvement of water-use efficiency during rice production should thus contribute significantly to agricultural water conservation and deserves much attention. Rice cultivars showing normal or even increased yield under drought stress conditions are expected to be closely related to those with high water-use efficiency.

Transgenic engineering approaches in plants have opened the door to the development of new cultivars with improved drought tolerance. Progress has been made in the generation of transgenic drought-tolerant rice plants. In this review, we begin with an overview of abiotic stress signaling pathways coordinated by a wide range of regulatory proteins, including key factors for the development of transgenic drought-tolerant rice plants, and then describe advances in the development of transgenic drought-tolerant rice. We also discuss growth regulatory mechanisms operating under water deficit stress conditions. Special attention is paid to transgenic rice plants showing improved drought tolerance under field conditions.

REGULATORY MECHANISMS OF RESPONSES TO ABIOTIC STRESSES, INCLUDING DROUGHT, IN *ARABIDOPSIS*

Abiotic stresses, such as drought, high salinity, and low temperature, induce the expression of a large number of genes. The induction of these genes under stress is regulated through complex transcriptional networks (Yamaguchi-Shinozaki and Shinozaki, 2006). The key genes functioning in these transcriptional networks have been revealed by molecular studies and are important candidates for the development of transgenic plants tolerant to abiotic stress. Here, we highlight two important pathways of transcriptional networks under abiotic stress conditions in *Arabidopsis*: an abscisic acid (ABA)-dependent signaling pathway and an ABA-independent regulatory network mediated by dehydration responsive element-binding (DREB)-type transcription factors (Figure 1A). Numerous excellent review articles on global abiotic stress regulatory networks have previously been published (Zhu, 2002; Bartels and Sunkar, 2005; Chinnusamy et al., 2007;Hua, 2009; Thomashow, 2010; Qin et al., 2011; Deinlein et al., 2014; Golldack et al., 2014; Yoshida et al., 2014).

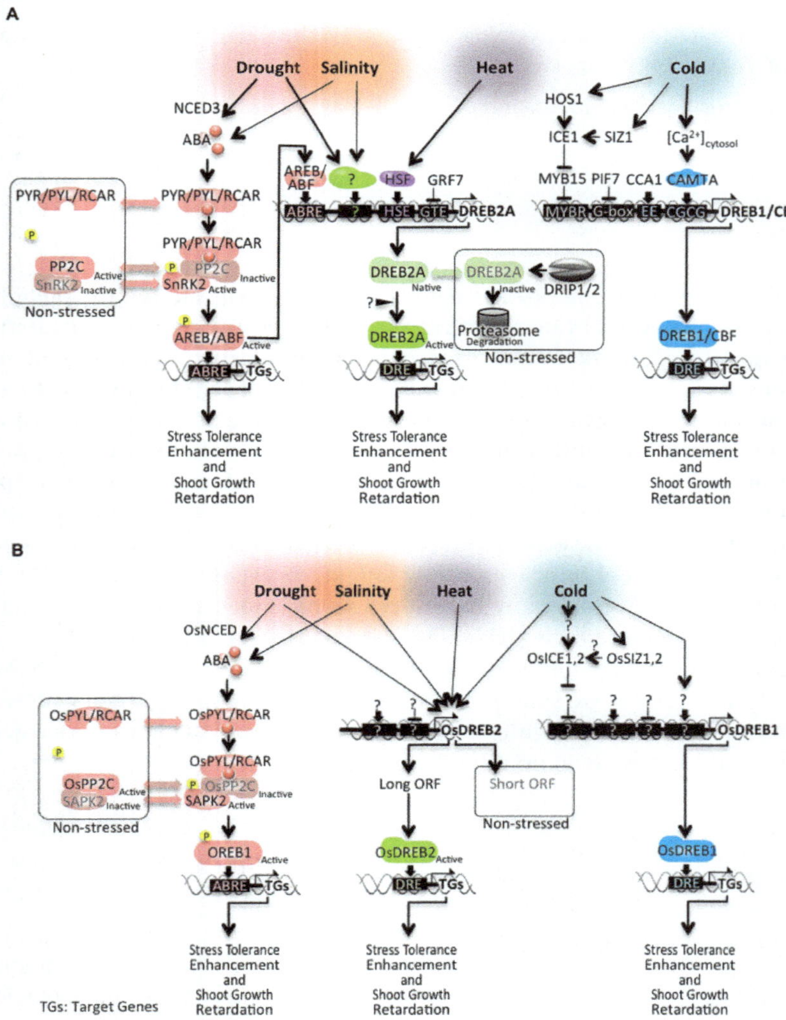

Figure 1: Abiotic stress signaling networks mediated by AREB, DREB1, and DREB2-type transcription factors (A) in *Arabidopsis* **and (B)** in rice.

The ABA-Dependent Signaling Pathway

In *Arabidopsis*, significant progress has been made in the elucidation of molecular mechanisms of transcriptional networks involved in abiotic stress response. The phytohormone ABA is a major molecule facilitating signal transduction during drought stress response. A key enzyme for ABA biosynthesis is 9-*cis*-epoxycarotenoid dioxygenase (NCED; Iuchi et al., 2001). Among five genes encoding NCED in*Arabidopsis*, expression of *NCED3* has been found to increase under water deficit conditions (Iuchi et al., 2001). A G-box recognition sequence located -2,248 bp from the *NCED3* transcriptional start site has recently been shown to be important for this gene's expression under water deficit conditions (Behnam et al., 2013). It has also emerged that ABA intercellular transport mechanisms are important for ABA-dependent drought responses. Kuromori et al. (2014) have demonstrated that specific cells in vascular tissue synthesize ABA and transport the molecule to target cells. Bauer et al. (2013) have proposed that ABA is autonomously synthesized in guard cells.

Synthesized or transported ABA is perceived by a receptor complex, which consists of PYR (PYRABACTIN RESISTANCE)/PYL (PYR1-LIKE)/RCAR (REGULATORY COMPONENT OF ABA RESPONSE), PP2C (protein phosphatase 2C), and SnRK2 (sucrose non-fermenting 1-related protein kinase 2; Cutler et al., 2010; Raghavendra et al., 2010; Umezawa et al., 2010; Weiner et al., 2010). A suite of studies has clarified the molecular structural changes that occur during ABA perception and in the ABA signaling cascade (Cutler et al., 2010; Raghavendra et al., 2010; Umezawa et al., 2010; Weiner et al., 2010;Miyakawa et al., 2013). In the absence of ABA, PP2Cs repress the ABA signaling pathway by dephosphorylation-triggered inactivation of SnRK2s. In the presence of ABA, ABA-bound PYL/PYR/RCARs recognize and bind to PP2Cs, thereby releasing SnRK2s from PP2C-dependent negative regulation. The activated SnRKs phosphorylate downstream proteins including AREB/ABF (ABA-responsive *cis*-element binding protein/ABA-responsive *cis*-element binding factor) transcription factors. The AREB/ABF transcription factors have a bZIP domain and four conserved domains containing SnRK2 phosphorylation sites. The phosphorylated AREB/ABFs are activated and bind to the ABA-responsive *cis*-element (ABRE; PyACGTGG/TC), which is enriched in promoter regions of drought-inducible genes. AREB/ABFs function as master transcriptional activators regulating ABRE-dependent gene expression in ABA signaling under drought stress conditions.

The ABA-Independent Signaling Pathway Mediated By DREB2 and Dreb1/Cbf

Evidence has revealed that ABA-independent signaling pathways are also important in abiotic stress response. DREB2 proteins are members of the AP2/ERF family of plant-specific transcription factors. Among the eight *DREB2* genes in *Arabidopsis*, *DREB2A* and *DREB2B* are highly induced by drought, high salinity, and heat stress, and function as transcriptional activators in the ABA-independent pathway. A negative regulatory domain has been identified in the DREB2A amino acid sequence and is reported to be involved in DREB2A protein stability (Sakuma et al., 2006a; Mizoi et al., 2012). Under non-stressed conditions, degradation of DREB2A proteins occurs via ubiquitination of DREB2A by the C3HC4 RING domain-containing proteins DRIP1 (DREB2A-interacting protein1) and DRIP2 (Qin et al., 2008; Morimoto et al., 2013). Kim et al. (2012a) have proposed that *DREB2A* expression is repressed by GRF7, a growth-regulating factor, to prevent growth inhibition under non-stressed conditions. DREB2A also plays a role in high temperature stress response and increased heat stress tolerance (Sakuma et al., 2006b). The use of HsfA1 multiple mutants has revealed that *DREB2A* expression under heat stress conditions is regulated by HsfA1 (heat shock factor A1) proteins (Yoshida et al., 2011).

DREB1/CBF transcription factors are another subfamily that regulates expression of many abiotic stress-responsive genes. DREB1/CBFs are key regulators in low-temperature stress-responsive gene expression. Transgenic *Arabidopsis* plants overexpressing *DREB1/CBF* genes have been found to improve low-temperature stress tolerance as well as drought and salinity stress tolerance (Kasuga et al., 1999; Yamaguchi-Shinozaki and Shinozaki, 2006). Several studies have reported that expression of *DREB1/CBF* genes is directly or indirectly modulated by regulatory factors such as HOS1 (Dong et al., 2006), ICE1 (Chinnusamy et al., 2003), SIZ1 (Miura et al., 2007), MYB15 (Agarwal et al., 2006), PIF7 (Kidokoro et al., 2009), CAMTA3 (Doherty et al., 2009), and a clock component (Dong et al., 2011). Recent excellent reviews byMedina et al. (2011) and Mizoi et al. (2012) are highly informative regarding DREB transcription factor proteins and the related signaling cascade in *Arabidopsis*.

MOLECULAR RESPONSES TO DROUGHT STRESS IN RICE

In rice, more than 5,000 genes are up-regulated and more than 6,000 are down-regulated by drought stress (Maruyama et al., 2014). A comparison between rice and *Arabidopsis* by Maruyama et al. (2014)demonstrated that different metabolites are accumulated under abiotic stress conditions. For example, high expression levels of genes encoding isocitrate lyase and malate synthase in the glyoxylate cycle along with glucose accumulation under abiotic stress conditions were observed in rice, but not in *Arabidopsis*. Additionally, reduced expression of the cytochrome P450 735A gene was correlated with decreased cytokinin levels in rice, but not in *Arabidopsis*. Similar comprehensive analyses of drought stress-responsive genes, proteins, and metabolites in rice have been performed as follows. Wang et al. (2011a) identified 5,284 drought stress-responsive genes. Lenka et al. (2011) compared drought-responsive genes in *indica* rice genotypes having contrasting drought tolerances. Up-regulation of the α-linolenic acid metabolic pathway was observed in the drought-tolerant genotype. Degenkolbe et al. (2009) also investigated comprehensive expression profiles of drought-responsive genes in both drought-sensitive and drought-tolerant rice genotypes. They found that senescence-related degradation processes and expression of photosynthesis-related genes were reduced in drought-tolerant cultivars compared with those in drought-sensitive ones. Using the comprehensive expression data, the authors also identified marker transcripts for selection of drought tolerance in a range of rice germplasm resources through integrated analyses of gene expression and stress tolerance (Degenkolbe et al., 2013). The marker transcripts showed a significant correlation between expression level and tolerance under drought stress conditions. One of the markers was a gene for cytosolic fructose-1,6-bisphosphatase, an enzyme that catalyzes a highly regulated step in C-metabolism. Ray et al. (2011) have reported that genes responsive to drought stress conditions significantly overlap with those expressed during panicle and seed development. Shu et al. (2011) performed integrated analyses of two-dimensional electrophoresis – mass spectrometry – mass spectrometry, cDNA microarray, and gas chromatography – mass spectrometry data from rice seedlings after drought stress treatment to, respectively, determine protein expression levels, gene expression levels, and metabolite contents. The authors speculated that energy conversion from carbohydrates and/or fatty acids to amino acids increased under drought stress conditions. Epigenetic research on drought response in rice has been also reported. Wang et al. (2011b)examined drought-induced genome-wide DNA methylation and its association with drought tolerance. Zong et al. (2013) investigated the genome-wide distribution pattern of histone H3 lysine 4 tri-methylation and found that methylation levels were positively correlated with expression levels of some of the evaluated drought-responsive genes.

While these large data sets have provided much information on drought responses in rice, studies on associated signaling cascades have been limited. Evidence indicates that *O. sativa NCED* transcripts are up-regulated along with ABA accumulation under drought stress conditions (Maruyama et al., 2014), and that a core ABRE sequence in the promoter regions of drought-inducible genes is enriched in rice, *Arabidopsis* and soybean (Maruyama et al., 2012). These results suggest that the ABA-dependent signaling pathway in rice is activated by drought stress, similar to that of *Arabidopsis* and other plant species. He et al. (2014) determined the crystal structure of the ABA–OsPYL2–OsPP2C06 complex and suggested that the complex has the potential to be an ABA receptor in rice. Kim et al. (2012b) identified a rice ABA signaling unit composed of OsPYL/RCAR5, OsPP2C30, SAPK2, and OREB1 for ABA-dependent gene regulation. They have reported that OsPYL/RCAR5 functions as a positive regulator of abiotic stress-responsive gene expression and that transgenic rice plants overexpressing *OsPYL/RCAR5* have improved drought tolerance. The OREB1 bZIP-type transcription factor, which is ortholog to *Arabidopsis* AREB, has been shown to regulate the ABA-dependent pathway in rice (Chae et al., 2007; Hong et al., 2011).

Experimental evidence has demonstrated that rice DREB transcription factors also function as important regulators in ABA-independent drought responses. The rice genome contains five DREB2-type genes, two of which—*OsDREB2A* and *OsDREB2B*—are up-regulated by abiotic stress (Matsukura et al., 2010). Transgenic rice plants overexpressing these two genes have been found to have increased drought tolerance (Chen et al., 2008; Cui et al., 2011). *OsDREB2B* generates two forms of the transcripts, *OsDREB2B1* and*OsDREB2B2*. *OsDREB2B1* encodes a non-functional protein (Short ORF in Figure1B). *OsDREB2B2*contains a coding region with AP2/ERF DNA binding domain (Long ORF in Figure 1B). *OsDREB2B2*transcripts were accumulated by heat, cold, drought, and high salinity stress treatments, while *OsDREB2B1*transcripts were not changed except for cold stress. These results suggest that OsDREB2B2 plays an important role in the abiotic stress response of rice through the alternative splicing regulatory system (Matsukura et al., 2010). Expression of *OsDREB1A* and *OsDREB1B* is up-regulated by low-temperature stress (Dubouzet et al., 2003), while expression of *OsDREB1F* and *OsDREB1G* is increased by water deficit stress (Chen et al., 2008; Wang et al., 2008). In contrast to other DREB1-type genes in rice, OsDREB1F likely participates as a regulatory factor in the ABA-dependent pathway. Transgenic rice plants overexpressing these genes also exhibit increased drought tolerance. It has been reported that OsICE1, OsICE2, OsSIZ1, and OsSIZ2 are involved in the cold stress signaling pathway that regulates *OsDREB1B*expression (Park et al., 2010; Nakamura et al., 2011).

TRANSGENIC RICE PLANTS THAT ENHANCE DROUGHT STRESS TOLERANCE

Genetic engineering has opened the door to the development of new cultivars with improved drought stress tolerance. Reports on transgenic rice plants that show increased drought stress tolerance are accumulating. A selected list of transgenic rice plants, which includes information on transgenes and promoters used for the transformations as well as plant stress tolerance and growth performance, is given in Table 1.

Table 1: Selective transgenic rice plants with improved drought stress tolerance

	Construct used for transformation			Type of transgenic engineering	Performance			
Gene function	Gene name	Gene source	Promoter		Stress tolerance	Field trial	Growth trait	Reference
Transcription factor								
bHLH	*OsbHLHU8*	Rice	Cytochrome*c*	COE*[1]	D↑*[5]	–	Not mentioned	Seo et al. (2011)
bZIP	*OsbZIP23*	Rice	Ubiquitin	COE	D↑, S↑*[6]	–	Hypersensitive to shoot growth inhibition by exogenous ABA	Xiang et al. (2008)
bZIP	*OsbZIP46*	Rice	Ubiquitin	COE	D↑, O↑*[7]	–	Hypersensitive to shoot growth inhibition by exogenous ABA	Tang et al. (2012)

bZIP	*OsbZIP71*	Rice	35S	COE	D↑, S↑,O↑	–	Not mentioned	Liu et al. (2014)
bZIP	*OsbZIP71*	Rice	RD29A	SIE*[2]	O↑	–	Not mentioned	Liu et al. (2014)
bZIP	*OsbZIP16*	Rice	Actin	COE	D↑	–	Hypersensitive to shoot growth inhibition by exogenous ABA	Chen et al. (2012)
bZIP	*ABF3*	*Arabidopsis*	Ubiquitin	COE	D↑	–	Normal shoot growth and seed production under control conditions	Oh et al. (2005)
AP2/ERF	*OSDERF1*	Rice	Actin	CRNAi*[3]	D↑	–	Not mentioned	Wan et al. (2011)
AP2/ERF	*OsERF4a*	Rice	Cytochrome*c*	COE	D↑	–	Increase in shoot growth under control conditions	Joo et al. (2013)
AP2/ERF	*OsERF4a*	Rice	Ai	AIE*[4]	D↑	–	Increase in shoot growth under control conditions	Joo et al. (2013)
AP2/ERF	*OsERF10a*	Rice	Cytochrome*c*	COE	D↑	–	Not mentioned	Joo et al. (2013)
AP2/ERF	*TSRF1*	Tomato	35S	COE	D↑, O↑	–	Increases in seedling weight and root length under drought conditions	Quan et al. (2010)
AP2/ERF	*JERF3*	Tomato	35S	COE	D↑, O↑	–	Normal shoot growth under control conditions	Zhang et al. (2010)
AP2/ERF	*OsDREB2A*	Rice	4ABRC	SIE	D↑, S↑	–	Increases in plant height and effective tillers under drought conditions	Cui et al. (2011)
AP2/ERF	*OsDREB2B*	Rice	4ABRC	SIE	D↑, S↑	–	Increases in plant height and effective tillers	Cui et al. (2011)
AP2/ERF	*HARDY*	*Arabidopsis*	35S	COE	D↑	–	Increases in shoot growth under control conditions and in root growth under drought conditions	Karaba et al. (2007)
AP2/ERF	*ZmCBF3*	Maize	Ubiquitin	COE	D↑, S↑, C↑*[8]	–	Normal yield under control conditions, hypersensitive to root growth inhibition by exogenous ABA	Xu et al. (2011)
AP2/ERF	*CBF3/ DREB1A*	*Arabidopsis*	Ubiquitin	COE	D↑, S↑	–	Normal shoot growth and seed production under non-stressed conditions	Oh et al. (2005)
AP2/ERF	*CBF3/ DREB1A*	*Arabidopsis*	RD29A	SIE	D↑	–	Increase in yield under drought conditions	Datta et al. (2012)
AP2/ERF	*CBF3/ DREB1A*	*Arabidopsis*	HVA22	SIE	D↑	Done	Increase in yield under drought conditions	Xiao et al. (2009)
AP2/ERF	*OsDREBIA*	Rice	Ubiquitin	COE	D↑, S↑, C↑	–	Shoot growth retardation under control conditions	Ito et al. (2006)
AP2/ERF	*OsDREBIB*	Rice	Ubiquitin	COE	D↑, S↑, C↑	–	Shoot growth retardation under control conditions	Ito et al. (2006)
AP2/ERF	*CBF3/ DREB1A*	*Arabidopsis*	Ubiquitin	COE	D↑, S↑, C↑	–	Shoot growth retardation under control conditions	Ito et al. (2006)

AP2/ERF	*CBF1/ DREB1B*	*Arabidopsis*	Ubiquitin	COE	D↑, S↑, C↑	–	Shoot growth retardation under control conditions	Ito et al. (2006)
AP2/ERF	*CBF2/ DREB1C*	*Arabidopsis*	Ubiquitin	COE	D↑, S↑, C↑	–	Shoot growth retardation under control conditions	Ito et al. (2006)
AP2/ERF	*HvCBF4*	Barley	Ubiquitin	COE	D↑, S↑, C↑	–	Normal shoot growth under control conditions	Oh et al. (2007)
AP2/ERF	*CBF2/ DREB1C*	*Arabidopsis*	Lip9	SIE	D↑	–	Increase in dry weight under drought conditions	Ishizaki et al. (2012)
AP2/ERF	*AP37*	Rice	OsCcI	COE	D↑, S↑, C↑	Done	Normal shoot growth under control conditions	Oh et al. (2009)
Homeodomain	*EDT1/HDG11*	*Arabidopsis*	Actin	COE	D↑	Done	Increases in shoot, root, and seed production under drought conditions	Yu et al. (2013)
Homeodomain	*Zmhdz10*	Maize	Ubiquitin	COE	D↑, S↑	–	Hypersensitive to shoot growth inhibition by exogenous ABA	Zhao et al. (2014)
MYB	*OsMYB2*	Rice	Ubiquitin	COE	D↑, S↑, C↑	–	Hypersensitive to shoot growth inhibition by exogenous ABA	Yang et al. (2012)
NAC	*OsNAC5*	Rice	RCc3	RSE*[10]	D↑	Done	Increases in yield and root diameter under drought conditions	Jeong et al. (2013)
NAC	*OsNAC9/ SNAC1*	Rice	RCc3	RSE	D↑	Done	Increases in yield and root diameter under drought conditions	Redillas et al. (2012)
NAC	*OsNAC10*	Rice	RCc3	RSE	D↑	Done	Increases in yield and root diameter under drought conditions	Jeong et al. (2010)
NAC	*OsNAC6*	Rice	Ubiquitin	COE	D↑	–	Shoot growth retardation under control conditions	Nakashima et al. (2007)
NAC	*SNAC1*	Rice	35S	COE	D↑, S↑	Done	Hypersensitive to shoot growth inhibition by exogenous ABA, increase in yield under drought conditions, normal shoot growth under control conditions	Hu et al. (2006)
WRKY	*OsWRKY30*	Rice	35S	COE	D↑	–	Not mentioned	Shen et al. (2012)
Zinc finger	*ZFP182*	Rice	35S	COE	D↑, S↑, C↑	–	Similar morphological traits under control conditions	Huang et al. (2012)
Zinc finger	*ZFP245*	Rice	35S	COE	D↑, C↑	–	Hypersensitive to shoot growth inhibition by exogenous ABA, normal shoot growth under control conditions	Huang et al. (2009)
Zinc finger	*ZFP252*	Rice	35S	COE	D↑, S↑	–	Not mentioned	Xu et al. (2008)
Zinc finger	*ZAT10*	*Arabidopsis*	Actin	COE	D↑	Done	Increases in yield under drought conditions	Xiao et al. (2009)

Zinc finger	*ZAT10*	*Arabidopsis*	HVA22	SIE	D↑	Done	Increases in yield under drought conditions	Xiao et al. (2009)
Kinases								
Calcium-dependent protein kinase	*OsCPK4*	Rice	Ubiquitin	COE	D↑, S↑	–	Not mentioned	Campo et al. (2014)
Calcium-dependent protein kinase	*OsCDPK1*	Rice	Ubiquitin	COE	D↑	–	Decreases in plant height and grain size under control conditions	Ho et al. (2013)
Calcium-dependent protein kinase	*OsCDPK7*	Rice	35S	COE	D↑, S↑, C↑	–	Normal growth under control conditions	Saijo et al. (2000)
Calcineurin B-like protein interacting protein kinase	*OsCIPK12*	Rice	Ubiquitin	COE	D↑	–	Normal growth under control conditions	Xiang et al. (2007)
Recetor-like kinase	*OsSIK1*	Rice	35S	COE	D↑, S↑	–	Normal growth under control conditions	Ouyang et al. (2010)
Recetor-like kinase	*OsSIK2*	Rice	35S	COE	D↑, S↑	–	Decrease in shoot growth under control conditions	Chen et al. (2013)
MAP kinase kinase kinase	*NPK1*	*Arabidopsis*	Actin	COE	D↑	Done	Increases in yield under drought conditions	Xiao et al. (2009)
MAP kinase kinase kinase	*NPK1*	*Arabidopsis*	HVA22	SIE	D↑	Done	Increases in yield under drought conditions	Xiao et al. (2009)
Phytohormones								
ABA receptor	*OsPYL/RCA R5*	Rice	Ubiquitin	COE	D↑, S↑	–	Hypersensitive to shoot growth inhibition by exogenous ABA, decrease in yield under control conditions	Kim et al. (2014)
Molybdenum cofactor sulfurase	*LOS5/ABA3*	*Arabidopsis*	Actin	COE	D↑	Done	Increases in yield under drought conditions	Xiao et al. (2009)
Molybdenum cofactor sulfurase	*LOS5/ABA3*	*Arabidopsis*	HVA22	SIE	D↑	Done	Increases in yield under drought conditions	Xiao et al. (2009)
Auxin efflux carrier	*OsPIN3t*	Rice	35S	COE	O↑	–	Increase in root growth under control conditions	Zhang et al. (2012)
Isopentenyl-transferase	*IPT*	*Agrobacterium*	SAPK	SMIE*11	D↑	–	Normal growth under control conditions	Peleg et al. (2011)
β-catotene Hydroxylase	*DSM2*	Rice	35S	COE	D↑	–	Increases in spikelet fertility under drought conditions	Du et al. (2010)
LEA proteins								
Group 3 LEA protein	*HVA1*	Barley	Actin	COE	D↑, S↑	–	Normal growth under control conditions	Xu et al. (1996)
Group 3 LEA protein	*OsLEA3-2*	Rice	2X35S	COE	D↑, S↑	–	Normal growth under control conditions	Duan and Cai (2012)
Group 3 LEA protein	*OsLEA3-1*	Rice	35S	COE	D↑	Done	Normal grain yield under control conditions	Xiao et al. (2007)
Group 3 LEA protein	*OsLEA3-1*	Rice	HVA1-like	SIE	D↑	Done	Normal grain yield under control conditions	Xiao et al. (2007)
Others								

Lipid transfer protein	*OsDIL*	Rice	Ubiquitin	COE	D↑	–	Increases in shoot growth under osmotic stress	Guo et al. (2013)
Heat shock protein	*OsHsp17.0*	Rice	35S	COE	D↑, S↑	–	Increases in shoot and root growth under osmotic stress	Zou et al. (2012)
Heat shock protein	*OsHsp23.7*	Rice	35S	COE	D↑, S↑	–	Increases in shoot and root growth under osmotic stress	Zou et al. (2012)
myo-Inositol oxygenase	*OsMIOX*	Rice	35S	COE	O↑	–	Increases in shoot growth under osmotic stress	Duan et al. (2012)
Ornithine δ-amino-transferase	*OsOAT*	Rice	Ubiquitin	COE	D↑, O↑	Done	Increases in shoot growth under osmotic stress	You et al. (2012)
His phosphotransfer protein	*OsAHP*	Rice	Ubiquitin	CRNAi	O↑	–	Decrease in internode length and increase in root growth under control conditions	Sun et al. (2014)
RING domain-containing protein	*OsRDCPl*	Rice	35S	COE	D↑	–	Not mentioned	Bae et al. (2011)
Ribosome-inactivating protein	*OsRIP18*	Rice	35S	COE	D↑, S↑	–	Not mentioned	Jiang et al. (2012)
RING-finger containing E3 ligase	*OsSDIRI*	Rice	Ubiquitin	COE	D↑	–	Normal growth at an adult stage under control conditions	Gao et al. (2011)
DNA and RNA helicase and ATPase	*OsSUV3*	Rice	35S	COE	O↑, S↑	–	Increases in shoot and root growth under high salinity	Tuteja et al. (2013)
Trehalose-6-phosphate synthase	*OsTPSI*	Rice	Actin	COE	D↑,S↑,C↑	–	Not mentioned	Li et al. (2011)
RNA binding protein	*GRP2*	*Arabidopsis*	Ubiquitin	COE	D↑	–	Normal shoot growth under control conditions	Yang et al. (2014)
RNA binding protein	*GRP7*	*Arabidopsis*	Ubiquitin	COE	D↑	–	Normal shoot growth under control conditions	Yang et al. (2014)
Harpin	*hrfl*	*Xanthomonas*	35S	COE	D↑	–	Not mentioned	Zhang et al. (2011)
Farnesyltransferase/ squalene synthase	*SQS*	Maize	Ubiquitin	CRNAi	D↑	–	Normal growth under control conditions and no inhibition of seedling growth by ABA treatment	Manavalan et al. (2012)
14-3-3 protein	*ZmGF14-6*	Maize	Ubiquitin	COE	D↑	–	Decreases in shoot growth and seed production under control conditions	Campo et al. (2012)
14-3-3 protein	*GF14c*	Rice	Ubiquitin	COE	D↑	–	Not mentioned	Ho et al. (2013)
Arginine decarboxylase	*Dadc*	*Datura*	Ubiquitin	COE	D↑	–	Normal growth under control conditions	Capell et al. (2004)
Manganese superoxide dimutase	*MnSOD*	Pea	SWPA2	SIE	O↑	–	Not mentioned	Wang et al. (2005)
Ski-interacting protein	*OsSKIPa*	Rice	Actin	COE	O↑	–	Normal growth under control conditions, less growth inhibition under ABA, salinity and osmotic stress condition;	Hou et al. (2009)

*[1]COE, constitutively overexpression; *[2]SIE, stress-inducible expression; *[3]CRNAi, constitutively triggering RNA interference; *[4]AIE, ABA-inducible expression; *[5]D↑, enhanced drought stress tolerance; *[6]S↑, enhanced salinity stress tolerance; *[7]O↑, enhanced osmotice stress tolerance; *[8]C↑, enhanced cold stress tolerance; *[10]RSE, root specific expression; *[11]SMIE, stress-and maturation-induced promoter.

bZIP-Type Transcription Factors

OsbZIP23, which is closely related to the *Arabidopsis* homologs ABF/AREB, is a major regulator of ABA-dependent pathways (Xiang et al., 2008). In the study of Xiang et al. (2008), *OsbZIP23*-overexpressing rice plants showed increased sensitivity to ABA at both germination and post-germination stages. The transgenic plants also exhibited enhanced tolerance to drought and salinity stresses. A transactivation assay indicated that OsbZIP23 functions as a transcriptional activator, with two regions of the OsbZIP23 amino acid sequence—at N- (1–59) and C- (210–240) termini—important for transcriptional activation. Microarray analysis detected hundreds of downstream genes of OsbZIP23 with diverse functions. These downstream genes included genes encoding stress-related transcription factors, protein kinases, dehydrins, and LEA proteins.

OsbZIP46 is a member of the subfamily that includes OsbZIP23. A transactivation assay conducted by Tang et al. (2012) revealed that an internal amino acid sequence (residues 122–219) of the OsbZIP46 protein had a negative affected on transactivation activity. A constitutively active form of OsbZIP46 (OsbZIP46CA1) was developed by deletion of this internal region. Transgenic rice plants overexpressing OsbZIP46CA1 showed increased drought tolerance. Microarray analysis was then used to detect up- or down-regulated genes in the*OsbZIP46CA1*-overexpressing transgenic rice plants. These differentially regulated genes were largely different from the OsbZIP23 downstream genes, suggesting that OsbZIP46CA1 regulates a different set of genes than does OsbZIP23. Although plant height under non-stressed conditions did not appear to differ between overexpressors and control plants, exogenous ABA application drastically decreased plant height of*OsbZIP46CA1* overexpressors. Similar growth inhibition was observed in *OsbZIP23* overexpressors. These results suggest that the ABA-dependent signaling pathway mediated by OsbZIP23 or OsbZIP46 is closely related to growth retardation mechanisms under drought stress conditions. This relationship implies that growth of transgenic rice plants overexpressing *OsbZIP23* or *OsbZIP46CA1* is decreased under drought stress conditions, even though transgenic plants exhibit increased stress tolerance.

Arabidopsis ABF3 belongs to the same bZIP subfamily as OsbZIP46 and OsbZIP23. Transgenic*Arabidopsis* plants overexpressing *ABF3* show improved drought tolerance (Kang et al., 2002). Under non-stressed conditions, overexpressors are morphologically identical to control plants. Transgenic rice plants overexpressing *ABF3* have also been found to exhibit improved drought tolerance without growth inhibition (Oh et al., 2005).

OsbZIP16 and OsbZIP71 are classified into group IV of the rice bZIP subfamily, which is different than the*Arabidopsis* ABF/AREB subfamily (Nijhawan et al., 2008). Transgenic rice plants overexpressing OsbZIP16 show improved drought tolerance and increased growth inhibition under exogenous ABA treatment (Chen et al., 2012), while OsbZIP71-overexpressing plants under the 35S promoter or a stress-inducible RD29A promoter exhibit improved drought, salt, and osmotic stress tolerance (Liu et al., 2014). Overexpression of these genes seems not to change plant architecture under non-stressed conditions.

These reports suggest that the bZIP-type transcription factors involved in the ABA signaling pathway are potentially useful for transgenic engineering to develop rice cultivars with enhanced drought tolerance. This notion is supported by the fact that ABA content in rice is increased by drought stress (Maruyama et al., 2014) and the finding that the ABRE *cis*-element is enriched in the promoter regions of drought-responsive genes in rice (Maruyama et al., 2012).

AP2/ERF-Type Transcription Factors

As mentioned above, the *Arabidopsis DREB1A* gene is a key regulator of abiotic stress response. At least three independent research groups have developed *DREB1A*-overexpressing transgenic rice plants. Oh et al. (2005) reported that overexpression of *DREB1A* driven by the ubiquitin promoter in rice plants enhances tolerance to drought and salinity stresses without growth retardation under non-stressed conditions. They observed up-regulated expression of several genes, including those responsive to stress, in transgenic rice plants. Datta et al. (2012) developed transgenic rice plants expressing *DREB1A* under the control of the stress-inducible *RD29* promoter. The yield of the transgenic rice plants under drought stress conditions was increased compared with that of non-transgenic plants. Ito et al. (2006) also reported that transgenic rice plants overexpressing *DREB1A* with the ubiquitin promoter showed enhanced tolerance to drought, cold, and salinity stresses. Elevated contents of osmoprotectants such as free proline and soluble sugars were also observed. A microarray analysis detected up-regulated genes in the transgenic rice plants; among the uncovered genes were genes for α-amylase and dehydrins, which were different from those identified by Oh et al. (2005). Unlike Oh et al. (2005), Ito et al. (2006) found that shoot growth retardation occurred in the transgenic plants under non-stressed conditions. Ito et al. (2006) also developed transgenic rice plants overexpressing *DREB1A* homologs, *Arabidopsis* paralogs *DREB1B* and *DREB1C* and rice orthologs*OsDREB1A* and *OsDREB1B* under the control of the ubiquitin promoter. These overexpressors showed enhanced tolerance to drought, salinity, and low temperature and displayed reduced growth under non-stressed conditions. Ishizaki et al. (2012) introduced *Arabidopsis DREB1C* into the upland rice cultivar NERICA1, an interspecific hybrid between stress-resistant *O. glaberrima* Steud. and high-yield *O. sativa*. The transgenic rice plants showed improved survival under drought stress conditions. *HvCBF4*, a member of the *Arabidopsis* DREB1A subfamily, has been isolated from barley as a low-temperature responsive gene. Overexpression of *HvCBF4* was found to enhance tolerance to drought, salinity, and low temperature while shoot growth was unaffected (Oh et al., 2007). ZmCBF3, a maize AP2/ERF-type transcription factor, is also a member of the *Arabidopsis* DREB1A subfamily. In a study by Xu et al. (2011), overexpression of*ZmCBF3* in transgenic rice plants enhanced tolerance to drought, salinity, and low-temperature stresses. Yields under control conditions were unchanged in the overexpressors compared with those in non-transgenic plants.

As mentioned above, the physiological role and molecular function of *Arabidopsis* DREB2A in abiotic stress responses have been vigorously studied (Mizoi et al., 2012). *Arabidopsis* DREB2A is widely recognized as a master regulator of both drought and heat stress responses and has a high potential to enhance drought and heat stress tolerance (Mizoi et al., 2012). The DREB2 regulatory mechanism seems to be well conserved in various plant species. In rice, there are five DREB2 family genes: *OsDREB2A*, *OsDREB2B*, *OsDREB2C*,*OsDREB2E*, and *OsABI4*. Transgenic *Arabidopsis* plants overexpressing *OsDREB2B* are reported to show increased expression of DREB2A target genes and enhanced tolerance to drought and heat stresses (Matsukura et al., 2010). Transgenic rice plants overexpressing *OsDREB2B* have also been found to increase drought tolerance (Chen et al., 2008). Overexpression of *OsDREB2A* under the control of an ABA-responsive promoter in rice plants increased contents of soluble sugars and proline at the seedling stage, resulting in increases in osmotic and salinity stress tolerance (Cui et al., 2011). The transgenic rice plants exhibited increased plant height and effective tiller numbers at the reproductive stage following drought treatment.

In addition to DREB1A and DREB2 subfamily members, several AP2/ERF-type and AP2/ERF-like transcription factors have been used to develop transgenic rice plants with enhanced abiotic stress tolerance. The *Arabidopsis HARDY* gene is an AP2/ERF-like transcription factor that enhances drought and salinity stress tolerance (Karaba et al., 2007). Transgenic rice plants overexpressing *HARDY* show enhanced photosynthetic assimilation and reduced transpiration, leading to increased shoot and root biomass. OsDERF1 is a protein that directly interacts with the GCC box in the promoter regions of *OsERF3* and*OsAP2-39* (Wan et al., 2011). Knockdown of *OsDERF1* increases ethylene biosynthesis and drought tolerance, suggesting that OsDERF1 modulates drought response via ethylene production. Transgenic rice plants overexpressing *OsERF3* show decreased drought tolerance (Zhang et al., 2013). In contrast, Joo et al. (2013) have reported that overexpression of OsERF3/OsERF4a decreases expression levels of a repressor involved in defense responses, leading to increased drought tolerance and seedling shoot growth. Overexpression of a tomato ERF gene, *TSRF1*, was found to improve osmotic and drought tolerance without growth retardation in rice seedlings. The improvement was attributed to corresponding increases in the expression of genes encoding MYC and MYB-type transcription factors and genes related to ABA synthesis, proline synthesis, and photosynthesis (Quan et al., 2010). Another tomato ERF gene, *JERF3*, has also been observed to increase drought tolerance when overexpressed in transgenic rice plants (Zhang et al., 2010). The studied transgenic rice plants showed higher contents of soluble sugars and proline compared with those of non-transgenic plants. Transgenic rice plants overexpressing *AP37*, a rice AP2/ERF transcription factor, showed enhanced tolerance to drought, cold and high salinity stresses (Oh et al., 2009). Microarray analysis identified *AP37* downstream genes, which included genes for PHD zinc finger and iron transporter.

Among the AP2/ERF transcription factor genes described here, *DREB1/CBF* genes have been widely used for the development of drought-tolerant transgenic crops. These developed crops include chrysanthemum (Hong et al., 2006), peanut (Bhatnagar-Mathur et al., 2013), soybean (Polizel et al., 2011), tobacco (Kasuga et al., 2004), tomato (Hsieh et al., 2002), tall fescue (Zhao et al., 2007), and wheat (Pellegrineschi et al., 2004).

NAC-Type Transcription Factors

NAC family proteins function in a wide variety of developmental processes and abiotic stress responses (Nakashima et al., 2012; Nuruzzaman et al., 2013). Constitutive overexpression of *OsNAC6* in rice plants was observed to increase tolerance to drought and salinity stresses (Nakashima et al., 2007). The transgenic rice plants showed decreased shoot growth under non-stressed conditions. When a stress-inducible promoter was used for the transgene expression, the transgenic plants showed normal growth under non-stressed conditions and improved salinity stress tolerance. In the same study, microarray analysis revealed OsNAC6 downstream genes including stress-related genes. The transgenic rice plants constitutively overexpressing*OsNAC6* also showed enhanced tolerance to blast disease, suggesting that OsNAC6 can act as a transcriptional regulator in both biotic and abiotic stress responses in rice.

In a study by Hu et al. (2006), *SNAC1* was demonstrated to be predominantly expressed in guard cells under drought conditions. Transgenic rice plants overexpressing *SNAC1* showed reduced water loss due to increased stomatal closure and enhanced expression of a large number of stress-related genes. As mentioned in the following section, the transgenic rice plants exhibited enhanced drought tolerance during field trials.

Transgenic rice plants overexpressing *OsNAC10* under the control of a root-specific promoter showed thicker roots and higher grain yields than those of control plants under drought stress conditions (Jeong et al., 2010). An accompanying microarray analysis identified various downstream genes, including *P450*, *Zn-finger*,*HAK5*, *2OG-Fe(II)*, *NCED*, *NAC*, and *KUP3* (Jeong et al., 2010). Similar to *OsNAC10* overexpressors, transgenic rice plants overexpressing *OsNAC5* (Jeong et al., 2013) and *OsNAC9/SNAC1* (Redillas et al., 2012) under root-specific promoter control have been shown to have thicker roots and higher grain yields than control plants under drought stress conditions. The microarray analysis of Jeong et al. (2013) identified 62 downstream genes of *OsNAC5*, including *NCED*, *calcium-transporting ATPase*, *germin-like protein*, and*meristem protein 5*. Only 17 of these downstream genes were up-regulated in *OsNAC10*-overexpressing transgenic rice plants (Jeong et al., 2013). With respect to *OsNAC9/SNAC1* overexpressors, identified downstream genes included *NCED* and *calcium-transporting ATPase* (Redillas et al., 2012). In addition to stress-responsive genes, OsNAC family downstream genes included genes involved in cell growth and development, suggesting that the OsNAC family is involved in regulatory mechanisms of stress responses and developmental processes. Song et al. (2011) observed that accumulations of proline and soluble sugars in*OsNAC5* overexpressors were higher than those of non-transgenic plants. Takasaki et al. (2010) have reported that OsNAC5 functions as a transcriptional activator and up-regulates expression of some stress-responsive genes in *OsNAC5* overexpressors. These authors also detected dimerization of OsNAC5 with OsNAC5, OsNAC5 with OsNAC6, and OsNAC5 with SNAC1.

Other Transcription Factors

Other transcription factors have also been applied in the successful development of transgenic rice plants with enhanced drought tolerance. Some of these plants have shown increased ABA sensitivity. *Arabidopsis EDT1/HDG11* encodes a homeodomain-leucine zipper transcription factor, which is likely involved in reproductive development (Yu et al., 2008). Overexpression of *EDT1/HDG11* in rice plants has been found to enhance root development, reduce stomatal density, and increase water-use efficiency (Yu et al., 2013). In the cited study, levels of ABA, proline and soluble sugars and activities of reactive oxygen species (ROS)-scavenging enzymes under drought stress conditions were higher in transgenic rice plants than in non-transgenic ones. Global gene expression analysis showed that stress-responsive genes, including *SNAC1*,*SNAC2*, *OsbZIP23*, and *OsNCED3*, were up-regulated in the transgenic rice plants. Under drought stress conditions, the transgenic rice plants exhibited higher grain yields compared with non-transgenic plants. Zhao et al. (2014) isolated a maize homeodomain-leucine zipper transcription factor gene, *Zmhdz10*, and generated transgenic rice plants overexpressing this gene. Overexpression of *Zmhdz10* enhanced tolerance to drought and salinity stresses and increased growth inhibition under exogenous ABA treatments. In another study, a rice R2R3-type MYB gene, *OsMYB2*, was overexpressed in rice plants (Yang et al., 2012). The transgenic overexpressors showed enhanced tolerance to drought, salinity and low-temperature stresses, and normal growth rates under non-stressed conditions. Exogenous ABA treatment resulted in greater growth inhibition of shoots of the overexpressors than those of non-transgenic plants.

Overexpression of *OsWRKY30* in rice plants has been found to enhance drought tolerance (Shen et al., 2012). In contrast, no improved drought tolerance due to overexpression of *OsWRKY30AA*, where all serine residues followed by proline are replaced by alanine residues in the encoded protein, has been observed. The observed interaction of these OsWRKY30 proteins with various MAP kinase proteins suggests that OsWRKY30 functions downstream of the MAPK cascades (Shen et al., 2012). A few studies have reported that overexpression of genes encoding C2H2-type zinc finger transcription factors improves drought tolerance in transgenic rice plants. For example, *ZFP182*-overexpressing transgenic rice plants exhibit increased expression levels of *OsDREB1A*, *OsDREB1B*, *OsP5CS*, and *OsLEA3* and show enhanced tolerance to drought, salinity and low-temperature stresses (Huang et al., 2012), suggesting that ZFP182 may function in the upstream pathway of OsDREB1. Overexpression of *ZFP245* in rice plants has been found to increase tolerance to drought and low-temperature stresses (Huang et al., 2009). Transgenic rice plants display elevated proline levels and ROS-scavenging enzyme activities. Overexpressing *ZFP252* in rice plants leads to enhanced tolerance to drought and salinity stresses and increased proline and soluble sugar contents (Xu et al., 2008). Increased drought tolerance has been observed in transgenic rice plants overexpressing*OsbHLH148*, a gene encoding MeJA-responsive transcription factor (Seo et al., 2011). In the cited study, expression of *OsDREB1* and *OsJAZ* family genes was up-regulated in the overexpressors, and OsJAZ and OsCOI1 proteins were demonstrated to interact with one another. These results suggest that OsbHLH148 acts on the JA signaling cascade with OsJAZ1 and OsCOI1 and functions as an upstream regulator of OsDREB1.

Protein Kinases

Transgenic rice plants overexpressing *OsCPK4*, a calcium-dependent protein kinase, showed enhanced tolerance to drought and salinity stresses in a study by Campo et al. (2014). In the overexpressors, genes related to lipid metabolism, such as those encoding proteins with lipid binding activities, lipid transfer proteins, and lipases, were up-regulated. Oxidative stress-responsive genes, including peroxidase, thioredoxin, GST, and laccase genes, were also up-regulated in the transgenic plants. These findings suggest that OsCPK4 is involved in the regulation of cellular membrane protection against oxidative damage (Campo et al., 2014). Ho et al. (2013) isolated *OsCDPK1* from sucrose-starved rice suspension cells and developed*OsCDPK1*-overexpressing transgenic rice plants. The transgenic rice plants displayed improved drought tolerance and activated (Ho et al., 2013) expression of a gene for a 14-3-3 protein, GF14c. Transgenic rice plants overexpressing *GF14c* also showed improved drought tolerance, suggesting that enhanced drought tolerance due to OsCDPK1 may be mediated by GF14c. Campo et al. (2012) have reported that transgenic rice plants overexpressing the gene encoding ZmGF14-6, a maize 14-3-3 protein, show enhanced drought tolerance. In their study, expression of stress-responsive genes, including *Rab21* and *Dip1*, was higher under drought stress conditions in transgenic rice plants than in non-transgenic ones, with transgenic plants also exhibiting a higher susceptibility to infection by fungal pathogens. These observations indicate that *ZmGF14-6* functions as a positive regulator in abiotic stress response, but as a negative regulator in biotic stress response. Saijo et al. (2000) discovered that transgenic rice plants overexpressing *OsCDPK7* showed elevated tolerance to drought, salinity, and low-temperature stresses. Overexpression of *OsCDPK7* increased the expression of several stress-responsive genes, suggesting that OsCDPK7 is a positive regulator of abiotic stress response. Finally, transgenic rice plants overexpressing *OsCIPK12* have been found to exhibit enhanced drought tolerance, with increased accumulation of proline and soluble sugars (Xiang et al., 2007).

Receptor-Like Kinases

OsSIK1 is a putative receptor-like kinase (RLK) with extracellular leucine-rich repeats (Ouyang et al., 2010). In the study by Ouyang et al. (2010), transgenic rice plants overexpressing *OsSIK1* showed enhanced tolerance to drought and salinity stresses. Leaves of the transgenic plants exhibited elevated peroxidase, superoxide dismutase and catalase activities and reduced accumulation of H_2O_2 compared with those of non-transgenic plants. Reduced stomatal density was also observed in the transgenic plants, suggesting that OsSIK1 may act as a negative regulator for stomatal development. Another rice RLK, OsSIK2, has been reported by Chen et al. (2013). In their study, OsSIK2 was predicted to be an S-domain RLK. Transgenic rice plants overexpressing *OsSIK2* showed enhanced tolerance to drought and salinity stresses, early leaf development, and a delayed dark-induced senescence phenotype. Their results suggest that OsSIK2 is involved in abiotic stress response and senescence processes.

LEA Proteins

Late embryogenesis abundant (LEA) proteins are important stress-inducible proteins involved in cellular protection against stresses (Hanin et al., 2011). Their protective roles in cells include cryoprotective (Bravo et al., 2003) and osmoprotective (Swire-Clark and Marcotte, 1999) behavior to stabilize proteins (Grelet et al., 2005), membranes (Koag et al., 2003), and glassy states (Wolkers et al., 2001). For example, recombinant pea LEA proteins have been shown to protect two mitochondrial matrix enzymes, fumarase, and rhodanese, during drying (Grelet et al., 2005). After transgenically introducing several LEA proteins into rice plants,Hong et al. (1988) investigated stress tolerance in the transgenic rice plants. A barley group-3 LEA protein, HVA1, was specifically accumulated in aleurone layers and embryos at the seed maturation stage. In another study, *HVA1*-overexpressing transgenic rice plants were found to have increased tolerance to drought and salinity stresses, with the increased stress tolerance correlated with HVA1 protein accumulation (Xu et al., 1996). Overexpression of *OsLEA3-1* or *OsLEA3-2* in rice plants leads to enhanced drought tolerance (Xiao et al., 2007; Duan and Cai, 2012). Overexpression of *OsLEA3-2* in yeast improved growth under salinity or osmotic stress conditions, with the OsLEA3-2 protein inhibiting protein aggregation in an *in vitro* assay (Duan and Cai, 2012).

Phytohormones

OsPYL/RCAR5 has been shown to be one of the ABA-signaling components in rice (Kim et al., 2012b). A protein-protein interaction assay and a transient gene expression assay performed by these authors identified an ABA-signaling unit composed of OsPYL/RCAR5, OsPP2C30, SAPK2, and OREB1. Kim et al. (2014)found that overexpression of *OsPYL/RCAR5* induced expression of numerous stress-responsive genes in rice and caused enhanced tolerance to drought and salinity stresses; however, field-grown transgenic rice plants had shorter heights and lower yields than their non-transgenic counterparts. *DSM2* encodes a chloroplast protein, a putative β-carotene hydroxylase involved in biosynthesis of the ABA precursor zeaxanthin (Du et al., 2010). Overexpression of *DSM2* in rice plants enhanced resistance to drought and oxidative stresses and increased xanthophyll levels and non-photochemical quenching.

Zhang et al. (2012) identified and investigated OsPIN3t, a putative auxin efflux carrier protein in rice. GFP proteins fused to OsPIN3t were expressed in the plasma membrane, while GUS activity in OsPIN3t promoter-driven GUS transgenic plants was detected in vascular tissues. These subcellular expressions and tissue-specific localization were changed by treatment with auxin transport inhibitors. Transgenic rice plants overexpressing *OsPIN3t* exhibited improved drought tolerance. These results suggest that OsPIN3t regulates polar auxin transport, thereby enhancing drought tolerance.

Isopentenyltransferase (IPT) is an enzyme that mediates cytokinin synthesis. Transgenic tobacco plants expressing the *IPT* gene under the control of a senescence-associated receptor kinase (SAPK), a maturation- and stress-inducible promoter, were developed by Rivero et al. (2007, 2009). The transgenic tobacco plants showed a drastic increase in plant productivity under drought stress conditions. The observed increased plant productivity was attributed to suppression of drought-induced leaf senescence (Rivero et al., 2007) and involvement in photorespiration (Rivero et al., 2009). Similarly, transgenic rice plants expressing the *IPT*gene under the control of the SAPK promoter were generated (Peleg et al., 2011). The developed transgenic rice plants displayed expression changes in genes involved in hormone homeostasis and resource mobilization, a delay in stress responses, and improvement of drought tolerance. In a study by Sun et al. (2014), two rice authentic histidine phosphotransfer proteins (OsAHP1 and OsAHP2) were knocked down simultaneously via RNA interference. The transgenic rice plants showed enhanced tolerance to osmotic stress and hyposensitivity to exogenous cytokinin, suggesting that OsAHPs function as positive regulators of the cytokinin signaling pathway in response to osmotic stress.

Osmoprotectants

Ornithine δ-aminotransferase is involved in proline and arginine metabolism. *OsOAT*, a rice gene encoding ornithine δ-aminotransferase, has been identified as a downstream gene of SNAC2 (Hu et al., 2008). You et al. (2012) demonstrated that SNAC2 can bind to the *OsOAT* promoter. In their study, overexpression of the*OsOAT* gene in rice plants enhanced δ-OAT activity and increased proline accumulation, glutachione content, and ROS-scavenging enzyme activity. The *OsOAT*-overexpressing transgenic rice plants displayed enhanced oxidative, drought, and osmotic stress tolerance. While seedling shoot lengths were similar between transgenic and non-transgenic plants under normal conditions, reduced inhibition of shoot growth was observed in transgenic plants under osmotic stress conditions compared with non-transgenic plants.

OsTPS1, a gene encoding a rice trehalose-6-phosphate synthase, acts as a key enzyme for trehalose biosynthesis. Overexpression of the gene in rice plants improved tolerance to drought, salinity, and low-temperature stresses in an investigation by Li et al. (2011). In the transgenic rice plants, trehalose and proline contents were increased and some stress-responsive genes, including *WSI18*, were up-regulated relative to those in non-transgenic plants.

Other Genes

Other genes encoding proteins with various characteristics have also been shown to enhance drought tolerance. Some of these proteins are stress responsive. The effect of overexpression of *O. sativa Drought-Induced LTP* (*OsDIL*), a lipid transfer protein gene, on drought stress tolerance in rice was investigated byGuo et al. (2013). The transgenic plants showed increased tolerance to drought stress at both vegetative and reproductive stages. Less severe tapetal defects and fewer defective anther sacs were observed in the transgenic plants. These results were consistent with data indicating that the *OsDIL* gene is expressed in anthers. Overexpression of the heat shock protein gene *OsHsp17.0*, or *OsHsp23.7*, has been found to improve tolerance to drought and salinity stresses in rice (Zou et al., 2012). In that study, the transgenic rice plants had lower relative electrical conductivities and malondialdehyde contents and higher proline contents compared with non-transgenic rice plants.

Modulation of ROS accumulation is also important for the enhancement of drought tolerance. Transgenic rice plants overexpressing the gene encoding manganese superoxide dismutase, an antioxidant enzyme, have improved osmotic stress tolerance (Wang et al., 2005). The cited authors found that electrolyte leakage in the transgenic plants was lower than in non-transgenic plants under osmotic stress conditions, and that photosynthetic rate was less affected by osmotic stress in the transgenic plants. The enzyme OsMIOX, a rice*myo*-inositol oxygenase, catalyzes the oxidation of myo-inositol to glucuronic acid. In a study by Duan et al. (2012), the *OsMIOX* gene was overexpressed in rice plants and caused increases in ROS-scavenging enzyme activities and proline content and enhancement of growth performance under osmotic stress conditions. Ski-interacting protein (SKIP), identified by yeast two-hybrid screening using the avian retrovirus oncogene v-Ski as bait (Dahl et al., 1998), has been well characterized as a transcriptional coregulator and a spliceosome component in humans (Figueroa and Hayman, 2004). Transgenic rice plants overexpressing *OsSKIPa* have shown improved drought tolerance and increased ROS-scavenging ability. Higher transcript levels of*SNAC1*, *OsCBF2*, *OsPP2C*, and *OsRD22* have been found in *OsSKIPa*-transformed rice plants compared with their non-transgenic counterparts (Hou et al., 2009).

Protein turnover via ubiquitin-dependent protein degradation and ribosomal protein synthesis has been shown to be involved in abiotic stress response regulatory networks. OsSDIR1 (*O. sativa* SALT-AND DROUGHT-INDUCED RING FINGER 1) is a functional RING-finger-containing E3 ligase, with the RING finger region required for its activity (Gao et al., 2011). Transgenic rice plants overexpressing *OsSDIR1* show enhanced drought tolerance and stomatal closure (Gao et al., 2011), while those overexpressing *OsRDCP1*, a rice RING domain-containing protein 1 gene, have improved drought tolerance (Bae et al., 2011). Molecular mechanisms underlying the improved drought tolerance of these transgenic rice plants remain largely unclear.Jiang et al. (2012) observed that transgenic rice plants overexpressing *OSRIP18*, a rice ribosome-inactivating protein 18 gene, exhibited improved drought and salinity tolerance. Microarray analysis detected differentially expressed genes, most of which were not regulated by abiotic stresses, in the transgenic rice plants.

RNA turnover may also be involved in abiotic stress-response regulatory networks. *OsSUV3* encodes an NTP-dependent RNA/DNA helicase (Tuteja et al., 2013). Transgenic rice plants overexpressing *OsSUV3*show reduced lipid peroxidation, electrolyte leakage, and H_2O_2 production, and enhanced antioxidant enzyme activities, thereby leading to enhanced tolerance to osmotic and salinity stresses (Tuteja et al., 2013). GRP encodes a glycine-rich RNA-binding protein. In a study by Yang et al. (2014), transgenic rice plants overexpressing *Arabidopsis GRP2* or *GRP7* displayed increased grain yield under drought conditions. The increased grain yield was caused by improved grain filling. Several stress-responsive genes, including *OSE2*,*Dip1*, and *PBZ1*, were up-regulated in the transgenic rice plants.

The application of genes encoding metabolic enzymes is also thought to be useful for the enhancement of drought tolerance. Squalene synthase (SQS) is one of several farnesyl-diphosphate farnesyltransferase proteins that catalyze the first reaction of the branch of the isoprenoid metabolic pathway involved in sterol biosynthesis (Tansley and Shechter, 2001). Disruption of *SQS* gene function by RNA interference has been found to improve drought tolerance in rice plants, with the transgenic plants showing increased root length, an elevated number of lateral roots and reduced stomatal conductance (Manavalan et al., 2012). Polyamines, such as putrescine, spermidine, and spermine, are compounds implicated in plant embryo development, stem elongation and stress response (Takahashi and Kakehi, 2010). Polyamine levels can be modulated by the regulation of metabolic enzymes, including arginine decarboxylase. Because rice plants overexpressing*Datura stramonium* arginine decarboxylase show improved drought tolerance along with increased putrescine content, Capell et al. (2004) have proposed a regulatory mechanism linking putrescine metabolism to drought tolerance. In contrast to putrescine, spermidine, and spermine are not involved in drought stress tolerance, as increased spermidine and spermine content has not been observed to enhance drought tolerance in rice plants (Peremarti et al., 2009).

Transgenes originating from non-plant species have also been used to enhance stress tolerance. Harpin proteins are secreted by the type-III protein secretion system of Gram-negative plant pathogenic bacteria (Wei et al., 1992). Harpin proteins trigger the hypersensitive response, a well-characterized defense response against various bacteria, fungi, nematodes, and viruses. Transgenic rice plants overexpressing the harpin-encoding gene hrf1 showed improved drought tolerance along with increased stomatal clousure and ABA, proline, and soluble sugar contents (Zhang et al., 2011). Increased expression of stress-responsive genes including *OsLEA3-1* was also observed in the transgenic rice plants. As reviewed by Sharma et al. (2013), the evidence that pathogenic-related genes can also improve abiotic stress tolerance suggests an overlapping regulatory cascade between biotic and abiotic stresses.

CHANGES IN SHOOT GROWTH OF DROUGHT-TOLERANT TRANSGENIC RICE PLANTS

Among the transgenic rice plants described in this review, 37% have been reported to display growth retardation under normal conditions or exogenous ABA application (Table 1). Such decreased shoot growth performance is also observed in non-transgenic plants subjected to drought stress conditions. Shoot growth retardation due to low soil water content is one of the earliest stress responses in plants, occurring even earlier than decreases in leaf water potential (Michelena and Boyer, 1982; Parent et al., 2010). This phenomenon suggests that plants actively decrease shoot growth instead of it being a consequence of decreased cell turgor (Claeys and Inzé, 2013). Growth regulation in proportion to soil water content is thus an important plant morphological response to water deficit. Molecular mechanisms underlying growth regulatory responses to water deficit have been investigated in *Arabidopsis*. DELLA proteins, which are negative regulators of gibberellic acid (GA) signaling, have been shown to integrate growth and abiotic stress tolerance in*Arabidopsis* (Achard et al., 2006). Skirycz et al. (2010) performed transcript profiling of expanding*Arabidopsis* leaves subjected to mild osmotic stress. Their results indicated that an ethylene- and gibberellin-dependent regulatory circuit modulated growth under the mild osmotic stress conditions, with no involvement from ABA.

Rapid accumulation of 1-aminocyclopropane-1-carboxylic acid (1-ACC), an ethylene precursor, was observed in the expanding leaf tissue under the mild osmotic stress conditions of their study. This accumulation has been proposed to activate a cascade of the growth regulatory circuit in *Arabidopsis* as follows (Claeys and Inzé, 2013). After activation by 1-ACC accumulation through a MAP kinase cascade, ethylene responsive factor 6 (ERF6) increases expression of *GA2OX6*, which encodes an enzyme that inactivates GAs. By the operation of GA2OX6, GAs are inactivated, with this inactivation stabilizing DELLA proteins. The DELLA proteins modulate the activity of ANAPHASE-PROMOTING COMPLEX/CYCLOSOME (APC/C), which controls the activity of CDK-cyclin complexes, through the repression of APC/C inhibitors DEL1 and UVI4. Finally, the modulated APC/C activity abolishes potential for cell proliferation and inhibits growth.

A similar growth regulatory circuit does not seem to hold for rice, as it has been generally accepted that ethylene and 1-ACC act as positive growth regulators under various conditions in rice (Bailey-Serres and Voesenek, 2008, 2010; Fukao and Xiong, 2013; Wang et al., 2013). In rice, growth regulatory mechanisms that are distinct from those in *Arabidopsis* should therefore be taken into consideration. We recently identified*O. sativa phytochrome interacting factor like 1* (*OsPIL1)*, a gene that regulates internode elongation under drought stress conditions in rice (Todaka et al., 2012). The bHLH-type transcription factor OsPIL1 functions as a transcriptional activator and modulates expression of cell elongation-related genes such as expansins. Increased expression of *OsPIL1* observed in the daytime under normal growth conditions was canceled under drought stress conditions. We proposed the following growth regulatory mechanistic model involving OsPIL1 in response to drought stress. Under normal growth conditions, OsPIL1 elevates expression of cell elongation-related genes such as expansins, causing normal shoot growth. When rice plants are exposed to drought stress, the canceled OsPIL1 expression leads to reduced expression of cell elongation-related genes, resulting in shoot growth reduction that likely conserves photosynthetic products and decreased shoot surface area. The saved energy may be used for activation of mechanisms involved in stress tolerance. Ogawa et al. (2011) have revealed that the rice protein RSS1 plays an important role in the maintenance of meristematic activity under salinity stress conditions. RSS1 proteins interact with protein phosphatase 1, a regulator of various cellular processes including the cell division cycle. RSS1 stability is regulated by the APC/C 26S proteasome pathway, which is responsible for degradation of mitotic cyclins.

Use of the growth regulatory genes described in this section may ameliorate the growth reduction observed in drought-tolerant transgenic rice plants. Although stress-inducible promoters are often useful for development, their efficacy seems to be limited under moderate long-term drought stress conditions. Long-term drought stress maintains a high level of transgene expression, thereby affecting growth performance.

EVALUATION OF ABIOTIC STRESS TOLERANCE UNDER FIELD CONDITIONS

When developing abiotic stress tolerant transgenic crops, plant productivity should be taken into consideration. Plant productivity is extensively affected by natural drought episodes under field conditions. Droughts are unpredictable events and vary in stress severity and duration. Simultaneously occurring stresses, such as drought and heat, are also observed. Results obtained under laboratory or greenhouse conditions are therefore not perfectly comparable to observations made under field conditions. Field trials are thus critical for the proper evaluation of stress-tolerant transgenic crops.

Xiao et al. (2007) analyzed drought tolerance of transgenic rice plants constitutively overexpressing*OsLEA3-1*, a gene encoding proteins that highly accumulate in water-stressed tissues, as well as plants expressing the transgene with a stress-inducible promoter under field conditions. Drought stress was initiated at the panicle development stage by draining surface water in paddy fields and halting irrigation until leaves were rolled. Although T_1 generations of both transgenic lines showed reduced yields under non-stressed conditions, T_2 and T_3 generations exhibited no yield penalty under non-stressed conditions and exhibited increased grain yield under drought conditions.

Xiao et al. (2009) also examined drought tolerance of transgenic rice plants overexpressing seven well-documented stress-related genes with an actin promoter under field conditions. These seven genes were*CBF3/DREB1A*, an AP2/ERF-type transcription factor; *SOS2*, a serine/threonine protein kinase; *NCED2*and *LOS5*, enzymes involved in ABA biosynthesis; *NPK1*, a mitogen-activated protein kinase kinase kinase;*ZAT10*, a C2H2-type zinc finger transcription factor; and *NHX1*, a vacuolar Na^+/H^+ antiporter. Although drought stress in the field decreased grain yield in these transgenic plants, grain yields in *LOS5*, *ZAT10*, and*NHX1* overexpressors were less affected. The authors also developed transgenic rice plants that expressed these genes with a stress-inducible promoter and field-tested their drought tolerance. Grain yields in these transgenic plants were similarly decreased by drought stress under field conditions. Grain yields in transgenic rice plants expressing *CBF3/DREB1A*, *SOS2*, *NPK1*, *LOS5*, *ZAT10*, and *NHX1* with the stress-inducible promoter were the least affected. Because absolute grain yields under normal growth conditions were lower in these transgenic rice plants than in non-transgenic ones, further improvement is needed for practical application.

Hu et al. (2006) subjected field-grown transgenic rice plants overexpressing *SNAC1*, a NAC-type transcription factor, to two different levels of drought stress treatments at the anthesis stage: severe stress with 15% soil moisture and moderate stress with 28% soil moisture. Both drought stress conditions increased spikelet fertility in the transgenic plants. Under non-stressed conditions, agronomic traits, including plant height, panicle number, spikelet number, spikelet fertility, and grain yield, were similar between transgenic plants and the controls. Drought resistance of transgenic rice plants overexpressing OsNAC5 (Jeong et al., 2013), OsNAC9/SNAC1 (Redillas et al., 2012), or OsNAC10 (Jeong et al., 2010) under the control of the root-specific promoter has also been examined in the field. In these studies, exposure to drought stress was performed at the panicle heading stage by draining surface water and halting irrigation until leaves were rolled. Similar results were observed among the three transgenic rice lines. Grain yield decreases under drought conditions were significantly smaller in all three transgenic lines than those observed in their non-transgenic counterparts. Drought tolerance of transgenic rice plants overexpressing *OsOAT* has also been investigated under field conditions (You et al., 2012). The OsOAT protein, an enzyme that increases proline content, is a direct target gene of the stress-responsive NAC transcription factor SNAC2. The field drought test was performed by stopping irrigation at the flowering stage in a refined paddy field covered with a movable rain-off shelter. Slower wilting, fewer withered leaves, and a higher rate of seed-setting were noted in the transgenic rice plants than in non-transgenic ones.

Increased grain yield was observed in transgenic rice plants overexpressing the *AP37* gene, an AP2/ERF-type transcription factor, when the transgenic plants were subjected to drought stress in the field (Oh et al., 2009). The field drought stress was performed at the panicle heading stage by draining the surface water and halting irrigation until leaves were rolled. The increased grain yield was due to the higher grain-filling rate in the drought-treated transgenic plants compared with the drought-treated non-transgenic plants. Finally, field evaluation of transgenic rice plants overexpressing *EDT1/HDG11*, a homeodomain-leucine zipper transcription factor, has also been carried out (Yu et al., 2013). The transgenic rice plants were grown in the field for a month after transplanting; irrigation was then stopped until the seed maturation stage. The drought-treated transgenic rice plants had higher grain yields than those observed in the drought-treated non-transgenic rice plants. The grain yield increase in the transgenic plants was a consequence of their larger panicle sizes and higher tiller numbers compared with the non-transgenic plants.

FUTURE DIRECTIONS IN THE DEVELOPMENT OF DROUGHT-TOLERANT TRANSGENIC RICE PLANTS

Although several studies have reported transgenic rice plants with improved drought tolerance during field trials, further research is needed to uncover the regulatory mechanism of drought response and tolerance under field conditions. Such investigations should lead to the discovery of new genes that increase drought tolerance without yield penalty even under drought conditions. Another approach to new gene exploration is to study stress tolerance mechanisms of stress-adapted extremophiles such as desert plants, halophilic plants, cold-water fishes, and thermophilic bacteria (Mittler and Blumwald, 2010). Even in well-characterized species, the functions of 18–38% of total proteins remain unknown (Gollery et al., 2006). The elucidation of these unknown function proteins should aid the discovery of new genes. Modification of root architecture is also important for the development of drought-tolerant rice plants. In this regard, Uga et al. (2013) reported that the QTL *Deeper Rooting 1 (DRO1)* increased the root growth angle in rice, leading to high-yield performance under drought conditions.

Rice has the highest potential of any crop to grow under submergence conditions. Studies of submergence-tolerance mechanisms and the development of submergence-tolerant rice cultivars have progressed significantly (Bailey-Serres and Voesenek, 2008, 2010; Hattori et al., 2009; Fukao and Xiong, 2013;Voesenek and Bailey-Serres, 2013). The results of these studies indicate that drought-tolerant rice plants with submergence-tolerant cultivar backgrounds are exceptional crops that can survive under both low and excessive soil–water content conditions. In the future, crops may be alternately exposed to drought and flood because of global climate change. Efforts to develop rice cultivars having high water usage flexibility should help solve this crisis.

ACKNOWLEDGMENTS

We are grateful to Etsuko Toma for skillful editorial assistance. This work was supported by a Grant-in-Aid for Young Scientists (B) (No. 25850007) and a Grant-in-Aid for Scientific Research on Innovative Areas (No. 22119004) from the Ministry of Education, Culture, Sports, Science and Technology of Japan (MEXT), the Program for the Promotion of Basic Research Activities for Innovative Biosciences (BRAIN) of Japan, and the Science and Technology Research Partnership for Sustainable Development (SATREPS) of the Japan Science and Technology Agency (JST)/Japan International Cooperation Agency (JICA).

REFERENCES

1. Achard P., Cheng H., De Grauwe L., Decat J., Schoutteten H., Moritz T., et al. (2006). Integration of plant responses to environmentally activated phytohormonal signals. *Science* 311 91–94 10.1126/science.1118642
2. Agarwal M., Hao Y., Kapoor A., Dong C.-H., Fujii H., Zheng X., et al. (2006). A R2R3 type MYB transcription factor is involved in the cold regulation of CBF genes and in acquired freezing tolerance.*J. Biol. Chem.* 281 37636–37645 10.1074/jbc.M605895200
3. Bae H., Kim S. K., Cho S. K., Kang B. G., Kim W. T. (2011). Over expression of OsRDCP1 a rice RING domain-containing E3 ubiquitin ligase, increased tolerance to drought stress in rice (*Oryza sativa* L.). *Plant Sci.* 180 775–782 10.1016/j.plantsci.2011.02.008
4. Bailey-Serres J., Voesenek L. A. (2008). Flooding stress: acclimations and genetic diversity. *Annu. Rev. Plant Biol.* 59 313–339 10.1146/annurev.arplant.59.032607.092752
5. Bailey-Serres J., Voesenek L. A. (2010). Life in the balance: a signaling network controlling survival of flooding. *Curr. Opin. Plant Biol.* 13 489–494 10.1016/j.pbi.2010.08.002
6. Bartels D., Sunkar R. (2005). Drought and salt tolerance in plants. *Crit. Rev. Plant Sci.* 24 23–58 10.1080/07352680590910410
7. Bauer H., Ache P., Lautner S., Fromm J., Hartung W., Al-Rasheid K. A. S., et al. (2013). The stomatal response to reduced relative humidity requires guard cell-autonomous ABA synthesis. *Curr. Biol.* 2353–57 10.1016/j.cub.2012.11.022
8. Behnam B., Iuchi S., Fujita M., Fujita Y., Takasaki H., Osakabe Y., et al. (2013). Characterization of the promoter region of an *Arabidopsis* gene for 9-cis-epoxycarotenoid dioxygenase involved in dehydration-

inducible transcription. *DNA Res.* 20 315–324 10.1093/dnares/dst012

9. Bhatnagar-Mathur P., Rao J. S., Vadez V., Dumbala S. R., Rathore A., Yamaguchi-Shinozaki K., et al. (2013). Transgenic peanut overexpressing the DREB1A transcription factor has higher yields under drought stress. *Mol. Breeding* 33 327–340 10.1007/s11032-013-9952-7

10. Bravo L. A., Gallardo J., Navarrete A., Olave N., Martınez J., Alberdi M., et al. (2003).Cryoprotective activity of a cold-induced dehydrin purified from barley. *Physiol. Plant.* 118 262–269 10.1034/j.1399-3054.2003.00060.x

11. Campo S., Baldrich P., Messeguer J., Lalanne E., Coca M., San Segundo B. (2014). Overexpression of a calcium-dependent protein kinase confers salt and drought tolerance in rice by preventing membrane lipid peroxidation. *Plant Physiol.* 165 688–704 10.1104/pp.113.230268

12. Campo S., Peris-Peris C., Montesinos L., Peñas G., Messeguer J., San Segundo B. (2012). Expression of the maize ZmGF14-6 gene in rice confers tolerance to drought stress while enhancing susceptibility to pathogen infection. *J. Exp. Bot.* 63 983–999 10.1093/jxb/err328

13. Capell T., Bassie L., Christou P. (2004). Modulation of the polyamine biosynthetic pathway in transgenic rice confers tolerance to drought stress. *Proc. Natl. Acad. Sci. U.S.A.* 101 9909–9914 10.1073/pnas.0306974101

14. Chae M.-J., Lee J.-S., Nam M.-H., Cho K., Hong J.-Y., Yi S.-A., et al. (2007). A rice dehydration-inducible SNF1-related protein kinase 2 phosphorylates an abscisic acid responsive element-binding factor and associates with ABA signaling. *Plant Mol. Biol.* 63 151–169 10.1007/s11103-006-9079-x

15. Chen H., Chen W., Zhou J., He H., Chen L., Chen H., et al. (2012). Basic leucine zipper transcription factor OsbZIP16 positively regulates drought resistance in rice. *Plant Sci.* 193–194, 8–17 10.1016/j.plantsci.2012.05.003

16. Chen J.-Q., Meng X.-P., Zhang Y., Xia M., Wang X.-P. (2008). Over-expression of OsDREB genes lead to enhanced drought tolerance in rice. *Biotechnol. Lett.* 30 2191–2198 10.1007/s10529-008-9811-5

17. Chen L.-J., Wuriyanghan H., Zhang Y.-Q., Duan K.-X., Chen H.-W., Li Q.-T., et al. (2013). An S-domain receptor-like kinase, OsSIK2, confers abiotic stress tolerance and delays dark-induced leaf senescence in rice. *Plant Physiol.* 163 1752–1765 10.1104/pp.113.224881

18. Chinnusamy V., Ohta M., Kanrar S., Lee B.-H., Hong X., Agarwal M., et al. (2003). ICE1: a regulator of cold-induced transcriptome and freezing tolerance in *Arabidopsis*. *Genes Dev.* 17 1043–1054 10.1101/gad.1077503

19. Chinnusamy V., Zhu J., Zhu J.-K. (2007). Cold stress regulation of gene expression in plants. *Trends Plant Sci.* 12 444–451 10.1016/j.tplants.2007.07.002

20. Claeys H., Inzé D. (2013). The agony of choice: how plants balance growth and survival under water-limiting conditions. *Plant Physiol.* 162 1768–1779 10.1104/pp.113.220921

21. Cui M., Zhang W., Zhang Q., Xu Z., Zhu Z., Duan F., et al. (2011). Induced over-expression of the transcription factor OsDREB2A improves drought tolerance in rice. *Plant Physiol. Biochem.* 49 1384–1391 10.1016/j.plaphy.2011.09.012

22. Cutler S. R., Rodriguez P. L., Finkelstein R. R., Abrams S. R. (2010). Abscisic acid: emergence of a core signaling network. *Annu. Rev. Plant Biol.* 61 651–679 10.1146/annurev-arplant-042809-112122

23. Dahl R., Wani B., Hayman M. J. (1998). The Ski oncoprotein interacts with Skip, the human homolog of *Drosophila* Bx42. *Oncogene* 16 1579–1586 10.1038/sj.onc.1201687

24. Datta K., Baisakh N., Ganguly M., Krishnan S., Yamaguchi Shinozaki K., Datta S. K. (2012). Overexpression of Arabidopsis and rice stress genes' inducible transcription factor confers drought and salinity tolerance to rice. *Plant Biotechnol. J.* 10 579–586 10.1111/j.1467-7652.2012.00688.x

25. Degenkolbe T., Do P. T., Kopka J., Zuther E., Hincha D. K., Köhl K. I. (2013). Identification of drought tolerance markers in a diverse population of rice cultivars by expression and metabolite profiling. *PLoS ONE* 8:e63637 10.1371/journal.pone.0063637

26. Degenkolbe T., Do P. T., Zuther E., Repsilber D., Walther D., Hincha D. K., et al. (2009). Expression profiling of rice cultivars differing in their tolerance to long-term drought stress. *Plant Mol. Biol.* 69133–153 10.1007/s11103-008-9412-7

27. Deinlein U., Stephan A. B., Horie T., Luo W., Xu G., Schroeder J. I. (2014). Plant salt-tolerance mechanisms. *Trends Plant Sci.* 19 371–379 10.1016/j.tplants.2014.02.001

28. Doherty C. J., Van Buskirk H. A., Myers S. J., Thomashow M. F. (2009). Roles for *Arabidopsis*CAMTA transcription factors in cold-regulated gene expression and freezing tolerance. *Plant Cell* 21972–984 10.1105/tpc.108.063958

29. Dong C.-H., Agarwal M., Zhang Y., Xie Q., Zhu J.-K. (2006). The negative regulator of plant cold responses, HOS1, is a RING E3 ligase that mediates the ubiquitination and degradation of ICE1. *Proc. Natl. Acad. Sci. U.S.A.* 103 8281–8286 10.1073/pnas.0602874103

30. Dong M. A., Farré E. M., Thomashow M. F. (2011). Circadian clock-associated 1 and late elongated hypocotyl regulate expression of the C-repeat binding factor (CBF) pathway in *Arabidopsis*. *Proc. Natl. Acad. Sci. U.S.A.* 108 7241–7246 10.1073/pnas.1103741108

31. Du H., Wang N., Cui F., Li X., Xiao J., Xiong L. (2010). Characterization of the beta-carotene hydroxylase gene DSM2 conferring drought and oxidative stress resistance by increasing xanthophylls and abscisic acid synthesis in rice. *Plant Physiol.* 154 1304–1318 10.1104/pp.110.163741

32. Duan J., Cai W. (2012). OsLEA3-2, an abiotic stress induced gene of rice plays a key role in salt and drought tolerance. *PLoS ONE* 7:e45117 10.1371/journal.pone.0045117

33. Duan J., Zhang M., Zhang H., Xiong H., Liu P., Ali J., et al. (2012). OsMIOX, a myo-inositol oxygenase gene, improves drought tolerance through scavenging of reactive oxygen species in rice (*Oryza sativa* L.). *Plant Sci.* 196 143–151 10.1016/j.plantsci.2012.08.003

34. Dubouzet J. G., Sakuma Y., Ito Y., Kasuga M., Dubouzet E. G., Miura S., et al. (2003). OsDREB genes in rice, *Oryza sativa* L., encode transcription activators that function in drought-, high-salt- , and cold-responsive gene expression. *Plant J.* 33 751–763 10.1046/j.1365-313X.2003.01661.x

35. Figueroa J. D., Hayman M. J. (2004). Differential effects of the Ski-interacting protein (SKIP) on differentiation induced by transforming growth factor-beta1 and bone morphogenetic protein-2 in C2C12 cells. *Exp. Cell Res.* 296 163–172 10.1016/j.yexcr.2004.01.025

36. Fukao T., Xiong L. (2013). Genetic mechanisms conferring adaptation to submergence and drought in rice: simple or complex? *Curr. Opin. Plant Biol.* 16 196–204 10.1016/j.pbi.2013.02.003

37. Gao T., Wu Y., Zhang Y., Liu L., Ning Y., Wang D., et al. (2011). OsSDIR1 overexpression greatly improves drought tolerance in transgenic rice. *Plant Mol. Biol.* 76 145–156 10.1007/s11103-011-9775-z

38. Golldack D., Li C., Mohan H., Probst N. (2014). Tolerance to drought and salt stress in plants: unraveling the signaling networks. *Front. Plant Sci.* 5:151 10.3389/fpls.2014.00151

39. Gollery M., Harper J., Cushman J., Mittler T., Girke T., Zhu J.-K., et al. (2006). What makes species unique? The contribution of proteins with obscure features. *Genome Biol.* 7 R57 10.1186/gb-2006-7-7-r57

40. Grelet J., Benamar A., Teyssier E., Avelange-Macherel M.-H., Grunwald D., Macherel D. (2005).Identification in pea seed mitochondria of a late-embryogenesis abundant protein able to protect enzymes from drying. *Plant Physiol.* 137 157–167 10.1104/pp.104.052480

41. Guo C., Ge X., Ma H. (2013). The rice OsDIL gene plays a role in drought tolerance at vegetative and reproductive stages. *Plant Mol. Biol.* 82 239–253 10.1007/s11103-013-0057-9

42. Hanin M., Brini F., Ebel C., Toda Y., Takeda S., Masmoudi K. (2011). Plant dehydrins and stress tolerance: versatile proteins for complex mechanisms. *Plant Signal. Behav.* 6 1503–1509 10.4161/psb.6.10.17088

43. Hattori Y., Nagai K., Furukawa S., Song X.-J., Kawano R., Sakakibara H., et al. (2009). The ethylene response factors SNORKEL1 and SNORKEL2 allow rice to adapt to deep water. *Nature* 460 1026–1030 10.1038/nature08258

44. He Y., Hao Q., Li W., Yan C., Yan N., Yin P. (2014). Identification and characterization of ABA receptors in *Oryza sativa*. *PLoS ONE* 9:e95246 10.1371/journal.pone.0095246

45. Ho S.-L., Huang L.-F., Lu C.-A., He S.-L., Wang C.-C., Yu S.-P., et al. (2013). Sugar starvation- and GA-inducible calcium-dependent protein kinase 1 feedback regulates GA biosynthesis and activates a 14-3-3 protein to confer drought tolerance in rice seedlings. *Plant Mol. Biol.* 81 347–361 10.1007/s11103-012-0006-z

46. Hong B., Tong Z., Ma N., Li J., Kasuga M., Yamaguchi-Shinozaki K., et al. (2006). Heterologous expression of the AtDREB1A gene in chrysanthemum increases drought and salt stress tolerance. *Sci. China C Life Sci.* 49 436–445 10.1007/s11427-006-2014-1

47. Hong B., Uknes S. J., Ho T.-H. (1988). Cloning and characterization of a cDNA encoding a mRNA rapidly induced by ABA in barley aleurone layers. *Plant Mol. Biol.* 11 495–506 10.1007/BF00039030

48. Hong J. Y., Chae M. J., Lee I. S., Lee Y. N., Nam M. H., Kim D. Y., et al. (2011). Phosphorylation-mediated regulation of a rice ABA responsive element binding factor. *Phytochemistry* 72 27–36 10.1016/j.phytochem.2010.10.005

49. Hou X., Xie K., Yao J., Qi Z., Xiong L. (2009). A homolog of human ski-interacting protein in rice positively regulates cell viability and stress tolerance. *Proc. Natl. Acad. Sci. U.S.A.* 106 6410–6415 10.1073/pnas.0901940106

50. Hsieh T. H., Lee J. T., Charng Y. Y., Chan M. T. (2002). Tomato plants ectopically expressing*Arabidopsis* cbf1 show enhanced resistance to water deficit stress. *Plant Physiol.* 130 618–626 10.1104/pp.006783

51. Hu H., Dai M., Yao J., Xiao B., Li X., Zhang Q., et al. (2006). Overexpressing a NAM, ATAF, and CUC (NAC) transcription factor enhances drought resistance and salt tolerance in rice. *Proc. Natl. Acad. Sci. U.S.A.* 103 12987–12992 10.1073/pnas.0604882103

52. Hu H., You J., Fang Y., Zhu X., Qi Z., Xiong L. (2008). Characterization of transcription factor gene SNAC2 conferring cold and salt tolerance in rice. *Plant Mol. Biol.* 67 169–181 10.1007/s11103-008-9309-5

53. Hua J. (2009). From freezing to scorching, transcriptional responses to temperature variations in plants.*Curr. Opin. Plant Biol.* 12 568–573 10.1016/j.pbi.2009.07.012

54. Huang J., Sun S., Xu D., Lan H., Sun H., Wang Z., et al. (2012). A TFIIIA-type zinc finger protein confers multiple abiotic stress tolerances in transgenic rice (*Oryza sativa* L.). *Plant Mol. Biol.* 80 337–350 10.1007/s11103-012-9955-5

55. Huang J., Sun S.-J., Xu D.-Q., Yang X., Bao Y.-M., Wang Z.-F., et al. (2009). Increased tolerance of rice to cold, drought and oxidative stresses mediated by the over expression of a gene that encodes the zinc finger protein ZFP245. *Biochem. Biophys. Res. Commun.* 389 556–561 10.1016/j.bbrc.2009.09.032

56. Ishizaki T., Maruyama K., Obara M., Fukutani A., Yamaguchi-Shinozaki K., Ito Y., et al. (2012).Expression of*Arabidopsis* DREB1C improves survival, growth, and yield of upland New Rice for Africa (NERICA) under drought. *Mol. Breed.* 31 255–264 10.1007/s11032-012-9785-9

57. Ito Y., Katsura K., Maruyama K., Taji T., Kobayashi M., Seki M., et al. (2006). Functional analysis of rice DREB1/CBF-type transcription factors involved in cold-responsive gene expression in transgenic rice. *Plant Cell Physiol.* 47 141–153 10.1093/pcp/pci230

58. Iuchi S., Kobayashi M., Taji T., Naramoto M., Seki M., Kato T., et al. (2001). Regulation of drought tolerance by gene manipulation of 9-cis-epoxycarotenoid dioxygenase, a key enzyme in abscisic acid biosynthesis in *Arabidopsis*. *Plant J.* 27 325–333 10.1046/j.1365-313x.2001.01096.x

59. Jeong J. S., Kim Y. S., Baek K. H., Jung H., Ha S.-H., Do Choi Y., et al. (2010). Root-specific expression of OsNAC10 improves drought tolerance and grain yield in rice under field drought conditions. *Plant Physiol.* 153 185–197 10.1104/pp.110.154773

60. Jeong J. S., Kim Y. S., Redillas M. C. F. R., Jang G., Jung H., Bang S. W., et al. (2013). OsNAC5 over expression enlarges root diameter in rice plants leading to enhanced drought tolerance and increased grain yield in the field. *Plant Biotechnol. J.* 11 101–114 10.1111/pbi.12011

61. Jiang S.-Y., Bhalla R., Ramamoorthy R., Luan H.-F., Venkatesh P. N., Cai M., et al. (2012). Over-expression of OSRIP18 increases drought and salt tolerance in transgenic rice plants. *Transgenic Res.*21 785–795 10.1007/s11248-011-9568-9

62. Joo J., Choi H. J., Lee Y. H., Kim Y.-K., Song S. I. (2013). A transcriptional repressor of the ERF family confers drought tolerance to rice and regulates genes preferentially located on chromosome 11.*Planta* 238 155–170 10.1007/s00425-013-1880-6

63. Kang J., Choi H., Im M., Kim S. Y. (2002). *Arabidopsis* basic leucine zipper proteins that mediate stress-responsive abscisic acid signaling. *Plant Cell* 14 343–357 10.1105/tpc.010362.tase

64. Karaba A., Dixit S., Greco R., Aharoni A., Trijatmiko K. R., Marsch-Martinez N., et al. (2007).Improvement of water use efficiency in rice by expression of HARDY, an *Arabidopsis* drought and salt tolerance gene. *Proc. Natl. Acad. Sci. U.S.A.* 104 15270–15275 10.1073/pnas.0707294104

65. Kasuga K., Liu Q., Miura S., Yamaguchi-Shinozaki K., Shinozaki K. (1999). Improving plant drought, salt, and freezing tolerance by gene transfer of a single stress-inducible transcription factor.*Nat. Biotechnol.* 17 287–291 10.1038/7036

66. Kasuga M., Miura S., Shinozaki K., Yamaguchi-Shinozaki K. (2004). A combination of the*Arabidopsis* DREB1A gene and stress-inducible rd29A promoter improved drought- and low-temperature stress tolerance in tobacco by gene transfer. *Plant Cell Physiol.* 45 346–350 10.1093/pcp/pch037

67. Kidokoro S., Maruyama K., Nakashima K., Imura Y., Narusaka Y., Shinwari Z. K., et al. (2009). The phytochrome-interacting factor PIF7 negatively regulates DREB1 expression under circadian control in *Arabidopsis*. *Plant Physiol.* 151 2046–2057 10.1104/pp.109.147033

68. Kim H., Lee K., Hwang H., Bhatnagar N., Kim D.-Y., Yoon I. S., et al. (2014). Over expression of PYL5 in rice enhances drought tolerance, inhibits growth, and modulates gene expression. *J. Exp. Bot.*65 453–464 10.1093/jxb/ert397

69. Kim J.-S., Mizoi J., Kidokoro S., Maruyama K., Nakajima J., Nakashima K., et al. (2012a).Arabidopsis growth-regulating factor7 functions as a transcriptional repressor of abscisic acid- and osmotic stress-responsive genes, including DREB2A. *Plant Cell* 24 3393–3405 10.1105/tpc.112.100933

70. Kim H., Hwang H., Hong J.-W., Lee Y.-N., Ahn I. P., Yoon I. S., et al. (2012b). A rice orthologue of the ABA receptor, OsPYL/RCAR5, is a positive regulator of the ABA signal transduction pathway in seed germination and early seedling growth. *J. Exp. Bot.* 63 1013–1024 10.1093/jxb/err338

71. Koag M.-C., Fenton R. D., Wilkens S., Close T. J. (2003). The binding of maize DHN1 to lipid vesicles. Gain of structure and lipid specificity. *Plant Physiol.* 131 309–316 10.1104/pp.011171

72. Kuromori T., Sugimoto E., Shinozaki K. (2014). Intertissue signal transfer of abscisic acid from vascular cells to guard cells. *Plant Physiol.* 164 1587–1592 10.1104/pp.114.235556

73. Lenka S. K., Katiyar A., Chinnusamy V., Bansal K. C. (2011). Comparative analysis of drought-responsive transcriptome in Indica rice genotypes with contrasting drought tolerance. *Plant Biotechnol. J.* 9 315–327 10.1111/j.1467-7652.2010.00560.x

74. Li H.-W., Zang B.-S., Deng X.-W., Wang X.-P. (2011). Over expression of the trehalose-6-phosphate synthase gene OsTPS1 enhances abiotic stress tolerance in rice. *Planta* 234 1007–1018 10.1007/s00425-011-1458-0

75. Liu C., Mao B., Ou S., Wang W., Liu L., Wu Y., et al. (2014). OsbZIP71, a bZIP transcription factor, confers salinity, and drought tolerance in rice. *Plant Mol. Biol.* 84 19–36 10.1007/s11103-013-0115-3

76. Manavalan L. P., Chen X., Clarke J., Salmeron J., Nguyen H. T. (2012). RNAi-mediated disruption of squalene synthase improves drought tolerance and yield in rice. *J. Exp. Bot.* 63 163–175 10.1093/jxb/err258

77. Maruyama K., Todaka D., Mizoi J., Yoshida T., Kidokoro S., Matsukura S., et al. (2012).Identification of cis-acting promoter elements in cold- and dehydration-induced transcriptional pathways in *Arabidopsis*, rice, and soybean. *DNA Res.* 19 37–49 10.1093/dnares/dsr040

78. Maruyama K., Urano K., Yoshiwara K., Morishita Y., Sakurai N., Suzuki H., et al. (2014). Integrated analysis of the effects of cold and dehydration on rice metabolites, phytohormones, and gene transcripts. *Plant Physiol.* 164 1759–1771 10.1104/pp.113.231720

79. Matsukura S., Mizoi J., Yoshida T., Todaka D., Ito Y., Maruyama K., et al. (2010). Comprehensive analysis of rice DREB2-type genes that encode transcription factors involved in the expression of abiotic stress-responsive genes. *Mol. Genet. Genomics* 283 185–196 10.1007/s00438-009-0506-y

80. Medina J., Catalá R., Salinas J. (2011). The CBFs: three *Arabidopsis* transcription factors to cold acclimate. *Plant Sci.* 180 3–11 10.1016/j.plantsci.2010.06.019

81. Michelena V. A., Boyer J. S. (1982). Complete turgor maintenance at low water potentials in the elongating

region of maize leaves. *Plant Physiol.* 69 1145–1149 10.1104/pp.69.5.1145

82. Mittler R., Blumwald E. (2010). Genetic engineering for modern agriculture: challenges and perspectives. *Annu. Rev. Plant Biol.* 61 443–462 10.1146/annurev-arplant-042809-112116

83. Miura K., Jin J. B., Lee J., Yoo C. Y., Stirm V., Miura T., et al. (2007). SIZ1-mediated sumoylation of ICE1 controls CBF3/DREB1A expression and freezing tolerance in *Arabidopsis*. *Plant Cell* 19 1403–1414 10.1105/tpc.106.048397

84. Miyakawa T., Fujita Y., Yamaguchi-Shinozaki K., Tanokura M. (2013). Structure and function of abscisic acid receptors. *Trends Plant Sci.* 18 259–266 10.1016/j.tplants.2012.11.002

85. Mizoi J., Shinozaki K., Yamaguchi-Shinozaki K. (2012). AP2/ERF family transcription factors in plant abiotic stress responses. *Biochim. Biophys. Acta* 1819 86–96 10.1016/j.bbagrm.2011.08.004

86. Morimoto K., Mizoi J., Qin F., Kim J.-S., Sato H., Osakabe Y., et al. (2013). Stabilization of*Arabidopsis* DREB2A is required but not sufficient for the induction of target genes under conditions of stress. *PLoS ONE* 8:e80457 10.1371/journal.pone.0080457

87. Nakamura J., Yuasa T., Huong T. T., Harano K., Tanaka S., Iwata T., et al. (2011). Rice homologs of inducer of *CBF* expression (OsICE) are involved in cold acclimation. *Plant Biotechnol.* 28 303–309 10.5511/plantbiotechnology.11.0421a

88. Nakashima K., Takasaki H., Mizoi J., Shinozaki K., Yamaguchi-Shinozaki K. (2012). NAC transcription factors in plant abiotic stress responses. *Biochim. Biophys. Acta* 1819 97–103 10.1016/j.bbagrm.2011.10.005

89. Nakashima K., Tran L.-S. P., Van Nguyen D., Fujita M., Maruyama K., Todaka D., et al. (2007).Functional analysis of a NAC-type transcription factor OsNAC6 involved in abiotic and biotic stress-responsive gene expression in rice. *Plant J.* 51 617–630 10.1111/j.1365-313X.2007.03168.x

90. Nijhawan A., Jain M., Tyagi A. K., Khurana J. P. (2008). Genomic survey and gene expression analysis of the basic leucine zipper transcription factor family in rice. *Plant Physiol.* 146 333–350 10.1104/pp.107.112821

91. Nuruzzaman M., Sharoni A. M., Kikuchi S. (2013). Roles of NAC transcription factors in the regulation of biotic and abiotic stress responses in plants. *Front. Microbiol.* 4:24810.3389/fmicb.2013.00248

92. Ogawa D., Abe K., Miyao A., Kojima M., Sakakibara H., Mizutani M., et al. (2011). RSS1 regulates the cell cycle and maintains meristematic activity under stress conditions in rice. *Nat. Commun.* 2 27810.1038/ncomms1279

93. Oh S.-J., Kim Y. S., Kwon C.-W., Park H. K., Jeong J. S., Kim J.-K. (2009). Overexpression of the transcription factor AP37 in rice improves grain yield under drought conditions. *Plant Physiol.* 1501368–1379 10.1104/pp.109.137554

94. Oh S.-J., Kwon C.-W., Choi D.-W., Song S. I., Kim J.-K. (2007). Expression of barley HvCBF4 enhances tolerance to abiotic stress in transgenic rice. *Plant Biotechnol. J.* 5 646–656 10.1111/j.1467-7652.2007.00272.x

95. Oh S., Song S. I., Kim Y. S., Jang H., Kim S. Y., Kim M., et al. (2005). *Arabidopsis* CBF3 / DREB1A and ABF3 in transgenic rice increased tolerance to abiotic stress without stunting growth.*Plant Physiol.* 138 341–351 10.1104/pp.104.059147.1

96. Ouyang S.-Q., Liu Y.-F., Liu P., Lei G., He S.-J., Ma B., et al. (2010). Receptor-like kinase OsSIK1 improves drought and salt stress tolerance in rice (*Oryza sativa*) plants. *Plant J.* 62 316–329 10.1111/j.1365-313X.2010.04146.x

97. Parent B., Suard B., Serraj R., Tardieu F. (2010). Rice leaf growth and water potential are resilient to evaporative demand and soil water deficit once the effects of root system are neutralized. *Plant Cell Environ.* 33 1256–1267 10.1111/j.1365-3040.2010.02145.x

98. Park H. C., Kim H., Koo S. C., Park H. J., Cheong M. S., Hong H., et al. (2010). Functional characterization of the SIZ/PIAS-type SUMO E3 ligases, OsSIZ1 and OsSIZ2 in rice. *Plant Cell Environ.* 33 1923–1934 10.1111/j.1365-3040.2010.02195.x

99. Peleg Z., Reguera M., Tumimbang E., Walia H., Blumwald E. (2011). Cytokinin-mediated source/sink modifications improve drought tolerance and increase grain yield in rice under water-stress. *Plant Biotechnol. J.* 9 747–758 10.1111/j.1467-7652.2010.00584.x

100. Pellegrineschi A., Reynolds M., Pacheco M., Brito R. M., Almeraya R., Yamaguchi-shinozaki K., et al. (2004). Stress-induced expression in wheat of the *Arabidopsis thaliana* DREB1A gene delays water stress symptoms under greenhouse conditions. *Genome* 47 493–500 10.1139/G03-140

101. Peremarti A., Bassie L., Christou P., Capell T. (2009). Spermine facilitates recovery from drought but does not confer drought tolerance in transgenic rice plants expressing *Datura stramonium* S-adenosylmethionine decarboxylase. *Plant Mol. Biol.* 70 253–264 10.1007/s11103-009-9470-5

102. Polizel A. M., Medri M. E., Nakashima K., Yamanaka N., Farias J. R. B., de Oliveira M. C. N., et al. (2011). Molecular, anatomical, and physiological properties of a genetically modified soybean line transformed with rd29A: AtDREB1A for the improvement of drought tolerance. *Genet. Mol. Res.* 103641–3656 10.4238/2011.October.21.4

103. Qin F., Sakuma Y., Tran L.-S. P., Maruyama K., Kidokoro S., Fujita Y., et al. (2008). *Arabidopsis*DREB2A-interacting proteins function as RING E3 ligases and negatively regulate plant drought stress-responsive gene expression. *Plant Cell* 20 1693–1707 10.1105/tpc.107.057380

104. Qin F., Shinozaki K., Yamaguchi-Shinozaki K. (2011). Achievements and challenges in understanding plant abiotic stress responses and tolerance. *Plant Cell Physiol.* 52 1569–1582 10.1093/pcp/pcr106

105. Quan R., Hu S., Zhang Z., Zhang H., Zhang Z., Huang R. (2010). Overexpression of an ERF transcription factor TSRF1 improves rice drought tolerance. *Plant Biotechnol. J.* 8 476–488 10.1111/j.1467-7652.2009.00492.x

106. Raghavendra A. S., Gonugunta V. K., Christmann A., Grill E. (2010). ABA perception and signalling. *Trends Plant Sci.* 15 395–401 10.1016/j.tplants.2010.04.006

107. Ray S., Dansana P. K., Giri J., Deveshwar P., Arora R., Agarwal P., et al. (2011). Modulation of transcription factor and metabolic pathway genes in response to water-deficit stress in rice. *Funct. Integr. Genomics* 11 157–178 10.1007/s10142-010-0187-y

108. Redillas M. C. F. R., Jeong J. S., Kim Y. S., Jung H., Bang S. W., Choi Y. D., et al. (2012). The overexpression of OsNAC9 alters the root architecture of rice plants enhancing drought resistance and grain yield under field conditions. *Plant Biotechnol. J.* 10 792–805 10.1111/j.1467-7652.2012.00697.x

109. Rivero R. M., Kojima M., Gepstein A., Sakakibara H., Mittler R., Gepstein S., et al. (2007). Delayed leaf senescence induces extreme drought tolerance in a flowering plant. *Proc. Natl. Acad. Sci. U.S.A.* 104 19631–19636 10.1073/pnas.0709453104

110. Rivero R. M., Shulaev V., Blumwald E. (2009). Cytokinin-dependent photorespiration and the protection of photosynthesis during water deficit. *Plant Physiol.* 150 1530–1540 10.1104/pp.109.139378

111. Saijo Y., Hata S., Kyozuka J., Shimamoto K., Izui K. (2000). Over-expression of a single Ca2+-dependent protein kinase confers both cold and salt/drought tolerance on rice plants. *Plant J.* 23 319–327 10.1046/j.1365-313x.2000.00787.x

112. Sakuma Y., Maruyama K., Osakabe Y., Qin F., Seki M., Shinozaki K. (2006a). Functional analysis of an *Arabidopsis* transcription factor, dreb2a, involved in drought-responsive gene expression. *Plant Cell* 18 1292–1309 10.1105/tpc.105.035881.1

113. Sakuma Y., Maruyama K., Qin F., Osakabe Y., Shinozaki K., Yamaguchi-Shinozaki K. (2006b). Dual function of an *Arabidopsis* transcription factor DREB2A in water-stress-responsive and heat-stress-responsive gene expression. *Proc. Natl. Acad. Sci. U.S.A.* 103 18822–18827 10.1073/pnas.0605639103

114. Seo J.-S., Joo J., Kim M.-J., Kim Y.-K., Nahm B. H., Song S. I., et al. (2011). OsbHLH148, a basic helix-loop-helix protein, interacts with OsJAZ proteins in a jasmonate signaling pathway leading to drought tolerance in rice. *Plant J.* 65 907–921 10.1111/j.1365-313X.2010.04477.x

115. Sharma R., De Vleesschauwer D., Sharma M. K., Ronald P. C. (2013). Recent advances in dissecting stress-regulatory crosstalk in rice. *Mol. Plant* 6 250–260 10.1093/mp/sss147

116. Shen H., Liu C., Zhang Y., Meng X., Zhou X., Chu C., et al. (2012). OsWRKY30 is activated by MAP kinases to confer drought tolerance in rice. *Plant Mol. Biol.* 80 241–253 10.1007/s11103-012-9941-y

117. Shu L., Lou Q., Ma C., Ding W., Zhou J., Wu J., et al. (2011). Genetic, proteomic and metabolic analysis of the regulation of energy storage in rice seedlings in response to drought. *Proteomics* 114122–4138 10.1002/pmic.201000485

118. Singh A. K., Choudhury B. U., Bouman B. A. M. (2002). "Water-wise rice production," in*Proceedings of the international workshop on water-wise rice production* eds Bouman B. A. M., Hengsdijk H., Hardy B., Bindraban P. S., Tuong T. P., Ladha J. K., editors. (Los Baños, Philippines: International Rice Research Institute) 237–248.

119. Skirycz A., De Bodt S., Obata T., De Clercq I., Claeys H., De Rycke R., et al. (2010). Developmental stage specificity and the role of mitochondrial metabolism in the response of *Arabidopsis* leaves to prolonged mild osmotic stress. *Plant Physiol.* 152 226–244 10.1104/pp.109.148965

120. Song S.-Y., Chen Y., Chen J., Dai X.-Y., Zhang W.-H. (2011). Physiological mechanisms underlying OsNAC5-dependent tolerance of rice plants to abiotic stress. *Planta* 234 331–345 10.1007/s00425-011-1403-2

121. Sun L., Zhang Q., Wu J., Zhang L., Jiao X., Zhang S., et al. (2014). Two rice authentic histidine phosphotransfer proteins, osahp1 and osahp2, mediate cytokinin signaling, and stress responses in rice.*Plant Physiol.* 165 335–345 10.1104/pp.113.232629

122. Swire-Clark G. A., Marcotte W. R. (1999). The wheat LEA protein Em functions as an osmoprotective molecule in *Saccharomyces cerevisiae*. *Plant Mol. Biol.* 39 117–128 10.1023/A:1006106906345

123. Takahashi T., Kakehi J.-I. (2010). Polyamines: ubiquitous polycations with unique roles in growth and stress responses. *Ann. Bot.* 105 1–6 10.1093/aob/mcp259

124. Takasaki H., Maruyama K., Kidokoro S., Ito Y., Fujita Y., Shinozaki K., et al. (2010). The abiotic stress-responsive NAC-type transcription factor OsNAC5 regulates stress-inducible genes and stress tolerance in rice. *Mol. Genet. Genomics* 284 173–183 10.1007/s00438-010-0557-0

125. Tang N., Zhang H., Li X., Xiao J., Xiong L. (2012). Constitutive activation of transcription factor OsbZIP46 improves drought tolerance in rice. *Plant Physiol.* 158 1755–1768 10.1104/pp.111.190389

126. Tansley T. R., Shechter I. (2001). Squalene synthase: structure and regulation. *Prog. Nucleic Acid Res. Mol. Biol.* 65 157–195 10.1016/S0079-6603(00)65005-5

127. Thomashow M. F. (2010). Molecular basis of plant cold acclimation: insights gained from studying the CBF cold response pathway. *Plant Physiol.* 154 571–577 10.1104/pp.110.161794

128. Todaka D., Nakashima K., Maruyama K., Kidokoro S., Osakabe Y., Ito Y., et al. (2012). Rice phytochrome-interacting factor-like protein OsPIL1 functions as a key regulator of internode elongation and induces a morphological response to drought stress. *Proc. Natl. Acad. Sci. U.S.A.* 10915947–15952 10.1073/pnas.1207324109

129. Tuteja N., Sahoo R. K., Garg B., Tuteja R. (2013). OsSUV3 dual helicase functions in salinity stress tolerance by maintaining photosynthesis and antioxidant machinery in rice (*Oryza sativa* L. cv. IR64).*Plant J.* 76 115–127 10.1111/tpj.12277

130. Uga Y., Sugimoto K., Ogawa S., Rane J., Ishitani M., Hara N., et al. (2013). Control of root system architecture by DEEPER ROOTING 1 increases rice yield under drought conditions. *Nat. Genet.* 451097–1102 10.1038/ng.2725

131. Umezawa T., Nakashima K., Miyakawa T., Kuromori T., Tanokura M., Shinozaki K., et al. (2010).Molecular basis of the core regulatory network in ABA responses: sensing, signaling and transport.*Plant Cell Physiol.* 51 1821–1839 10.1093/pcp/pcq156

132. USDA. (2014). *Crop Production 2013 Summary.* Chesterfield, MO: National Statistics for Corn.

133. Voesenek L. A. C. J., Bailey-Serres J. (2013). Flooding tolerance: O2 sensing and survival strategies.*Curr. Opin. Plant Biol.* 16 647–653 10.1016/j.pbi.2013.06.008

134. Wan L., Zhang J., Zhang H., Zhang Z., Quan R., Zhou S., et al. (2011). Transcriptional activation of OsDERF1 in OsERF3 and OsAP2-39 negatively modulates ethylene synthesis and drought tolerance in rice. *PLoS ONE* 6:e25216 10.1371/journal.pone.0025216

135. Wang D., Pan Y., Zhao X., Zhu L., Fu B., Li Z. (2011a). Genome-wide temporal-spatial gene expression profiling of drought responsiveness in rice. *BMC Genomics* 12:149 10.1186/1471-2164-12-149

136. Wang F.-Z., Wang Q.-B., Kwon S.-Y., Kwak S.-S., Su W.-A. (2005). Enhanced drought tolerance of transgenic rice plants expressing a pea manganese superoxide dismutase. *J. Plant Physiol.* 162 465–472 10.1016/j.jplph.2004.09.009

137. Wang Q., Guan Y., Wu Y., Chen H., Chen F., Chu C. (2008). Overexpression of a rice OsDREB1F gene increases salt, drought, and low temperature tolerance in both *Arabidopsis* and rice. *Plant Mol. Biol.* 67 589–602 10.1007/s11103-008-9340-6

138. Wang Q., Zhang W., Yin Z., Wen C.-K. (2013). Rice CONSTITUTIVE TRIPLE-RESPONSE2 is involved in the ethylene-receptor signalling and regulation of various aspects of rice growth and development. *J. Exp. Bot.* 64 4863–4875 10.1093/jxb/ert272

139. Wang W.-S., Pan Y.-J., Zhao X.-Q., Dwivedi D., Zhu L.-H., Ali J., et al. (2011b). Drought-induced site-specific DNA methylation and its association with drought tolerance in rice (*Oryza sativa* L.). *J. Exp. Bot.* 62 1951–1960 10.1093/jxb/erq391

140. Wei Z., Laby R. J., Zumoff C. H., Bauer D. W., He S. Y., Colimer A., et al. (1992). Harpin, elicitor of the hypersensitive response produced by the plant pathogen *Erwinia Amylovora. Sci.* 257 85–88 10.1126/science.1621099

141. Weiner J. J., Peterson F. C., Volkman B. F., Cutler S. R. (2010). Structural and functional insights into core ABA signaling. *Curr. Opin. Plant Biol.* 13 495–502 10.1016/j.pbi.2010.09.007

142. Wolkers W. F., McCready S., Brandt W. F., Lindsey G. G., Hoekstra F. A. (2001). Isolation and characterization of a D-7 LEA protein from pollen that stabilizes glasses in vitro. *Biochim. Biophys. Acta* 1544 196–206 10.1016/S0167-4838(00)00220-X

143. Xiang Y., Huang Y., Xiong L. (2007). Characterization of stress-responsive CIPK genes in rice for stress tolerance improvement. *Plant Physiol.* 144 1416–1428 10.1104/pp.107.101295

144. Xiang Y., Tang N., Du H., Ye H., Xiong L. (2008). Characterization of OsbZIP23 as a key player of the basic leucine zipper transcription factor family for conferring abscisic acid sensitivity and salinity and drought tolerance in rice. *Plant Physiol.* 148 1938–1952 10.1104/pp.108.128199

145. Xiao B.-Z., Chen X., Xiang C.-B., Tang N., Zhang Q.-F., Xiong L.-Z. (2009). Evaluation of seven function-known candidate genes for their effects on improving drought resistance of transgenic rice under field conditions. *Mol. Plant* 2 73–83 10.1093/mp/ssn068

146. Xiao B., Huang Y., Tang N., Xiong L. (2007). Over-expression of a LEA gene in rice improves drought resistance under the field conditions. *Theor. Appl. Genet.* 115 35–46 10.1007/s00122-007-0538-9

147. Xu D., Duan X., Wang B., Hong B., Ho T. D., Wu R. (1996). Expression of a late embryogenesis abundant protein cene, HVA7, from barley confers tolerance to water deficit and salt stress in transgenic rice. *Plant Physiol.* 110 249–257.

148. Xu D.-Q., Huang J., Guo S.-Q., Yang X., Bao Y.-M., Tang H.-J., et al. (2008). Overexpression of a TFIIIA-type zinc finger protein gene ZFP252 enhances drought and salt tolerance in rice (*Oryza sativa*L.). *FEBS Lett.* 582 1037–1043 10.1016/j.febslet.2008.02.052

149. Xu M., Li L., Fan Y., Wan J., Wang L. (2011). ZmCBF3 over expression improves tolerance to abiotic stress in transgenic rice (*Oryza sativa*) without yield penalty. *Plant Cell Rep.* 30 1949–1957 10.1007/s00299-011-1103-1

150. Yamaguchi-Shinozaki K., Shinozaki K. (2006). Transcriptional regulatory networks in cellular responses and tolerance to dehydration and cold stresses. *Annu. Rev. Plant Biol.* 57 781–803 10.1146/annurev.arplant.57.032905.105444

151. Yang A., Dai X., Zhang W.-H. (2012). A R2R3-type MYB gene, OsMYB2, is involved in salt, cold, and dehydration tolerance in rice. *J. Exp. Bot.* 63 2541–2556 10.1093/jxb/err431

152. Yang D. H., Kwak K. J., Kim M. K., Park S. J., Yang K.-Y., Kang H. (2014). Expression of*Arabidopsis* glycine-rich RNA-binding protein AtGRP2 or AtGRP7 improves grain yield of rice (*Oryza sativa*) under drought stress conditions. *Plant Sci.* 214 106–112 10.1016/j.plantsci.2013.10.006

153. Yoshida T., Mogami J., Yamaguchi-Shinozaki K. (2014). ABA-dependent and ABA-independent signaling in response to osmotic stress in plants. *Curr. Opin. Plant Biol.* 21 133–139 10.1016/j.pbi.2014.07.009

154. Yoshida T., Ohama N., Nakajima J., Kidokoro S., Mizoi J., Nakashima K., et al. (2011). *Arabidopsis*HsfA1 transcription factors function as the main positive regulators in heat shock-responsive gene expression. *Mol. Genet. Genomics* 286 321–332 10.1007/s00438-011-0647-7

155. You J., Hu H., Xiong L. (2012). An ornithine δ-aminotransferase gene OsOAT confers drought and oxidative stress tolerance in rice. *Plant Sci.* 197 59–69 10.1016/j.plantsci.2012.09.002

156. Yu H., Chen X., Hong Y.-Y., Wang Y., Xu P., Ke S.-D., et al. (2008). Activated expression of an*Arabidopsis* HD-START protein confers drought tolerance with improved root system and reduced stomatal density. *Plant Cell* 20 1134–1151 10.1105/tpc.108.058263

157. Yu L., Chen X., Wang Z., Wang S., Wang Y., Zhu Q., et al. (2013). *Arabidopsis* enhanced drought tolerance1/HOMEODOMAIN GLABROUS11 confers drought tolerance in transgenic rice without yield penalty. *Plant Physiol.* 162 1378–1391 10.1104/pp.113.217596

158. Zhang H., Liu W., Wan L., Li F., Dai L., Li D., et al. (2010). Functional analyses of ethylene response factor JERF3 with the aim of improving tolerance to drought and osmotic stress in transgenic rice.*Transgenic Res.* 19 809–818 10.1007/s11248-009-9357-x

159. Zhang H., Zhang J., Quan R., Pan X., Wan L., Huang R. (2013). EAR motif mutation of rice OsERF3 alters the regulation of ethylene biosynthesis and drought tolerance. *Planta* 237 1443–1451 10.1007/s00425-013-1852-x

160. Zhang L., Xiao S., Li W., Feng W., Li J., Wu Z., et al. (2011). Overexpression of a Harpin-encoding gene hrf1 in rice enhances drought tolerance. *J. Exp. Bot.* 62 4229–4238 10.1093/jxb/err131

161. Zhang Q., Li J., Zhang W., Yan S., Wang R., Zhao J., et al. (2012). The putative auxin efflux carrier OsPIN3t is involved in the drought stress response and drought tolerance. *Plant J.* 72 805–816 10.1111/j.1365-313X.2012.05121.x

162. Zhao J., Ren W., Zhi D., Wang L., Xia G. (2007). *Arabidopsis* DREB1A/CBF3 bestowed transgenic tall fescue increased tolerance to drought stress. *Plant Cell Rep.* 26 1521–1528 10.1007/s00299-007-0362-3

163. Zhao Y., Ma Q., Jin X., Peng X., Liu J., Deng L., et al. (2014). A novel maize homeodomain-leucine zipper (HD-Zip) I gene, Zmhdz10, positively regulates drought, and salt tolerance in both rice and*Arabidopsis*. *Plant Cell Physiol.* 55 1142–1156 10.1093/pcp/pcu054

164. Zhu J.-K. (2002). Salt and drought stress signal transduction in plants. *Annu. Rev. Plant Biol.* 53 247–273 10.1146/annurev.arplant.53.091401.143329

165. Zong W., Zhong X., You J., Xiong L. (2013). Genome-wide profiling of histone H3K4-tri-methylation and gene expression in rice under drought stress. *Plant Mol. Biol.* 81 175–188 10.1007/s11103-012-9990-2

166. Zou J., Liu C., Liu A., Zou D., Chen X. (2012). Over expression of OsHsp17.0 and OsHsp23.7 enhances drought and salt tolerance in rice. *J. Plant Physiol.* 169 628–635 10.1016/j.jplph.2011.12.014

CHAPTER 13

RECENT ADVANCES IN UTILIZING TRANSCRIPTION FACTORS TO IMPROVE PLANT ABIOTIC STRESS TOLERANCE BY TRANSGENIC TECHNOLOGY

Hongyan Wang[1], Honglei Wang[1], Hongbo Shao[2, 3], and Xiaoli Tang[3]

[1]Laboratory of Plant Molecular Physiology, Graduate School of Agricultural and Life Sciences, The University of Tokyo, Tokyo, Japan

[2]RIKEN Center for Sustainable Resource Science, Yokohama, Japan

ABSTRACT

Agricultural production and quality are adversely affected by various abiotic stresses worldwide and this will be exacerbated by the deterioration of global climate. To feed a growing world population, it is very urgent to breed stress-tolerant crops with higher yields and improved qualities against multiple environmental stresses. Since conventional breeding approaches had marginal success due to the complexity of stress tolerance traits, the transgenic approach is now being popularly used to breed stress-tolerant crops. So identifying and characterizing the critical genes involved in plant stress responses is an essential prerequisite for engineering stress-tolerant crops. Far beyond the manipulation of single functional gene, engineering certain regulatory genes has emerged as an effective strategy now for controlling the expression of many stress-responsive genes. Transcription factors (TFs) are good candidates for genetic engineering to breed stress-tolerant crop because of their role as master regulators of many stress-responsive genes. Many TFs belonging to families AP2/EREBP, MYB, WRKY, NAC, bZIP have been found to be involved in various abiotic stresses and some TF genes have also been engineered to improve stress tolerance in model and crop plants. In this review, we take five large families of TFs as examples and review the recent progress of TFs involved in plant abiotic stress responses and their potential utilization to improve multiple stress tolerance of crops in the field conditions.

INTRODUCTION

Agricultural production and quality are adversely affected by a broad range of abiotic stresses including drought, salinity, heat, and cold. Especially when these stresses occur in combination, it can have devastating effects on plant growth and productivity. It is estimated that more than 50% of worldwide yield loss for major crop are caused by abiotic stresses (Shao et al., 2009; Ahuja et al., 2010; Lobell et al., 2011). According to the current climate prediction models, the deterioration of global climate will inevitably cause an increased frequency of drought, heat wave, and salinization (Easterling et al., 2000; Ipcc, 2007, 2008). This means that agricultural productivity will face a greater challenge in fighting against environmental stresses. Meanwhile, the growing world population will reach close to ten billion by the year 2050 and then almost two times of current agricultural productivity is needed to feed the large population (Bengtsson et al., 2006; United Nations, 2015). Moreover, such a tremendous increase of crop productivity must be achieved with no increase in arable land and in the face of multiple environmental stresses. Where is the way to solve this problem? Many scholars and experts worldwide

Citation: Hongyan Wang, Honglei Wang, Hongbo Shao, and Xiaoli Tang, Recent Advances in Utilizing Transcription Factors to Improve Plant Abiotic Stress Tolerance by Transgenic Technology, doi: 10.3389/fpls.2016.00067.

have reached a consensus that breeding stress-tolerant crops with higher yields and improved qualities against multiple environmental stresses is an effective strategy, as well as one of the greatest challenges faced by modern agriculture (Takeda and Matsuoka, 2008; Newton et al., 2011; Liu J.-H. et al., 2014). In the past few decades, a great deal of efforts has been devoted to breeding crop cultivars with various stress-tolerant traits. Two main approaches have been employed to this process. One is traditional breeding methods such as wide-cross hybridization and mutation breeding, which often brings about unpredictable results. Another is modern transgenic technology by introducing novel exogenous genes or altering the expression levels of endogenous genes to improve stress tolerance. Since conventional breeding approaches have marginal success due to the complexity of stress tolerance traits, the transgenic approach is now being popularly used to develop transgenic crops tolerant to abiotic stresses (Yamaguchi and Blumwald, 2005). Therefore, deciphering the molecular mechanisms by which plants perceive and transduce stress signals to cellular machinery to initiate adaptive responses is an essential prerequisite for identification of the key genes and pathways to engineer stress-tolerant crop plants (Ray et al., 2009; Heidarvand and Amiri, 2010; Sanchez et al., 2011).

Substantial progress has been made to unravel the molecular mechanisms of abiotic stress responses in plants by means of high throughput sequencing and functional genomics tools. To date, a number of critical genes involved in abiotic stress tolerance have been identified and validated, which are generally classified into two types: functional genes and regulatory genes (Shinozaki et al., 2003). The former encodes important enzymes and metabolic proteins (functional proteins), such as detoxification enzyme, water channel, ion transporter, heat shock protein (HSP), and late embryogenesis abundant (LEA) protein, which directly function to protect cells from stresses. The latter encodes various regulatory proteins including TFs, protein kinases and protein phosphatases, which regulate signal transduction and gene expression in the stress responses. Although there have been numerous studies on functional genes, most of these studies pay more attention to single gene or several genes encoding enzymes and protective proteins by imposing a given stress. Due to the complexity of stress responses regulated by multi-genes, little success has been achieved by a single functional gene approach to significantly enhance plant stress tolerance (Mittler and Blumwald, 2010; Varshney et al., 2011). Given the complexity and variability of field conditions where crops are often simultaneously subjected to multiple abiotic stresses or some in combination (Ahuja et al., 2010), more and more studies have paid close attention to regulatory genes and found that some regulatory genes including TFs play essential roles in multiple abotic stress responses by regulating a large spectrum of downstream stress-responsive genes. Thus, genetically modifying the expression of certain regulatory genes can greatly influence plant stress tolerance because it mimics or enhances stress signals to regulate many downstream stress-responsive genes at a time (Century et al., 2008; Yang et al., 2011). Among the regulatory genes, stress-responsive TFs have attracted particular attention due to their important roles in plant stress responses (Chen and Zhu, 2004; Xu et al.,2008a). In this paper, we mainly review the recent progress of TFs involved in plant abiotic stress responses and their potential utilization to improve multiple stress tolerance of crops in the field conditions.

THE GENERIC SIGNALING PATHWAY INVOLVED IN PLANT ABIOTIC STRESS RESPONSES

As sessile organisms, plants have evolved various defense mechanisms at multiple levels to respond to unfavorable environment including diverse abiotic stresses. As stated before, it is imperative to dissect regulatory mechanisms of stress response and identify the key regulators involved in this process to breed or genetically engineer stress-tolerant plants. With the availability of plant genomes and various omics tools including genomics, transcriptomics, and proteomics tools, major progress has been made in deciphering the stress signaling pathways and relevant components involved in plant abiotic stress response, but there is still much more to be determined (Liu J.-H. et al., 2014). According to our current knowledge about stress signaling pathways, the generic signaling pathway for any given abiotic stress can be divided into the following major steps: signal perception, signal transduction, stress responsive gene expression, in turn the activation of physiological, and metabolic responses (Chaves et al., 2003; Yamaguchi-Shinozaki and Shinozaki, 2006; Pérez-Clemente et al., 2013). In this process, plant cells first perceive stress stimulus through sensors or receptors located in the cell wall or membrane. Then the captured extracellular signals are converted into intracellular ones through second messengers including calcium ions, inositol phosphate, reactive oxygen species (ROS), cyclic nucleotides (cAMP and cGMP), sugars, and nitric oxide. Subsequently, these second messengers initiate the corresponding signaling pathways to transduce the signals (Chaves et al., 2009; Bhargava and Sawant, 2013). In many signal transduction pathways, the phosphorylation, and dephosphorylation of proteins mediated by protein kinase and phosphatases, respectively, is an important and effective mechanism for signal relay (Singh et al., 2003). For example, the mitogen activated protein kinases (MAPKs) pathway and calcium-dependent protein kinases (CDPKs) pathway are known to be involved in plant abiotic stress responses (Schaller et al., 2008; Huang G.T. et al.,2012). At the end of the phosphorylation cascade, TFs are activated or suppressed by protein kinases or phosphatases, and they further bind specifically to cis-elements in the promoters of stress-responsive genes and regulate their transcription (Danquah et al., 2014). Meanwhile, TFs themselves are regulated at the transcription level by other upstream components (Hirayama and Shinozaki, 2010) and also subjected to various tiers of modifications at the post-transcription level, such as ubiquitination and sumoylation, thus forming a complex regulatory network to modulate the expression of stress responsive genes, which in turn determine the activation of physiological and metabolic responses (Dong et al., 2006; Miura et al., 2007; Mizoi et al., 2013). All the components mentioned above, from the foremost receptors to the downstream functional genes, constitute the generic pathway for plant abiotic stress signal transduction (Figure1). As one of the most important regulators, TFs function as terminal transducers and directly regulate the expression of an array of downstream genes by interacting with the specific cis-elements in their promoter region (Yamaguchi-Shinozaki and Shinozaki, 2006). In the last few decades, considerable research has been conducted to identify and characterize various TFs involved in plant abiotic stress responses either in abscisic acid (ABA)-dependent pathway or ABA-independent pathway, such as *AP2/EREBP, MYB, WRKY, NAC, bZIP*, and so on (Vinocur and Altman, 2005; Umezawa et al., 2006; Golldack et al., 2011). Numerous efforts have been also made to improve plant stress tolerance by engineering these TF genes, and some promising results have been reported in succession (Table1).1). In the following sections, we mainly summarize current information on several major TF families including their features, roles, and biotechnological uses for improving the abiotic stress tolerance in plants.

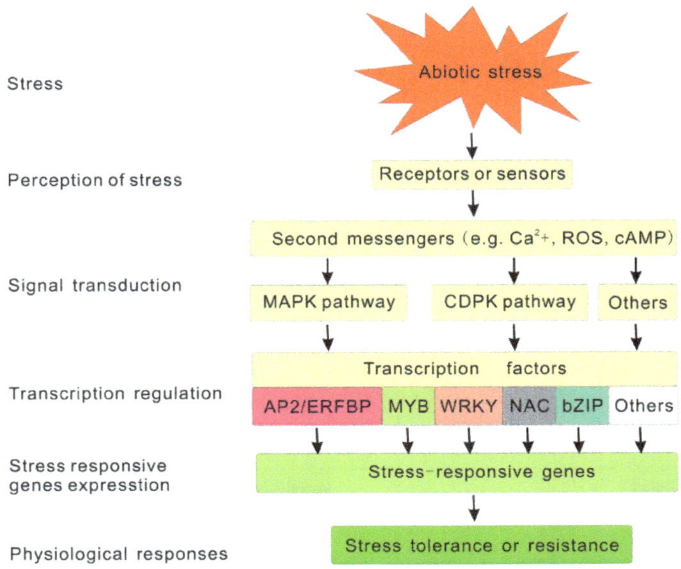

Figure 1: Generic signaling pathway involved in plant abiotic stress responses.

Table 1: Some examples of transgenic plants over-expressing transcription factor genes in recent years

Family	Gene	Donor	Acceptor	Enhanced tolerance	References
AP2/ERFBP	*LcDREB3a*	*Leymus chinensis*	*Arabidopsis*	Drought and salinity↑	Peng et al., 2013
	LcDREB2	*Leymus chinensis*	*Arabidopsis*	Salinity↑	Peng et al., 2013
	LcERF054	*Lotus corniculatus*	*Arabidopsis*	Salinity↑	Sun et al., 2014
	VrDREB2A	*Vigna radiata*	*Arabidopsis*	Drought and salinity↑	Chen et al., 2015
	GmERF3	*Glycine max*	*Tobacco*	Drought and salinity↑	Zhang et al., 2009
	GmERF7	*Glycine max*	Tobacco	Salinity↑	Zhai et al., 2013
	SsDREB	*Suaeda salsa*	Tobacco	Drought and salinity↑	Zhang X. et al., 2015
	JERF3	*Solanum lycopersicum*	Rice	Drought↑	Zhang et al., 2010
	OsDREB2A	*Oryza sativa*	Rice	Drought and salinity↑	Mallikarjuna et al., 2011
	OsERF4a	*Oryza sativa*	Rice	Drought↑	Joo et al., 2013
	AtDREB1A	*Arabidopsis*	Rice	Drought↑	Ravikumar et al., 2014
	TaERF3	*Triticum aestivum*	Wheat	Drought and salinity↑	Rong et al., 2014
	TaPIE1	*Triticum aestivum*	Wheat	Cold↑	Zhu et al., 2014
	EaDREB2	*Erianthus arundinaceus*	Sugarcane	Drought and salinity↑	Augustine et al., 2015
	StDREB1	*Solanum tuberosum*	Potato	Salinity↑	Bouaziz et al., 2013
MYB	*AtMYB15*	*Arabidopsis*	*Arabidopsis*	Drought and salinity↑	Ding et al., 2009
	LcMYB1	*Leymus chinensis*	*Arabidopsis*	Salinity↑	Cheng et al., 2013b
	GmMYBJ1	*Glycine max*	*Arabidopsis*	Drought and cold↑	Su et al., 2014

	TaMYB3R1	*Triticum aestivum*	*Arabidopsis*	Drought and salinity↑	Cai et al., 2015
	TaPIMP1	*Triticum aestivum*	Tobacco	Drought and salinity↑	Liu et al., 2011b
	SbMYB2	*Scutellaria baicalensis*	Tobacco	NaCl, mannitol, and ABA stresses↑	Qi et al., 2015
	SbMYB7	*Scutellaria baicalensis*	Tobacco	NaCl, mannitol, and ABA stresses↑	Qi et al., 2015
	LeAN2	*Lycopersicum esculentum*	Tobacco	Chilling and oxidative stresses↑	Meng et al., 2014
	LeAN2	*Lycopersicum esculentum*	Tomato	Heat↑	Meng et al., 2015
	AtMYB44	*Arabidopsis*	Soybean	Drought and salinity↑	Seo et al., 2012
	OsMYB2	*Oryza sativa*	Rice	Drought, cold, and salinity↑	Yang et al., 2012
	OsMYB91	*Oryza sativa*	Rice	Salinity↑	Zhu et al., 2015
	MdSIMYB1	*Malus × domestica*	Apple	Drought, cold, and salinity↑	Wang et al., 2014
WRKY	*AtWRKY28*	*Arabidopsis*	*Arabidopsis*	Salinity↑	Babitha et al., 2013
	OsWRKY45	*Oryza sativa*	*Arabidopsis*	Drought and salinity↑	Qiu and Yu, 2009
	TaWRKY79	*Triticum aestivum*	*Arabidopsis*	Drought↑	Qin et al., 2013
	VvWRKY11	*Vitis vinifera*	*Arabidopsis*	Drought↑	Liu et al., 2011a
	ZmWRKY33	*Zea may*	*Arabidopsis*	Salinity↑	Li et al., 2013
	GhWRKY34	*Gossypium hirsutum*	*Arabidopsis*	Salinity↑	Zhou et al., 2015
	GsWRKY20	*Glycine soja*	*Arabidopsis*	Drought↑	Luo et al., 2013
	TaWRKY79	*Triticum aestivum*	*Arabidopsis*	Salinity and ionic stress↑	Qin et al., 2013
	TaWRKY93	*Triticum aestivum*	*Arabidopsis*	Salinity, drought, and low temperature↑	Qin et al., 2015
	TaWRKY10	*Triticum aestivum*	Tobacco	Drought and salinity↑	Wang et al., 2013
	GhWRKY39	*Gossypium hirsutum*	Tobacco	Salinity↑	Shi et al., 2014
	BdWRKY36	*Brachypodium distachyon*	Tobacco	Drought↑	Sun et al., 2015
	ZmWRKY58	*Zea may*	Rice	Drought and salinity↑	Cai et al., 2014
	MtWRKY76	*Medicago truncatula*	*Medicago truncatula*	Drought and salinity↑	Liu et al., 2016
NAC	*ANAC019*	*Arabidopsis*	*Arabidopsis*	Cold↑	Jensen et al., 2010
	ONAC063	*Oryza sativa*	*Arabidopsis*	Salinity and osmotic tolerance↑	Yokotani et al., 2009
	GmNAC20	*Glycine max*	*Arabidopsis*	Salinity and freezing tolerance↑	Hao et al., 2011
	ZmSNAC1	*Zea may*	*Arabidopsis*	Cold, salinity, and drought↑	Lu et al., 2012
	BnNAC5	*Brassica napus*	*Arabidopsis*	Salinity↑	Zhong et al., 2012
	TaNAC67	*Triticum aestivum*	*Arabidopsis*	Cold, salinity, and drought↑	Mao et al., 2014
	TaNAC29	*Triticum aestivum*	*Arabidopsis*	Drought and salinity↑	Huang et al., 2015
	MLNAC5	*Miscanthus lutarioriparius*	*Arabidopsis*	Drought and cold↑	Yang et al., 2015
	TaNAC2a	*Triticum aestivum*	Tobacco	Drought	Tang et al., 2012

	AhNAC3	*Arachis hypogaea*	Tobacco	Drought↑	Liu et al., 2013
	SNAC1	*Oryza sativa*	Wheat	Drought and salinity↑	Saad et al., 2013
	OsNAP	*Oryza sativa*	Rice	Cold, salinity, and drought↑	Chen et al., 2014
bZIP	*ABP9*	*Zea may*	*Arabidopsis*	Drought, salinity, and cold↑	Zhang et al., 2011
	GmbZIP1	*Glycine max*	*Arabidopsis*	Drought, salinity, and cold↑	Gao et al., 2011
	ZmbZIP72	*Zea may*	*Arabidopsis*	Drought and salinity↑	Ying et al., 2012
	TabZIP60	*Triticum aestivum*	*Arabidopsis*	Drought, salt, and freezing tolerance↑	Zhang L. et al., 2015
	PtrABF	*Poncirus trifoliata*	Tobacco	Drought↑	Huang et al., 2010
	GmbZIP1	*Glycine max*	Tobacco	Drought, salinity, and cold↑	Gao et al., 2011
	LrbZIP	*Nelumbo nucifera*	Tobacco	Salinity↑	Cheng et al., 2013a
	OsbZIP71	*Oryza sativa*	Rice	Drought and salinity↑	Liu C. et al., 2014

AP2/EREBP TRANSCRIPTION FACTORS

AP2/ERFBP family includes a large group of plant-specific TFs and is characterized by the presence of the highly conserved *AP2*/ethylene-responsive element-binding factor (ERF) DNA-binding domain that directly interact with GCC box and/or dehydration-responsive element (DRE)/C-repeat element (CRT) cis-acting elements at the promoter of downstream target genes (Riechmann and Meyerowitz, 1998). AP2/ERFBP TFs perform a variety of roles in plant developmental processes and stress responses, such as vegetative and reproductive development, cell proliferation, abiotic and biotic stress responses, and plant hormone responses (Nakano et al., 2006; Licausi et al., 2010; Sharoni et al., 2011). Presently, a multitude of AP2/ERFBP members have been identified in various species by means of genome-wide analysis, such as 145 in*Arabidopsis* (Riechmann and Meyerowitz, 1998), 163 in rice (Sharoni et al., 2011), 200 in poplar (Zhuang et al., 2008), 291 in *Chinese cabbage* (Song et al., 2013), 171 in *foxtail millet* (Lata et al., 2014), 116 in moso bamboo (Wu et al., 2015). Based on the number and similarity of *AP2/ERF* domains, these *AP2/EREBP* TFs are grouped into four major subfamilies: *AP2* (Apetala 2), *RAV* (related to *ABI3/VP1*), *DREB* (dehydration-responsive element-binding protein), and *ERF* (Sakuma et al., 2002; Sharoni et al., 2011). Among these, both*ERF* and *DREB* subfamilies have been extensively studied due to their involvement in plant responses to biotic and abiotic stresses.

The DREB subfamily can regulate the expression of multiple dehydration/cold-regulated (*RD/COR*) genes by interacting with *DRE/CRT* cis-elements (A/GCCGAC) located in the promoters of *RD/COR* genes that are responsive to water deficit and low-temperature, such as *COR15A, RD29A/COR78*, and *COR6.6*(Stockinger et al., 1997; Liu et al., 1998; Lucas et al., 2011). A lot of DREB-type TFs have been tested in many plants including *Arabidopsis*, wheat, tomato, soybean, rice, maize, and barley (Agarwal et al., 2006; Lata and Prasad, 2011; Mizoi et al., 2012). According to the variation in some conserved motifs and biological functions in divergent species, *DREB* TFs are further classified into two major subgroups:*DREB1/C*-repeat-binding factor (*DREB1/CBF*) and *DREB2*, and each of them is involved in separate signal transduction pathway under abiotic stresses (Dubouzet et al., 2003). Commonly, *DREB1/CBF* genes are involved in low temperature stress responses in *Arabidopsis* and rice, while *DREB2* genes respond to dehydration, high salinity and heat shock (Liu et al., 1998; Sakuma et al., 2002; Lucas et al., 2011). For example, three major *DREB1/CBF* members in *Arabidopsis, DREB1A/CBF3, DREB1B/CBF1* and*DREB1C/CBF2* are rapidly induced in response to cold stress (Stockinger et al., 1997; Gilmour et al., 1998; Liu et al., 1998; Shinwari et al., 1998). Over-expressing any one of these three *DREB1s/CBFs* displayed significantly improved tolerance to freezing, drought, and high salinity in transgenic *Arabidopsis* (Gilmour et al., 1998; Jaglo-Ottosen et al., 1998; Liu et al., 1998). Further, over-expressing *Arabidopsis DREB1/CBF*genes improved freezing tolerance in oilseed rape (Jaglo et al., 2001) and chilling tolerance in tomato, tobacco and rice (Tsai-Hung et al., 2002; Kasuga et al., 2004; Ito et al., 2006). Some *DREB1/CBF*homologous genes have also been isolated from many other plant species including tomato, oilseed rape, wheat, rye, rice, and maize, and some of them have been used to produce transgenic plants with improved tolerance to abiotic stress (Jaglo et al., 2001; Dubouzet et al., 2003; Qin et al., 2004). In contrast, *DREB2*genes have been studied in a limited number of plant species, but the existing studies have shown that*DREB2* genes are also involved in abiotic stress responses in plants. In *Arabidopsis, DREB2A* and *DREB2B*are major *DREB2s* induced by dehydration, high salinity, and heat, while *DREB2C* is induced by heat later than them (Liu et al., 1998; Nakashima et al., 2000; Sakuma et al., 2006b; Lim et al., 2007). Over-expression of the constitutively active form of *AtDREB2A* from *Arabidopsis* improved the tolerance to drought and osmotic stress in transgenic *Arabidopsis* plants (Sakuma et al., 2006a; Xu et al., 2008b). Over-expression of*ZmDREB2A* from maize enhanced drought tolerance in transgenic *Arabidopsis* plants (Qin et al., 2007). The transgenic plants harboring *GmDREB2* from soybean also showed enhanced tolerance to drought and high salinity without any growth retardation (Chen et al., 2007), as was observed in transgenic rice by over-expressing *OsDREB2A* under control of stress-inducible *RD29A* promoter (Mallikarjuna et al., 2011).

The ERF subfamily is the largest group of the AP2/EREBP TF family (Dietz et al., 2010) and functions in plant stress tolerance by regulating the stress-responsive genes through interacting with the cis-element GCC boxes with core sequence of AGCCGCC (Ohme-Takagi and Shinshi, 1995; Hao et al., 1998). An array of ERF genes are induced by various abiotic stresses, such as drought, high salinity, osmotic stress, and cold (Xu et al., 2008a). Over-expression of these ERF genes resulted in improvement of abiotic stress tolerance in transgenic plants, as summarized in Table1. It is worth mentioning that some ERFs function in both biotic and abiotic stress tolerance, and this is partly due to their involvement in various hormonal signaling pathways including ethylene, JA, or SA (Liang et al., 2008). For example, over-expressing *TaPIE1* in wheat significantly enhanced resistance to both pathogen and freezing stress (Zhu et al., 2014). Over-expressing*GmERF3* in tobacco not only enhanced resistance against infection by pathogen and tobacco mosaic virus (TMV) but also improved tolerance to high salinity and dehydration (Zhang et al., 2009). So far, functions of a limited number of *ERFs* have been well characterized, but most of *ERF* family members have yet to be identified.

MYB TRANSCRIPTION FACTORS

The MYB TFs are widely distributed in plants and form a large family characterized by a highly conserved MYB domain for DNA-binding, which contains from 1 to 4 imperfect repeats (*MYB* repeat) at the N-terminus. In contrast, the activation domain is located at the C-terminus and varies significantly among MYBs, leading to versatile regulatory roles of *MYB* family. According to the number of MYB domain repeats, the MYB TFs are divided into four groups: 1R-MYB (MYB-related type), R2R3-MYB, R1R2R3-MYB, and 4R-MYB, containing one, two, three, and four *MYB* repeats, respectively. Among them, the R2R3-MYBs are more prevalent in plants (Dubos et al., 2010; Ambawat et al., 2013; Li et al., 2015). So far, large numbers of MYB members have been identified in different plant species, such as 198 in *Arabidopsis*(Yanhui et al., 2006), 183 in rice (Yanhui et al., 2006), 229 in apple (Cao et al., 2013), 177 in sweet orange (Hou et al., 2014), 209 in *foxtail millet* (Muthamilarasan et al., 2014).

Numerous MYB TFs have been found to function in many significant physiological and biochemical processes including cell development and cell cycle, primary and secondary metabolism, hormone synthesis and signal transduction, as well as in plant responses to various biotic and abiotic stresses (Allan et al., 2008; Dubos et al., 2010; Ambawat et al., 2013). Recently, some abiotic stress-responsive *MYB* TFs in *Arabidopsis*and other plants have been well summarized by Li (Li et al., 2015). For example, *AtMYB15* improved freezing tolerance by regulating *CBF* expression (Agarwal et al., 2006); *AtMYB44*, *AtMYB60*, and *AtMYB61*improved drought tolerance by regulating stomatal movement (Cominelli et al., 2005; Liang et al., 2005; Jung et al., 2008). Especially, *AtMYB96* improved drought tolerance either by integrating ABA and auxin signals (Seo et al., 2009) or by activating cuticular wax biosynthesis (Seo et al., 2011), and also improved freezing and drought tolerance by regulating a lipid-transfer protein LTP3. This fact shows that a MYB factor can regulate diverse target genes involved in various physiological processes under abiotic stresses. In addition, *OsMYB2* from rice was induced by salt, cold, and dehydration stress. The transgenic plants with over-expressing *OsMYB2* exhibited enhanced tolerance to various stresses by the alteration of expression levels of numerous genes involving diverse functions in stress response (Yang et al., 2012). Salt and freezing tolerance in *Arabidopsis* was significantly enhanced by over-expressing either *GmMYB76* or *GmMYB177*from soybean (Liao et al., 2008).

WRKY TRANSCRIPTION FACTORS

The WRKY family is also extensively distributed in plants and contains many members. WRKY TFs are characterized by the presence of one or two highly conserved WRKY domains of about 60 amino acid residues, which contains a conserved WRKYGQK motif at the N-terminus and a C2H2 or C2HC zinc-finger motif at the C-terminus The WRKY domains can specifically bind to W-box cis-elements with a core sequence of TTGACC/T, located at the promoters of many target genes. According to the number of WRKY domains and the feature of the zinc finger motif, the WRKY TFs can be categorized into three groups. Group I members contain two WRKY domains and a C2H2 zinc-finger motif, whereas group II and III members contain one WRKY domain with a C2H2 and C2HC zinc-finger motif, respectively (Eulgem et al., 2000; Ulker and Somssich, 2004; Pandey and Somssich, 2009; Rushton et al., 2010). Since the cloning of the first cDNA encoding a WRKY protein (SPF1) from sweet potato (Ishiguro and Nakamura, 1994), an increasing number of WRKY TFs have been identified in various plants, such as 74 in *Arabidopsis* (Ulker and Somssich, 2004), 102 in rice (Wu et al., 2005), 104 in poplar (He et al., 2012), 86 in *Brachypodium distachyon* (Wen et al., 2014), 182 in soybean (Bencke-Malato et al., 2014), and 116 and 102 genes in two different species of cotton (Dou et al., 2014).

WRKY TFs have been shown to participate in various processes in plants, including plant growth, seed development, leaf senescence, and responses to biotic and abiotic stresses (Rushton et al., 2010). Accumulating evidence has demonstrated that WRKY TFs play key roles in plant responses to a variety of abiotic stresses such as drought, salt, heat, cold, and osmotic pressure, and these topics have been extensively reviewed recently (Chen et al., 2012; Rushton et al., 2012; Tripathi et al., 2014; Banerjee and Roychoudhury,2015). Over-expression of some stress-responsive *WRKY* genes showed enhanced tolerance to abiotic stresses in transgenic plants. For example, transgenic rice plants harboring *OsWRKY11* gene showed significant heat and drought tolerance (Wu et al., 2009). Transgenic *Arabidopsis* plants over-expressing*GmWRKY21* gene exhibited improved tolerance to cold stress, while over-expressing *GmWRKY54* gene improved tolerance to drought and salt stress (Zhou et al., 2008). Transgenic *Arabidopsis* plants over-expressing *VvWRKY11* improved to tolerance mannitol-induced osmotic stress (Liu et al., 2011a). Although some WRKYs in several plants have been functionally characterized, the vast majority of WRKYs in many plants, especially in non-model plants, are far from being functionally elucidated.

NAC TRANSCRIPTION FACTORS

Like the transcription factor families mentioned above, the NAC TFs also comprise a large plant-specific superfamily present in a wide range of plant species. The typical features of a NAC TF contain a highly conserved NAC domain in the N-terminal region and a variable transcriptional regulatory region in the C-terminal region. The NAC domain is associated with DNA binding, nucleus-oriented localization, and the formation of homodimers or heterodimers with other NAC proteins, while the transcriptional regulatory functions as a transcriptional activator or repressor (Olsen et al., 2005; Puranik et al., 2012). NAC TFs can regulate the transcription of downstream target genes by interacting with *NAC* recognition sequence (NACRS) with the CACG core-DNA binding motif in the promoter of these genes. NAC TFs have been found to participate in various processes including flower development, formation of secondary walls and cell division, shoot apical meristem formation, leaf senescence, as well as biotic and abiotic stress responses (Olsen et al., 2005; Tran et al., 2010; Nakashima et al., 2012; Nuruzzaman et al., 2013; Banerjee and Roychoudhury, 2015). To date, a lot of putative NAC TFs have been identified in many sequenced species at genome-wide scale, such as 117 in *Arabidopsis* and 151 in rice (Nuruzzaman et al., 2010), 152 in soybean (Le et al., 2011), 204 in Chinese cabbage (Liu T.K. et al., 2014), 152 in maize (Shiriga et al., 2014), tomato (Su et al., 2015). Moreover, quite a lot of them have been found to be involved in abiotic stress responses. For instance, 33 *NAC* genes changed significantly in *Arabdopsis* under salt stress (Jiang and Deyholos,2006), 40 *NAC* genes changed under drought or salt stress in rice (Fang et al., 2008), 38 *NAC* genes changed in soybean under drought stress (Le et al., 2011), 32 *NAC* genes responded to at least two kinds of treatments in *Chrysanthemum lavandulifolium* (Huang H. et al., 2012). These stress-responsive *NAC* genes showed differential expression patterns such as tissue-specific, developmental stage- or stress-specific expression, indicating their involvement in the complex signaling networks during plant stress responses. Some of these stress-responsive *NAC* genes have been over-expressed in *Arabidopsis*, rice and other plants and displayed positive effects, summarized in Table1.

BZIP TRANSCRIPTION FACTORS

The basic leucine zipper (bZIP) family contains a conserved bZIP domain which is composed of a highly basic region for nuclear localization and DNA binding at the N-terminus and a leucine-rich motif for dimerization at the C-terminus (Landschulz et al., 1988; Hurst, 1994). Like other TFs, the bZIP TFs not only play pivotal roles in developmental processes but also respond to various abiotic stresses such as drought, high salinity, and cold stresses (Jakoby et al., 2002). Now, many members of the bZIP TF family have been identified or predicted at genome-wide level in some species. For example, it has been reported 75 in*Arabidopsis* (Jakoby et al., 2002), 89 in rice (Nijhawan et al., 2008), 125 in maize (Wei et al., 2012), 89 in barley (Pourabed et al., 2015), 55 in grapevine (Liu J. et al., 2014), 96 in *Brachypodium distachyon* (Liu and Chu, 2015). However, only a small portion of bZIP TFs has been well studied and most studies on their involvement in stress responses have shown that bZIP TFs are induced by ABA and regulate the expression of stress-related genes in ABA-dependent manner through interaction with specific ABA-responsive cis-acting elements (ABRE) in their promoter region (Uno et al., 2000; Kim et al., 2006; Zou et al., 2008). A lot of efforts have been made to improve abiotic stress tolerance in transgenic plants by over-expressing some stress-responsive bZIP genes and some successful example have been achieved, as listed in Table1.

CONCLUSIONS AND PERSPECTIVES

Taking five large families of TFs as examples, this review emphasizes the promising roles of TFs as tools to improve plant responses to multiple abiotic stresses. In addition to the above-mentioned several TF families, there are still other TF families such as DNA binding with one finger (Dof) TFs, basic helix-loop-helix (bHLH) TFs, homeodomain-leucine zipper (HD-Zip) TFs, heat shock TFs (HSFs), etc. How to select the key TFs in such a huge gene families and fully display its potential is still an important issue before us. Although a great deal of information about TFs has been accumulated on their involvement in response to diverse abiotic stresses and a good number of promising candidate TF genes have been validated, but there are still some problems to be solved. First, functional redundancy between different TF members may hinder the dissection of the functions of an individual member. Second, most of transgenic studies based on TFs focused on plant growth and tolerance to a given stress at a given developmental stage rather than whole stage. Moreover, the evaluation of transgenic plants was conducted in controlled laboratory or greenhouse conditions rather than field conditions. Third, the constitutive over-expression of some TF genes may improve the stress tolerance, but occasionally lead to negative effects in transgenic plants such as dwarfing, late flowering, and lower yields. Finally, the complete regulation mechanism of individual transcription factor including its upstream and downstream co-regulators, as well as their interactions remains largely unknown.

Abiotic stress response in plants is an extremely complicated process because of the huge gene families and the complex interactions between TFs and cis-elements on the promoters of target genes. Moreover, one transcription factor may regulate a vast array of target genes with the corresponding cis-elements on the promoters, whereas one gene with several types of cis-elements may be regulated by different families of TFs. Thus, the stress-responsive TFs not only function independently but also cross-talk between each other in response to various abiotic stress responses, which indicates the complexity of signaling networks involved in plant stress responses. In the future research, we should first identify multiple stress-responsive TF genes by comparing their expression patterns and the identification of commonly regulated genes which have been proposed to be required for universal stress responses or represent points of cross-talk between signaling pathways (Prasch and Sonnewald, 2015). Genetic manipulation of these identified genes should be a more powerful approach for improving plant tolerance to multiple stresses than manipulation of individual functional genes. Then, the selected TF genes should be validated not only in model plant species but also in crop plants by use of stress-inducible promoter which can minimize the negative effects caused by over-expressing some TF genes. Furthermore, the critical field trials are required to evaluate the transgenetic plants, especially focusing on their growth and tolerance in the whole life period. which is a necessary step in many strategies to develop stress-tolerant crops. Taken together, we still need to struggle for a complete understanding the precise regulatory mechanisms involved in plant abiotic stress responses, which helps to obtain the promising candidate TF genes for breeding multiple abiotic stress-tolerant crops with better yields and qualities.

AUTHOR CONTRIBUTIONS

HYW, HLW, XT wrote the paper. HS provided the paper frame and revised the final paper. All authors reviewed the final manuscript.

ACKNOWLEDGMENTS

This work has been jointly supported by the National Natural Science Foundation of China (41171216), Jiangsu Autonomous Innovation of Agricultural Science and Technology [CX(15)1005], National Basic Research Program of China (2013CB430403; 2013CB430400), Open Foundation of Jiangsu Key Laboratory for Bioresources of Saline Soils (JKLBS2014005), Jiangsu Province Science and Technology Support Plan (BE2013429), and Yantai Double-hundred Talent Plan (XY-003-02), and Shuangchuang Talent Plan of Jiangsu Province.

REFERENCES

1. Agarwal M., Hao Y., Kapoor A., Dong C. H., Fujii H., Zheng X., et al. . (2006). A R2R3 type MYB transcription factor is involved in the cold regulation of CBF genes and in acquired freezing tolerance.J. Biol. Chem. 281, 37636–37645. 10.1074/jbc.M605895200
2. Ahuja I., de Vos R. C., Bones A. M., Hall R. D. (2010). Plant molecular stress responses face climate change. Trends Plant Sci. 15, 664–674. 10.1016/j.tplants.2010.08.002
3. Allan A. C., Hellens R. P., Laing W. A. (2008). MYB transcription factors that colour our fruit. Trends Plant Sci. 13, 99–102. 10.1016/j.tplants.2007.11.012
4. Ambawat S., Sharma P., Yadav N. R., Yadav R. C. (2013). MYB transcription factor genes as regulators for plant responses: an overview. Physiol. Mol. Biol. Plants 19, 307–321. 10.1007/s12298-013-0179-1
5. Augustine S. M., Ashwin Narayan J., Syamaladevi D. P., Appunu C., Chakravarthi M., Ravichandran V., et al. . (2015). Overexpression of EaDREB2 and pyramiding of EaDREB2 with the pea DNA helicase gene (PDH45) enhance drought and salinity tolerance in sugarcane (Saccharum spp. hybrid).Plant Cell Rep. 34, 247–263. 10.1007/s00299-014-1704-6
6. Babitha K. C., Ramu S. V., Pruthvi V., Mahesh P., Nataraja K. N., Udayakumar M. (2013). Co-expression of AtbHLH17 and AtWRKY28 confers resistance to abiotic stress in *Arabidopsis*.Transgenic Res. 22, 327–341. 10.1007/s11248-012-9645-8
7. Banerjee A., Roychoudhury A. (2015). WRKY proteins: signaling and regulation of expression during abiotic stress responses. Sci. World J. 2015:807560. 10.1155/2015/807560
8. Bencke-Malato M., Cabreira C., Wiebke-Strohm B., Bücker-Neto L., Mancini E., Osorio M. B., et al. . (2014). Genome-wide annotation of the soybean WRKY family and functional characterization of genes involved in response to *Phakopsora pachyrhizi* infection. BMC Plant Biol. 14:236. 10.1186/s12870-014-0236-0
9. Bengtsson M., Shen Y., Oki T. (2006). A SRES-based gridded global population dataset for 1990–2100. Popul. Environ. 28, 113–131. 10.1007/s11111-007-0035-8
10. Bhargava S., Sawant K. (2013). Drought stress adaptation: metabolic adjustment and regulation of gene expression. Plant Breeding 132, 21–32. 10.1111/pbr.12004
11. Bouaziz D., Pirrello J., Charfeddine M., Hammami A., Jbir R., Dhieb A., et al. . (2013).Overexpression of StDREB1 transcription factor increases tolerance to salt in transgenic potato plants.Mol. Biotechnol. 54, 803–817. 10.1007/s12033-012-9628-2
12. Cai H., Tian S., Dong H., Guo C. (2015). Pleiotropic effects of TaMYB3R1 on plant development and response to osmotic stress in transgenic *Arabidopsis*. Gene 558, 227–234. 10.1016/j.gene.2014.12.066
13. Cai R., Zhao Y., Wang Y., Lin Y., Peng X., Li Q., et al. (2014). Overexpression of a maize WRKY58 gene enhances drought and salt tolerance in transgenic rice. Plant Cell Tiss. Org. Cult. 119, 565–577. 10.1007/s11240-014-0556-7
14. Cao Z. H., Zhang S. Z., Wang R. K., Zhang R. F., Hao Y. J. (2013). Genome wide analysis of the apple MYB transcription factor family allows the identification of MdoMYB121 gene confering abiotic stress tolerance in plants. PLoS ONE 8:e69955. 10.1371/journal.pone.0069955
15. Century K., Reuber T. L., Ratcliffe O. J. (2008). Regulating the regulators: the future prospects for transcription-factor-based agricultural biotechnology products. Plant Physiol. 147, 20–29. 10.1104/pp.108.117887
16. Chaves M. M., Flexas J., Pinheiro C. (2009). Photosynthesis under drought and salt stress: regulation mechanisms from whole plant to cell. Ann. Bot. 103, 551–560. 10.1093/aob/mcn125
17. Chaves M. M., Maroco J., Joao S. P. (2003). Understanding plant responses to drought - from genes to the whole plant. Funct. Plant Biol. 30, 239–264. 10.1071/FP02076
18. Chen H., Liu L., Wang L., Wang S., Cheng X. (2015). VrDREB2A, a DREB-binding transcription factor from *Vigna radiata*, increased drought and high-salt tolerance in transgenic *Arabidopsis thaliana*. J. Plant Res. [Epub ahead of print] 10.1007/s10265-015-0773-0
19. Chen L., Song Y., Li S., Zhang L., Zou C., Yu D. (2012). The role of WRKY transcription factors in plant abiotic stresses. Biochim. Biophys. Acta 1819, 120–128. 10.1016/j.bbagrm.2011.09.002
20. Chen M., Wang Q. Y., Cheng X. G., Xu Z. S., Li L. C., Ye X. G., et al. . (2007). GmDREB2, a soybean DRE-binding transcription factor, conferred drought and high-salt tolerance in transgenic plants. Biochem. Biophys. Res. Commun. 353, 299–305. 10.1016/j.bbrc.2006.12.027 [PubMed]

21. Chen W. J., Zhu T. (2004). Networks of transcription factors with roles in environmental stress response. Trends Plant Sci. 9, 591–596. 10.1016/j.tplants.2004.10.007

22. Chen X., Wang Y., Lv B., Li J., Luo L., Lu S., et al. . (2014). The NAC family transcription factor OsNAP confers abiotic stress response through the ABA pathway. Plant Cell Physiol. 55, 604–619. 10.1093/pcp/pct204

23. Cheng L., Li S., Hussain J., Xu X., Yin J., Zhang Y., et al. . (2013a). Isolation and functional characterization of a salt responsive transcriptional factor, LrbZIP from lotus root (*Nelumbo nucifera*Gaertn). Mol. Biol. Rep. 40, 4033–4045. 10.1007/s11033-012-2481-3

24. Cheng L., Li X., Huang X., Ma T., Liang Y., Ma X., et al. . (2013b). Overexpression of sheepgrass R1-MYB transcription factor LcMYB1 confers salt tolerance in transgenic *Arabidopsis*. Plant Physiol. Biochem. 70, 252–260. 10.1016/j.plaphy.2013.05.025

25. Cominelli E., Galbiati M., Vavasseur A., Conti L., Sala T., Vuylsteke M., et al. . (2005). A guard-cell-specific MYB transcription factor regulates stomatal movements and plant drought tolerance. Curr. Biol. 15, 1196–1200. 10.1016/j.cub.2005.05.048

26. Danquah A., de Zelicourt A., Colcombet J., Hirt H. (2014). The role of ABA and MAPK signaling pathways in plant abiotic stress responses. Biotechnol. Adv. 32, 40–52. 10.1016/j.biotechadv.2013.09.006

27. Dietz K. J., Vogel M. O., Viehhauser A. (2010). AP2/EREBP transcription factors are part of gene regulatory networks and integrate metabolic, hormonal and environmental signals in stress acclimation and retrograde signalling. Protoplasma 245, 3–14. 10.1007/s00709-010-0142-8

28. Ding Z., Li S., An X., Liu X., Qin H., Wang D. (2009). Transgenic expression of MYB15 confers enhanced sensitivity to abscisic acid and improved drought tolerance in *Arabidopsis thaliana*. J. Genet. Genomics 36, 17–29. 10.1016/S1673-8527(09)60003-5

29. Dong C. H., Agarwal M., Zhang Y., Xie Q., Zhu J. K. (2006). The negative regulator of plant cold responses, HOS1, is a RING E3 ligase that mediates the ubiquitination and degradation of ICE1. Proc. Natl. Acad. Sci. U.S.A. 103, 8281–8286. 10.1073/pnas.0602874103

30. Dou L., Zhang X., Pang C., Song M., Wei H., Fan S., et al. . (2014). Genome-wide analysis of the WRKY gene family in cotton. Mol. Genet. Genomics 289, 1103–1121. 10.1007/s00438-014-0872-y

31. Dubos C., Stracke R., Grotewold E., Weisshaar B., Martin C., Lepiniec L. (2010). MYB transcription factors in *Arabidopsis*. Trends Plant Sci. 15, 573–581. 10.1016/j.tplants.2010.06.005

32. Dubouzet J. G., Sakuma Y., Ito Y., Kasuga M., Dubouzet E. G., Miura S., et al. . (2003). OsDREB genes in rice, *Oryza sativa* L., encode transcription activators that function in drought-, high-salt- and cold-responsive gene expression. Plant J. 33, 751–763. 10.1046/j.1365-313X.2003.01661.x

33. Easterling D. R., Meehl G. A., Parmesan C., Changnon S. A., Karl T. R., Mearns L. O. (2000).Climate extremes: observations, modeling, and impacts. Science 289, 2068–2074. 10.1126/science.289.5487.2068

34. Eulgem T., Rushton P. J., Robatzek S., Somssich I. E. (2000). The WRKY superfamily of plant transcription factors. Trends Plant Sci. 5, 199–206. 10.1016/S1360-1385(00)01600-9

35. Fang Y., You J., Xie K., Xie W., Xiong L. (2008). Systematic sequence analysis and identification of tissue-specific or stress-responsive genes of NAC transcription factor family in rice. Mol. Genet. Genomics 280, 547–563. 10.1007/s00438-008-0386-6

36. Gao S. Q., Chen M., Xu Z. S., Zhao C. P., Li L., Xu H. J., et al. . (2011). The soybean GmbZIP1 transcription factor enhances multiple abiotic stress tolerances in transgenic plants. Plant Mol. Biol. 75, 537–553. 10.1007/s11103-011-9738-4

37. Gilmour S. J., Zarka D. G., Stockinger E. J., Salazar M. P., Houghton J. M., Thomashow M. F. (1998). Low temperature regulation of the *Arabidopsis* CBF family of AP2 transcriptional activators as an early step in cold-induced COR gene expression. Plant J. 16, 433–442. 10.1046/j.1365-313x.1998.00310.x

38. Golldack D., Luking I., Yang O. (2011). Plant tolerance to drought and salinity: stress regulating transcription factors and their functional significance in the cellular transcriptional network. Plant Cell Rep. 30, 1383–1391. 10.1007/s00299-011-1068-0

39. Hao D., Ohme-Takagi M., Sarai A. (1998). Unique mode of GCC box recognition by the DNA-binding domain of ethylene-responsive element-binding factor (ERF domain) in plant. J. Biol. Chem.273, 26857–26861. 10.1074/jbc.273.41.26857

40. Hao Y. J., Wei W., Song Q. X., Chen H. W., Zhang Y. Q., Wang F., et al. . (2011). Soybean NAC transcription factors promote abiotic stress tolerance and lateral root formation in transgenic plants.Plant J. 68, 302–313. 10.1111/j.1365-313X.2011.04687.x

41. He H., Dong Q., Shao Y., Jiang H., Zhu S., Cheng B., et al. . (2012). Genome-wide survey and characterization of the WRKY gene family in *Populus trichocarpa*. Plant Cell Rep. 31, 1199–1217. 10.1007/s00299-012-1241-0

42. Heidarvand L., Amiri R. M. (2010). What happens in plant molecular responses to cold stress? Acta Physiol. Plant. 32, 419–431. 10.1007/s11738-009-0451-8

43. Hirayama T., Shinozaki K. (2010). Research on plant abiotic stress responses in the post-genome era: past, present and future. Plant J. 61, 1041–1052. 10.1111/j.1365-313X.2010.04124.x

44. Hou X. J., Li S. B., Liu S. R., Hu C. G., Zhang J. Z. (2014). Genome-wide classification and evolutionary and expression analyses of citrus MYB transcription factor families in sweet orange.PLoS ONE 9:e112375. 10.1371/journal.pone.0112375

45. Huang G. T., Ma S. L., Bai L. P., Zhang L., Ma H., Jia P., et al. . (2012). Signal transduction during cold, salt, and drought stresses in plants. Mol. Biol. Rep. 39, 969–987. 10.1007/s11033-011-0823-1

46. Huang H., Wang Y., Wang S., Wu X., Yang K., Niu Y., et al. . (2012). Transcriptome-wide survey and expression analysis of stress-responsive NAC genes in *Chrysanthemum lavandulifolium*. Plant Sci.193–194, 18–27. 10.1016/j.plantsci.2012.05.004

47. Huang Q., Wang Y., Li B., Chang J., Chen M., Li K., et al. . (2015). TaNAC29, a NAC transcription factor from wheat, enhances salt and drought tolerance in transgenic *Arabidopsis*. BMC Plant Biol.15:268. 10.1186/s12870-015-0644-9

48. Huang X. S., Liu J. H., Chen X. J. (2010). Overexpression of PtrABF gene, a bZIP transcription factor isolated from *Poncirus trifoliata*, enhances dehydration and drought tolerance in tobacco via scavenging ROS and modulating expression of stress-responsive genes. BMC Plant Biol. 10:230. 10.1186/1471-2229-10-230

49. Hurst H. C. (1994). Transcription factors. 1: bZIP proteins. Protein Profile 1, 123–168.

50. Ipcc (2007). Climate change 2007: the physical science basis, in Contribution of Working Group I to the Fourth Assessment Report of the Intergovernmental Panel on Climate Change, eds Solomon S., Qin D., Manning M., Chen Z., Marquis M., Averyt K. B., Tignor M., Miller H. L., editors. (Geneva: Ipcc Secretariat).

51. Ipcc (2008). Climate change and water, in Technical Paper of the Intergovernmental Panel on Climate Change, eds Bates B. C., Kundzewicz Z. W., Palutikof J., Wu S., editors. (Geneva: Ipcc Secretariat;).

52. Ishiguro S., Nakamura K. (1994). Characterization of a cDNA encoding a novel DNA-binding protein, SPF1, that recognizes SP8 sequences in the 5' upstream regions of genes coding for sporamin and beta-amylase from sweet potato. Mol. Gen. Genet. 244, 563–571. 10.1007/BF00282746

53. Ito Y., Katsura K., Maruyama K., Taji T., Kobayashi M., Seki M., et al. . (2006). Functional analysis of rice DREB1/CBF-type transcription factors involved in cold-responsive gene expression in transgenic rice. Plant Cell Physiol. 47, 141–153. 10.1093/pcp/pci230

54. Jaglo K. R., Kleff S., Amundsen K. L., Zhang X., Haake V., Zhang J. Z., et al. . (2001). Components of the *Arabidopsis* C-repeat/dehydration-responsive element binding factor cold-response pathway are conserved in *Brassica napus* and other plant species. Plant Physiol. 127, 910–917. 10.1104/pp.010548

55. Jaglo-Ottosen K. R., Gilmour S. J., Zarka D. G., Schabenberger O., Thomashow M. F. (1998).*Arabidopsis* CBF1 overexpression induces COR genes and enhances freezing tolerance. Science 280, 104–106. 10.1126/science.280.5360.104

56. Jakoby M., Weisshaar B., Dröge-Laser W., Vicente-Carbajosa J., Tiedemann J., Kroj T., et al. . (2002).bZIP transcription factors in *Arabidopsis*. Trends Plant Sci. 7, 106–111. 10.1016/S1360-1385(01)02223-3

57. Jensen M. K., Kjaersgaard T., Nielsen M. M., Galberg P., Petersen K., O'shea C., et al. . (2010). The*Arabidopsis thaliana* NAC transcription factor family: structure-function relationships and determinants of ANAC019 stress signalling. Biochem. J. 426, 183–196. 10.1042/BJ20091234

58. Jiang Y., Deyholos M. K. (2006). Comprehensive transcriptional profiling of NaCl-stressed*Arabidopsis* roots reveals novel classes of responsive genes. BMC Plant Biol. 6:25. 10.1186/1471-2229-6-25

59. Joo J., Choi H. J., Lee Y. H., Kim Y. K., Song S. I. (2013). A transcriptional repressor of the ERF family confers drought tolerance to rice and regulates genes preferentially located on chromosome 11.Planta 238, 155–170. 10.1007/s00425-013-1880-6

60. Jung C., Seo J. S., Han S. W., Koo Y. J., Kim C. H., Song S. I., et al. . (2008). Overexpression of AtMYB44 enhances stomatal closure to confer abiotic stress tolerance in transgenic *Arabidopsis*. Plant Physiol. 146, 623–635. 10.1104/pp.107.110981

61. Kasuga M., Miura S., Shinozaki K., Yamaguchi-Shinozaki K. (2004). A combination of the*Arabidopsis* DREB1A gene and stress-inducible rd29A promoter improved drought- and low-temperature stress tolerance in tobacco by gene transfer. Plant Cell Physiol. 45, 346–350. 10.1093/pcp/pch037

62. Kim Y. S., Kim S. G., Park J. E., Park H. Y., Lim M. H., Chua N. H., et al. . (2006). A membrane-bound NAC transcription factor regulates cell division in *Arabidopsis*. Plant Cell 18, 3132–3144. 10.1105/tpc.106.043018

63. Landschulz W. H., Johnson P. F., McKnight S. L. (1988). The leucine zipper: a hypothetical structure common to a new class of DNA binding proteins. Science 240, 1759–1764. 10.1126/science.3289117

64. Lata C., Mishra A. K., Muthamilarasan M., Bonthala V. S., Khan Y., Prasad M. (2014). Genome-wide investigation and expression profiling of AP2/ERF transcription factor superfamily in foxtail millet (*Setaria italica* L.). PLoS ONE 9:e113092. 10.1371/journal.pone.0113092

65. Lata C., Prasad M. (2011). Role of DREBs in regulation of abiotic stress responses in plants. J. Exp. Bot. 62, 4731–4748. 10.1093/jxb/err210

66. Le D. T., Nishiyama R., Watanabe Y., Mochida K., Yamaguchi-Shinozaki K., Shinozaki K., et al. . (2011). Genome-wide survey and expression analysis of the plant-specific NAC transcription factor family in soybean during development and dehydration stress. DNA Res. 18, 263–276. 10.1093/dnares/dsr015

67. Li C., Ng C. K. Y., Fan L.-M. (2015). MYB transcription factors, active players in abiotic stress signaling. Environ. Exp. Bot. 114, 80–91. 10.1016/j.envexpbot.2014.06.014

68. Li H., Gao Y., Xu H., Dai Y., Deng D., Chen J. (2013). ZmWRKY33, a WRKY maize transcription factor conferring enhanced salt stress tolerances in *Arabidopsis*. Plant Growth Regul. 70, 207–216. 10.1007/s10725-013-9792-9

69. Liang H., Lu Y., Liu H., Wang F., Xin Z., Zhang Z. (2008). A novel activator-type ERF of*Thinopyrum intermedium*, TiERF1, positively regulates defence responses. J. Exp. Bot. 59, 3111–3120. 10.1093/jxb/ern165

70. Liang Y. K., Dubos C., Dodd I. C., Holroyd G. H., Hetherington A. M., Campbell M. M. (2005).AtMYB61,

an R2R3-MYB transcription factor controlling stomatal aperture in *Arabidopsis thaliana*.Curr. Biol. 15, 1201–1206. 10.1016/j.cub.2005.06.041

71. Liao Y., Zou H. F., Wang H. W., Zhang W. K., Ma B., Zhang J. S., et al. . (2008). Soybean GmMYB76, GmMYB92, and GmMYB177 genes confer stress tolerance in transgenic *Arabidopsis*plants. Cell Res. 18, 1047–1060. 10.1038/cr.2008.280
72. Licausi F., Giorgi F. M., Zenoni S., Osti F., Pezzotti M., Perata P. (2010). Genomic and transcriptomic analysis of the AP2/ERF superfamily in *Vitis vinifera*. BMC Genomics 11:719. 10.1186/1471-2164-11-719
73. Lim C. J., Hwang J. E., Chen H., Hong J. K., Yang K. A., Choi M. S., et al. . (2007). Over-expression of the *Arabidopsis* DRE/CRT-binding transcription factor DREB2C enhances thermotolerance. Biochem. Biophys. Res. Commun. 362, 431–436. 10.1016/j.bbrc.2007.08.007
74. Liu C., Mao B., Ou S., Wang W., Liu L., Wu Y., et al. . (2014). OsbZIP71, a bZIP transcription factor, confers salinity and drought tolerance in rice. Plant Mol. Biol. 84, 19–36. 10.1007/s11103-013-0115-3
75. Liu H., Yang W., Liu D., Han Y., Zhang A., Li S. (2011a). Ectopic expression of a grapevine transcription factor VvWRKY11 contributes to osmotic stress tolerance in *Arabidopsis*. Mol. Biol. Rep. 38, 417–427. 10.1007/s11033-010-0124-0
76. Liu H., Zhou X., Dong N., Liu X., Zhang H., Zhang Z. (2011b). Expression of a wheat MYB gene in transgenic tobacco enhances resistance to *Ralstonia solanacearum*, and to drought and salt stresses.Funct. Integr. Genomics 11, 431–443. 10.1007/s10142-011-0228-1
77. Liu J., Chen N., Chen F., Cai B., Dal Santo S., Tornielli G. B., et al. . (2014). Genome-wide analysis and expression profile of the bZIP transcription factor gene family in grapevine (*Vitis vinifera*). BMC Genomics 15:281. 10.1186/1471-2164-15-281
78. Liu J.-H., Peng T., Dai W. (2014). Critical cis-acting elements and interacting transcription factors: key players associated with abiotic stress responses in plants. Plant Mol. Biol. Report. 32, 303–317. 10.1007/s11105-013-0667-z
79. Liu L., Zhang Z., Dong J., Wang T. (2016). Overexpression of MtWRKY76 increases both salt and drought tolerance in *Medicago truncatula*. Environ. Exp. Bot. 123, 50–58. 10.1016/j.envexpbot.2015.10.007
80. Liu T. K., Song X. M., Duan W. K., Huang Z. N., Liu G. F., Li Y., et al. (2014). Genome-wide analysis and expression patterns of NAC transcription factor family under different developmental stages and abiotic stresses in Chinese Cabbage. Plant Mol. Biol. Report. 32, 1041–1056. 10.1007/s11105-014-0712-6
81. Liu Q., Kasuga M., Sakuma Y., Abe H., Miura S., Yamaguchi-Shinozaki K., et al. . (1998). Two transcription factors, DREB1 and DREB2, with an EREBP/AP2 DNA binding domain separate two cellular signal transduction pathways in drought- and low-temperature-responsive gene expression, respectively, in *Arabidopsis*. Plant Cell 10, 1391–1406. 10.1105/tpc.10.8.1391
82. Liu X., Chu Z. (2015). Genome-wide evolutionary characterization and analysis of bZIP transcription factors and their expression profiles in response to multiple abiotic stresses in *Brachypodium distachyon*. BMC Genomics 16:227. 10.1186/s12864-015-1457-9
83. Liu X., Liu S., Wu J., Zhang B., Li X., Yan Y., et al. . (2013). Overexpression of *Arachis hypogaea*NAC3 in tobacco enhances dehydration and drought tolerance by increasing superoxide scavenging.Plant Physiol. Biochem. 70, 354–359. 10.1016/j.plaphy.2013.05.018
84. Lobell D. B., Schlenker W., Costa-Roberts J. (2011). Climate trends and global crop production since 1980. Science 333, 616–620. 10.1126/science.1204531
85. Lu M., Ying S., Zhang D. F., Shi Y. S., Song Y. C., Wang T. Y., et al. . (2012). A maize stress-responsive NAC transcription factor, ZmSNAC1, confers enhanced tolerance to dehydration in transgenic *Arabidopsis*. Plant Cell Rep. 31, 1701–1711. 10.1007/s00299-012-1284-2
86. Lucas S., Durmaz E., Akpinar B. A., Budak H. (2011). The drought response displayed by a DRE-binding protein from *Triticum dicoccoides*. Plant Physiol. Biochem. 49, 346–351. 10.1016/j.plaphy.2011.01.016
87. Luo X., Bai X., Sun X., Zhu D., Liu B., Ji W., et al. . (2013). Expression of wild soybean WRKY20 in *Arabidopsis* enhances drought tolerance and regulates ABA signalling. J. Exp. Bot. 64, 2155–2169. 10.1093/jxb/ert073
88. Mallikarjuna G., Mallikarjuna K., Reddy M. K., Kaul T. (2011). Expression of OsDREB2A transcription factor confers enhanced dehydration and salt stress tolerance in rice (*Oryza sativa* L.).Biotechnol. Lett. 33, 1689–1697. 10.1007/s10529-011-0620-x
89. Mao X., Chen S., Li A., Zhai C., Jing R. (2014). Novel NAC transcription factor TaNAC67 confers enhanced multi-abiotic stress tolerances in *Arabidopsis*. PLoS ONE 9:e84359. 10.1371/journal.pone.0084359
90. Meng X., Wang J. R., Wang G. D., Liang X. Q., Li X. D., Meng Q. W. (2015). An R2R3-MYB gene, LeAN2, positively regulated the thermo-tolerance in transgenic tomato. J. Plant Physiol. 175, 1–8. 10.1016/j.jplph.2014.09.018
91. Meng X., Yin B., Feng H. L., Zhang S., Liang X. Q., Meng Q. W. (2014). Overexpression of R2R3-MYB gene leads to accumulation of anthocyanin and enhanced resistance to chilling and oxidative stress. Biol. Plant. 58, 121–130. 10.1007/s10535-013-0376-3
92. Mittler R., Blumwald E. (2010). Genetic engineering for modern agriculture: challenges and perspectives. Annu. Rev. Plant Biol. 61, 443–462. 10.1146/annurev-arplant-042809-112116
93. Miura K., Jin J. B., Lee J., Yoo C. Y., Stirm V., Miura T., et al. . (2007). SIZ1-mediated sumoylation of ICE1 controls CBF3/DREB1A expression and freezing tolerance in *Arabidopsis*. Plant Cell 19, 1403–1414. 10.1105/tpc.106.048397
94. Mizoi J., Ohori T., Moriwaki T., Kidokoro S., Todaka D., Maruyama K., et al. . (2013).GmDREB2A;2, a canonical DEHYDRATION-RESPONSIVE ELEMENT-BINDING PROTEIN2-type transcription

factor in soybean, is posttranslationally regulated and mediates dehydration-responsive element-dependent gene expression. Plant Physiol. 161, 346–361. 10.1104/pp.112.204875

95. Mizoi J., Shinozaki K., Yamaguchi-Shinozaki K. (2012). AP2/ERF family transcription factors in plant abiotic stress responses. Biochim. Biophys. Acta 1819, 86–96. 10.1016/j.bbagrm.2011.08.004

96. Muthamilarasan M., Khandelwal R., Yadav C. B., Bonthala V. S., Khan Y., Prasad M. (2014).Identification and molecular characterization of MYB Transcription Factor Superfamily in C4 model plant foxtail millet (*Setaria italica* L.). PLoS ONE 9:e109920. 10.1371/journal.pone.0109920

97. Nakano T., Suzuki K., Fujimura T., Shinshi H. (2006). Genome-wide analysis of the ERF gene family in *Arabidopsis* and rice. Plant Physiol. 140, 411–432. 10.1104/pp.105.073783

98. Nakashima K., Shinwari Z. K., Sakuma Y., Seki M., Miura S., Shinozaki K., et al. . (2000).Organization and expression of two *Arabidopsis* DREB2 genes encoding DRE-binding proteins involved in dehydration- and high-salinity-responsive gene expression. Plant Mol. Biol. 42, 657–665. 10.1023/A:1006321900483

99. Nakashima K., Takasaki H., Mizoi J., Shinozaki K., Yamaguchi-Shinozaki K. (2012). NAC transcription factors in plant abiotic stress responses. Biochim. Biophys. Acta 1819, 97–103. 10.1016/j.bbagrm.2011.10.005

100. Newton A. C., Johnson S. N., Gregory P. J. (2011). Implications of climate change for diseases, crop yields and food security. Euphytica 179, 3–18. 10.1007/s10681-011-0359-4

101. Nijhawan A., Jain M., Tyagi A. K., Khurana J. P. (2008). Genomic survey and gene expression analysis of the basic leucine zipper transcription factor family in rice. Plant Physiol. 146, 333–350. 10.1104/pp.107.112821

102. Nuruzzaman M., Manimekalai R., Sharoni A. M., Satoh K., Kondoh H., Ooka H., et al. . (2010).Genome-wide analysis of NAC transcription factor family in rice. Gene 465, 30–44. 10.1016/j.gene.2010.06.008

103. Nuruzzaman M., Sharoni A. M., Kikuchi S. (2013). Roles of NAC transcription factors in the regulation of biotic and abiotic stress responses in plants. Front. Microbiol. 4:248. 10.3389/fmicb.2013.00248

104. Ohme-Takagi M., Shinshi H. (1995). Ethylene-inducible DNA binding proteins that interact with an ethylene-responsive element. Plant Cell 7, 173–182. 10.1105/tpc.7.2.173

105. Olsen A. N., Ernst H. A., Leggio L. L., Skriver K. (2005). NAC transcription factors: structurally distinct, functionally diverse. Trends Plant Sci. 10, 79–87. 10.1016/j.tplants.2004.12.010

106. Pandey S. P., Somssich I. E. (2009). The role of WRKY transcription factors in plant immunity. Plant Physiol. 150, 1648–1655. 10.1104/pp.109.138990

107. Peng X., Zhang L., Zhang L., Liu Z., Cheng L., Yang Y., et al. (2013). The transcriptional factor LcDREB2 cooperates with LcSAMDC2 to contribute to salt tolerance in *Leymus chinensis*. Plant Cell Tissue Organ Cult. 113, 245–256. 10.1007/s11240-012-0264-0

108. Pérez-Clemente R. M., Vives V., Zandalinas S. I., López-Climent M. F., Munoz V., Gomez-Cadenas A. (2013). Biotechnological approaches to study plant responses to stress. Biomed Res. Int.2013:654120. 10.1155/2013/654120

109. Pourabed E., Ghane Golmohamadi F., Soleymani Monfared P., Razavi S. M., Shobbar Z. S. (2015). Basic leucine zipper family in barley: genome-wide characterization of members and expression analysis. Mol. Biotechnol. 57, 12–26. 10.1007/s12033-014-9797-2

110. Prasch C. M., Sonnewald U. (2015). Signaling events in plants: stress factors in combination change the picture. Environ. Exp. Bot. 114, 4–14. 10.1016/j.envexpbot.2014.06.020

111. Puranik S., Sahu P. P., Srivastava P. S., Prasad M. (2012). NAC proteins: regulation and role in stress tolerance. Trends Plant Sci. 17, 369–381. 10.1016/j.tplants.2012.02.004

112. Qi L., Yang J., Yuan Y., Huang L., Chen P. (2015). Overexpression of two R2R3-MYB genes from*Scutellaria baicalensis* induces phenylpropanoid accumulation and enhances oxidative stress resistance in transgenic tobacco. Plant Physiol. Biochem. 94, 235–243. 10.1016/j.plaphy.2015.06.007

113. Qin F., Kakimoto M., Sakuma Y., Maruyama K., Osakabe Y., Tran L. S., et al. . (2007). Regulation and functional analysis of ZmDREB2A in response to drought and heat stresses in *Zea mays* L. Plant J. 50, 54–69. 10.1111/j.1365-313X.2007.03034.x

114. Qin F., Sakuma Y., Li J., Liu Q., Li Y. Q., Shinozaki K., et al. . (2004). Cloning and functional analysis of a novel DREB1/CBF transcription factor involved in cold-responsive gene expression in*Zea mays* L. Plant Cell Physiol. 45, 1042–1052. 10.1093/pcp/pch118

115. Qin Y., Tian Y., Han L., Yang X. (2013). Constitutive expression of a salinity-induced wheat WRKY transcription factor enhances salinity and ionic stress tolerance in transgenic *Arabidopsis thaliana*.Biochem. Biophys. Res. Commun. 441, 476–481. 10.1016/j.bbrc.2013.10.088

116. Qin Y., Tian Y., Liu X. (2015). A wheat salinity-induced WRKY transcription factor TaWRKY93 confers multiple abiotic stress tolerance in *Arabidopsis thaliana*. Biochem. Biophys. Res. Commun.464, 428–433. 10.1016/j.bbrc.2015.06.128

117. Qiu Y., Yu D. (2009). Over-expression of the stress-induced OsWRKY45 enhances disease resistance and drought tolerance in *Arabidopsis*. Environmental and experimental botany 65. Environ. Exp. Bot.65, 35–47. 10.1016/j.envexpbot.2008.07.002

118. Ravikumar G., Manimaran P., Voleti S. R., Subrahmanyam D., Sundaram R. M., Bansal K. C., et al. . (2014). Stress-inducible expression of AtDREB1A transcription factor greatly improves drought stress tolerance in transgenic indica rice. Transgenic Res. 23, 421–439. 10.1007/s11248-013-9776-6

119. Ray S., Dansana P. K., Bhaskar A., Giri J., Kapoor S., Khurana J. P., et al. (2009). Emerging trends in functional genomics for stress tolerance in crop plants, in Plant Stress Biology, ed Hirt H., editor. (Weinheim: Wiley-VCH Verlag GmbH & Co. KGaA;), 37–63.

120. Riechmann J. L., Meyerowitz E. M. (1998). The AP2/EREBP family of plant transcription factors.Biol. Chem. 379, 633–646.

121. Rong W., Qi L., Wang A., Ye X., Du L., Liang H., et al. . (2014). The ERF transcription factor TaERF3 promotes tolerance to salt and drought stresses in wheat. Plant Biotechnol. J. 12, 468–479. 10.1111/pbi.12153

122. Rushton D. L., Tripathi P., Rabara R. C., Lin J., Ringler P., Boken A. K., et al. . (2012). WRKY transcription factors: key components in abscisic acid signalling. Plant Biotechnol. J. 10, 2–11. 10.1111/j.1467-7652.2011.00634.x

123. Rushton P. J., Somssich I. E., Ringler P., Shen Q. J. (2010). WRKY transcription factors. Trends Plant Sci. 15, 247–258. 10.1016/j.tplants.2010.02.006

124. Saad A. S., Li X., Li H. P., Huang T., Gao C. S., Guo M. W., et al. . (2013). A rice stress-responsive NAC gene enhances tolerance of transgenic wheat to drought and salt stresses. Plant Sci. 203–204, 33–40. 10.1016/j.plantsci.2012.12.016

125. Sakuma Y., Liu Q., Dubouzet J. G., Abe H., Shinozaki K., Yamaguchi-Shinozaki K. (2002). DNA-binding specificity of the ERF/AP2 domain of *Arabidopsis* DREBs, transcription factors involved in dehydration- and cold-inducible gene expression. Biochem. Biophys. Res. Commun. 290, 998–1009. 10.1006/bbrc.2001.6299

126. Sakuma Y., Maruyama K., Osakabe Y., Qin F., Seki M., Shinozaki K., et al. . (2006a). Functional analysis of an *Arabidopsis* transcription factor, DREB2A, involved in drought-responsive gene expression. Plant Cell 18, 1292–1309. 10.1105/tpc.105.035881

127. Sakuma Y., Maruyama K., Qin F., Osakabe Y., Shinozaki K., Yamaguchi-Shinozaki K. (2006b). Dual function of an *Arabidopsis* transcription factor DREB2A in water-stress-responsive and heat-stress-responsive gene expression. Proc. Natl. Acad. Sci. U.S.A. 103, 18822–18827. 10.1073/pnas.0605639103

128. Sanchez D. H., Pieckenstain F. L., Szymanski J., Erban A., Bromke M., Hannah M. A., et al. . (2011).Comparative functional genomics of salt stress in related model and cultivated plants identifies and overcomes limitations to translational genomics. PLoS ONE 6:e17094. 10.1371/journal.pone.0017094

129. Schaller G. E., Kieber J. J., Shiu S. H. (2008). Two-component signaling elements and histidyl-aspartyl phosphorelays. Arabidopsis Book 6:e0112. 10.1199/tab.0112

130. Seo J., Sohn H., Noh K., Jung C., An J., Donovan C., et al. (2012). Expression of the *Arabidopsis*AtMYB44 gene confers drought/salt-stress tolerance in transgenic soybean. Mol. Breed. 29, 601–608. 10.1007/s11032-011-9576-8

131. Seo P. J., Lee S. B., Suh M. C., Park M. J., Go Y. S., Park C. M. (2011). The MYB96 transcription factor regulates cuticular wax biosynthesis under drought conditions in *Arabidopsis*. Plant Cell 23, 1138–1152. 10.1105/tpc.111.083485

132. Seo P. J., Xiang F., Qiao M., Park J. Y., Lee Y. N., Kim S. G., et al. . (2009). The MYB96 transcription factor mediates abscisic acid signaling during drought stress response in *Arabidopsis*.Plant Physiol. 151, 275–289. 10.1104/pp.109.144220

133. Shao H. B., Chu L. Y., Jaleel C. A., Manivannan P., Panneerselvam R., Shao M. A. (2009).Understanding water deficit stress-induced changes in the basic metabolism of higher plants - biotechnologically and sustainably improving agriculture and the ecoenvironment in arid regions of the globe. Crit. Rev. Biotechnol. 29, 131–151. 10.1080/07388550902869792

134. Sharoni A. M., Nuruzzaman M., Satoh K., Shimizu T., Kondoh H., Sasaya T., et al. . (2011). Gene structures, classification and expression models of the AP2/EREBP transcription factor family in rice.Plant Cell Physiol. 52, 344–360. 10.1093/pcp/pcq196

135. Shi W., Liu D., Hao L., Wu C. A., Guo X., Li H. (2014). GhWRKY39, a member of the WRKY transcription factor family in cotton, has a positive role in disease resistance and salt stress tolerance.Plant Cell Tissue Organ Cult. 118, 17–32. 10.1007/s11240-014-0458-8

136. Shinozaki K., Yamaguchi-Shinozakiy K., Sekiz M. (2003). Regulatory network of gene expression in the drought and cold stress responses. Curr. Opin. Plant Biol. 6, 410–417. 10.1016/S1369-5266(03)00092-X

137. Shinwari Z. K., Nakashima K., Miura S., Kasuga M., Seki M., Yamaguchi-Shinozaki K., et al. . (1998). An *Arabidopsis* gene family encoding DRE/CRT binding proteins involved in low-temperature-responsive gene expression. Biochem. Biophys. Res. Commun. 250, 161–170. 10.1006/bbrc.1998.9267

138. Shiriga K., Sharma R., Kumar K., Yadav S. K., Hossain F., Thirunavukkarasu N. (2014). Genome-wide identification and expression pattern of drought-responsive members of the NAC family in maize. Meta Gene 2, 407–417. 10.1016/j.mgene.2014.05.001

139. Singh R., Usha Rizvi S. M. H., Sonia Jaiwal P. (2003). Genetic engineering for enhancing abiotic stress tolerance, in Improvement Strategies of Leguminosae Biotechnology, eds Jaiwal P., Singh R., editors. (Dordrecht: Springer;), 223–243.

140. Song X., Li Y., Hou X. (2013). Genome-wide analysis of the AP2/ERF transcription factor superfamily in Chinese cabbage (*Brassica rapa* ssp. pekinensis). BMC Genomics 14:573. 10.1186/1471-2164-14-573

141. Stockinger E. J., Gilmour S. J., Thomashow M. F. (1997). *Arabidopsis thaliana* CBF1 encodes an AP2 domain-containing transcriptional activator that binds to the C-repeat/DRE, a cis-acting DNA regulatory element that stimulates transcription in response to low temperature and water deficit. Proc. Natl. Acad. Sci. U.S.A. 94, 1035–1040. 10.1073/pnas.94.3.1035

142. Su H. Y., Zhang S. Z., Yin Y. L., Zhu D. Z., Han L. Y. (2015). Genome-wide analysis of NAM-ATAF1, 2-CUC2 transcription factor family in *Solanum lycopersicum*.

J. Plant Biochem. Biotechnol.24, 176–183. 10.1007/s13562-014-0255-9

143. Su L. T., Li J. W., Liu D. Q., Zhai Y., Zhang H. J., Li X. W., et al. . (2014). A novel MYB transcription factor, GmMYBJ1, from soybean confers drought and cold tolerance in *Arabidopsis thaliana*. Gene 538, 46–55. 10.1016/j.gene.2014.01.024

144. Sun J., Hu W., Zhou R., Wang L., Wang X., Wang Q., et al. . (2015). The *Brachypodium distachyon*BdWRKY36 gene confers tolerance to drought stress in transgenic tobacco plants. Plant Cell Rep. 34, 23–35. 10.1007/s00299-014-1684-6

145. Sun Z. M., Zhou M. L., Xiao X. G., Tang Y. X., Wu Y. M. (2014). Genome-wide analysis of AP2/ERF family genes from *Lotus corniculatus* shows LcERF054 enhances salt tolerance. Funct. Integr. Genomics 14, 453–466. 10.1007/s10142-014-0372-5

146. Takeda S., Matsuoka M. (2008). Genetic approaches to crop improvement: responding to environmental and population changes. Nat. Rev. Genet. 9, 444–457. 10.1038/nrg2342

147. Tang Y., Liu M., Gao S., Zhang Z., Zhao X., Zhao C., et al. . (2012). Molecular characterization of novel TaNAC genes in wheat and overexpression of TaNAC2a confers drought tolerance in tobacco.Physiol. Plant. 144, 210–224. 10.1111/j.1399-3054.2011.01539.x

148. Tran L. S., Nishiyama R., Yamaguchi-Shinozaki K., Shinozaki K. (2010). Potential utilization of NAC transcription factors to enhance abiotic stress tolerance in plants by biotechnological approach. GM Crops 1, 32–39. 10.4161/gmcr.1.1.10569

149. Tripathi P., Rabara R. C., Rushton P. J. (2014). A systems biology perspective on the role of WRKY transcription factors in drought responses in plants. Planta 239, 255–266. 10.1007/s00425-013-1985-y

150. Tsai-Hung H., Jent-Turn L., Pei-Tzu Y., Li-Hui C., Yee-Yung C., Yu-Chie W., et al. . (2002).Heterology expression of the *Arabidopsis*C-repeat/dehydration response element binding Factor 1 gene confers elevated tolerance to chilling and oxidative stresses in transgenic tomato. Plant Physiol.129, 1086–1094. 10.1104/pp.003442

151. Ulker B., Somssich I. E. (2004). WRKY transcription factors: from DNA binding towards biological function. Curr. Opin. Plant Biol. 7, 491–498. 10.1016/j.pbi.2004.07.012

152. Umezawa T., Fujita M., Fujita Y., Yamaguchi-Shinozaki K., Shinozaki K. (2006). Engineering drought tolerance in plants: discovering and tailoring genes to unlock the future. Curr. Opin. Biotechnol. 17, 113–122. 10.1016/j.copbio.2006.02.002

153. United Nations (2015). The World Population Prospects. New York, NY: United Nations Department of Economic and Social Affairs.

154. Uno Y., Furihata T., Abe H., Yoshida R., Shinozaki K., Yamaguchi-Shinozaki K. (2000). *Arabidopsis*basic leucine zipper transcription factors involved in an abscisic acid-dependent signal transduction pathway under drought and high-salinity conditions. Proc. Natl. Acad. Sci. U.S.A. 97, 11632–11637. 10.1073/pnas.190309197

155. Varshney R. K., Bansal K. C., Aggarwal P. K., Datta S. K., Craufurd P. Q. (2011). Agricultural biotechnology for crop improvement in a variable climate: hope or hype? Trends Plant Sci. 16, 363–371. 10.1016/j.tplants.2011.03.004

156. Vinocur B., Altman A. (2005). Recent advances in engineering plant tolerance to abiotic stress: achievements and limitations. Curr. Opin. Biotechnol. 16, 123–132. 10.1016/j.copbio.2005.02.001

157. Wang C., Deng P., Chen L., Wang X., Ma H., Hu W., et al. . (2013). A wheat WRKY transcription factor TaWRKY10 confers tolerance to multiple abiotic stresses in transgenic tobacco. PLoS ONE8:e65120. 10.1371/journal.pone.0065120

158. Wang R. K., Cao Z. H., Hao Y. J. (2014). Overexpression of a R2R3 MYB gene MdSIMYB1 increases tolerance to multiple stresses in transgenic tobacco and apples. Physiol. Plant. 150, 76–87. 10.1111/ppl.12069

159. Wei K., Chen J., Wang Y., Chen Y., Chen S., Lin Y., et al. . (2012). Genome-wide analysis of bZIP-encoding genes in maize. DNA Res. 19, 463–476. 10.1093/dnares/dss026

160. Wen F., Zhu H., Li P., Jiang M., Mao W., Ong C., et al. . (2014). Genome-wide evolutionary characterization and expression analyses of WRKY family genes in *Brachypodium distachyon*. DNA Res. 21, 327–339. 10.1093/dnares/dst060

161. Wu H., Lv H., Li L., Liu J., Mu S., Li X., et al. . (2015). Genome-wide analysis of the AP2/ERF transcription factors family and the expression patterns of DREB genes in Moso bamboo (*Phyllostachys edulis*). PLoS ONE 10:e0126657. 10.1371/journal.pone.0126657

162. Wu K. L., Guo Z. J., Wang H. H., Li J. (2005). The WRKY family of transcription factors in rice and*Arabidopsis* and their origins. DNA Res. 12, 9–26. 10.1093/dnares/12.1.9

163. Wu X., Shiroto Y., Kishitani S., Ito Y., Toriyama K. (2009). Enhanced heat and drought tolerance in transgenic rice seedlings overexpressing OsWRKY11 under the control of HSP101 promoter. Plant Cell Rep. 28, 21–30. 10.1007/s00299-008-0614-x

164. Xu Z.-S., Chen M., Li L.-C., Ma Y.-Z. (2008a). Functions of the ERF transcription factor family in plants. Botany 86, 969–977. 10.1139/B08-041

165. Xu Z.-S., Ni Z. Y., Liu L., Nie L. N., Li L. C., Chen M., et al. . (2008b). Characterization of the TaAIDFa gene encoding a CRT/DRE-binding factor responsive to drought, high-salt, and cold stress in wheat. Mol. Genet. Genomics 280, 497–508. 10.1007/s00438-008-0382-x

166. Yamaguchi T., Blumwald E. (2005). Developing salt-tolerant crop plants: challenges and opportunities. Trends Plant Sci. 10, 615–620. 10.1016/j.tplants.2005.10.002

167. Yamaguchi-Shinozaki K., Shinozaki K. (2006). Transcriptional regulatory networks in cellular responses and tolerance to dehydration and cold stresses. Annu. Rev. Plant Biol. 57, 781–803. 10.1146/annurev.arplant.57.032905.105444

168. Yang A., Dai X., Zhang W. H. (2012). A R2R3-type MYB gene, OsMYB2, is involved in salt, cold, and dehydration tolerance in rice. J. Exp. Bot. 63, 2541–2556. 10.1093/jxb/err431

169. Yang W., Liu X. D., Chi X. J., Wu C. A., Li Y. Z., Song L. L., et al. . (2011). Dwarf apple MbDREB1 enhances plant tolerance to low temperature, drought, and salt stress via both ABA-dependent and ABA-independent pathways. Planta 233, 219–229. 10.1007/s00425-010-1279-6

170. Yang X., Wang X., Lu J., Yi Z., Fu C., Ran J., et al. . (2015). Overexpression of a *Miscanthus lutarioriparius* NAC gene MlNAC5 confers enhanced drought and cold tolerance in *Arabidopsis*.Plant Cell Rep. 34, 943–958. 10.1007/s00299-015-1756-2

171. Yanhui C., Xiaoyuan Y., Kun H., Meihua L., Jigang L., Zhaofeng G., et al. . (2006). The MYB transcription factor superfamily of *Arabidopsis*: expression analysis and phylogenetic comparison with the rice MYB family. Plant Mol. Biol. 60, 107–124. 10.1007/s11103-005-2910-y

172. Ying S., Zhang D. F., Fu J., Shi Y. S., Song Y. C., Wang T. Y., et al. . (2012). Cloning and characterization of a maize bZIP transcription factor, ZmbZIP72, confers drought and salt tolerance in transgenic *Arabidopsis*. Planta 235, 253–266. 10.1007/s00425-011-1496-7

173. Yokotani N., Ichikawa T., Kondou Y., Matsui M., Hirochika H., Iwabuchi M., et al. . (2009).Tolerance to various environmental stresses conferred by the salt-responsive rice gene ONAC063 in transgenic *Arabidopsis*. Planta 229, 1065–1075. 10.1007/s00425-009-0895-5

174. Zhai Y., Wang Y., Li Y., Lei T., Yan F., Su L., et al. . (2013). Isolation and molecular characterization of GmERF7, a soybean ethylene-response factor that increases salt stress tolerance in tobacco. Gene513, 174–183. 10.1016/j.gene.2012.10.018

175. Zhang G., Chen M., Li L., Xu Z., Chen X., Guo J., et al. . (2009). Overexpression of the soybean GmERF3 gene, an AP2/ERF type transcription factor for increased tolerances to salt, drought, and diseases in transgenic tobacco. J. Exp. Bot. 60, 3781–3796. 10.1093/jxb/erp214

176. Zhang H., Liu W., Wan L., Li F., Dai L., Li D., et al. . (2010). Functional analyses of ethylene response factor JERF3 with the aim of improving tolerance to drought and osmotic stress in transgenic rice. Transgenic Res. 19, 809–818. 10.1007/s11248-009-9357-x

177. Zhang L., Zhang L., Xia C., Zhao G., Liu J., Jia J., et al. . (2015). A novel wheat bZIP transcription factor, TabZIP60, confers multiple abiotic stress tolerances in transgenic *Arabidopsis*. Physiol. Plant.153, 538–554. 10.1111/ppl.12261

178. Zhang X., Liu X., Wu L., Yu G., Wang X., Ma H. (2015). The SsDREB transcription factor from the succulent halophyte suaeda salsa enhances abiotic stress tolerance in transgenic tobacco. Int. J. Genomics 2015, 875497. 10.1155/2015/875497

179. Zhang X., Wang L., Meng H., Wen H., Fan Y., Zhao J. (2011). Maize ABP9 enhances tolerance to multiple stresses in transgenic *Arabidopsis* by modulating ABA signaling and cellular levels of reactive oxygen species. Plant Mol. Biol. 75, 365–378. 10.1007/s11103-011-9732-x

180. Zhong H., Guo Q. Q., Chen L., Ren F., Wang Q. Q., Zheng Y., et al. . (2012). Two *Brassica napus*genes encoding NAC transcription factors are involved in response to high-salinity stress. Plant Cell Rep. 31, 1991–2003. 10.1007/s00299-012-1311-3

181. Zhou L., Wang N. N., Gong S. Y., Lu R., Li Y., Li X. B. (2015). Overexpression of a cotton (*Gossypium hirsutum*) WRKY gene, GhWRKY34, in *Arabidopsis* enhances salt-tolerance of the transgenic plants. Plant Physiol. Biochem. 96, 311–320. 10.1016/j.plaphy.2015.08.016

182. Zhou Q. Y., Tian A. G., Zou H. F., Xie Z. M., Lei G., Huang J., et al. . (2008). Soybean WRKY-type transcription factor genes, GmWRKY13, GmWRKY21, and GmWRKY54, confer differential tolerance to abiotic stresses in transgenic *Arabidopsis* plants. Plant Biotechnol. J. 6, 486–503. 10.1111/j.1467-7652.2008.00336.x

183. Zhu N., Cheng S., Liu X., Du H., Dai M., Zhou D. X., et al. . (2015). The R2R3-type MYB gene OsMYB91 has a function in coordinating plant growth and salt stress tolerance in rice. Plant Sci. 236, 146–156. 10.1016/j.plantsci.2015.03.023

184. Zhu X., Qi L., Liu X., Cai S., Xu H., Huang R., et al. . (2014). The wheat ethylene response factor transcription factor pathogen-induced ERF1 mediates host responses to both the necrotrophic pathogen*Rhizoctonia cerealis* and freezing stresses. Plant Physiol. 164, 1499–1514. 10.1104/pp.113.229575

185. Zhuang J., Cai B., Peng R. H., Zhu B., Jin X. F., Xue Y., et al. . (2008). Genome-wide analysis of the AP2/ERF gene family in *Populus trichocarpa*. Biochem. Biophys. Res. Commun. 371, 468–474. 10.1016/j.bbrc.2008.04.087

186. Zou M., Guan Y., Ren H., Zhang F., Chen F. (2008). A bZIP transcription factor, OsABI5, is involved in rice fertility and stress tolerance. Plant Mol. Biol. 66, 675–683. 10.1007/s11103-008-9298-4

CHAPTER 14

RECENT ADVANCES IN SOYBEAN TRANSFORMATION AND THEIR APPLICATION TO MOLECULAR BREEDING AND GENOMIC ANALYSIS

Tetsuya Yamada[1], Kyoko Takagi[2] and Masao Ishimoto[2]

[1] Research Faculty of Agriculture, Hokkaido University, Kita 9 Nishi 9, Kita, Sapporo, Hokkaido 060-8589, Japan

[2] Soybean Applied Genomics Research Unit, National Institute of Agrobiological Sciences, 2-1-2 Kannondai, Tsukuba, Ibaraki 305- 8602, Japan

ABSTRACT

Herbicide-resistant transgenic soybean plants hold a leading market share in the USA and other countries, but soybean has been regarded as recalcitrant to transformation for many years. The cumulative and, at times, exponential advances in genetic manipulation have made possible further choices for soybean transformation. The most widely and routinely used transformation systems are cotyledonary node–*Agrobacterium*-mediated transformation and somatic embryo–particle-bombardment-mediated transformation. These ready systems enable us to improve seed qualities and agronomic characteristics by transgenic approaches. In addition, with the accumulation of soybean genomic resources, convenient or promising approaches will be requisite for the determination and use of gene function in soybean. In this article, we describe recent advances in and problems of soybean transformation, and survey the current transgenic approaches for applied and basic research in Japan.

INTRODUCTION

Soybean [*Glycine max* (L.) Merrill] is an important crop, with food, nutritional, industrial, and pharmaceutical uses. Soybean seeds contain about 40% protein and about 20% oil. They are also abundant in physiologically active metabolites such as isoflavones, lecithins, tocopherols and saponins, in addition to functional proteins and are used as an affordable source of foods that promote and maintain health (Sugano 2005). Soybean production has increased the most among major crops in response to recent increases in demand for vegetable protein, oil and other constituents (Hartman *et al.* 2011). Therefore, soybean improvement is crucial to meeting demand.

The genomic era is now under way for soybean, as for other many crops. Recently, a soybean genomics database has been developed from the whole genome sequence (Schmutz *et al.* 2010), and a large number of genomic, transcriptional, and functional annotated sequences can be retrieved from Phytozome (http://www.phytozome.net/search.php). In addition to efforts to sequence the whole genome, several resources have been developed, including an expressed sequence tag (EST) database, full-length cDNAs and cDNA microarrays (Stacey *et al.* 2004, Umezawa *et al.* 2008). These resources provide a range of opportunities for soybean improvement by marker-assisted breeding and transgenic approaches, and for understanding gene function by map-

Citation: Tetsuya Yamada, Kyoko Takagi and Masao Ishimoto, Recent advances in soybean transformation and their application to molecular breeding and genomic analysis, doi:10.1270/jsbbs.61.480.

based cloning and reverse genetic approaches. An efficient and stable transformation system is essential to these goals.

Roundup Ready soybean cultivars are an example of transgenic soybean (Padgette *et al.* 1995), and have been planted on the majority of soybean fields in the world since 2004 (ISAAA, http://www.isaaa.org/). However, soybean remains recalcitrant to routine genetic transformation. The first fertile transgenic soybeans were produced nearly simultaneously by *Agrobacterium tumefaciens* infection with cotyledonary node plant regeneration (Hinchee *et al.* 1988), and by particle bombardment of meristems of immature soybean seeds (McCabe *et al.* 1988). The system was successfully adapted to embryogenic suspension cultures for the regeneration of fertile transgenic soybeans (Finer and McMullen 1991). Since then, these two methods have continued to be improved and have produced most transgenic soybeans to date.

In this review, we describe recent advances in and problems of soybean transformation, with a focus on the methods that generate fertile transgenic plants (Table 1). We discuss the convenience and prospects of transgenic approaches for the identification of gene function and the improvement of agronomic characteristics (Table 2), and survey the recent transgenic research in Japan.

Table 1: Summary of representative soybean transformation systems

Transformation method	Explant	Soybean genotype	Strain of *A. tumefaciens*	Selection		References
				Marker	Agent	
Agrobacterium	Cotyledonary explant	Peking, Maple Prest	A208	*npt II*	kanamycin	Hinchee *et al.* (1988)
		A3237	EHA105	*bar*	glufosinate	Zhang *et al.* (1999)
		AC Colibri	EHA105	*npt II*	kanamycin	Donaldson and Simmonds (2000)
		Bert	AGL1	*bar*	phosphinothricin	Olhoft and Somers (2001)
		Bert	LBA4404, EHA105	*hpt*	hygromycin	Olhoft *et al.* (2003)
		Williams 82	EHA101	*bar*	glufosinate	Zeng *et al.* (2004)
		Williams, Williams 79, Peking, Thorne	EHA101	*bar*	glufosinate or bialaphos	Paz *et al.* (2004)
		Thorne, Williams, Williams 79, Williams 82	EHA101	*bar*	glufosinate	Paz *et al.* (2006)
		Jungery	LBA4404	*bar*	phosphinothricin	Xue *et al.* (2006)
		Kariyutaka	EHA105	*bar*	glufosinate	Sato *et al.* (2007), Yamada *et al.* (2010)
		Hefeng 25, Dongnong 42, Heinong 37, Jilin 39, Jiyu 58	EHA105	*hpt*	hygromycin	Liu *et al.* (2008)
	Somatic embryo	Peking, PI 283332	LBA4404, EHA101	*npt II*	G418	Parrott *et al.* (1989a)
		Chapman	EHA105	*hpt*	hygromycin	Trick and Finer (1998)
	Embryonic tip	Hefeng 25, Hefeng 35, Hefeng 39, Heinong 37, Heinong 43, Dongnong 42, Lefeng 39	KYRT1	*bar*	phosphinothricin	Dang and Wei (2007)
Particle bombardment	Embryonic axis	Williams 82, Mandarin Ottawa	-	*npt II*	Undefined	McCabe *et al.* (1988)
		BR-16, Doko RC, BR-91, Conquista	-	*ahas*	imazapyr	Arãgao *et al.* (2000)
	Somatic embryo	Fayette	-	*hpt*	hygromycin	Finer and McMullen (1991)
		Fayette	-	*npt II*	G418	Sato *et al.* (1993)
		Jack and its derivative line	-	*hpt*	hygromycin	Parrott *et al.* (1994), Stewart *et al.* (1996), Maughan *et al.* (1999), Reddy *et al.* (2003), El-Shemy *et al.* (2004), Furutani and Hidaka (2004), Khalafalla *et al.* (2005), Kita *et al.* (2007) etc.

Table 2: Summary of transgenic approaches for improvement of seed components, agronomic traits, and functional genomics in soybean

Target traits	Target gene	Origin of target gene	Target tissue	Promoter	Effect	Transformation method[1)]	Soybean genotype	References
Seed component								
Seed protein	β-casein gene	bovine	seed	soybean lectin	Accumulation of β-casein protein	PB	Jack	Maughan *et al.* (1999)
	15-kDa zein gene	maize	seed	common bean β-phaseolin	Accumulation of zein protein	PB & AG	Jack, F173	Dinkins *et al.* (2001), Reddy *et al.* (2003)
	Gly m Bd 30K gene	soybean	seed	soybean α-subunit of β-conglycinin	Reduction of allergen (Gly m Bd 30K protein)	PB	Jack	Herman *et al.* (2003)
	Modified glycinin gene (*V3-1*)	soybean	seed	soybean glycinin (*gy2*)	Accumulation of V3-1 protein	PB	Jack	El-Shemy *et al.* (2004)
	11-kDa δ-zein gene	maize	seed	soybean α' subunit of β-conglycinin	Accumulation of zein protein	AG	Williams 82	Kim and Krishnan (2004)
	27-kDa γ-zein gene	maize	seed	soybean α' subunit of β-conglycinin	Accumulation of zein protein	PB	Jack	Li *et al.* (2005)
	K99 fimbrial subunit gene (*fanC*)	*Escherichia coli*	seed	cauliflower mosaic virus (CaMV) 35S	Accumulation of FanC	AG	Thorne	Piller *et al.* (2005)
	Basic fibroblast growth factor (*bFGF*) gene	human	seed	CaMV 35S or soybean glycinin (*gy1*)	Accumulation of bFGF	AG	Sichuan	Ding *et al.* (2006)
	Modified β-conglycinin α' subunit gene containing bioactive peptide (Novokinin, LPYPR, Rubiscolin)	modified materials from soybean	seed	soybean α' subunit of β-conglycinin	Accumulation of bioactive peptides	PB	Jack	Nishizawa *et al.* (2008)
	Modified β-conglycinin α' subunit gene (*4novokinin-α'*)	modified materials from soybean	seed	soybean α' subunit of β-conglycinin	Accumulation of bioactive peptides	Whisker	Jack	Yamada *et al.* (2008)
	β-conglycinin α' subunit gene	soybean	seed	soybean glycinin (*gy2*)	Reduction of β-conglycinin	PB	Jack	Nishizawa *et al.* (2010)
	Human growth hormone gene (*hgh*)	human	seed	soybean α' subunit of β-conglycinin	Accumulation of mature form of hGH	PB	BR-16	Cunha *et al.* (2011)
Oil	Δ12 fatty acid desaturase gene (*FAD2-1*), Palmitoyl-thioesterase gene (*Fat*B)	soybean	seed	common bean β-phaseolin or soybean β-conglycinin	Increase of oleic acid and decrease of saturated fatty acid	AG	A3237, Thorne	Buhr *et al.* (2002)
	Δ6 desaturase gene	*Arabidopsis thaliana*	seed	soybean β-conglycinin	Production of γ-linolenic acid (GLA) and stearidonic acid (STA)	AG	A3237, Thorne, NE3001	Sato *et al.* (2004)
	Δ5 desaturase gene, Δ6 desaturase gene, GLELO elongase gene, Δ15 desaturase gene	Mortierella alpina 1S-4 (Δ5 and 6 desaturase,GLELO), soybean (Δ15 desaturase)	seed	soybean α' subunit of β-conglycinin	Production of arachidonic acid	PB	Jack	Chen *et al.* (2006)
	Δ6 desaturase gene, Δ15 desaturase gene (*fad3*)	*B.officinalis* (Δ6 desaturase gene), *A. thaliana* (*fad3*)	seed	soybean β-conglycinin	High accumulation of stearidonic acid (STA)	AG	Thorne, NE3001, 420-5	Eckert *et al.* (2006)
	ω-3 fatty acid desaturase gene (*GmFAD3*)	soybean	seed	soybean glycinin	Reduction of α-linolenic acids (18:3)	AG	Jack	Flores *et al.* (2008)
	Δ6 desaturase gene (*MpDES6*), Δ6 elongase gene (*MpELO1*), Δ5 desaturase gene (*MpDES5*)	*Marchantia polymorpha*	seed	soybean α' subunit of β-conglycinin	Production of of C20-LCPUFAs (long-chain polyunsaturated fatty acids)	PB	Jack	Kajikawa *et al.* (2008)
	Diacylglycerol acyltransferase 2A gene (*UrDGAT2A*)	*Umbelopsis ramanniana*	seed	soybean α' subunit of β-conglycinin	Increase of oil content	AG	Undefined	Lardizabal *et al.* (2008)
	Δ12 fatty acid desaturase gene (*GmFAD2-1*)	soybean	seed	soybean lectin	Increase of oleic acid	AG	Heinong44	Wang and Xu (2008)
	Sphingolipid compensation gene (*SLC1*)	*Saccharomyces cerevisiae*	seed	common bean phaseolin	Increase of oil content	PB	Jack	Rao and Hildebrand (2009)
	Fatty acid ω-6 desaturase 2 gene (*FAD2*), Acyl-acyl carrier protein thioesterase 2 genes (*FATB-4* and *FATB-5*), Diacylglycerol acyltransferase gene (*DGAT1*), Dihydrodipicolinate synthetase gene (*DHPS*), High-lysine protein gene (*BHL8*), truncated cysteine synthase gene (*CGS*)	soybean (*FAD2*, *FATB4* and *5*, *CGS*), *Yarrowia lipolytica* (*DGAT1*), *Corynebacterium glutamicum* (*DHPS*), barley (*BHL8*),	seed	soybean KTI3 (Kunitz trypsin inhibitor 3)	Improvement of oil content and composition	PB	Undefined	Li Z. *et al.* (2010)
	Epoxygenase gene (*SlEPX*), Diacylglycerol acyltransferase genes (*VgDGAT1* and *VgDGAT2*)	*Stokesia laevis* (*SlEPX*), *Vernonia galamensis* (*VgDGAT1and 2*)	seed	common bean phaseolin	Increase of epoxy fatty acid	PB	Jack	Li R. *et al.* (2010)

TWO COMMON PLATFORMS FOR SOYBEAN TRANSFORMATION

Cotyledonary node–Agrobacterium-mediated transformation

A biological vector, *Agrobacterium tumefaciens*, is used to transfer desirable genes placed in the T-DNA region into a host plant genome (Beijersbergen *et al.* 1992, Horsch *et al.* 1985). The advantages of*Agrobacterium*-meditated transformation include its straightforward methodology, familiarity to researchers, minimal equipment cost and reliable insertion of a single transgene, or a low copy number (Hansen and Wright 1999). *Agrobacterium*-mediated transformation of soybean in co-cultivation has been followed by organogenesis from cotyledonary nodes (Hinchee *et al.* 1988), immature cotyledons (Parrott *et al.* 1989a,1994), and embryogenic suspension cultures (Trick and Finer 1998). Originally the method relied on a soybean genotype that conferred susceptibility to *A. tumefaciens* infection and on the availability of plant regeneration (Delzer *et al.* 1990, Hinchee *et al.* 1988, Owens and Cress 1985). However, recent advances, as described below, overcome some of these shortcomings (Dinkins and Collins 2008, Olhoft and Somers 2007,Somers *et al.* 2003).

The successful and repeatable production of transgenic soybean has been achieved by using cotyledonary node explants from young seedlings and imbibed mature seeds (Donaldson and Simmonds 2000, Hinchee *et al.* 1988, Olhoft *et al.* 2003, Paz *et al.* 2006, Zhang *et al.* 1999) for *Agrobacterium*-mediated transformation. Cotyledonary node regions contain axillary meristems at the junction between cotyledon and hypocotyl. The axillary meristems proliferate and regenerate through the formation of multiple adventitious shoots on culture medium containing the cytokinin benzylaminopurine. The degree of shoot formation depends on the genotype of an explant, most types of which can form adventitious shoots at the cotyledonary nodes. In general, cotyledonary nodes are pre-wounded mechanically with a scalpel (Olhoft *et al.* 2001) or a small needle (Xue *et al.* 2006), but it requires practiced skill to prepare enough target tissue for bacterial infection (Zhang *et al.* 1999). In contrast, scratching with a stainless steel micro-brush enables any technician to wound the tissues easily and uniformly, regardless of skill (Yamada *et al.* 2010).

The addition of reducing agents such as l-cysteine and thiol compounds in the solidified co-cultivation medium significantly increases the efficiency of transformation of cotyledonary node cells (Olhoft *et al.* 2001,Olhoft and Somers 2001) and the production of fertile transgenic plants (Olhoft *et al.* 2003). The reducing agents seem to inhibit wound- and pathogen-induced responses, thereby increasing the capacity for*Agrobacterium*-mediated transformation (Olhoft *et al.* 2001). The combination of the reducing agents, a super-binary vector, and acetosyringone has increased transformation efficiencies and the competency of soybean genotypes for transformation (Dang and Wei 2007, Liu *et al.* 2008, Sato *et al.* 2007). The first transgenic soybeans were produced using the *nptII* gene, which detoxifies kanamycin as a selectable marker (Hinchee *et al.* 1988). Now transgenic cells are selected exclusively by the combination of the *bar* gene and the herbicide phosphinothricin (glufosinate) (Zeng *et al.* 2004, Zhang *et al.* 1999). The concentration of the selection agent greatly affects the transformation frequency (Zeng *et al.* 2004), so the appropriate selection schemes are varied among soybean genotypes.

These improved protocols have been widely applied to several Japanese soybean cultivars, including Kariyutaka, Kinusayaka, Tamahomare, and Suzuyutaka (Sato *et al.* 2007, Sayama *et al.* unpublished data). Kariyutaka, with an early maturity genotype, produces a small number of T_1 seeds about 5 months after co-cultivation with *A. tumefaciens* (Sato *et al.* 2007). Its short life span might be useful in the rapid development of transgenic soybean lines. Transformation frequencies range from 0.2% to around 10% (Olhoft *et al.* 2003,Paz *et al.* 2004, 2006, Zeng *et al.* 2004), indicating that the transformation efficiency still relies on the skill of the practitioner and on the soybean genotype. The frequency of transformation is still low in comparison with somatic embryo–particle-bombardment-mediated transformation.

In the USA, public facilities, including the Plant Transformation Facility at Iowa State University and the Plant Transformation Core Facility at the University of Missouri, provide transgenic plants for public research, mainly by cotyledonary node–*Agrobacterium*-mediated transformation. A similar facility needs to be launched in Japan.

Somatic embryo–particle-bombardment-mediated transformation

Particle bombardment, otherwise known as gene gun or biolistic technology, directs small tungsten or gold particles coated with the desired genes toward the target plant cells (Christou *et al.* 1988). Since an electrical-discharge gene gun was first used in soybean (McCabe *et al.* 1988), transformation by particle bombardment has been achieved in immature seed meristem (McCabe *et al.* 1988), somatic embryogenic tissue (Finer and McMullen 1991), and apical meristem (Aragão *et al.* 2000).

Somatic embryos were initially used as a target for *Agrobacterium*-mediated transformation (Parrott *et al.*1989a), and later found to be amenable to transformation by particle bombardment (Finer and McMullen 1991, Maughan *et al.* 1999, Sato *et al.* 1993). Somatic embryogenesis in soybean was first reported byChristianson *et al.* (1983). Somatic embryos are induced from immature cotyledons cultured on medium containing moderately high concentrations of an auxin such as 2,4-dichlorophenoxyacetic acid (2,4-D), and are used to generate proliferative embryogenic cultures and to recover whole plants (Finer and Nagasawa 1988, Lazzeri *et al.* 1985, 1987, Parrott *et al.* 1988, Ranch *et al.* 1985). As the formation of proliferative embryogenic tissue depends on genotype, the use of transformation has been limited to a few soybean cultivars. On the basis of its capacity for induction of primary somatic embryos, proliferative embryogenic cultures, and recovery of whole plants, cultivar Jack has been recognized as a competent genotype for transformation and has been exclusively used to generate transgenic soybeans (Meurer *et al.* 2001, Stewart *et al.* 1996, Tomlin *et al.* 2002), because modification of tissue culture protocols have only partially overcome the effects of genotype (Bailey *et al.* 1993a, 1993b). The limitation has often precluded the functional analysis of transgenes in combination with a specific genotype, and the direct improvement of leading cultivars by transformation. Somatic embryogenesis is a heritable trait and can be improved by hybridization breeding (Parrott *et al.* 1989b); the competence for somatic embryogenesis was successfully transferred and combined in other genotypes (Kita *et al.* 2007, 2010).

Physical procedures for transformation tend to result in the integration of large complexes, fragmentation, and reconstitution of transgenes, which sometimes lead to the silencing of transgenes or homologous endogenous genes (El-Shemy *et al.* 2004, Kinney *et al.* 2001, Reddy *et al.* 2003). The use of a reporter gene such as*sGFP(S65T)* or *DsRed2* in addition to a selectable marker gene could help to reduce the problem of gene silencing associated with physical transformation systems and facilitate the recovery of transgenic plants that stably express the target gene between the two marker genes (El-Shemy *et al.* 2004, Nishizawa *et al.* 2006). As shown in rice transformation (Fu *et al.* 2000), linearized transgene constructs lacking vector backbone sequences might also generate transgenic soybean plants with a low transgene copy number by the simple integration of the constructs.

Soybean somatic embryos have attracted additional attention as a model of zygotic embryos. Proliferative somatic embryos can retain regenerative properties for more than a year, with differentiation and development being readily induced when required (Finer and Nagasawa 1988, Parrott *et al.* 1988). Mature somatic embryos accumulate seed storage proteins with the same temporal and spatial regulation as developing seeds (Dahmer *et al.* 1992, Nishizawa and Ishimoto 2009), and their fatty acid composition is similar to that of seeds (Dahmer *et al.* 1991, Shoemaker and Hammond 1988). Transgenic embryos have usually been obtained within 7 weeks after the introduction of exogenous genes by particle bombardment (Khalafalla *et al.*2005), and homogeneous masses of transgenic embryos can be readily and repeatedly induced to differentiate. Somatic embryos have therefore been used to assess transgenic seed traits before recovery of whole plants, and then selected clones are recovered as whole transgenic plants (Cahoon *et al.* 2000, 2002,Chen *et al.* 2006, Herman *et al.* 2003, Nishizawa *et al.* 2010).

The improved and refined protocols for somatic embryo–particle-bombardment-mediated transformation are widely reproducible across laboratories, even though there are still some limitations as previously noted (El-Shemy *et al.* 2004, Furutani and Hidaka 2004, Furutani *et al.* 2006, 2007, Ishimoto *et al.* 2010, Khalafalla *et al.* 2005, Kita *et al.* 2009, 2010, Nishizawa *et al.* 2008, Takagi *et al.* 2011, Tougou *et al.* 2006, 2007,Yamada *et al.* 2008). The RIKEN Plant Science Center has supported the Transformation Network Consortium (TRANSNET) to enhance both basic and applied research in plant biology in Japan since 2008. Under a collaborative research agreement, staff at the National Agricultural Research Center for Hokkaido Region will create transgenic soybeans by particle-bombardment-mediated transformation on request from academic researchers in Japan.

TRANSGENIC APPROACHES TO IMPROVEMENT OF SEED COMPONENTS AND AGRONOMIC TRAITS

Modification of seed components

Protein and amino acid compositions

The abundant proteins and oil in soybean seeds are attractive targets for improvement by transformation. Soybean protein is the nutritional equivalent of meat and eggs except for its deficiency of sulfur amino acids, especially methionine (FAO/WHO 1990, Young 1991). High-methionine proteins such as bovine β-casein and maize zein were induced to accumulate in soybean seed under the regulation of seed-expression promoters (Dinkins *et al.* 2001, Kim and Krishnan 2004, Li *et al.* 2005, Maughan *et al.* 1999), but not enough for nutritional improvement. The accumulation of these methionine-rich proteins may be limited by the absence of the proper maturation process in soybean or by the availability of sulfur-containing amino acids or of sulfur itself. Although there is no information about the increase of free sulfur-containing amino acids in soybean, three other essential amino acids, lysine, tryptophan and threonine, substantially increased in soybean seeds by the expression of genes for feedback-insensitive enzymes involved in their synthesis (Falco *et al.* 1995, Ishimoto *et al.* 2010, Kita *et al.* 2010, Qi *et al.* 2011). Improvement of the pool of soluble amino acids would seem to be a reliable approach to improving the nutritional quality of soybean.

Soybean is also considered one of the most efficient protein bioreactors for plant molecular farming. Pharmaceutical proteins such as human growth hormone, fibroblast growth factor, and an edible vaccine were accumulated in stable transgenic soybean seeds (Cunha *et al.* 2011, Ding *et al.* 2006, Piller *et al.* 2005). Although bioactive proteins comprised up to 3% of the total seed protein content, the content of pharmaceutical proteins is nowhere near the content of endogenous storage proteins. Instead, another strategy was devised to use the major storage proteins, β-conglycinin and glycinin, as carriers for bioactive peptides (Nishizawa *et al.* 2008, Yamada *et al.* 2008). A bioactive hexa-peptide, novokinin, was incorporated into the α′ subunit of β-conglycinin at four sites by minimum replacement of amino acids constituting analogous sequences, and transgenic soybean seeds accumulated the modified protein with the intended properties (Yamada *et al.* 2008). So far, however, the levels of modified storage proteins have not come close to the amount of the original protein. Mutant lines lacking all subunits of glycinin and β-conglycinin may prove more amenable to the accumulation of modified storage proteins, and of foreign proteins (Kita *et al.* 2007,Takahashi *et al.* 2003), since a decrease in the abundance of the endogenous storage proteins prolamine and globulin in rice was compensated for by the enrichment of foreign proteins, resulting in an almost equivalent total amount of seed storage proteins (Tada *et al.* 2003).

Although soy proteins are highly nutritious, some are recognized as allergens in some people (Ogawa *et al.*2000). Among them, Gly m Bd 30K, also called P34, is regarded as the major or immunodominant allergen in soybean seed. Transgene-induced gene silencing (co-suppression) could be used to remove allergens from soybean seeds without any compositional, developmental, or structural changes (Herman *et al.* 2003).

Oil composition

Almost three-fourths of global vegetable oil production comes from oil palm, soybean, rape-seed and sunflower, in that order. Soybean oil is widely used in food and in industry in printing ink, lubricants and bio-diesel. Improvement of the oil content and its composition has been a goal in the use of transformation technology. As vegetable oil is stored in seeds in the triacylglycerol form, exotic acyltransferase genes were introduced into soybean to enhance the biosynthesis of triacylglycerol, resulting in a maximum increase of 3.2% (by weight) in seed oil content in mature seeds (Lardizabal *et al.* 2008, Li Z. *et al.* 2010, Rao and Hildebrand 2009).

Oil composition determines the performance of an oil. Transgenic approaches could provide many options to tailor soybean oil for specific uses. Typically, soybean oil is composed of palmitic, stearic, oleic, linoleic and linolenic acids (Yadav 1996). The high level of polyunsaturated fatty acids in natural soybean oil renders the oil unstable and thus susceptible to the development of disagreeable odors and flavors. Therefore, soybean oil with decreased polyunsaturated fatty acids would be ideal for use in food. Down-regulation of the desaturation of fatty acids by ribozyme termination of RNA transcripts or RNA interference (RNAi) gene silencing (see Kasai and Kanazawa 2012) decreased the content of polyunsaturated fatty acids or increased that of oleic acid (Buhr *et al.* 2002, Flores *et al.* 2008, Li R. *et al.* 2010, Wang and Xu 2008). On the other hand, ectopic expression of heterogeneous genes involved in fatty acid modification could generate other fatty acids such as γ-linolenic, stearidonic, arachidonic, eicosapentaenoic and vernolic acids, which are undetectable or minor fatty acids in non-transgenic soybean seeds (Chen *et al.* 2006, Eckert *et al.* 2006,Kajikawa *et al.* 2008, Li R. *et al.* 2010, Sato *et al.* 2004).

The vitamin E family comprises tocopherols and tocotrienols (α, β, γ and δ forms). All isoforms possess lipid antioxidant activity, and α-tocopherol possesses the highest vitamin E activity in mammals (Bramley *et al.*2000, Herbers 2003). Vitamin E is widely used as an antioxidant in foods and oils, as a nutrient additive in poultry and cattle feeds and as a supplement in the human diet to help prevent diseases. In soybean processing, tocopherols are extracted with the oil. Their content is only about 1.5% of the total oil component, yet they are critical for oxidative stability of the oil (Hoppe and Krennrich 2000). Enhancing the key step in the conversion of γ-tocopherol to α-tocopherol elevated the α-tocopherol content to 95% of the total tocopherol content in transgenic soybean seeds (Kim *et al.* 2005, Tavva *et al.* 2007, Van Eenennaam *et al.* 2003).

Other compounds

Isoflavones are an important group of compounds that are synthesized in legumes. In addition to their role in the mediation of plant–microbe interactions (Ebel 1986, Rivera-Vargas *et al.* 1993, Subramanian *et al.* 2005,van Rhijn and Vanderleyden 1995), isoflavones are known as phytoestrogens and biologically active substances associated with human health benefits such as anti-cancer effects and decreased risk of coronary heart disease (Setchell 1998). The soybean isoflavones daidzein, genistein, and glycitein are synthesized through the phenylpropanoid pathway, modified by legume-specific enzymes, and stored in the vacuole as glycosidic conjugates (Graham 1991). Activation of the phenylpropanoid pathway by the maize C1 and R transcription factors combined with blockage of the competing pathway by co-suppression of flavanone 3-hydroxylase increased isoflavone accumulation by up to four times that in wild-type seed (Yu *et al.* 2003). In contrast, transgenic soybeans containing three gene cassettes encoding chalcone synthase, isoflavone synthase, and phenylalanine ammonia lyase produced seeds with 70% less isoflavone (Zernova *et al.* 2009). These results indicate that regulation of the expression of genes for phenylpropanoid biosynthesis enzymes and isoflavone-specific enzymes can alter the content and composition of isoflavones.

Saponins are a group of structurally diverse molecules that include glycosylated triterpenic or steroidal compounds, and are widely distributed among plant species. In soybean, a number of triterpenoid saponins have been identified, and have been classified into four groups (A, B, E and DDMP) on the basis of the chemical structure of the aglycone (Kudou *et al.* 1992, Shiraiwa *et al.* 1991a, 1991b). Soybean saponins have various pharmacological effects such as anti-lipidemic effects (Topping *et al.* 1980) and antiproliferative effects against human colon cancer cells (Ellington *et al.* 2005, 2006). On the other hand, they are considered unwanted components in foods, because they are the main cause of undesirable flavors and of foaming in tofu production. The biosynthesis of saponins in transgenic seeds was almost completely suppressed by RNAi silencing of β-amyrin synthase, a key enzyme in the synthesis of a common aglycone of soybean saponins (Takagi *et al.* 2011).

Soybean seeds contain large quantities of phytic acid (phytate), which releases phosphorus (P) and myoinositol during seed germination. Monogastric animals lack phytase, the digestive enzyme required to remove phosphate from the inositol in phytate, and therefore P in phytate is not available to them. Fertile transgenic soybean plants containing phytase showed a nearly threefold increase in P availability as well as a reduction of phytate (Chiera *et al.* 2004). Myoinositol-1-phosphate is synthesized from glucose 6-phosphate in a reaction catalyzed by myoinositol-1-phosphate synthase, and then converted into phytate. RNAi gene silencing drastically reduced phytate and inhibited seed development (Nunes *et al.* 2006). Suppressing a multidrug-resistance–associated protein (MRP) ATP-binding cassette (ABC) transporter gene in maize and soybean generated low-phytic-acid seed (Shi *et al.* 2007).

Enhancement of biotic and abiotic resistance

Insect and nematode resistance

Insecticidal crystal proteins (cry proteins or δ-endotoxins) are an active component of *Bacillus thuringiensis*(Bt) toxin, a biological insecticide (Tabashnik 1994). Expression of the Bt *cry* gene in soybean has proven highly effective for controlling insect pests (Dufourmantel *et al.* 2005, Miklos *et al.* 2007, Parrott *et al.* 1994,Stewart *et al.* 1996), and the resistance to lepidopteran pests in a transgenic line expressing Bt *cry1A* was confirmed under field conditions (Walker *et al.* 2000). However, the discovery that insects can adapt to Bt cry proteins raises concerns about long-term or high-dose use (McGaughey and Whalon 1992). Strategies suggested for managing the development of resistance to Bt cry proteins include the combination of the Bt*cry* gene and defoliating insect resistance QTLs or other insecticidal proteins (Macrae *et al.* 2005, Walker *et al.* 2002, Zhu *et al.* 2008).

Soybean cyst nematode (SCN; *Heterodera glycines* Ichinohe) is a primary pest of soybean production. Effective management of SCN relies on the combination of resistant cultivars and crop rotation. Resistance to SCN is controlled by multiple loci, but diverse nematode populations have broken down the elaborate resistance. Therefore, other strategies for SCN resistance are needed. $Hs1^{pro-1}$, a gene from wild beet for resistance to the closely related beet cyst nematode, enhanced SCN resistance in soybean (McLean *et al.*2007).

Disease resistance

Soybean mosaic virus (SMV) is endemic in virtually all regions where soybeans are grown in the presence of vector insects. SMV can cause serious yield losses (Ross 1969), so virus resistance is an essential trait for introduction. There have been some efforts to improve virus resistance in soybean by transgenic approaches. Overexpression of a coat protein gene and the 3-UTR region from SMV resulted in high resistance to SMV in transgenic soybean plants (Furutani *et al.* 2006, Wang *et al.* 2001). In addition, resistance to bean pod mottle virus and soybean dwarf virus has been introduced into susceptible soybean by transgenic approaches (Di *et al.* 1996, Reddy *et al.* 2001, Tougou *et al.* 2006, 2007).

Sclerotinia stem rot (white mould) is serious fungal disease of soybean. As oxalic acid is an important pathogenicity factor of the fungus (Godoy *et al.* 1990), the introduction of a gene to degrade oxalic acid would provide an effective defense against the fungus in soybean. Overexpression of heterogeneous genes encoding oxalate oxidase or oxalate decarboxylase reduced disease progression and lesion length after inoculation of leaves and stems with the fungus (Cunha *et al.* 2010, Donaldson *et al.* 2001).

Abiotic stress tolerance

Drought stress is one of the major environmental limitations on crop production. Transgenic soybean expressing *P5CR*, encoding l-Δ^1-pyrroline-5-carboxylate reductase, which catalyzes the final step in proline biosynthesis, under the control of an inducible heat shock promoter was more tolerant to drought and high temperature than non-transgenic plants (De Ronde *et al.* 2004a, 2004b). Furthermore, overexpression of an endogenous gene encoding ER-resistant molecular chaperon binding protein from soybean (*soyBiPD*) delayed leaf senescence during drought (Valente *et al.* 2009).

Iron is abundant in soil, but its availability is sometimes limited in aerated soil. Ectopic expression of the*Arabidopsis* ferric chelate reductase gene conferred tolerance to iron deficiency chlorosis, but constitutive expression decreased productivity (Vasconcelos *et al.* 2006).

Herbicide resistance

The most successful trans-genic trait introduced into soybean is resistance to the non-selective herbicide glyphosate (*N*-phosphonomethyl-glycine; Roundup) (Padgette *et al.* 1995). Roundup Ready soybean cultivars were introduced into commercial production in 1996 and have been planted on most soybean fields since 2004 (ISAAA, http://www.isaaa.org/). Glyphosate binds to and blocks the activity of 5-enolpyruvylshikimate-3-phosphate synthase (EPSPS), an enzyme of the shikimic acid pathway, which produces aromatic amino acids. A glyphosate-tolerant EPSPS was introduced into soybean to confer a high level of glyphosate tolerance (Hinchee *et al.* 1988, Padgette *et al.* 1995). In addition, the introduction of genes for acetohydroxyacid synthase (AHAS) from Arabidopsis, 4-hydroxyphenylpyruvate dioxygenase (HPPD) from *Pseudomonas fluorescens*, and phosphinothricin *N*-acetyltransferase (PAT) from bialaphos-resistant soil bacteria conferred tolerance to, respectively, imazapyr, isoxaflu-tole and phosphinothricin (Aragão *et al.* 2000, Dufourmantel *et al.* 2007, Kita *et al.* 2009). These herbicide resistance genes are also used as markers to allow the selection of transgenic soybeans (Rech *et al.* 2008).

TRANSGENIC APPROACHES TO SOYBEAN GENOMICS RESEARCH

Soybean genes have often been evaluated for their function in heterogeneous plants such as *A. thaliana* or tobacco because soybean has remained recalcitrant to routine transformation. However, they should also be evaluated in the genetic background of a soybean with a null mutant or recessive allele for the target gene. Therefore, the functional analysis of target genes requires the transformation of a wide range of soybean genotypes. *Agrobacterium*-mediated transformation has now been successfully used in a wide range of soybean genotypes and been simplified (Table 1). This transformation system could provide a sophisticated method of gene functional analysis for soybean genomics research. There is one example of the complementation of an isolated gene by the transgenic approach. The habit of stem growth is an important agronomic trait. A recessive allele, *dt1*, decreases plant height and number of nodes. The *Dt1* gene of soybean was isolated as a *TFL1* orthologue of*A. thaliana* (Liu et al. 2010). The genomic region of the *Dt1*allele was introduced into the genetic background of the *dt1* allele by *Agrobacterium*-mediated transformation to complement the *dt1* allele (Liu *et al.* 2010), revealing that the *Dt1* locus exactly controls stem growth habit in soybean.

Agrobacterium tumefaciens is commonly used for DNA delivery. An alternative system using*Agrobacterium rhizogenes* is termed hairy root transformation. This system, which inserts the T-DNA region into the genome of host plant root cells (Chilton *et al.* 1982), has been optimized to the study of symbiotic and pathogenic interactions in roots (Kereszt *et al.* 2007). Hairy root transformation offers the advantage over*A. tumefaciens*-mediated transformation that as every transgenic root represents an independent transformation event, high numbers of transformants can be obtained and analyzed in a relatively short period of time. This system has contributed to elucidating the molecular mechanism of nodulation in soybean root (Indrasumunar *et al.* 2011, Kasai and Kanazawa 2012, Yang *et al.* 2010).

The process of soybean transformation is sometimes integrated into systems of gene-tagging or mutagenesis. Transformation mediated by *A. tumefaciens* or *A. rhizogenes* has been used to develop gene-tagging by transposon elements or site-direct mutagenesis using zinc-finger nucleases (Curtin *et al.* 2011, Mathieu *et al.*2009). These combination systems are appropriate for soybean genomics research.

CONCLUDING REMARKS

Transformation procedures have been simplified and optimized for various soybean genotypes. The techniques provide soybean breeders and researchers with opportunities to use transgenic plants for the improvement of agronomic traits as well as the analysis of gene function. Indeed, herbicide-resistant transgenic soybeans have been successfully released and planted in many countries. If a transgenic soybean were developed with agronomically important traits such as high yielding ability and multiple stress resistance which could not be achieved by current genetic resources, transgenic approaches might be more widely accepted in soybean breeding. In addition, transformation is an essential approach for genomics research in many crops, not only soybean. Target genes are readily isolated by map-based cloning or database information through well-organized genomic resources, which provide information on a large number of genomic, transcriptional, and functionally annotated sequences in soybean. Transgenic approaches are likely to become routine for the elucidation of gene function by over-expression, suppression, or complementation testing in the appropriate genetic background.

ACKNOWLEDGEMENTS

We wish to thank all of our past and present colleagues who have worked on the establishment of soybean transformation systems in Japan. This work was supported in part by Grants-in-Aid for Scientific Research from the Ministry of Education, Culture, Sports, Science and Technology of Japan to T. Yamada; and by a grant from the Ministry of Agriculture, Forestry and Fisheries of Japan (Genomics for Agricultural Innovation, GMZ-1004) and TRANSNET, organized by the RIKEN Plant Science Center, to M. Ishimoto.

REFERENCES

1. Aragão,F.J.L., L.Sarokin, G.R.Vianna and E.L.Rech (2000) Selection of transgenic meristematic cells utilizing a herbicidal molecule results in the recovery of fertile transgenic soybean [Glycine max (L.) Merril] plants at a high frequency. Theor. Appl. Genet. 101: 1–6.
2. Bailey,M.A., H.R.Boerma and W.A.Parrott (1993a) Genotype-specific optimization of plant regeneration from somatic embryos of soybean. Plant Sci. 93: 117–120.
3. Bailey,M.A., H.R.Boerma and W.A.Parrott (1993b) Genotype effects on proliferative embryogenesis and plant regeneration of soybean. In Vitro Cell. Dev. Biol. 29P: 102–108.
4. Beijersbergen, A., A.D. Dulk-Ras, R.A. Schilperoort and P.J.J. Hooykaas (1992) Conjugative transfer by the virulence system of Agrobacterium tumefaciens. Science 256: 1324–1327.
5. Bramley,P.M., I.Elmadfa, A.Kafatos, F.J.Kelly, Y.Manios, H.E. Roxborough, W.Schuch, P.J.A.Sheehy and K.-H.Wagner (2000) Vitamin E. J. Sci. Food Agric. 80: 913–938.
6. Buhr,T., S.Sato, F.Ebrahim, A.Xing, Y.Zhou, M.Mathiesen, B. Schweiger, A. Kinney, P. Staswick and T. Clemente (2002) Ribozyme termination of RNA transcripts down-regulate seed fatty acid genes in transgenic soybean. Plant J. 30: 155–163.
7. Cahoon,E.B., E.-F.Marillia, K.L.Stecca, S.E.Hall, D.C.Taylor and A.J.Kinney (2000) Production of fatty acid components of meadowfoam oil in somatic soybean embryos. Plant Physiol. 124: 243– 251.
8. Cahoon,E.B., K.G.Ripp, S.E.Hall and B.McGonigle (2002) Transgenic production of epoxy fatty acids by expression of a cytochrome P450 enzyme from Euphorbia lagascae seed. Plant Physiol. 128: 615–624.
9. Chen,R., K.Matsui, M.Ogawa, M.Oe, M.Ochiai, H.Kawashima, E.Sakuradani, S.Shimizu, M.Ishimoto, M.Hayashi et al. (2006) Expression of Δ6, Δ5 desaturase and GLELO elongase genes from Mortierella alpina for production of arachidonic acid in soybean [Glycine max (L.) Merrill] seeds. Plant Sci. 170: 399–406.
10. Chiera,J.M., J.J.Finer and E.A.Grabau (2004) Ectopic expression of a soybean phytase in developing seeds of Glycine max to improve phosphorus availability. Plant Mol. Biol. 56: 895–904.
11. Chilton,M.-D., D.A.Tepfer, A.Petit, C.David, F.Casse-Delbart and J.Temp (1982) Agrobacterium rhizogenes inserts T-DNA into the genomes of the host plant root cells. Nature 295: 432–434.
12. Christianson,M.L., D.A.Warnick and P.S.Carlson (1983) A morphogenetically competent soybean suspension culture. Science 222: 632–634.
13. Christou,P., D.E.McCabe and W.F.Swain (1988) Stable transformation of soybean callus by DNA-coated gold particles. Plant Physiol. 87: 671–674.
14. Cunha, N.B., A.M. Murad, T.M. Cipriano, A.C.G. Araújo, F.J.L. Aragão, A.Leite, G.R.Vianna, T.R.McPhee, G.H.M.F.Souza, M.J. Waters et al. (2011) Expression of functional recombinant human growth hormone in transgenic soybean seeds. Transgenic Res. 20: 811–826.
15. Cunha,W.G., M.L.P.Tinoco, H.L.Pancoti, R.E.Ribeiro and F.J.L. Aragão (2010) High resistance to Sclerotinia sclerotiorum in transgenic soybean plants transformed to express an oxalate decarboxylate gene. Plant Pathol. 59: 654–660.
16. Curtin,S.J., F.Zhang, J.D.Sander, W.J.Haun, C.Starker, N.J.Baltes, D.Reyon, E.J.Dahlborg, M.J.Goodwin, A.P.Coffman et al. (2011) Targeted mutagenesis of duplicated genes in soybean with zincfinger nucleases. Plant Physiol. 156: 466–473.
17. Dahmer,M.L., G.B.Collins and D.F.Hildebrand (1991) Lipid content and composition of soybean somatic embryos. Crop Sci. 31: 741– 746.

18. Dahmer,M.L., D.F.Hildebrand and G.B.Collins (1992) Comparative protein accumulation patterns in soybean somatic and zygotic embryos. In Vitro Cell. Dev. Biol. 28P: 106–114.

19. Dang,W. and Z.Wei (2007) An optimized Agrobacterium-mediated transformation for soybean for expression of binary insect resistance genes. Plant Sci. 173: 381–389.

20. DeRonde,J.A., R.N.Laurie, T.Caetano, M.M.Greyling and I.Kerepesi (2004a) Comparative study between transgenic and non-transgenic soybean lines proved transgenic lines to be more drought tolerant. Euphytica 138: 123–132.

21. De Ronde,J.A., W.A.Cress, G.H.J.Krüger, R.J.Strasser and J.Van Staden (2004b) Photosynthetic response of transgenic soybean plants, containing an Arabidopsis P5CR gene, during heat and drought stress. J. Plant Physiol. 161: 1211–1224.

22. Delzer, B.W., D.A. Somers and J.H. Orf (1990) Agrobacterium tumefaciens susceptibility and plant regeneration of 10 soybean genotypes in maturity groups 00 to II. Crop Sci. 30: 320–322.

23. Di,R., V.Purcell, G.B.Collins and S.A.Ghabrial (1996) Production of transgenic soybean lines expressing the bean pod mottle virus coat protein precursor gene. Plant Cell Rep. 15: 746–750.

24. Ding,S.-H., L.-Y.Huang, Y.-D.Wang, H.-C.Sun and Z.-H.Xiang (2006) High-level expression of basic fibroblast growth factor in transgenic soybean seeds and characterization of its biological activity. Biotechnol. Lett. 28: 869–875.

25. Dinkins,R.D., M.S.S.Reddy, C.A. Meurer, B. Yan, H. Trick, F. Thibaud-Nissen, J.J.Finer, W.A.Parrott and G.B.Collins (2001) Increased sulfur amino acids in soybean plants overexpressing the maize 15 kDa zein protein. In Vitro Cell. Dev. Biol. Plant 37: 742– 747.

26. Dinkins,R.D. and G.B.Collins (2008) Agrobacterium-mediated genetic transformation of soybean. In: Kirti,P.D. (ed.) Handbook of New Technologies for Genetic Improvement of Legumes, CRC Press, Florida, pp. 89–102.

27. Donaldson, P.A. and D.H. Simmonds (2000) Susceptibility to Agrobacterium tumefaciens and cotyledonary node transformation in short-season soybean. Plant Cell Rep. 19: 478–484.

28. Donaldson,P.A., T.Anderson, B.G.Lane, A.L.Davidson and D.H. Simmonds (2001) Soybean plants expressing an active oligomeric oxalate oxidase from the wheat gf-2.8 (germin) gene are resistant to the oxalate-secreting pathogen Sclerotina sclerotiorum. Physiol. Mol. Plant Pathol. 59: 297–307.

29. Dufourmantel, N., G.Tissot, F.Goutorbe, F. Garçon, C. Muhr, S. Jansens, B.Pelissier, G.Peltier and M.Dubald (2005) Generation and analysis of soybean plastid transformants expressing Bacillus thuringiensis Cry1Ab protoxin. Plant Mol. Biol. 58: 659–668.

30. Dufourmantel,N., M.Dubald, M.Matringe, H.Canard, F.Garcon, C. Job, E.Kay, J.-P.Wisniewski, J.-M.Ferullo, B.Pelissier et al. (2007) Generation and characterization of soyben and marker-free tobacco plastid transformants over-expressing a bacterial 4- hydroxyphenylpyruvate dioxygenase which provides strong herbicide tolerance. Plant Biotechnol. J. 5: 118–133.

31. Ebel,J. (1986) Phytoalexin synthesis: the biochemical analysis of the induction process. Ann. Rev. Phytopathol. 24: 235–264.

32. Eckert,H., B.L.Vallee, B.J.Schweiger, A.J.Kinney, E.B.Cahoon and T.Clemente (2006) Co-expression of the borage Δ^6 desaturase and the Arabidopsis Δ^{15} desaturase results in high accumulation of steridonic acid in the seeds of transgenic soybean. Planta 224: 1050–1057.

33. El-Shemy, H.A., M.Teraishi, M.M. Khalafalla, T. Katsube-Tanaka, S.Utsumi and M.Ishimoto (2004) Isolation of soybean plants with stable transgene expression by visual selection based on green fluorescent protein. Mol. Breed. 14: 227–238.

34. Ellington,A.A., M.Berhow and K.W.Singletary (2005) Induction of macroautophagy in human colon cancer cells by soybean B-group triterpenoid saponins. Carcinogenesis 26: 159–167.

35. Ellington,A.A., M.A.Berhow and K.W.Singletary (2006) Inhibition of Akt signaling and enhanced ERK1/2 activity and involved in induction of macroautophagy by triterpenoid B-group soyasaponins in colon cancer cells. Carcinogenesis 27: 298–306.

36. Falco,S.C., T.Guida, M.Locke, J.Mauvais, C.Sanders, R.T.Ward and P.Webber (1995) Transgenic canola and soybean seeds with increased lysine. Nat. Biotechnol. 13: 577–582.

37. FAO/WHO (1990) Expert consultation on protein quality evaluation. Food and Agriculture Organization of the United Nations, Rome.

38. Finer,J.J. and A.Nagasawa (1988) Development of an embryogenic suspension culture of soybean (Glycine max Merrill.). Plant Cell Tissue Organ Cult. 15: 125–136.

39. Finer,J.J. and M.D.McMullen (1991) Transformation of soybean via particle bombardment of embryogenic suspension culture tissue. In Vitro Cell. Dev. Biol. 27P: 175–182.

40. Flores,T., O.Karpova, X.Su, P.Zeng, K.Bilyeu, D.A.Sleper, H.T. Nguyen and Z.J.Zhang (2008) Silencing of GmFAD3 gene by siRNA leads to low α-linolenic acids (18: 3) of fad3-mutant phenotype in soybean [Glycine max (Merr.)]. Transgenic Res. 17: 839– 850.

41. Fu,X., L.T.Duc, S.Fontana, B.B.Bong, P.Tinjuangjun, D.Sudhakar, R.M.Twyman, P.Christou and A.Kohli (2000) Linear transgene constructs lacking vector backbone sequences generate low-copynumber transgenic plants with simple integration patterns. Transgenic Res. 9: 11–19.

42. Furutani,N. and S.Hidaka (2004) Efficient production of transgenic soybean using a co-transformation method. Breed. Sci. 54: 91–98.

43. Furutani,N., S.Hidaka, Y.Kosaka, Y.Shizukawa and S.Kanematsu (2006) Coat protein gene-mediated resistance to soybean mosaic virus in transgenic soybean. Breed. Sci. 56: 119–124.

44. Furutani,N., N.Yamagishi, S.Hidaka, Y.Shizukawa, S.Kanematsu and Y.Kosaka (2007) Soybean mosaic virus resistance in transgenic soybean caused by post-transcriptional gene silencing. Breed. Sci. 57: 123–128.

45. Godoy,G., J.R.Steadman, M.B.Dickman and R.Dam (1990) Use of mutants to demonstrate the role of oxalic acid in pathogenicity of Sclerotinia sclerotiorum on Phaseolus vulgaris. Physiol. Mol. Plant Pathol. 37: 179–191.

46. Graham,T.L. (1991) Flavonoid and isoflavonoid distribution in developing soybean seedling tissues and in seed and root exudates. Plant Physiol. 95: 594–603.

47. Hansen,G. and M.S.Wright (1999) Recent advances in the transformation of plants. Trends Plant Sci. 4: 226–231.

48. Hartman,G.L., E.D.West and T.K.Herman (2011) Crops that feed the World 2. Soybean—worldwide production, use, and constraints caused by pathogens and pests. Food Sec. 3: 5–17.

49. Herbers,K. (2003) Vitamin production in transgenic plants. J. Plant Physiol. 160: 821–829.

50. Herman,E.M., R.M.Helm, R.Jung and A.J.Kinney (2003) Genetic modification removes an immunodominant allergen from soybean. Plant Physiol. 132: 36–43.

51. Hinchee,M.A.W., D.V.Conner-Ward, C.A.Newell, R.E.McDonnell, S.J.Sato, C.S.Gasser, D.A.Fischhoff, D.B.Re, R.T.Fraley and R.B.Horsch (1988) Production of transgenic soybean plants using Agrobacterium-mediated DNA transfer. Nat. Biotechnol. 6: 915– 922.

52. Hoppe,P.P. and G.Krennrich (2000) Bioavailability and potency of natural-source and all-racemic α-tocopherol in the human: a dispute. Eur. J. Nutr. 39: 183–193.

53. Horsch,R.B., J.E.Fry, N.L.Hoffmann, D.Eichholtz, S.G.Rogers and R.T.Fraley (1985) A simple and general method for transferring genes into plants. Science 227: 1229–1231.

54. Inaba,Y., J.E.Brotherton, A.Ulanov and J.M.Widholm (2007) Expression of a feedback insensitive anthranilate synthase gene from tobacco increases free tryptophan in soybean plants. Plant Cell Rep. 26: 1763–1771.

55. Indrasumunar,A., I.Searle, M.-H.Lin, A.Kereszt, A.Men, B.J.Carroll and P.M.Gresshoff (2011) Nodulation factor receptor kinase 1α controls nodule organ number in soybean (Glycine max L. Merr). Plant J. 65: 39–50.

56. Ishimoto,M., S.M.Rahman, M.S.Hanafy, M.M.Khalafalla, H.A. El-Shemy, Y.Nakamoto, Y.Kita, K.Takanashi, F.Matsuda, Y. Murano et al. (2010) Evaluation of amino acid content and nutritional quality of transgenic soybean seeds with high-level tryptophan accumulation. Mol. Breed. 25: 313–326.

57. Kajikawa,M., K.Matsui, M.Ochiai, Y.Tanaka, Y.Kita, M.Ishimoto, Y.Kohzu, S.Shoji, K.T.Yamato, K.Ohyama et al. (2008) Production of arachidonic and eicosapentaenoic acids in plants using bryophyte fatty acid Δ6-desaturase, Δ6-elongase, and Δ5-desaturase genes. Biosci. Biotechnol. Biochem. 72: 435–444.

58. Kasai,M. and A.Kanazawa (2012) RNA silencing as a tool to uncover gene function and engineer novel traits in soybean. Breed. Sci. 61: 468–479.

59. Kereszt, A., D. Li, A. Indrasumunar, C.D.T. Nguyen, S. Nontachaiyapoom, M. Kinkema and P.M. Gresshoff (2007) Agrobacterium rhizogenes-mediated transformation of soybean to study root biology. Nat. Protoc. 2: 948–952.

60. Khalafalla,M.M., S.M.Rahman, H.A.El-Shemy, Y.Nakamoto, K. Wakasa and M.Ishimoto (2005) Optimization of particle bombardment conditions by monitoring of transient sGFP(S65T) expression in transformed soybean. Breed. Sci. 55: 257–263.

61. Kim,W.-S. and H.B. Krishnan (2004) Expression of an 11kDa methionine-rich delta-zein in transgenic soybean results in the formation of two types of novel protein bodies in transitional cells situated between the vascular tissue and storage parenchyma cells. Plant Biotechnol. J. 2: 199–210.

62. Kim,Y.H., Y.Y.Lee, Y.H.Kim, M.S.Choi, K.H.Jeong, S.K.Lee, M.J. Seo, H.T.Yun, C.K.Lee, W.H.Kim et al. (2011) Antioxidant activity and inhibition of lipid peroxidation in germinating seeds of transgenic soybean expressing OsHGGT. J. Agric. Food Chem. 59: 584–591.

63. Kim,Y.J., H.Y.Seo, T.I.Park, S.H.Baek, W.C.Shin, H.S.Kim, J.G. Kim, Y.E.Choi and S.J.Yun (2005) Enhanced biosynthesis of α- tocopherol in transgenic soybean by introducing γ-TMT gene. J. Plant Biotechnol. 7: 203–209.

64. Kinney,A.J., R.Jung and E.M.Herman (2001) Cosuppression of the α subunits of β-conglycinin in transgenic soybean seeds induces the formation of endoplasmic reticulum-derived protein bodies. Plant Cell 13: 1165–1178.

65. Kita,Y., K.Nishizawa, M.Takahashi, M.Kitayama and M.Ishimoto (2007) Genetic improvement of the somatic embryogenesis and regeneration in soybean and transformation of the improved breeding lines. Plant Cell Rep. 26: 439–447.

66. Kita,Y., M.S.Hanafy, M.Deguchi, H.Hasegawa, T.Terakawa, K. Kitamura and M.Ishimoto (2009) Generation and characterization of herbicide-resistant soybean plants expressing novel phosphinothricin N-acetyltransferase genes. Breed. Sci. 59: 245–251.

67. Kita,Y., Y.Nakamoto, M.Takahashi, K.Kitamura, K.Wakasa and M.Ishimoto (2010) Manipulation of amino acid composition in soybean seeds by the combination of deregulated tryptophan biosynthesis and storage protein deficiency. Plant Cell Rep. 29: 87– 95.

68. Kudou,S., M.Tonomura, C.Tsukamoto, M.Shimoyamada, T.Uchida and K.Okubo (1992) Isolation and structural elucidation of the major genuine soybean saponin. Biosci. Biotech. Biochem. 56: 142– 143.

69. Lardizabal,K., R.Effertz, C.Levering, J.Mai, M.C.Pedroso, T.Jury, E.Aasen, K.Gruys and K.Bennett (2008) Expression of Umbelopsis ramanniana DGAT2A in seed increases oil in soybean. Plant Physiol. 148: 89–96.

70. Lazzeri,P.A., D.F.Hildebrand and G.B.Collins (1985) A procedure for plant regeneration from immature cotyledon tissue of soybean. Plant Mol. Biol. Rep. 3: 160–167.

71. Lazzeri,P.A., D.F.Hildebrand and G.B.Collins (1987) Soybean somatic embrygenesis: Effects of hormones and culture munipulations. Plant Cell Tissue Organ Cult. 10: 197–208.

72. Li,R., K.Yu, T.Hatanaka and D.F.Hildebrand (2010) Vernonia DGATs increase accumulation of epoxy fatty acids in oil. Plant Biotechnol. J. 8: 184–195.

73. Li,Z., S.Meyer, J.S.Essig, Y.Liu, M.A.Schapaugh, S.Muthukrishnan, B.E.Hainline and H.N.Trick (2005) High-level expression of maize γ-zein protein in transgenic soybean (Glycine max). Mol. Breed. 16: 11–20.

74. Li,Z., B.P.Moon, A.Xing, Z.-B.Liu, R.P.McCardell, H.G.Damude and S.C.Falco (2010) Stacking multiple transgenes at a selected genomic site via repeated recombinase-mediated DNA cassette exchanges. Plant Physiol. 154: 622–631.

75. Liu,B., S.Watanabe, T.Uchiyama, F.Kong, A.Kanazawa, Z.Xia, A.Nagamatsu, M.Arai, T.Yamada, K.Kitamura et al. (2010) The soybean stem growth habit gene Dt1 is an ortholog of Arabidopsis TERMINAL FLOWER1. Plant Physiol. 153: 198–210.

76. Liu, S.-J., Z.-M.Wei and J.-Q.Huang (2008) The effect of cocultivation and selection parameters on Agrobacterium-mediated transformation of Chinese soybean varieties. Plant Cell Rep. 27: 489–498.

77. Macrae,T.C., M.E.Baur, D.J.Boethel, B.J.Fitzpatrick, A.-G.Gao, J.C.Gamundi, L.A.Harrison, V.T.Kabuye, R.M.Mcpherson, J.A.

78. Miklos et al. (2005) Laboratory and field evaluations of transgenic soybean exhibiting high-dose expression of a synthetic Bacillus thuringiensis cry1A gene for control of lepidoptera. J. Econ. Entomol. 98: 577–587.

79. Mathieu,M., E.K.Winters, F.Kong, J.Wan, S.Wang, H.Eckert, D. Luth, M.Paz, C.Donovan, Z.Zhang et al. (2009) Establishment of a soybean (Glycine max Merr. L) transposon-based mutagenesis repository. Planta 229: 279–289.

80. Maughan,P.J., R.Philip, M.-J.Cho, J.M.Widholm and L.O.Vodkin (1999) Biolistic transformation, expression, and inheritance of bovine β-casein in soybean (Glycine max). In Vitro Cell. Dev. Biol. Plant 35: 344–349.

81. McCabe,D.E., W.F.Swain, B.J.Martinell and P.Christou (1988) Stable transformation of soybean (Glycine max) by particle acceleration. Nat. Biotechnol. 6: 923–926.

82. McGaughey,W.H. and M.E.Whalon (1992) Managing insect resistance to Bacillus thuringiensis toxins. Science 258: 1451–1455.

83. McLean,M.D., G.J. Hoover, B.Bancroft, A.Makhmoudova, S.M. Clark, T.Welacky, D.H.Simmonds and B.J.Shelp (2007) Identification of the full-length Hs1^{pro-1} coding sequence and preliminary evaluation of soybean cyst nematode resistance in soybean transformed with Hs1^{pro-1} cDNA. Can. J. Bot. 85: 437–441.

84. Meurer,C.A., R.D.Dinkins, C.T.Redmond, K.P.McAllister, D.T. Tucker, D.R.Walker, W.A.Parrott, H.N.Trick, J.S.Essig, H.M. Frantz et al. (2001) Embryogenic response of multiple soybean [Glycine max (L.) Merrill] cultivars across three locations. In Vitro Cell. Dev. Biol. 37: 62–67.

85. Miklos,J.A., M.F.Alibhai, S.A.Bledig, D.C.Connor-Ward, A.-G.Gao, B.A. Holmes, K.H. Kolacz, V.T. Kabuye, T.C. MacRae, M.S. Paradise et al. (2007) Characterization of soybean exhibiting high expression of a synthetic Bacillus thuringiensis cry1A transgene that confers a high degree of resistance to lepidopteran pests. Crop Sci. 47: 148–157.

86. Nishizawa,K., Y.Kita, M.Kitayama and M.Ishimoto (2006) A red fluorescent protein, DsRed2, as a visual reporter for transient expression and stable transformation in soybean. Plant Cell Rep. 25: 1355–1361.

87. Nishizawa,K., A.Kita, C.Doi, Y.Yamada, K.Ohinata, M.Yoshikawa and M.Ishimoto (2008) Accumulation of the bioactive peptides, novokinin, LPYPR and rubiscolin, in seeds of genetically modified soybean. Biosci. Biotechnol. Biochem. 72: 3301–3305.

88. Nishizawa,K. and M.Ishimoto (2009) Maturation of somatic embryos as a model for soybean seed development. Plant Biotechnol. 26: 543–550.

89. Nishizawa,K., K.Takagi, M.Teraishi, A.Kita and M.Ishimoto (2010) Application of somatic embryos to rapid and reliable analysis of soybean seed components by RNA interference-mediated gene silencing. Plant Biotechnol. 27: 409–420.

90. Nunes, A.C.S., G.R. Vianna, F. Cuneo, J. Amaya-Farfán, G. de Capdeville, E.L.Rech and F.J.L.Aragão (2006) RNAi-mediated silencing of the myo-inositol-1-phosphate synthase gene (GmMIPS1) in transgenic soybean inhibited seed development and reduced phytate content. Planta 224: 125–132.

91. Ogawa,T., M.Samoto and K.Takahashi (2000) Soybean allergens and hypoallergenic soybean products. J. Nutr. Sci. Vitaminol. 46: 271– 279.

92. Olhoft,P.M., K.Lin, J.Galbraith, N.C.Nielsen and D.A.Somers (2001) The role of thiol compounds in increasing Agrobacteriummediated transformation of soybean cotyledonary-node cells. Plant Cell Rep. 20: 731–737.

93. Olhoft, P.M. and D.A. Somers (2001) L-Cysteine increases Agrobacterium-mediated T-DNA delivery into soybean cotyledonarynode cells. Plant Cell Rep. 20: 706–711.

94. Olhoft,P.M., L.E.Flagel, C.M.Donovan and D.A.Somers (2003) Efficient soybean transformation using hygromycin B selection in the cotyledonary-node method. Planta 216: 723–735.

95. Olhoft,P.M. and D.A.Somers (2007) Soybean. In: Pua,E.C. and M.R. Davey (eds.) Biotechnology in Agriculture and Forestry. Vol.61, Springer, Berlin, pp. 3–27

96. Owens,L.D. and D.E.Cress (1985) Genotypic variability of soybean response to Agrobacterium strains harboring the Ti or Ri plasmids. Plant Physiol. 77: 87–94

97. Padgette,S.R., K.H.Kolacz, X.Delannay, D.Re, B.J.LaVallee, C.N. Tinius, W.K.Rhodes, Y.I.Otero, G.F.Barry, D.A.Eichholtz et al. (1995) Development, identification, and characterization of a glyphosate-tolerant soybean line. Crop Sci. 35: 1451–1461.

98. Parrott,W.A., G.Dryde, S.Vogt, D.F.Hildebrand, G.B.Collins and E.G.Williams (1988) Optimization of somatic embryogenesis and embryo germination in soybean. In Vitro Cell. Dev. Biol. 24: 817– 820.

99. Parrott,W.A., L.M.Hoffman, D.F.Hildebrand, E.G.Williams and G.B. Collins (1989a) Recovery of primary transformants of soybean. Plant Cell Rep. 7: 615–617.

100. Parrott,W.A., E.G.Williams, D.F.Hildebrand and G.B.Collins (1989b) Effect of genotype on somatic embryogenesis from immature cotyledons of soybean. Plant Cell Tissue Organ Cult. 16: 15–21.

101. Parrott,W.A., J.N.All, M.J.Adang, M.A.Bailey, H.R.Boerma and C.N. Stewart (1994) Recovery and evaluation of soybean plants transgenic for a Bacillus thuringiensis var. Kurstaki insecticidal gene. In Vitro Cell. Dev. Biol. Plant 30: 144–149.

102. Paz,M.M., H.Shou, Z.Guo, Z.Zhang, A.K.Banerjee and K.Wang (2004) Assessment of conditions affecting Agrobacterium-mediated soybean transformation using the cotyledonary node explant. Euphytica 136: 167–179.

103. Paz,M.M., J.C.Martinez, A.B.Kalvig, T.M.Fonger and K.Wang (2006) Improved cotyledonary node method using an alternative explant derived from mature seed for efficient Agrobacteriummediated soybean transformation. Plant Cell Rep. 25: 206–213

104. Piller,K.J., T.E.Clemente, S.M.Jun, C.C.Petty, S.Sato, D.W.Pascual and K.L.Bost (2005) Expression and immunogenicity of an Escherichia coli K99 fimbriae subunit antigen in soybean. Planta 222: 6–18.

105. Qi,Q., J.Huang, J.Crowley, L.Ruschke, B.S.Goldman, L.Wen and W.D.Rapp (2011) Metabolically engineered soybean seed with enhanced threonine levels: biochemical characterization and seedspecific expression of lysine-insensitive variants of aspartate kinases from the enteric bacterium Xenorhabdus bovienii. Plant Biotechnol. J. 9: 193–204.

106. Ranch,J.P., L.Oglesby and A.C.Zielinski (1985) Plant regeneration from embryo-derived tissue cultures of soybeans. In Vitro Cell. Dev. Biol. 21: 653–658.

107. Rao,S.S. and D.Hildebrand (2009) Changes in oil content of transgenic soybeans expressing the yeast SLC1 gene. Lipids 44: 945–951.

108. Rech,E.L., G.R.Vianna and F.J.Aragão (2008) High-efficiency transformation by biolistics of soybean, common bean and cotton transgenic plants. Nat. Protoc. 3: 410–418.

109. Reddy,M.S.S., S.A.Ghabrial, C.T.Redmond, R.D.Dinkins and G.B. Collins (2001) Resistance to Bean pod mottle virus in transgenic soybean lines expressing the capsid polyprotein. Phytopathology 91: 831–838.

110. Reddy,M.S.S., R.D.Dinkins and G.B.Collins (2003) Gene silencing in transgenic soybean plants transformed via particle bombardment. Plant Cell Rep. 21: 676–683.

111. Rivera-Vargas,L.I., A.F.Schmitthenner and T.L.Graham (1993) Soybean flavonoid effects on and metabolism by Phytophthora sojae. Phytochemistry 32: 851–857.

112. Ross,J.P. (1969) Effect of time and sequence of inoculation of soybeans with soybean mosaic and bean pod mottle viruses on yields and seed characters. Phytopathology 59: 1404–1408.

113. Sato,H., T.Yamada, Y.Kita, M.Ishimoto and K.Kitamura (2007) Production of transgenic plants and their early seed set in Japanese soybean variety, Kariyutaka. Plant Biotechnol. 24: 533–536.

114. Sato,S., C.Newell, K.Kolacz, L.Tredo, J.Finer and M.Hinchee (1993) Stable transformation via particle bombardment in two different soybean regeneration systems. Plant Cell Rep. 12: 408–413.

115. Sato,S., A.Xing, X.Ye, B.Schweiger, A.Kinney, G.Graef and T. Clemente (2004) Production of γ-linolenic acid and stearidonic acid in seeds of marker-free transgenic soybean. Crop Sci. 44: 646–652.

116. Schmutz,J., S.B.Cannon, J.Schlueter, J.Ma, T.Mitros, W.Nelson, D.L. Hyten, Q.Song, J.J.Thelen, J.Cheng et al. (2010) Genome sequence of the palaeopolyploid soybean. Nature 463: 178–183.

117. Setchell,K.D. (1998) Phytoestrogens: the biochemistry, physiology, and implications for human health of soy isoflavones. Am. J. Clin. Nutr. 68: 1333S–1346S.

118. Shi,J., H.Wang, K.Schellin, B.Li, M.Faller, J.M.Stoop, R.B.Meeley, D.S.Ertl, J.P.Ranch and K.Glassman (2007) Embryo-specific silencing of a transporter reduces phytic acid content of maize and soybean seeds. Nat. Biotechnol. 25: 930–937.

119. Shiraiwa,M., K.Harada and K.Okubo (1991a) Composition and structure of "group B saponin" in soybean seed. Agric. Biol. Chem. 55: 911–917.

120. Shiraiwa,M., S.Kudo, M.Shimoyamada, K.Harada and K.Okubo (1991b) Composition and structure of "group A saponin" in soybean seed. Agric. Biol. Chem. 55: 315–322.

121. Shoemaker,R.C.and E.G.Hammond (1988) Fatty acid composition of soybean (Glycine max (L.) Merr.) somatic embryos. In Vitro Cell. Dev. Biol. 24: 829–832.

122. Somers,D.A., D.A.Samac and P.M.Olhoft (2003) Recent advances in legume transformation. Plant Physiol. 131: 892–899.

123. Stacey,G., L.Vodkin, W.A.Parrott and R.C.Shoemaker (2004) National Science Foundation-sponsored workshop report. Draft plan for soybean genomics. Plant Physiol. 135: 59–70.

124. Staswick,P.E., Z.Zhang, T.E.Clemente and J.E.Specht (2001) Efficient down-regulation of the major vegetative strage protein genes in transgenic soybean does not compromise plant productivity. Plant Physiol. 127: 1819–1826.

125. Stewart,C.N.J., M.J.Adang, J.N.All, H.R.Boerma, G.Cardineau, D. Tucker and W.A.Parrott (1996) Genetic transformation, recovery, and characterization of fertile soybean transgenic for a synthetic Bacillus thuringiensis cryIAc gene. Plant Physiol. 112: 121–129.

126. Subramanian,S., M.Y.Graham, O.Yu and T.L.Graham (2005) RNA interference of soybean isoflavone synthase genes leads to silencing in tissues distal to the transformation site and to enhanced susceptibility to Phytophthora sojae. Plant Physiol. 137: 1345–1353.

127. Sugano,M. (2005) Soy in Health and Disease Prevention, CRC Press, New York, USA.

128. Tabashnik,B.E. (1994) Evolution of resistance to Bacillus thuringiensis. Annu. Rev. Entomol. 39: 47–79.

129. Tada,Y., S.Utsumi and F.Takaiwa (2003) Foreign gene products can be enhanced by introduction into low storage protein mutants. Plant Biotechnol. J. 1: 411–422.

130. Takagi,K., K.Nishizawa, A.Hirose, A.Kita and M.Ishimoto (2011) Manipulation of saponin biosynthesis by RNA interferencemediated silencing of β-amyrin synthase gene expression in soybean. Plant Cell Rep. 30: 1835–1846.

131. Takahashi,M., Y.Uematsu, K.Kashiwaba, K.Yagasaki, M.Hajika, R.Matsunaga, K.Komatsu and M.Ishimoto (2003) Accumulation of high levels of free amino acids in soybean seeds through integration of mutations conferring seed protein deficiency. Planta 217: 577–586.

132. Tavva,V.S., Y.H.Kim, I.A.Kagan, R.D.Dinkins, K.H.Kim and G.B. Collins (2007) Increased α-tocopherol content in soybean seed overexpressing the Perilla frutescens γ-tocopherol methyltransferase gene. Plant Cell Rep. 26: 61–70.

133. Tomlin,E.S., S.R.Branch, D.Chamberlain, H.Gabe, M.S.Wright and C.N.Stewart (2002) Screening of soybean, Glycine max (L.) Merrill, lines for somatic embryo induction and maturation capability from immature cotyledons. In Vitro Cell. Dev. Biol. Plant 38: 543–548.

134. Topping,D.L., G.B.Storer, G.D.Calvert, R.J.Illman, D.G.Oakenfull and R.A.Weller (1980) Effect of dietary saponins on fecal bile acids and neutral sterols, plasma lipids, and lipoprotein turnover in the pig. Am. J. Clin. Nutr. 33: 783–786.

135. Tougou,M., N.Furutani, N.Yamagishi, Y.Shizukawa, Y.Takahata and S.Hidaka (2006) Development of resistant transgenic soybeans with inverted repeat-coat protein genes of soybean dwarf virus. Plant Cell Rep. 25: 1213–1218.

136. Tougou,M., N.Yamagishi, N.Furutani, Y.Shizukawa, Y.Takahata and S.Hidaka (2007) Soybean dwarf virus-resistant transgenic soybeans with the sense coat protein gene. Plant Cell Rep. 26: 1967– 1975.

137. Trick,H.N. and J.J.Finer (1998) Sonication-assisted Agrobacteriummediated transformation of soybean [Glycine max (L.) Merrill] embryogenic suspension culture tissue. Plant Cell Rep. 17: 482–488.

138. Umezawa,T., T.Sakurai, Y.Totoki, A.Toyoda, M.Seki, A.Ishiwata, K.Akiyama, A.Kurotani, T.Yoshida, K.Mochida et al. (2008) Sequencing and analysis of approximately 40000 soybean cDNA clones from a full-length-enriched cDNA library. DNA Res. 15: 333–346.

139. Valente,M.A.S., J.A.Q.A. Faria, J.R.L. Soares-Ramos, P.A.B.Reis, G.L.Pinheiro, N.D.Piovesan, A.T.Morais, C.C.Menezes, M.A.O. Cano and L.G.Fietto et al. (2009) The ER luminal binding protein (BiP) mediates an increase in drought tolerance in soybean and delays drought-induced leaf senescence in soybean and tobacco. J. Exp. Bot. 60: 533–546.

140. VanEenennaam,A.L., K.Lincoln, T.P.Durrett, H.E.Valentin, C.K. Shewmaker, G.M.Thorne, J.Jiang, S.R.Baszis, C.K.Levering and E.D.Aasen (2003) Engineering vitamin E content: from Arabidopsis mutant to soy oil. Plant Cell 15: 3007–3019.

141. vanRhijn,P. and J.Vanderleyden (1995) The Rhizobium-plant symbiosis. Microbiol. Mol. Biol. Rev. 59: 124–142.

142. Vasconcelos,M., H.Eckert, V.Arahama, G.Graef, M.A.Grusak and T.Clemente (2006) Molecular and phenotypic characterization of transgenic soybean expressing the Arabidopsis ferric chelate reductase gene, FRO2. Planta 224: 1116–1128.

143. Walker,D., H.R.Boerma, J.All and W.Parrott (2002) Combining cry1Ac with QTL alleles from PI 229358 to improve soybean resistance to lepidopteran pests. Mol. Breed. 9: 43–51.

144. Walker,D.R., J.N.All, R.M.McPherson, H.R.Boerma and W.A.Parrott (2000) Field evaluation of soybean engineered with a synthetic cry1Ac transgene for resistance to corn earworm, soybean looper, velvetbean caterpillar (Lepidoptera: Noctuidae), and lesser cornstalk borer (Lepidoptera: Pyralidae). J. Econ. Entomol. 93: 613–

145. Wang,G. and Y.Xu (2008) Hypocotyl-based Agrobacterium-mediated transformation of soybean (Glycine max) and application for RNA interference. Plant Cell Rep. 27: 1177–1184.

146. Wang,X.Y., A.L.Eggenberger, F.W.NutterJr. and J.H.Hill (2001) Pathogen-derived transgenic resistance to soybean mosaic virus in soybean. Mol. Breed. 8: 119–127.

147. Xue,R.G., H.F.Xie and B.Zhang (2006) A multi-needle-assisted transformation of soybean cotyledonary node cells. Biotechnol. Lett. 28: 1551–1557

148. Yadav,N.S. (1996) Genetic modification of soybean oil quality. In: Verma,D.P.S. and R.C. Shoemaker (eds.) Soybean: Genetics, Molecular Biology and Biotechnology, CAB INTERNATIONAL, USA, pp. 165–188.

149. Yamada,T., S.Watanabe, M.Arai, K.Harada and K.Kitamura (2010) Cotyledonary node pre-wounding with a micro-brush increased frequency of Agrobacterium-mediated transformation in soybean. Plant Biotechnol. 27: 217–220.

150. Yamada,Y., K.Nishizawa, M.Yokoo, H.Zhao, K.Onishi, M.Teraishi, S.Utsumi, M.Ishimoto and M.Yoshikawa (2008) Anti-hypertensive activity of genetically modified soybean seeds accumulating novokinin. Peptides 29: 331–337.

151. Yang,S.M., F.Tang, M.Q.Gao, H.B.Krishnan and H.Y.Zhu (2010) R gene-controlled host specificity in the legume-rhizobia symbiosis. Proc. Natl. Acad. Sci. USA 107: 18735–18740.

152. Young,V.R. (1991) Soy protein in relation to human protein and amino acid nutrition. J. Am. Diet. Assoc. 91: 828–835.

153. Yu,O., J.Shi, A.O.Hession, C.A.Maxwell, B.McGonigle and J.T. Odell (2003) Metabolic engineering to increase isoflavone biosynthesis in soybean seed. Phytochemistry 63: 753–763.

154. Zeng,P., D.A.Vadnais, Z.Zhang and J.C.Polacco (2004) Refined glufosinate selection in Agrobacterium-mediated transformation of soybean [Glycine max (L.) Merrill]. Plant Cell Rep. 22: 478–482.

155. Zernova,O.V., A.V.Lygin, J.M.Widholm and V.V.Lozovaya (2009) Modification of isoflavones in soybean seeds via expression of multiple phenolic biosynthetic genes. Plant Physiol. Biochem. 47: 769–777.

156. Zhang,Z., A.Xing, P.Staswick and T.E.Clemente (1999) The use of glufosinate as a selective agent in Agrobacterium-mediated transformation of soybean. Plant Cell Tissue Organ Cult. 56: 37–46.

157. Zhu,S., D.R.Walker, H.R.Boerma, J.All and W.A.Parrott (2008) Effects of defoliating insect resistance QTLs and a cry1Ac transgene in soybean near-isogenic lines. Theor. Appl. Genet. 116: 455–463.

CHAPTER 15

RNA INTERFERENCE FOR FUNCTIONAL GENOMICS AND IMPROVEMENT OF COTTON (GOSSYPIUM SP.)

Ibrokhim Y. Abdurakhmonov,[1,] Mirzakamol S. Ayubov[1] Khurshida A. Ubaydullaeva[1] Zabardast T. Buriev[1] Shukhrat E. Shermatov[1] Haydarali S. Ruziboev,[1] Umid M. Shapulatov,[1,2] Sukumar Saha[3] Mauricio Ulloa[4] John Z. Yu[5] Richard G. Percy[5] Eric J. Devor[6] Govind C. Sharma[7] Venkateswara R. Sripathi[7] Siva P. Kumpatla[8] Alexander van der Krol[2]Hake D. Kater[9] Khakimdjan Khamidov[1] Shavkat I. Salikhov[1] Johnie N. Jenkins[3] Abdusattor Abdukarimov[1] and Alan E. Pepper[10]

[1] Center of Genomics and Bioinformatics, Structural and Functional Genomics, Academy of Sciences the Republic of Uzbekistan, Ministry of Agriculture and Water Resources the Republic of Uzbekistan and "Uzpakhtasanoat" Association, Kibray, Uzbekistan,

[2] Laboratory of Plant Physiology, Wageningen University, Wageningen, Netherlands,

[3] Crop Science Research Laboratory, United States Department of Agriculture – Agricultural Research Service, Starkville, MS, USA,

[4] Plant Stress and Germplasm Development Research, United States Department of Agriculture – Agricultural Research Service, Lubbock, TX, USA

[5] Crop Germplasm Research Unit, United States Department of Agriculture – Agricultural Research Service, College Station, TX, USA

[6] Department of Obstetrics and Gynecology, University of Iowa Carver College of Medicine, Iowa City, IA, USA

[7] Department of Biological and Environmental Sciences, Alabama A&M University, Normal, AL, USA

[8] Dow AgroSciences LLC, Indianapolis, IN, USA

[9] Agricultural and Environmental Research, Cary, NC, USA

[10] Department of Biology, Texas A&M University, Colleges Station, TX, USA

ABSTRACT

RNA interference (RNAi), is a powerful new technology in the discovery of genetic sequence functions, and has become a valuable tool for functional genomics of cotton (*Gossypium* sp.). The rapid adoption of RNAi has replaced previous antisense technology. RNAi has aided in the discovery of function and biological roles of many key cotton genes involved in fiber development, fertility and somatic embryogenesis, resistance to important biotic and abiotic stresses, and oil and seed quality improvements as well as the key agronomic traits including yield and maturity. Here, we have comparatively reviewed seminal research efforts in previously used antisense approaches and currently applied breakthrough RNAi studies in cotton, analyzing developed RNAi methodologies, achievements, limitations, and future needs in functional characterizations of cotton genes. We also highlighted needed efforts in the development of RNAi-based cotton cultivars, and their safety and risk assessment, small and large-scale field trials, and commercialization.

INTRODUCTION

Today, the world demands and uses more cotton (*Gossypium* sp.) fiber than ever before. Global business revenue stimulated by cotton production is estimated at US$500 billion per year (Chen et al., 2007;Abdurakhmonov et al., 2012a) making cotton the most economically important value-added crop. Cotton is grown worldwide in more than 80 countries (Sunilkumar et al., 2006; Abdurakhmonov et al., 2012a) on 32–34 million

Citation: Ibrokhim Y. Abdurakhmonov, Mirzakamol S. Ayubov, Khurshida A. Ubaydullaeva, Zabardast T. Buriev, Shukhrat E. Shermatov, Haydarali S. Ruziboev, Umid M. Shapulatov, Sukumar Saha, Mauricio Ulloa, John Z. Yu, Richard G. Percy, Eric J. Devor, Govind C. Sharma, Venkateswara R. Sripathi, Siva P. Kumpatla, Alexander van der Krol, Hake D. Kater, Khakimdjan Khamidov, Shavkat I. Salikhov, Johnie N. Jenkins, Abdusattor Abdukarimov and Alan E. Pepper, RNA Interference for Functional Genomics and Improvement of Cotton (Gossypium sp.), doi: 10.3389/fpls.2016.00202.

hectares with annual total production of 25.65 million metric tons (MMTs; United States Department of Agriculture [USDA] Report, 2013). Uses of cotton fiber vary widely (Chen et al., 2007;Zhang et al., 2008; Campbell et al., 2010) from a large clothing and household usage, to numerous industrial items that account for many thousands of bales. In addition to the cotton fiber, cottonseed products (meal and hulls) are used for livestock feed, and industrial lubricants (Campbell et al., 2010) as well as a fertilizer. Cottonseed oil is used as an ingredient in food products as well as a premium cooking oil and salad dressing. Cotton stalks and leaves are used as organic matter to enrich soil and cotton stalks have been widely used as a firewood for primary energy in cooking process in rural areas as well as conversion to biogas or composting to bio-fertilizers (Federation of Indian Chambers of Commerce and Industry [FICCI] Report, 2012).

Genome of allopolyploid cotton (*Gossypium* sp.) is very complex and poorly studied. To date, cotton lags behind many other crops in studies and gains related to genomics and genetics as well as applications of marker-assisted selection (MAS) due to existence of low level of molecular polymorphisms (Abdurakhmonov et al., 2008b, 2009, 2012a) among cultivar germplasm caused by a "genetic bottleneck" during cotton domestication (Iqbal et al., 2001). The complexity of cultivated allotetraploid cotton genomes is due to the high levels of gene redundancy because of several genome duplications events before and after allopolyploidization (Adams et al., 2003). Allopolyploidization had induced a natural gene silencing, organ specific, and homoeologously biased expression of genes regulated at developmental and epigenetic level (Adams et al., 2003, 2004; Adams and Wendel, 2005; Keyte et al., 2006; Chaudhary et al., 2009). Recent studies reported that a whole genome duplication (WGD) of ancestor-like diploid cotton *Gossypium raimondii* (D5) had occurred approximately 60 million years ago (MYA) with five- to six-fold ploidy increase (Renny-Byfield et al., 2014). The gene expression analyses of this first sequenced cotton genome demonstrated complex, tissue-specific differential gene expression patterns and the diversification of gene expression level before divergence (Renny-Byfield et al., 2014) suggesting that even more complexities may be evident among cultivated allotetraploid genomes.

The allotetraploidization of A and D genome diploids occurred about 1.5-2 MYA that resulted in five distinct genomes (Adams et al., 2003; Chen et al., 2007; Zhang et al., 2008), out of which there are two allotetraploid species of cultivated cotton *Gossypium hirsutum* (so called Upland cotton) and *Gossypium barbadense* [so called Extra Long Staple (ELS) or Pima/Sea Island cotton; Abdurakhmonov et al., 2012a]. Being grown in more than 90% of world cotton growing area (Campbell et al., 2010), Upland cotton cultivars are demanded for their combination of exceptional yield, early maturity and other agronomic traits as well as moderately good fiber properties. In contrast, Pima or ELS cotton is grown in less than 5% of worldwide cotton growing area, and it is well-known for its exceptionally good fiber qualities (Liu et al., 2015a) with relatively lower yield and less desirable agronomic traits such as longer growing time, susceptibility to various diseases, and requirement of more water. One primary objective of worldwide cotton breeding effort has been to transfer ELS fiber quality into Upland cotton while keeping their early maturity, resistance, and productivity of Upland genotypes (Abdurakhmonov et al., 2012b, 2014). However, traditional breeding successes have been minimal over past century of worldwide breeding, which suffered from linkage drag and distorted segregation that occurred in interspecific hybrid progenies from Upland and Pima sexual crosses (Saha et al., 2012). Besides, simultaneous improvement of fiber quality, early flowering, early maturity, and productivity in Upland cotton (*G. hirsutum*) is a very difficult task using conventional breeding methods due to existence of negative correlations between major fiber quality and yield or maturity traits (Miller and Rawlings, 1967;Zhang et al., 2011; Abdurakhmonov et al., 2014). This challenge perhaps is even more severe under above-mentioned complexities of cotton genome, including higher rates of transgressive, differential and novel gene expression patterns, and homoeologous gene silencing (Yoo et al., 2013) that make it difficult to achieve a breeding goal.

Despite this complexity, there is a need to develop Upland cotton cultivars with increased yield, early maturity, and disease and pest resistance while producing longer and stronger fibers to be competitive with synthetic fibers in the global market. Cotton must overcome challenges from manmade fiber and become sustainable and environmentally "friendly" with reduced costs and hazards associated with chemical control methods. Novel genes must be discovered and deployed to combat salt, drought, and heat stresses as well as ever-evolving pathogens and pests such as *Verticillium/Fusarium* fungi, nematodes, and viruses (Hake, 2012;Abdurakhmonov, 2013). This prompted the cotton research community to develop an 'innovative new generation crop technology' to address this largely eluded and fundamentally longstanding challenge in worldwide cotton improvement programs. One of such innovative approaches is the expanded utilization of gene silencing techniques such as RNA interference (RNAi). RNAi helps to identify functions of genes with agricultural importance, and thus, address the cotton farming challenges via creating "biotechnology-derived" cotton cultivars ("biotech cotton") with the suppression of undesirable genes but improved expression of desired trait(s) of interest. A survey of literature revealed that to date, RNAi has been applied for the functional studies of many agronomically, biologically, and physiologically important cotton (*Gossypium* sp.) genes related to the cotton fiber development; early maturity and flowering; increased yield potential; fertility and embryogenesis; viral, fungal, and insect pest resistance; tolerance to various abiotic stresses; and cottonseed and oil quality improvements.

Application of RNAi is rapidly expanding due to the advances made in cotton genomics. Efforts should be expanded further with the completed genome sequencing efforts of two ancestor-like diploid D_5 (Paterson et al., 2012; Wang et al., 2012) and A_2 (Li et al., 2014) cottons as well as two cultivated allotetraploid Upland [*G. hirsutum* L. $(AD)_1$ cv. Texas Marker-1 (TM-1); Li et al., 2015a; Zhang et al., 2015] and ELS [*G.barbadense* $(AD)_2$ cv. Xinhai21; Liu et al., 2015a] cotton genotypes. Searching PubMed (http://www.ncbi.nlm.nih.gov/pubmed) database with the unquoted keywords of "*Gossypium gene silencing*" and "*Gossypium RNA interference*" found almost 70 refereed journal publications that experimentally cover the above-mentioned RNAi-related topics (Figure 1). Here, we highlighted these efforts in detail, discussing successes, existing methods, cultivar development, field trials, safety, risks, limitations, and future perspectives of RNAi in cotton research and breeding.

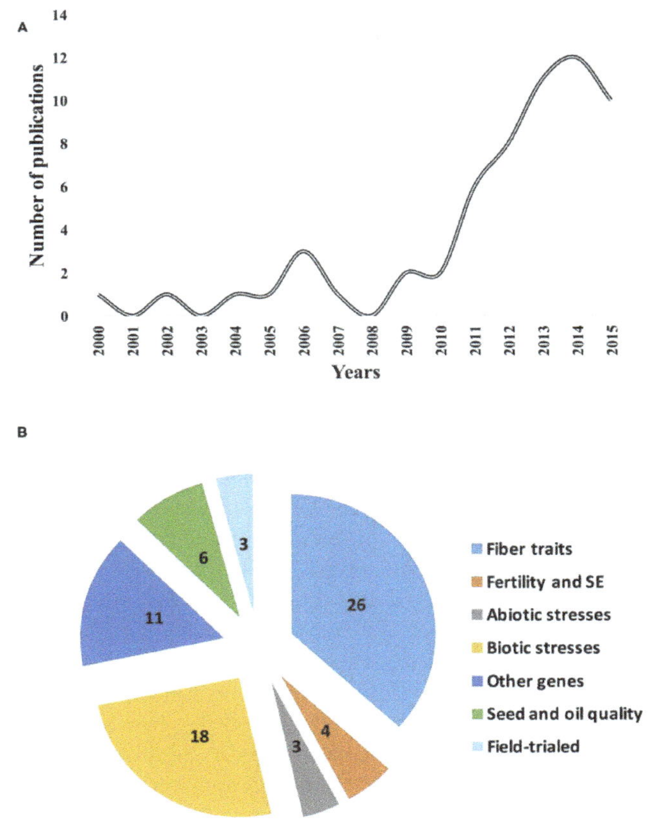

Figure 1:

Dynamics of publications and thematic coverage of RNA interference (RNAi) studies in *Gossypium* **sp.** (A) Number of publications for the past 15-years period; and (B) RNAi studies on the targeted cotton genes or traito:

RNAI METHODS IN COTTON

RNA interference, known as "co-suppression," "quelling," and/or "post-transcriptional gene silencing (PTGS)," is an evolutionarily conserved, double-stranded RNA-dependent, universal eukaryotic process to silence the expression of genes in a sequence-specific manner (Napoli et al., 1990; Romano and Macino, 1992; Hannon, 2002; Roberts et al., 2015). RNAi is induced by exo- or endogenous (i.e., micro RNAs) double stranded RNA (dsRNA) molecules. RNAi is part of the normal cellular function as well as an immune response against foreign nucleic acid signatures from viral infections (Roberts et al., 2015). In the cells, the dsRNAs are recognized by the Dicer family of enzymes and cleaved into short double stranded fragments of ~21–25 bp long siRNAs (Hannon, 2002; Roberts et al., 2015). Further, siRNAs are separated into two single-stranded 'passenger' and 'guide' RNAs. The guide strand is incorporated into the RNA-induced silencing complex (RISC), which triggers the recognition and digestion of complementary RNA sequence signatures, whereas the passenger strand gets degraded. RISC has enzymatic digestion activity, which consists of the key components of argonaute (AGO) and P-element induced wimpy testis (PIWI) proteins (Hannon, 2002; Jakymiw et al., 2005). AGO/PIWI proteins are considered to be the critical sites for RNAi and these proteins localize within the specific P-body regions in the cytoplasm (Jakymiw et al., 2005;Sen and Blau, 2005). RISC and downstream RNAi machinery are common for exogenous and endogenous dsRNA recognition and target sequences digestion; however, there are clear differences in distinctly processing and handling of exogenous and endogenous dsRNAs (Hannon, 2002).

The regulation of gene activity of cells at the PTGS level is the main biological function of RNAi, which occurs either by the inhibition of translation of mRNA or by direct degradation of the mRNA (Hannon, 2002). Moreover, RNAi pathway components (i.e., Dicer, siRNA, and RISC) may contribute to maintenance of genome organization and structure through RNA-induced histone modification, which may silence gene activity at the pre-transcriptional level (Holmquist and Ashley, 2006). At the same time, small dsRNAs may also possibly up-regulate expression of genes through binding into a promoter region and histone demethylation (Li et al., 2006).

Usually, RNAi can be induced by the expression of antisense RNA, dsRNA and by virus induced gene silencing (VIGS) constructs. A stable RNAi in cotton is achieved by employing hairpin (HP) RNAi binary*Agrobacterium* vector constructs, which produce dsRNAs that induce RNAi machinery. The majority of cotton RNAi studies highlighted here utilized HP-mediated gene silencing. For constructing HP vectors, usually 200–500 bp long fragments of gene of interest are used. One limitation of these HP constructs is that they sometimes could generate "off-target" gene silencing due to the generation of multiple variants of short interfering RNA (siRNA) from inserted target gene fragments. This is particularly challenging when there are several paralogous, orthologous, and highly similar gene family members in an allotetraploid genome like cultivated cotton. To address this issue, researchers have developed short synthetic oligonucleotide cassette binary vectors, consisting of 19–24 bp highly specific target-gene or sRNA/miRNA sequence with intronic loop size of 7–9 bp. These synthetic oligonucleotide-based vectors could efficiently and selectively silence target genes in plants (Higuchi et al., 2009) including cotton (Abdukarimov et al., 2011). However, both HP and synthetic oligonucleotide vectors require genetic transformation where many genotypes of cotton are recalcitrant to genetic transformation. Therefore, there is a need for rapid *in vitro* and *in vivo* testing of RNAi effects for targeted genes.

Tang et al. (2004) experimentally tested the silencing power of siRNA designed for targeting the green fluorescent protein (*GFP*) gene. The silencing effects of two *GFP* derived siRNAs were tested *in vitro*cultured *GFP* transgenic cells of rice, cotton, Fraser fir, and Virginia pine. These efforts resulted in efficient silencing of *GFP* marker gene expression, which is useful for large-scale screening of gene function and drug target validation (Tang et al., 2004). Due to the limitation of siRNA delivery *in vitro* experiments, that affect the efficiency of RNAi, researchers later developed the first efficient delivery system of siRNA to plant cells including cotton by a nanosecond pulsed laser-induced stress wave (LISW). This LISW-mediated siRNA delivery system was found to be a reliable and effective method for inducing PTGS efforts *in vitro* cultured cells (Tang et al., 2006).

Further, to perform a high throughput, rapid functional validation of cotton genes and phenotypic effects researchers developed potent RNAi assays using VIGS, facilitating transient PTGS. For example, because of the important role in viral disease symptom modulation in cotton leaf curl virus (CLCV) disease, VIGS-mediated RNAi was utilized to affect betasatellite DNA of CLCV of Multan (CLCuMV). When inoculated, such VIGS system showed efficient silencing of the target genes in tobacco, *Arabidopsis* and Upland cotton plants (Kumar et al., 2014).

A variety of *Agrobacterium*-mediated VIGS vectors bearing various marker genes to monitor RNAi efficiency were developed rapidly to test the gene function of interest from cotton genome. Examples include tobacco rattle virus (TRV) vector with *GrCLA1*, *GaPDS*, and *GaANR* or fused *GaPDS*/*GaANR* marker genes (Gao et al., 2011b; Gao and Shan, 2013; Pang et al., 2013) resulting in albino or brownish plant phenotypes. In addition to the TRV-VIGS vector, a cotton leaf crumple virus (CLCrV)-based vector was developed recently and was shown as an effective RNAi method for gene discovery in cotton (Gu et al., 2014). As described above and extensively referenced herein, VIGS has been widely used for the discovery and identification of many useful genes in cotton. For instance, RNAi of cotton phytoene synthase (*GhPSY*) using TRV-induced VIGS caused a uniform bleaching of the red color in newly emerged leaves, suggesting its role in controlling red plant phenotype (Cai et al., 2014). Similarly, the VIGS induced RNAi was instrumental to discover the plant phenotypes of anthocyanidin reductase (*GhANR*) pathway genes of cotton involved in the biosynthesis of proanthocyanidins (PAs). In this study, the gene silencing effort has resulted in a significant increase in anthocyanins and a decrease in the PAs, (-)-epicatechin, and (-)-catechin in the stems and leaves of VIGS-infected plants. Results demonstrated the role of ANR pathway in the biosynthesis of flavan-3-ols and PAs in cotton (Zhu et 15).

ANTISENSE GENE SILENCING FOR EARLY FUNCTIONAL STUDIES OF COTTON GENES

It is noteworthy to mention that the first pioneering attempts on silencing of cotton genes were performed by using antisense constructs. The pioneering effort is dated to John (1996) who studied two E6 genes isolated from Upland and Pima cottons. A 60–98% suppression of E6 activity using antisense transgenic cotton plants revealed no noticeable phenotypic changes in fiber development. This research demonstrated that E6 was not critical to the fiber development process (John, 1996). Antisense suppression of cotton small GTPase Rac (*RAC13*) genes decreased the levels of H_2O_2 in fiber cells, which in turn affected secondary wall formation of cotton fibers (Potikha et al., 1999). Ruan et al. (2003) successfully constructed antisense vectors for a 70% suppression of cotton sucrose synthase (*SUS*) gene that resulted in fiberless and shrunken seed phenotypes. Later, the functions of several cotton genes such as cotton steroid 5-alpha-reductase (GhDET2; Luo et al., 2007), cotton myeloblastosis (*GhMYB109*; Pu et al., 2008), and *GhPEL* gene, encoding a pectate lyase (Wang et al., 2010a) were studied using antisense technology, which resulted in significant reduction of fiber elongation and shorter fibers. These results have suggested important roles of these genes in fiber development.

Antisense technology was also utilized to regulate other aspects of cotton besides fiber genes. Inverted-repeat-based gene constructs designed for two key cotton seed-specific fatty acid desaturase genes, *GhSAD-1*and *GhFAD2-1*, resulted in increased levels of stearic and oleic acids in RNAi cotton lines, respectively. Results also showed that the content of palmitic acid significantly decreased in both high-stearic and high-oleic lines, providing a promising opportunity for the development of nutritionally improved cottonseed oil (Liu et al., 2000). CLCV significantly reduces boll formation and development in cotton. Antisense vectors were constructed for important genes of CLCV with the aim of affecting the virulence of this virus. Such targeted genes were 5′ and 3′ fragments of DNA replication gene (*AC1*) as well as a transcription activator (*AC2*) and replication enhancer (*AC3*) genes. Transformation of these antisense constructs into tobacco plant helped to obtain virus resistant transgenic plants (Asad et al., 2003).

Gossypol is a general biocide that provides protection from insect predation, but it restricts the usefulness of cottonseed as a feed and protein source for human and monogastric animals. In the efforts to manipulate gossypol, Martin et al. (2003) developed and transformed cotton with an antisense construct of *CDN1-Cl*, a member of a complex gene family of delta-(+) cadinene (CDN) synthase. These efforts resulted in a reduction of *CDN* synthase gene activity and decreased up to 70% of gossypol content. Further analyses of the first generation transgenic cotton plants demonstrated a significant amount of reduction of gossypol (92%), hemigossypolone (83%) and heliocides (68%) in leaves (Martin et al., 2003) and seeds (Benedict et al., 2004), negatively influencing the biosynthesis of cadinene sesquiterpenoids and heliocides in cotton plants. Later, Townsend et al. (2005) reported that the *Agrobacterium*-mediated genetic transformation of constitutive or seed-specific antisense constructs demonstrated the suppression of *CDN1-C4* genes in a response to bacterial blight infection of cotyledons in the constitutive antisense plants, suggesting a specific role of particular cotton sesquiterpene in the bacterial blight pathogenesis in cotton (Townsend et al., 2005). Similar antisense technology was also used for improvement of cottonseed oil by Sunilkumar et al. (2005)where authors developed the homologous alpha-globulin B promoter driven antisense construct for *FAD2*gene that reduced the expression of delta -12 desaturase in cottonseed. Efforts resulted in two-fold increase of the oleic acid level with an accompanying decrease of linoleic acids in transgenic cottonseeds.

Further, with the discovery and understanding of the mechanisms, RNAi has been applied widely in cotton research, and it has become a major research tool for studying gene functions and breeding of novel cotton cultivars. For the past 15-years, more than 60 RNAi related studies have been published in cotton where we see dramatic increase of efforts after 2010. These studies have targeted functional genomics of important cotton genes including, but not limited to the fiber development, fertility and somatic embryogenesis (SM), abiotic and biotic stress, cottonseed and oil quality genes of cg**tary Table S1**). These efforts have resulted in the development of efficient RNAi methodologies, and RNAi-derived novel biotech cotton lines and cultivars that were subsequently utilized for the functional studies, breeding, and farming of cotton, which we have reviewed belotail.

RNAI FOR FUNCTIONAL GENOMICS OF FIBER GENES AND QUALITY IMPROVEMENT

Majority (37%) of RNAi published studies have targeted the study of functional aspects of fiber-related genes in coure Figure1B1B). Cell elongation and cell cytoskeleton factors are important for fiber development. One of the first fiber-related RNAi efforts targeted the actin cytoskeleton (*GhACT*) genes that express during fiber development. RNAi of *GhACT1* significantly repressed the expression of the target and disrupted the actin cytoskeleton network in fibers, resulting in the inhibition of fiber elongation. Researchers concluded that*GhACT1* plays an important role in fiber elongation but not fiber initiation (Li et al., 2005). Because cotton fibers are seed epidermal cells and resemble epidermal trichomes of leaves, researchers targeted the characterization of orthologs of plant leaf trichome initiation factors in cotton fibers. Two highly similar cotton *GhMYB25* genes were characterized in allotetraploid cotton, which were acquired from A and D genome ancestors. RNAi silencing of both *GhMYB25* genes using single hairpin (HP) construct altered the timing of rapid fiber elongation and resulted in short fibers, dramatic reduction in trichomes on other parts of the plant, and a reduction in seed production. These results demonstrated the important functional role of*MYB* genes in cotton fiber and seed development (Machado et al., 2009).

Later, Walford et al. (2011) characterized *GhMYB25-like* cotton *MYB* gene with *R2R3* domain from a fiberless mutant of cotton (Xu142 *fl*). RNAi of *GhMYB25-like* gene caused fiberless cottonseeds in transformed cotton plants. However, normal trichome development occurred in other parts of the cotton plant, similar to the Xu142*fl* mutant lines. Like Xu142*fl* mutants, *GhMYB25-like* RNAi plants had a lowered expression of fiber-specific *MYBs* including *GhMYB25* and *GhMYB109*; however, RNAi of both *GhMYB25*and *GhMYB109* did not change the expression level of *GhMYB25-like*. This suggested that *GhMYB25-like* is upstream from those *MYBs* and important factor for cotton fiber development (Walford et al., 2011). Another trichome and early fiber expressed *MYB*-like gene with a putative homeodomain leucine zipper (HD-ZIP) transcription factor (*GhHD-1*), functioning downstream of *GhMYB25-like* gene, was characterized in cotton. RNAi suppression of *GhHD-1* decreased trichome formation and delayed the timing of fiber initiation.*GhHD-1*-derived RNAi plant analyses proved that this gene conditions the levels of ethylene and reactive oxidation species (ROS) through a WRKY transcription factor and calcium-signaling pathway genes (Walford et al., 2012). Similarly, RNAi of another cotton HD-ZIP transcription factor (*GhHOX3*) gene greatly (>80%) reduced fiber length, whereas its overexpression led to longer fibers. Results of this recent report elucidated a novel mechanism of transduction of gibberellic acid signal by a homeodomain protein to promote fiber cell elongation (Shan et al., 2014).

Arabinogalactan proteins (AGPs) are involved in many aspects of plant development. Researchers reported the essential roles of fasciclin-like arabinoglactan (*FLAs*) genes in cotton fiber development using *GhAGP4*RNAi-derived transgenic plants. Results revealed significant suppression of target genes as well as partial suppression of the expression of other *FLA* related genes, resulting in an inhibition of fiber initiation and fiber elongation processes. The resulting RNAi plants had significantly shorter fibers and inferior fiber quality parameters due to disruption of the cytoskeleton network and the cellulose deposition in fiber cells (Li et al., 2010). Subsequently, *GhFLA1* was identified in cotton, and it was demonstrated that its suppression slowed down fiber initiation and elongation. As a consequence, plants produced significantly shorter fibers compared to the wild type controls (Huang et al., 2013). These results demonstrated the importance of cotton *FLA*genes in fiber length and quality improvements.

Another cell elongation and expansion factor of cotton, characterized using RNAi approach, was vacuolar invertase (*GhVIN1*). *GhVIN1* activity was significantly higher in fiber compared to leaf, stem, and root. It expressed at subtle level in fiberless cottonseed epidermis. Interestingly this gene started expressing in initiating fibers and reached a high-level in elongation phase, but dramatically dropped after the elongation. RNAi plants for *GhVIN1* gene efficiently suppressed the transcript and negatively affected the elongation, while overexpression of *GhVIN1* showed positive changes in fiber elongation rates. It was demonstrated that*GhVIN1* regulates fiber elongation in an osmotic dependent manner (Wang et al., 2010b). Further, RNAi- induced down-regulation of *GhVIN1* expression has generated a dosage-dependent fiberless cottonseed phenotype without influencing seed development, which was due to the markedly reduction of VIN activity (Wang et al., 2014). The results of Wang et al. (2014) have suggested that *GhVIN1* mediates hexose signaling, which is important for the regulation of key regulatory genes of fiber cell initiation and differentiation from the ovule tissue in early stage of fiber development.

RNA interference approach was also instrumental in analyzing the function of a cotton *PROTODERMAL FACTOR1* gene (*GbPDF1*), where its RNAi caused a retardation of fiber initiation, produced shorter fibers and lowered the lint percentage compared to the wild-type controls. These results demonstrated that*GbPDF1* is an important genetic factor, positively conditioning cotton fiber development. The detailed characterization of *GbPDF1* RNAi plants revealed a higher accumulation of hydrogen peroxide, and reduced expression of ethylene and pectin synthesis or sugar transport related genes during fiber development. These RNAi study results suggested that "*GbPDF1 plays a critical role together with interaction partners in hydrogen peroxide homeostasis and steady biosynthesis of ethylene and pectin during fiber development via the core cis-element HDZIP2ATATHB2*" (Deng et al., 2012).

Because of their single celled-expansion-model property, developing cotton fibers have been used to elucidate cell wall formation using various transgenomic tools, including VIGS and high throughput modern experimental technologies (Haigler et al., 2012). Researchers extended TRV triggered VIGS in cotton leaves to study the function and biological role of a microtubule-severing protein (KATANIN) and a positive regulator of lipid biosynthesis (*WRINKLED1*) genes in cotton fiber development. RNAi plants having a decreased expression of *KATANIN* gene had shorter fibers and increased cottonseed oil in endosperm, whereas RNAi suppression of *WRINKLED1* resulted in increased fiber length with reduced cottonseed oil level. These results not only demonstrated the possibility to increase fiber length by repartitioning carbon flow but also highlighted the effectiveness of the TRV-VIGS method for rapid functional analysis of genes involved in cotton fiber development (Qu et al., 2012). A recent study has demonstrated that the use of VIGS constructs with several cell wall membrane lipid genes of cotton encoding fatty acid desaturases [*Gh (15)FAD*], phosphatidylinositol synthase (*GhPIS*), or phosphatidylinositol kinase (*GhPIK*) resulted in significant decrease of fiber length (Liu et al., 2015b).

RNA interference has also been utilized to annotate the function of several transcription factor genes involved in fiber development. For instance, RNAi suppression of a class I TCP transcription factor (*GbTCP*) yielded shorter fiber, a reduced lint percentage, and a lower fiber quality than the wild-type plants. Results from this study suggested that the modulation of jasmonic acid (JA) biosynthesis/response as well as other pathways such as ROS, calcium channel, and ethylene signaling by *GbTCP* gene is important for fiber and root hair development (Hao et al., 2012). Early attempts of studying the roles of sucrose synthase (*SUS*) genes using suppression of its activity resulted in shorter cotton fibers (Ruan et al., 2003). Recently, a novel cotton sucrose synthase (*GhSUSA1*) was characterized using germplasm analysis and genetic mapping strategies, where its RNAi-mediated suppression in transgenic cotton reduced fiber quality and decreased the boll size and seed weight sharply contrasting *GhSUSA1* overexpression events (Jiang et al., 2012). The interesting role of some metabolites in cotton fiber was recently elucidated in both overexpression and RNAi silencing experiments of flavanone 3-hydroxylase (*F3H*). *F3H* uses naringenin (NAR) metabolite as its substrate. An excess of NAR negatively affects fiber development; therefore, RNAi of *F3H* gene significantly increased the NAR content of fiber cells that caused a delay in fiber development. RNAi phenotype was transferable by hybridization to other genotypes, and appeared to be more severe in the genetic background of high-flavonoid brown fibered *G. hirsutum* T586 line. Results suggested the importance and prospective utilization of flavonoid metabolism as a novel pathway for cotton fiber improvement (Tan et al., 2013).

RNA interference experiments, targeting the *REPRESSOR OF SILENCING-1* (*ROS1*) gene with a conserved DNA demethylase domain, revealed the promotion of DNA methylation in resulting transgenic RNAi cotton lines, accompanied by a significantly reduced fiber growth (Jin et al., 2013). These results suggested that fiber development is greatly affected and under the control of epigenetic regulations. RNAi was a great help in annotating the exact function of several 'candidate' cotton genes in the fiber development, particularly those involved in various transport and ion exchange proteins that play a vital role. Two of these proteins are water transporter aquaporins (*GhPIP1* and *GhPIP2*) that condition fiber elongation. There is a complex hetero-oligomer *GhPIP2* protein interaction network, and RNAi cotton plants with reduced expression of *PIP2* transcript markedly delayed fiber elongation and produced shorter fibers. Results elucidated the molecular mechanism of *GhPIP2* involvement in fiber development. In conclusion, *GhPIP2*proteins are the primary aquaporin isoforms in fibers, proving their importance for rapid fiber elongation (Li et al., 2013). Similarly, RNAi of cotton Annexins [*GhAnn2*; Ca(2+)- and phospholipid-binding proteins] significantly retarded fiber elongation and secondary cell wall synthesis, resulting in shorter and thinner mature fibers in RNAi plants. This primarily was due to decreased level of the rate of Ca(2+) influx at the fiber cell apex of *GhAnn2* RNAi lines suggesting possible fiber development modulation through Ca(2+) fluxes and signaling (Tang et al., 2014). Further, RNAi of cotton α-expansin genes *GbEXPA2* and its truncated version *GbEXPATR* also resulted in shorter fibers with thicker cell walls sharply contrasting the results of their overexpression. These results fueled a hope for development of cultivars with improved fiber (Li et al., 2015b).

In contrast, RNAi of a proline-rich protein of cotton (*GhPRP5*) enhanced fiber development resulting in longer fibers in RNAi cotton plants compared to wild type controls (Xu et al., 2013) opening a promising venue to develop agriculturally valuable superior fiber RNAi cotton cultivars. Abdurakhmonov et al. (2014)reported a similar effect when cotton phytochrome A1 (*PHYA1*) gene was repressed by using RNAi, which resulted in improvement of fiber quality. Because of known influence of red/far-red light ratio on the fiber length researchers investigated RNAi of the cotton *PHYA1* gene that drastically reduced *PHYA1* gene transcript up to ~70%, but caused a compensatory overexpression of remaining phytochrome genes by up to ~20-fold in somatically regenerated *PHYA1* RNAi cotton plants. RNAi of *PHYA1* gene significantly enhanced the upper half-mean fiber length and improved other major fiber characteristics without adverse effects to the other agronomically important traits such as maturity and yield. RNAi phenotype expression related to fiber quality, root and vegetative growth, and early flowering was genetically stable, heritable, and transferable. Importantly, RNAi of *PHYA1* gene increased seed cotton by 10–17% when compared with the controls in replicated field plot experimental trials (Abdurakhmonov et al., 2012b, 2014). As a result of this effort, the first generation RNAi cultivars were developed using state-of-the-art RNAi technology, which are the world's first biotech cotton cultivars with improved fiber quality and other key agronomic traits without adversely affecting the yield (Abdurakhmonov et al., 2014; Cotton Outlook, 2014n, 2014).

RNAI FOR BIOTIC STRESS STUDIES

As for fiber development studies, RNAi has been a golden tool to study biotic resistance aspects in cotton. Twenty five percent of all published RNAi(Figure Figure1B1B) have resulted in many successful applications for functional analyses of viral, fungal, and insect resistance genes as well as for developing improved cotton cultivars.

RNAi for Functional Genomics of Fungal Resistance

Verticillium dahliae and *Fusarium oxysporum* are two species of wilt disease-causing fungal pathogens of cotton that globally pose a huge threat for cotton farming and production. RNAi has been successfully applied to study wilt disease resistance aspects and develop wilt-resistant cotton lines. Using VIGS-mediated RNAi approach, researchers demonstrated silencing of *GhNDR1* and *GhMKK2* genes that decreased*Verticillium* wilt resistance in cotton (Gao et al., 2011a). Similarly, VIGS of *Verticillium* wilt-induced *GbVE1*gene isolated from wilt resistant Pima (*G. barbadense*) decreased resistance to this fungal pathogen in contrast to its overexpression (Zhang et al., 2012). Results suggested that the *GbVE1* gene is one of the candidate genes to develop wilt resistant cotton cultivars through modern biotechnology and conventional breeding. VIGS applied for subfamily of SM receptor kinases genes *GhSERK/GhBAK1* (Gao et al., 2013a), a key gossypol biosynthesis enzyme gene *GbCAD1*, and an important regulatory *GbSSI2* gene (functioning in the crosstalk between salicylic acid and JA signal pathways; Gao et al., 2013b) revealed functional importance of these genes in *Verticillium* wilt resistance of cotton with decreased resistance when they were silenced (Gao et al., 2013a).

Zhu et al. (2013) characterized the ghr-miR482 family in cotton and demonstrated its down-regulation in cotton seedlings of a susceptible line after infection with *Verticillium* wilt, which caused up-regulation of several nucleotide binding site-leucine-rich-repeat (NBS-LRR) target genes. This provided evidence that like other plants, cotton has similar miRNA-mediated gene regulation mechanism that release control over the expression of NBS-LRR defense genes upon fungal pathogen attack. Heterologous VIGS, using a conserved fragment from the enhanced disease susceptibility 1 (*GbEDS1*) gene of cotton in *Nicotiana benthamiana*plants, increased susceptibility to *Verticillium* wilt. Therefore, *GbEDS1* could be a novel biotechnological tool in the regulation of cotton defense against *V. dahliae* (Su et al., 2014). Mitogen-activated protein kinase (*MAPK*) cascade genes are considered important in plant growth, development, and stress responses. The*MAPK* gene family of cotton, consisting of 28 putative members, has been characterized and phylogenetically clustered into the four known A, B, C and D groups, with more *MAPKs* containing the TEY phosphorylation site (18 members) than the TDY motif (10 members). Furthermore, TRV-VIGS approach has helped to discover important roles of *MAPKs* in diverse functions in cotton, including various developmental stages of vegetative and reproductive growth and in the biotic stress response. It was shown that VIGS of *MPK9*, *MPK13*, and *MPK25* could significantly enhance cotton susceptibility to the*Verticillium* wilt pathogen (Zhang et al., 2014).

Availability of detailed plant genome information and its combination with a rapid gene function detection using VIGS and heterologous expression facilitated a global level of gene function screening in plants including cotton. Recently, using this approach, Xu et al. (2014) screened 38 cotton genes for their response to *V. dahliae* and jasmonate treatment. Twenty-four of these genes were found to be differentially regulated by pathogen inoculation and most of these have positively responded to both *Verticillium* infection and JA stimuli. The results of functional analysis suggest that reactive oxygen species, salicylic acid- and JA-signaling pathways are involved in disease resistance response of cotton against *V. dahliae*, highlighting the power of VIGS in cotton functional genomics and data mining (Xu et al., 2014).

Very limited information is available on RNAi directed studies in understanding the resistance of *Fusarium oxysporum* (FOV) wilting pathogen that has been a serious threat to cotton production across the world (Abdullaev et al., 2015). With the target to develop novel RNAi-based cotton cultivars against FOV,Shapulatov et al. (2015) recently characterized a "top-layer" of small RNA/miRNA profile in FOV race 3 infected versus uninfected cotton seedlings of resistant and susceptible cotton genotypes. This study identified many small RNA and miRNA signatures that expressed differentially and targeted key proteins associated with disease resistance (Shapulatov et al., 2015) that could be useful for future RNAi studies.

RNAi for Insect Resistance and Pest Control

Conventionally, cotton pests are controlled by using chemically derived insecticides, which have been later expanded with the *Bacillus thuringiensis* (Bt) *Cry* toxin gene expressed in transgenic crops (Qiu et al., 2015). Bt-crops significantly helped to combat major agricultural insects and decreased the pesticide utilization in many crops including cotton. There is no doubt that Bt-cotton brought significant economic and environmental benefits; however, currently the examples exist where insect resistance has arisen against the Bt toxins in several crops, including cotton. This prompted the plant research community to explore alternative approaches for insect pest control.

In this context, RNAi appeared as an alternative approach to control insect pests based on findings in*Caenorhabditis elegans* where feeding nematodes with dsRNAs, produced by *Escherichia coli*, could cause silencing of expression of targeted gene (Roberts et al., 2015). Although the mechanism of dsRNA uptake, respond, and efficiency in insects are distinct and variable compared to *C. elegans*, a constitutive expression of dsRNAs for target insect genes through the plant-mediated RNAi could trigger pest target gene RNAi when fed with plant tissues, expressing insect gene-specific dsRNAs (Roberts et al., 2015). Several important factors should be taken into consideration to apply RNAi via insect-resistant transgenic plants. These include: (1) detection of vital and specific gene of target insect whose RNAi must cause developmental retardation or lethality to the preferentially targeted insect species; and (2) dsRNAs must be uptaken by the insect midgut system from plant tissues, and they must induce silencing of the target gene, amplify rapidly and spread systemically.

The first proof of the concept of RNAi-plant mediated pest control of cotton bollworm was demonstrated byMao et al. (2007) using dsRNA of insect-derived cytochrome P450 monooxygenase gene (dsCYP6AE14). Consequently, feeding the insects with dsCYP6AE14 expressing plant tissues increased the levels of dsCYP6AE14 transcript in the midgut of insects and decreased CYP6AE14 expression in insects. This resulted in retardation of larval growth due to intolerance to toxic gossypol as well as more sensitivity to the insecticides used (Mao et al., 2007, 2011; Tao et al., 2012). This research highlighted the power of RNAi cotton plants expressing dsCYP6AE14 that efficiently controlled cotton bollworms and developed insect-proof cotton cultivars (Mao et al., 2011; Tao et al., 2012). Further, the combined usage of dsCYP6AE14 and plant cysteine proteases, such as *GhCP1* from cotton (*G. hirsutum*) and *AtCP2* from *Arabidopsis*, have provided the increased protection from bollworm than either of the single-transgenic events. The increased effect was due to the improvement of dsCYP6AE14 uptake by insect cells because of increasing the permeability of the insect peritrophic matrix by cysteine proteases. This provided a potent plant-mediated RNAi approach against herbivorous insects (Mao et al., 2013).

A similar effort reported the potential of an alternative gene for plant RNAi-mediated insect pest control. *In vivo* RNAi of the 3-hydroxy-3-methylglutaryl coenzyme A reductase [HMG-CoA reductase (*HMGR*)] gene, a key enzyme in the mevalonate pathway in insects, effectively inhibited the fecundity and oviposition of the females, and significantly reduced vitellogenin (Vg) mRNA levels, demonstrating its potential against*Helicoverpa armigera* and other insect pests (Wang et al., 2013). In order to better understand RNAi mechanism of insects and to identify target genes for RNAi, the transcriptome database was developed for the cotton boll weevil, *Anthonomus grandis*. Several key components of RNAi machinery (PAZ domain and SID-like proteins) and key insect genes (e.g., chitin synthase 1) for inducing RNAi have been characterized. Further, *A. grandis* female adults exposed to dsRNA of a chitin synthase gene produced non-viable eggs and malformed alive larvae (Firmino et al., 2013). Further, RNAi has been shown to be a very valuable tool to identify key molecular receptors providing efficiency of widely used Bt-crops through improvement of susceptibility to Cry toxins. An oral RNAi inducer for cadherin *SeCad1b* gene significantly decreased susceptibility to Cry1Ac and Cry2Aa in *Spodoptera exigua* larvae (Qi., 2015).

RNAI FOR ABIOTIC STRESS GENES

An understanding of tolerance mechanisms to the abiotic stress (including salinity, drought, and heat), and their molecular basis have been major goals of plant research community, and are imperative to the improvement of stress tolerance, and environmental adaptation of crop plants, including cotton. RNAi methodology has been productive in characterizing the biological function of many cotton genes in abiotic stress tolerance. For instance, a VIGS-mediated sucrose non-fermenting 1- related protein kinase 2 (*GhSnRK2*) gene silencing mitigated drought tolerance in cotton plants, indicating that *GhSnRPK2* positively condition drought stress and low temperature tolerance (Bello et al., 2014).

Another VIGS-mediated RNAi study with an R2R3-type *GbMYB5* transcription factor, sharply contrasting its overexpression, decreased the tolerance of cotton plantlets to drought due to decreased proline content, antioxidant enzyme activities, and increased malondialdehyde (MDA) content in cotton under drought stress. Results from this study suggested that *GbMYB5* is an important positive factor contributing to the environmental adaptation of plants under drought stress condition (Chen et al., 2015). Further, as mentioned-above, RNAi of cotton *PHYA1* genes (Abdurakhmonov et al., 2014) generated improved drought, salt, and heat tolerance compared to wild type plant that might be due to increased photosynthesis, regulation of plant salt tolerance genes and longer, better developed root systems of *PHYA1* RNAi cotton plants (Abdurakhmono., 2014).

RNAI FOR FERTILITY AND EMBRYOGENESIS STUDIES

Somatic embryogenesis, pollen/anther development, fertility and heterosis questions are the topics of central interest to cotton physiology, genetics and breeding as well as modern biotechnology studies. In particular, an understanding of SE in cotton is very important for genetic engineering and modern cotton breeding programs due to limitations imposed by a high degree of genotype-dependency and only a very few genotypes having regenerative capability from single cells. An RNAi approach has been particularly helpful in the study of high mobility group box (*HmgB*) family genes of cotton involved in SE. RNAi of the*GhHmgB3* gene (Hu et al., 2011) decreased level of the *GhHmgB3* in hypocotyl-derived cotton somatic cells, which rapidly dedifferentiated but did not develop into embryogenic callus (EC) tissues. When the*GhHmgB3* RNAi construct was transformed into EC cells, tissue proliferation and differentiation in transformed samples were significantly improved, but embryonic tissues did not develop into complete plantlets. Further, researchers found that RNAi of *GhHmgB3* caused a series of β-catenin-related mechanisms that might condition the deregulation of proliferation and differentiation of cells in cotton SE (Hu et al., 2011).

RNA interference approach to studying the *GhSERK1* gene, encoding a putative leucine-rich repeat receptor protein kinase (LRR-RLK) with 11 domains, generated a series of male-sterile cotton lines, including mutants with normal fertility, semi-sterility, and complete sterility. Sterility was strongly correlated with the level of RNAi. Results demonstrated that the *GhSERK1* gene has an important role in the development of anthers, especially in the formation of pollen grains (Shi et al., 2014). Recent RNAi and co-suppression mediated down regulation study toward investigating the potential function of cotton KORRIGAN (*GhKOR1*), a highly conserved membrane-bound endoglucanase, has resulted in smaller filial tissue and reduced seed weight, leading to various abnormalities in endosperm formation and delayed embryo development. Results collectively suggested that the *GhKOR1* gene is required for the development of viable cottonseed with normal seed germination and seedling growth (Shang et al., 2015). Another interesting RNAi gene silencing study targeting cotton acyl-CoA N-acyltransferase (EC 2.3; *GhACNAT*) using VIGS resulted in abnormal anther formation and sterile plants. *GhACNAT* is involved in lipid metabolism and JA biosynthesis; therefore, treatments with exogenous methyl jasmonate "*rescued anther dehiscence and pollen release in GhACNAT-silenced plants and caused self-fertility*" (F., 2015).

RNAI FOR SEED AND OIL QUALITY

Following the inverted-repeat-based gene constructs study (Liu et al., 2000), a hairpin RNA-mediated gene silencing was applied by the same research group to genetically modify the fatty acid composition of cottonseed oil (Liu et al., 2002). The target genes were two key fatty acid desaturase genes, *GhSAD-1*encoding stearoyl-acyl-carrier protein delta 9-desaturase and *GhFAD2-1* encoding oleoyl-phosphatidylcholine omega 6-desaturase. RNAi of these two genes revealed similar down-regulation effects as reported earlier (Liu et al., 2000), i.e., 40 and 77% increase of stearic and oleic acids, respectively, with substantial decrease of palmitic acid content in both dsSAd-1 and dsFAD2 expressing lines. However, down-regulation was observed at lower frequencies for the antisense constructs as compared to the HP RNAi approach. Interestingly, researchers demonstrated "stacking" of both dsSAd-1 and dsFAD2 HP constructs into one cotton genotype via sexual intercrossing of independently silenced RNAi lines that resulted in the same degree of down-regulation of the target genes as observed in the individually silenced parental lines. Further, RNAi modulation of these key fatty acid genes (single or stacked) to various degrees provided an opportunity to produce cottonseed oils with novel combinations of palmitic, stearic, oleic, and linoleic acid contents. These results should help to increase cottonseed oil quality and consumer satisfaction (Liu et al., 2002).

Besides oil production, cottonseed is a rich source of proteins that "*can potentially provide the protein requirements for half a billion people per year*" (Sunilkumar et al., 2006), provided if cottonseed is free of the toxic gossypol present within the seed glands. Conventional geneticist developed gossypol-free glandless cultivars with mutations of gossypol biosynthesis pathway in entire cotton plant. Therefore, such glandless cotton cultivars were highly susceptible to insect pests limiting their wide farming. However, RNAi was of a particular help in providing an opportunity to develop seed tissue-specific, highly stable and heritable silencing of the key delta-cadinene synthase gene of gossypol biosynthesis during seed development, leaving gossypol intact in other green parts of the cotton. This breakthrough research in cotton offered a novel solution to increase food sources for sufficient nutrition for hundreds of millions of people (Sunilkumar et al., 2006). Rathore et al. (2012) further fine-tuned RNAi-knockdown of δ-cadinene synthase gene(s) and developed ultra-low gossypol cottonseed (ULGCS) cotton lines that exhibited multi-generational stab effects.

NATURAL RNAI FOR COTTON GENES AND TRAITS

Besides the use of RNAi for targeted inhibition of protein coding genes, it is important to understand the function of natural miRNAs present in cotton. The first 'wet-bench' small RNA characterization from developing cotton ovule was reported by Abdurakhmonov et al. (2008a) followed by the characterization of small RNAs in fiber development in a large scale deep sequencing efforts (Kwak et al., 2009; Pang et al., 2009; Ruan et al., 2009). Deep sequencing of small RNAs isolated from different cotton tissues has revealed the presence of evolutionarily conserved miRNAs, which target common developmental programs like organ and tissue differentiation and timing of development. Many cotton-specific miRNAs have been identified but their functions are still unclear and need experimental validation (Wang and Zhang, 2015). Natural cotton miRNAs have been studied in Upland cotton (*G. hirsutum*, AADD), Sea Island cotton (*G. barbadense*, AADD), *G. herbaceum* (A_1), *G. arboretum* (A_2), and *G. raimondii* (D_5). Some miRNAs show differential expression under salt stress; they could therefore be potential targets to improve salt stress responses (Yin et al., 2012). Most interestingly, miRNAs may play an important role in cotton fiber development as 140 targets of 30 conserved miRNAs and 38 targets of five candidate miRNAs were identified through degradome sequencing of cotton fiber RNA (Zhang et al., 2013a; Liu et al., 2014). Targeting of one species of miRNA (miRNA156/157) resulted in the reduction of mature fiber length, illustrating the importance of natural miRNAs in cotton.

Recently, several cotton micro RNA (miRNA) sequence signatures have been identified (Shweta and Khan, 2014) that efficiently target and stop the replication of Allahabad CLCV (CLCuAV) genome. The ghr-miR2950 was capable of targeting all the viral genes while ghr-miR408 targeted overlapping transcripts of*AC1* and *AC2* genes. Further, researchers identified that ghr-miR394, ghr-miR395a and miR395d could bind overlapping transcripts of *AC1* and *AC4* genes, which are involved in CLCuAV viral replication; therefore, these RNAi inducers could be used as potential tools to develop virus resistant cotton cultivars (Shweta and Khan, 2014).

In addition to natural miRNAs, Wang et al. (2015a) reported the first comprehensive identification of 30,550 long intergenic non-coding RNA (lincRNA) loci and 4,718 long non-coding natural antisense transcript (lncNAT) loci in *Gossypium* sp., demonstrating their biased expression patterns toward subgenomes and overall higher methylation levels compared to protein coding genes. Several functional long non-coding RNA (lncRNA) candidates, involved in cotton fiber initiation and elongation were identified. For instance, the miR397 generating lncRNAs have an important role in regulating lignin metabolism in domesticated tetraploid cotton fibers (Wang et al., 2015a). All of these natural miRNAs, lncRNA, and lncNAT are important candidate loci to elucidate many important functional questions in cotton that will serve as a base for designing novel RNAi approaches and studiesr future.

RNAI CULTIVARS, FIELD TRIALS, AND COMMERCIALIZATION

Although RNAi has been extensively used and will remain a valuable experimental tool for functional genomics of cotton, RNAi-based cultivar development and its commercialization for cotton farming is in its very early stage. There is limited information on whether the above-mentioned RNAi studies have resulted in a cultivar development, subjected to field trials, and are being targeted for commercialization or not. To our best knowledge, such efforts were covered by only ~5% of all RNAi studies reviewe(Figure Figure1B1B), as detailed below.

Field trials for two selected *FAD2* cotton RNAi lines were conducted in Narrabri, Australia during 2003/2004 revealing that RNAi lines had high oleic acid content in cottonseed without adverse effects on any of tested key agronomic traits (Liu et al., 2009). There is no information on the commercialization target of these cottonseed oil lines. Rathore and his colleagues (Rathore et al., 2012; Palle et al., 2013) have targeted the commercialization of the ULGCS cotton RNAi lines mentioned above. Small-scale field trials of these ULGCS lines over a period of 3 years demonstrated that the RNAi phenotype was stable under field conditions and plants exhibited sufficient insect pest resistance due to wild-type levels of gossypol and related terpenoids in other plant organs. Most importantly, there were no negative effects of RNAi on either the yield or quality of the fiber with significantly higher (4–8%) oil content in cottonseeds of RNAi lines compared to wild type parental plants. Researchers concluded that "*field trial results confirmed the stability and specificity of the ULGCS trait suggesting that this RNAi-based product has the potential to be commercially viable*" (Stokstad, 2006; Rathore et al., 2012; Palle et al., 2013). Field and laboratory trials were conducted in 2015, 2014, and 2013 by Texas A&M University to develop the data package necessary to seek food, feed, and cultivation deregulation of ULGCS biotech events from all relevant U.S. authorities (United States Department of Agriculture [USDA] Portal, 2015).

Abdurakhmonov et al. (2014) successfully applied state-of-the-art cotton *PHYA1* RNAi concept and results to the Uzbekistan cotton improvement program. They have demonstrated the transfer of phytochrome-associated RNAi phenotypes from somatically regenerated RNAi Coker-312 to four different (recipient) commercial Upland cotton cultivars via sexual crosses. Efforts effectively converted recipient Upland genotypes to the superior well-adapted RNAi cotton cultivars. As a result, the first generation of novel RNAi cotton cultivar series "Porloq-1," "Porloq-2," "Porloq-3," and "Porloq-4" (translates to "Shiny" or "Great future") has been deFigure Figures22 and 3**3**). These novel RNAi cultivars successfully passed 3 years (2012–2014) of extended field trials across 13 different soil-climatic regions in Uzbekistan (Cotton Outlo; Figure Figure33). Field trials demonstrated superiority of RNAi cultivars to any traditional Uzbekistan cultivars in terms of fiber quality, adaptiveness to harsh environmental conditions across Uzbekistan, early maturity and significant increase in seed cotton yield and production of increased lint fiber per hectare (Abdurakhmonovt al., 2012b).

Figure 2: The plant architecture of PHYA1 gene-derived and field grown (in 2013, Tashkent farm) novel RNAi cotton cultivar "Porloq-1" (translates as "Shiny-1" or "Great furture-1", Cotton Outlook, 2014). This cultivar has been developed through genetic hybridization and multi-generation individual selection from self-pollinated F2–5 progeny of the sexual cross between somatically single-cell regenerated RNAi Coker-312 (T-1-7 RNAi family) and Uzbek cotton cultivar AN-Boyovut-2. RNAi of PHYA1 genes resulted in vigorous shoot and root development, bushy plant architecture, early flowering and increased boll number, early synchronous boll opening and early plant senescence with superior fiber quality parameters among other important morphological and agronomic improvements (refer to Abdurakhmonov et al., 2014 for fiber quality and yield inicultivar ...

Figure 3: A general view of large-field trial experimental plot in 2013 for novel Uzbek RNAi cotton cultivar "Porloq-2" derived from the sexual cross between RNAi Coker 312 (T-31 family) and Uzbek commercial Upland cultivar "C-6524" ("Mashhura-N" cotton, Namangan region, Uzbekistan). This field view demonstrates high-yield potential, early autumn induced rapid-senescence, and simultaneous boll maturity accompanied by increased anthocyanin pigmentation on stems and leaves of RNAi cotton plants under normal solar ar "C-6524" ...

There are several positive ramifications associated with the anticipated usage of *PHYA1*-dervied RNAi cultivars by farmers locally and globally that include: (1) opportunity to produce superior, novel RNAi cotton fiber that should have a premium price increase and income per acre, and (2) opportunity to spin a fine-count cotton yarn from any production zone. Furthermore, (3) early flowering and maturity of *PHYA1*-derived RNAi varieties should provide an opportunity for early and quality crop harvest and on-time planting of rotation crops before harsh weather arrives, perfectly suitable and demanding for northernmost cotton growing country like Uzbekistan. RNAi cultivars have (4) increased seed cotton and lint yield per acre – the foremost target and interest of farmers. Most importantly, early observations and ongoing field experiments revealed (5) a better utilization of fertilizers and nutrients (due to increased nitrate reductase activity and strong root system), increased photosynthesis rate, and salt, heat and drought tolerance are added advantages of novel *PHYA1* RNAi varieties. This sufficiently addresses the issue of problematic shortages in irrigation and water deficiencies triggered by forecasted global warming that remains a priority danger for the Central Asian regions (Abdurakhmonov, 2013). The use of cotton's own gene unlike other existing transgenic technologies of cotton, utilizing foreign genes, this state-of-the-art *PHYA1*-mediated RNAi technology assures (6) ecological safety of RNAi cotton cultivars. A high premium price of cotton fiber and increased yield should allow the expanded cotton production (7) on marginal land and create a new cotton fiber/cloth market; and (8) decrease in cotton planting area provides an opportunity to grow more food crops/plants that help to sustain world agricultural, food security and environment at the regional and global levels.

Production of 'not yet existed in world fiber market' novel Upland cotton fiber with 38–41 code (versus possibility of production of fiber with only maximum 35–37 code from existing ordinary Upland cottons) would have a premium price and at least $100 increase in income per acre in Uzbekistan and worldwide. This means that Uzbekistan could earn additional $250 million from its current annual production of one MMT of lint fiber. From this high quality Upland fiber it is possible to spin a fine count 50–70 Ne cotton yarn (against 30-40Ne from ordinary Uplands that Uzbekistan produces). The yarn quality difference itself provides a 10% higher income from these novel RNAi cultivars. Reducing the cotton acreages due to yield and fiber quality improvement of RNAi cultivars would provide opportunity of planting more other food crops ensuring food security for people and sustainability of the environment suffering from cotton production and application of agricultural chemicals.

In 2014, all four RNAi cultivars of "Porloq" series mentioned above were approved for larger field trials in Uzbekistan. RNAi cultivars have been registered through the State List of Approved Agricultural Crops of Uzbekistan. With the purpose of conducting large-scale field trial, RNAi cultivars were planted in over 60,000 hectares (5% of total cotton growing area of Uzbekistan) in the 2015 season in the six major provinces (Namangan, Fergana, Tashkent, Jizzakh, Samarkand, Surkhandarya, and Khorazm) of Uzbekistan. This is to increase RNAi cultivar seeds for upcoming seasons as well as produce larger quantities of "first bales" of RNAi fiber for internal market to test textile quality parameters in larger scale and gather consumer's opinion (Cotton Outlook, 2014). Further, a collaborative effort between USDA/ARS and Uzbekistan made it possible to transfer the RNAi seeds from Uzbekistan to the USA where USDA partners are field-trialing RNAi lines in the USA environment in 2014/15. Efforts are in progress to mobilize *PHYA1*RNAi effects into several Upland cotton cultivars of the USA and conduct extended field evaluations in 2015/16. These efforts will help in technology transfer among the USA farmers and eventually to other stakeholders of all cotton-grountries.

SAFETY AND RISK ASSESSMENTS

Although RNAi technology is generally perceived as a safer technology for obtaining desired traits, currently RNAi-derived traits are considered as GE products and, therefore, they are also subjected to safety assessment like the products derived from other methods of transgenesis. RNAi activity may not only target genes of interest but could also affect genes with sufficient sequence homology leading to the 'off-target' gene silencing with the possible adverse effects on human and animal health or the environment including non-target organisms (NTOs; Casacuberta et al., 2014; Ramon et al., 2014; Roberts et al., 2015). Particularly, the utilization of plant-mediated RNAi as a biological insecticide has raised debates and discussions on its impact to NTOs "*including natural enemies, pollinators, soil decomposers, leaf shredders, wildlife, and fish*" that could possibly be exposed to the plant expressed dsRNAs (Roberts et al., 2015).

Moreover, it should be noted that despite the use of cottons own gene(s) or gene fragment(s), all RNAi hairpin binary vectors currently used for plant transformation, including cotton, have the antibiotic resistance marker (ARM) genes that lately raised great concerns with the use of ARMs in the GMO and its products. However, there is already a long history of safe utilization of ARMs in food and non-food plant products of biotechnology, including cotton. Moreover, there are confirmed evidences of negligibility of the practical impacts of consumption of ARM-containing plant products by humans and animals and horizontal transfer of ARMs from GM crops to environment in natural conditions (Goldstein et al., 2005). Furthermore, cotton is a technical crop and grown mainly for its fiber, making it a more suitable crop for RNAi modifications compared to food crops.

For the assessments of possible risks of RNAi, cotton cultivars and their products there are pre and post-RNAi development tasks. During pre-RNAi cultivar development, an important step is that the designed dsRNA fragment, to be cloned into a hairpin construct, must be thoroughly analyzed *in silico* for the production of all possible siRNA species as well as their putative targets in key consumer or readily exposed organism genomes using the target-searching algorithms against available genome databases. Although there are some limitations due to the lack of genome sequences for non-models and NTOs (Casacuberta et al., 2014), this step could identify the first step risks possibly caused by unintended function of 'candidate' dsRNA from a gene of interest. If any such unintended targeting or highly specific off-target matching in consumer genomes is detected, then RNAi development must be stopped, or candidate dsRNA must be changed, fine-tuned/modified and optimized to eliminate the unintended effects. If dsRNA is targeting only the intended function, then the next step is to develop the RNAi cultivar and begin post-RNAi cultivar testing. It includes: (1) the evaluation of the RNAi cultivar for desired trait expression; (2) experimental quantification of the gene-knockdown level of targeted gene expression; and (3) comparative sequencing of siRNA sequence profile before and after RNAi. Although currently not validated within the regulatory framework for food safety assessment, additionally, comparative studying of proteome and metabolome profiles of RNAi cultivar/product may be considered (Ricroch et al., 2011; Clarke et al., 2013; Simó et al., 2014; Wang et al., 2015b). These steps could identify any "off-target" phenotypes, visible side effects, production of unexpected novel siRNA or/and secondary RNA signatures as well as unknown proteome/metabolome components in RNAi cultivar genome and its products. If such novel, unintended, and secondary biomolecules are identified, it should be a warning signal to be cautious with the future use of RNAi product/cultivar. Further, RNAi cultivar products such as cell extracts containing products of RNAi are suggested to be tested (5) in cell/tissue culture and further in animal feeding experiments in model organisms including mammals to assure the safety of RNAi products *in vivo* level. These steps would verify possible biological risks on the utilization of RNAi cultivar and its products or prove its complete safety.Heinemann et al. (2013), has outlined the risk assessment steps, highlighted here.

In Uzbekistan, following the above mentioned risk assessment criteria/guidelines, we started a large, coordinated mega-project for the risk assessment of *PHYA1*

RNAi cotton cultivars with involvement of responsible organizations including Cotton Industry Associations, Food Industry Companies, Ministry of Agriculture and Water Resources, Academy of Sciences, Cambridge Center for High Technologies, and Ministry of Health of Uzbekistan. Cotton-derived products including cottonseed oil and feeding products are being comparatively evaluated biochemically and toxicologically via animal feeding expe in mice.

CONCLUSION AND FUTURE PERSPECTIVES

Thus, replacing the anti-sense technology, RNAi has been proven to be a valuable tool for functional genomics of cotton for the past decade. RNAi revolutionized the discovery of many key functions and biological roles of cotton genes involved in fiber development, fertility and SM, resistance to key biotic and abiotic stresses, oil and seed quality and other important cotton genes as well as in the improvement of key agronomic traits including yield and maturity. Future studies most likely will target the identification and RNAi of more complex, multi-functional, and developmental genes with cross-talking features among many interconnected networks and biochemical pathways of cotton ontogenesis (Abdurakhmonov, 2013) that generate simultaneous improvements of key agronomic traits (Abdurakhmonov et al., 2014). Some efforts highlighted herein demonstrate early examples to show the power of RNAi for cotton improvement that needs to be extended in the future.

The development of RNAi cotton lines targeting core RNAi machinery including proteins with important DNA methylase and demethylase activity [such as DOMAINS REARRANGED METHYLASE 1/2 (DRM1/2), CHROMOMETHYLASE 2/3 (CMT 2/3), ROS1, and DEMETER (DME)] in Upland cotton will aid in elucidating the key regulatory mechanisms in WGD, chromosomal rearrangements, dosage compensation, and evolutionary advantage of being polyploids. Further, screening the epigenetic modulators for specific traits such as fiber and yield and comparing against the genetic standard TM-1 will aid in understanding the epigenetic landscape of Upland cotton.

There is limited information available on downstream stage of RNAi cotton cultivar development, conducted field trials, or targeting for its commercialization. This indicates that possibly RNAi-based "biotech cotton" cultivar development is in its very early stage that requires more attention, effort, investment and partnership of the cotton research community and private seed companies. However, there are some commercially viable and already ongoing, targeted applications of RNAi technologies, which are passing the successful small- or large-scale field trials, and safety/risk assessment studies. These RNAi-based cotton cultivars, highlighted in this paper, present substantial benefit in cotton production with increased seed and oil quality, fiber, and key agronomic trait improvements. These efforts are believed to boost cotton production and its sustainability worldwide in the era of global climate change and increased crop biosecurity threats.

Despite general safety, RNAi based cultivars are subject for risk assessment before commercialization as per available (Heinemann et al., 2013) and highlighted guidelines herein. The existence and use of ARM genes in current RNAi-based cotton cultivars is subject for rising "unjustified" warnings and requests for removal of ARM genes from RNAi cultivars, and this perhaps continues to be a main roadblock for future commercialization of RNAi cotton cultivars. Therefore, there is a need for designing ARM-free RNAi transformation experiments and development of ARM-free RNAi cotton cultivars.

RNA interference studies, reviewed herein, mostly have utilized stable HP or transient VIGS vector constructs. There is a need for application of novel genome modification and editing tools such as artificial microRNA (amiRNA; Liang et al., 2012), short synthetic interfering siRNA oligonucleotides (Higuchi et al., 2009; Abdukarimov et al., 2011), Transcription Activator-like Effector Nucleases (TALENs; Zhang et al., 2013b), and Clustered Regularly Interspaced Short Palindromic Repeats (CRISPRs/Cas9; Larson et al., 2013) system to generate more effective, fine-tuned, native knockdowns/knockouts than currently used RNAi methods. There is no doubt that the cotton research community is already targeting these objectives, having several diploid (Paterson et al., 2012; Wang et al., 2012; Li et al., 2014) and key allotetraploid (Li et al., 2015a; Liu et al., 2015a; Zhang et al., 2015) genome sequences in hand. All of these will require more coordinated efforts, wider international collaborations, larger investment, and understandings of regulatory and stakeagencies.

AUTHOR CONTRIBUTIONS

IA – coordinated, wrote, revised manuscript; MA, KU, ZB, SES, HR, and US -collected and analyzed world literature, prepared tables and figures, drafted subsections; SS, MU, JY, RP, ED, VS, and SK – critically read and edited manuscript, drafted subsections; GS, SK, AK, HK, KK, SIS, JJ, AA, and AP rigourously edited and apprnuscript.

CONFLICT OF INTEREST STATEMENT

Cotton PHYA1 RNAi study mentioned in this review article has been filed for patenting in Uzbekistan (IAP: 20120069), USA (USPTO:13/445696), and internationally (PCT/US13/27801).

The reviewer AD-P and handling Editor declared their shared affiliation, and the handling Editor states that the process nevertheless met the standards of a fair and oe review.

ACKNOWLEDGMENTS

We thank Ms. Melanie Peterson, Office of International Research Programs of USDA-ARS, and Drs. Robert Griesbach, Evelyn Rabin, Gail Poulos, and Thomas Valco, Office of Technology Transfers, USDA-ARS for being instrumental on technology transfer. We also thank Uzbekistan Government, "CottonIndustry Associations," Ministry of Agriculture and Water Resources, and Ministry of Foreign Economic Relations, Trade and Investments of Uzbekistan for capacity funding of the Center of Genomics and Bioinformatics, Uzbekistan. Disclaimer: Mention of trade names or commercial products in this article is solely for the purpose of providing specific information and does not imply recommendation or endorsement by the U. S. Department of Agriculture. The U. S. Department of Agriculture is an equal opportunity providrence research.

REFERENCES

1. Abdukarimov A., Abdurakhmonov I. Y., Buriev Z. T.,

Bozorov T. A. (2011). *Small Interfering RNA (siRNA) for Knocking Down Gene Expression in Plant Cells*, Uzbekistan patent No. IAP 04300 Official bulletins of State Patents 2-6.

2. Abdullaev A., Salakhutdinov I. B., Egamberdiev S. S., Kuryazov Z., Glukhova L. A., Adilova A. T., et al. (2015). Analyses of *Fusarium* wilt race 3 resistance in upland cotton (*Gossypium hirsutum* L.).*Genetica* 143 385–392. 10.1007/s10709-015-9837-2
3. Abdurakhmonov I. Y. (2013). “Role of genomic studies in boosting yield,” in *Proceedings of International Cotton Advisory Board (ICAC)*, ed. Townsend T., editor. (Washington DC: ICAC Press;), 7–22.
4. Abdurakhmonov I. Y., Buriev Z. T., Saha S., Jenkins J. N., Abdukarimov A., Pepper A. E. (2014).Phytochrome RNAi enhances major fibre quality and agronomic traits of the cotton *Gossypium hirsutum* L. *Nat. Comm.* 5 3062 10.1038/ncomms4062
5. Abdurakhmonov I. Y., Buriev Z. T., Saha S., Jenkins J. N., Abdukarimov A., Pepper A. E. (2012b).*Cotton PHYA1 RNAi Improves Fibre Quality, Root Elongation, Flowering, Maturity and Yield Potential in Gossypium hirsutum L*, U.S. Patent Application No. 13445696 Washington, DC: U.S. Patent and Trademark Office.
6. Abdurakhmonov I. Y., Buriev Z. T., Shermatov S. S., Abdullaev A. A., Urmonov K., Kushanov F., et al. (2012a). “Genetic diversity in *Gossypium* genus,” in *Genetic Diversity in Plants*, ed. Galiskan M., editor. (Rijeka: InTech Press;), 331–338. 10.5772/35384
7. Abdurakhmonov I. Y., Devor E. J., Buriev Z. T., Huang L., Makamov A., Shermatov S. E., et al. (2008a). Small RNA regulation of ovule development in the cotton plant, *G. hirsutum* L. *BMC Plant Biol.* 8:93 10.1186/1471-2229-8-93
8. Abdurakhmonov I. Y., Kohel R. J., Yu J. Z., Pepper A. E., Abdullaev A. A., Kushanov F. N., et al. (2008b). Molecular diversity and association mapping of fiber quality traits in exotic *G. hirsutum* L. germplasm. *Genomics* 92 478–487. 10.1016/j.ygeno.208.07.013
9. Abdurakhmonov I. Y., Saha S., Jenkins J. N., Buriev Z. T., Shermatov S. E., Scheffler B. E., et al. (2009). Linkage disequilibrium based association mapping of fiber quality traits in *G. hirsutum* L. variety germplasm. *Genetica* 136 401–417. 10.1007/s10709-008-9337-8
10. Adams K. L., Cronn R., Percifield R., Wendel J. F. (2003). Genes duplicated by polyploidy show unequal contributions to the transcriptome and organ-specific reciprocal silencing. *Proc. Natl. Acad. Sci. U.S.A.* 100 4649–4654. 10.1073/pnas.0630618100
11. Adams K. L., Percifield R., Wendel J. F. (2004). Organ-specific silencing of duplicated genes in a newly synthesized cotton allotetraploid. *Genetics* 168 2217–2226. 10.1534/genetics.104.033522
12. Adams K. L., Wendel J. F. (2005). Allele-specific, bidirectional silencing of an alcohol dehydrogenase gene in different organs of interspecific diploid cotton hybrids. *Genetics* 171 2139–2142. 10.1534/genetics.105.047357
13. Asad S., Haris W. A., Bashir A., Zafar Y., Malik K. A., Malik N. N., et al. (2003). Transgenic tobacco expressing geminiviral RNAs are resistant to the serious viral pathogen causing cotton leaf curl disease. *Arch. Virol.* 148 2341–2352. 10.1007/s00705-003-0179-5
14. Bello B., Zhang X., Liu C., Yang Z., Yang Z., Wang Q., et al. (2014). Cloning of *Gossypium hirsutum* sucrose non-fermenting 1-related protein kinase 2 gene (GhSnRK2) and its overexpression in transgenic *Arabidopsis* escalates drought and low temperature tolerance. *PLoS ONE* 9:e11226910.1371/journal.pone.0112269
15. Benedict C. R., Martin G. S., Liu J., Puckhaber L., Magill C. W. (2004). Terpenoid aldehyde formation and lysigenous gland storage sites in cotton: variant with mature glands but suppressed levels of terpenoid aldehydes. *Phytochemistry* 65 1351–1359. 10.1016/j.phytochem.2004.03.032
16. Cai C., Zhang X., Niu E., Zhao L., Li N., Wang L., et al. (2014). GhPSY, a phytoene synthase gene, is related to the red plant phenotype in upland cotton (*Gossypium hirsutum* L.). *Mol. Biol. Rep.* 414941–4952. 10.1007/s11033-014-3360-x
17. Campbell B. T., Saha S., Percy R., Frelichowski J., Jenkins J. N., Park W., et al. (2010). Status of global cotton germplasm resources. *Crop Sci.* 50 1161–1179. 10.2135/cropsci2009.09.0551
18. Casacuberta J. M., Devos Y., du Jardin P., Ramon M., Vaucheret H., Nogué F. (2014).Biotechnological uses of RNAi in plants: risk assessment considerations. *Trends Biotechnol.* 33 145–147. 10.1016/j.tibtech.2014.12.003
19. Chaudhary B., Flagel L., Stupar R. M., Udall J. A., Verma N., Springer N. M., et al. (2009).Reciprocal silencing, transcriptional bias and functional divergence of homeologs in polyploid cotton (*Gossypium*). *Genetics* 182 503–517. 10.1534/genetics.109.102608
20. Chen T., Li W., Hu X., Guo J., Liu A., Zhang B. (2015). A cotton MYB transcription factor, GbMYB5, is positively involved in plant adaptive response to drought stress. *Plant Cell Physiol.* 56917–929. 10.1093/pcp/pcv019
21. Chen Z. J., Scheffler B. E., Dennis E., Triplett B. A., Zhang T., Guo W., et al. (2007). Toward sequencing cotton (*Gossypium*) genomes. *Plant Physiol.* 145 1303–1310. 10.1104/pp.107.107672
22. Clarke J. D., Alexander D. C., Ward D. P., Ryals J. A., Mitchell M. W., Wulff J. E., et al. (2013).Assessment of genetically modified soybean in relation to natural variation in the soybean seed metabolome. *Sci. Rep.* 3:3082 10.1038/srep03082
23. Cotton Outlook (2014). *Improving Cotton Yield – Uzbekistan Expands New Variety.* Available at:https://www.cotlook.com/2014/10/21/improving-cotton-yield-uzbekistan-expands-new-variety
24. Deng F., Tu L., Tan J., Li Y., Nie Y., Zhang X. (2012). GbPDF1 is involved in cotton fiber initiation via the core cis-element HDZIP2ATATHB2. *Plant Physiol.* 158 890–904. 10.1104/pp.111.186742
25. Federation of Indian Chambers of Commerce and Industry [FICCI] Report (2012). *Cotton 2020: Roadmap of Sustainable Production.* Available at: www.tierraseedscience.com/pdf/cotton-2020-roadmap-india.pdf

26. Firmino A. A., Fonseca F. C., de Macedo L. L., Coelho R. R., Antonino de Souza J. D., Jr., Togawa R. C., et al. (2013). Transcriptome analysis in cotton boll weevil (*Anthonomus grandis*) and RNA interference in insect pests. *PLoS ONE* 8:e85079 10.1371/journal.pone.0085079

27. Fu W., Shen Y., Hao J., Wu J., Ke L., Wu C., et al. (2015). Acyl-CoA N-acyltransferase influences fertility by regulating lipid metabolism and jasmonic acid biogenesis in cotton. *Sci. Rep.* 5:1179010.1038/srep11790

28. Gao W., Long L., Zhu L. F., Xu L., Gao W. H., Sun L. Q., et al. (2013b). Proteomic and virus-induced gene silencing (VIGS) analyses reveal that gossypol, brassinosteroids, and jasmonic acid contribute to the resistance of cotton to *Verticillium dahliae*. *Mol. Cell. Proteomics* 12 3690–3703. 10.1074/mcp.M113.031013

29. Gao X., Britt R. C., Jr., Shan L., He P. (2011b). Agrobacterium-mediated virus-induced gene silencing assay in cotton. *J. Vis. Exp.* 54 pii2938 10.3791/2938

30. Gao X., Li F., Li M., Kianinejad A. S., Dever J. K., Wheeler T. A., et al. (2013a). Cotton GhBAK1 mediates *Verticillium* wilt resistance and cell death. *J. Integr. Plant Biol.* 55 586–596. 10.1111/jipb.12064

31. Gao X., Shan L. (2013). Functional genomic analysis of cotton genes with agrobacterium-mediated virus-induced gene silencing. *Methods Mol. Biol.* 975 157–165. 10.1007/978-1-62703-278-0_12

32. Gao X., Wheelerm T., Lim Z., Kenerley C. M., He P., Shan L. (2011a). Silencing GhNDR1 and GhMKK2 compromises cotton resistance to *Verticillium* wilt. *Plant J.* 66 293–305. 10.1111/j.1365-13X.2011.04491.x

33. Goldstein D. A., Tinland B., Gilbertson L. A., Staub J. M., Bannon G. A., Goodman R. E., et al. (2005). Human safety and genetically modified plants: a review of antibiotic resistance markers and future transformation selection technologies. *J. Appl. Microbiol.* 99 7–23. 10.1111/j.1365-2672.2005.02595.x

34. Gu Z., Huang C., Li F., Zhou X. (2014). A versatile system for functional analysis of genes and microRNAs in cotton. *Plant Biotechnol. J.* 12 638–649. 10.1111/pbi.12169100

35. Haigler C. H., Betancur L., Stiff M. R., Tuttle J. R. (2012). Cotton fiber: a powerful single-cell model for cell wall and cellulose research. *Front. Plant Sci.* 3:104 10.3389/fpls.2012.00104

36. Hake K. (2012). *Climate Disruptions to Fiber Yield Growth. Cotton Research Journal July-December.* 160–166. Available at: http://www.cicr.org.in/isci-image/3.pdf

37. Hannon G. J. (2002). RNA interference. *Nature* 418 244–251. 10.1038/418244a

38. Hao J., Tu L., Hu H., Tan J., Deng F., Tang W., et al. (2012). GbTCP, a cotton TCP transcription factor, confers fiber elongation and root hair development by a complex regulating system. *J. Exp. Bot.*63 6267–6281. 10.1093/jxb/ers278

39. Heinemann J. A., Agapito-Tenfen S. Z., Carman J. A. (2013). A comparative evaluation of the regulation of GM crops or products containing dsRNA and suggested improvements to risk assessments. *Environ. Int.* 55 43–55. 10.1016/j.envint.2013.02.010

40. Higuchi M., Yoshizumi T., Kuriyama T., Hara H., Akagi C., Shimada H., et al. (2009). Simple construction of plant RNAi vectors using long oligonucleotides. *J. Plant Res.* 122 477–482. 10.1007/s10265-009-0228-6

41. Holmquist G., Ashley T. (2006). Chromosome organization and chromatin modification: influence on genome function and evolution. *Cytogenet. Genome Res.* 114 96–125. 10.1159/000093326

42. Hu L., Yang X., Yuan D., Zeng F., Zhang X. (2011). GhHmgB3 deficiency deregulates proliferation and differentiation of cells during somatic embryogenesis in cotton. *Plant Biotechnol. J.* 9 1038–1048. 10.1111/j.1467-7652.2011.00617.x

43. Huang G. Q., Gong S. Y., Xu W. L., Li W., Li P., Zhang C. J., et al. (2013). A fasciclin-like arabinogalactan protein, GhFLA1, is involved in fiber initiation and elongation of cotton. *Plant Physiol.* 161 1278–1290. 10.1104/pp.112.203760

44. Iqbal J., Reddy O. U., El-Zik K. M., Pepper A. M. (2001). A genetic bottleneck in the 'evolution under domestication' of upland cotton *Gossypium hirsutum* L. examined using DNA fingerprinting.*Theor. Appl. Genet.* 103 547–554. 10.1007/PL00002908

45. Jakymiw A., Lian S., Eystathioy T., Li S., Satoh M., Hamel J., et al. (2005). Disruption of P bodies impairs mammalian RNA interference. *Nat. Cell Biol.* 7 1267–1274. 10.1038/ncb1334

46. Jiang Y., Guo W., Zhu H., Ruan Y. L., Zhang T. (2012). Overexpression of GhSusA1 increases plant biomass and improves cotton fiber yield and quality. *Plant Biotechnol. J.* 10 301–312. 10.1111/j.1467-7652.2011.00662.x

47. Jin X., Pang Y., Jia F., Xiao G., Li Q., Zhu Y. (2013). A potential role for CHH DNA methylation in cotton fiber growth patterns. *PLoS ONE* 12:e60547 10.1371/journal.pone.0060547

48. John M. E. (1996). Structural characterization of genes corresponding to cotton fiber mRNA, E6: reduced E6 protein in transgenic plants by antisense gene. *Plant Mol. Biol.* 30 297–306. 10.1007/BF00020115

49. Keyte A. L., Percifield R., Liu B., Wendel J. F. (2006). Infraspecific DNA methylation polymorphism in cotton (*Gossypium hirsutum* L.). *J. Hered.* 97 444–450. 10.1093/jhered/esl023

50. Kumar J., Gunapati S., Kumar J., Kumari A., Kumar A., Tuli R., et al. (2014). Virus-induced gene silencing using a modified betasatellite: a potential candidate for functional genomics of crops. *Arch. Virol.* 159 2109–2113. 10.1007/s00705-014-2039-x

51. Kwak P. B., Wang Q. Q., Chen X. S., Qiu C. X., Yang Z. M. (2009). Enrichment of a set of microRNAs during the cotton fiber development. *BMC Genomics* 10:457 10.1186/1471-2164-10-457

52. Larson M. H., Gilbert L. A., Wang X., Lim W. A., Weissman J. S., Qi L. S. (2013). CRISPR interference (CRISPRi) for sequence-specific control of gene

expression. *Nat. Protocols* 8 2180–2196. 10.1038/nprot.2013.132

53. Li D. D., Ruan X. M., Zhang J., Wu Y. J., Wang X. L., Li X. B. (2013). Cotton plasma membrane intrinsic protein 2s (PIP2s) selectively interact to regulate their water channel activities and are required for fiber development. *New Phytol.* 199 695–707. 10.1111/nph.12309
54. Li F., Fan G., Lu C., Xiao G., Zou C., Kohel R. J., et al. (2015a). Genome sequence of cultivated Upland cotton (*Gossypium hirsutum* TM-1) provides insights into genome evolution. *Nat. Biotechnol.*33 524–530. 10.1038/nbt.3208
55. Li F., Fan G., Wang K., Sun F., Yuan Y., Song G., et al. (2014). Genome sequence of the cultivated cotton *Gossypium arboreum*. *Nat. Genet.* 46 567–572. 10.1038/ng.2987
56. Li L. C., Okino S. T., Zhao H., Pookot D., Place R. F., Urakami S., et al. (2006). Small dsRNAs induce transcriptional activation in human cells. *Proc. Natl. Acad. Sci. U.S.A.* 103 17337–17342. 10.1073/pnas.0607015103
57. Li X. B., Fan X. P., Wang X. L., Cai L., Yang W. C. (2005). The cotton ACTIN1 gene is functionally expressed in fibers and participates in fiber elongation. *Plant Cell* 17 859–875. 10.1105/tpc.104.029629
58. Li Y., Liu D., Tu L., Zhang X., Wang L., Zhu L., et al. (2010). Suppression of GhAGP4 gene expression repressed the initiation and elongation of cotton fiber. *Plant Cell Rep.* 29 193–202. 10.1007/s00299-009-0812-1
59. Li Y., Tu L., Pettolino F. A., Ji S., Hao J., Yuan D., et al. (2015b). GbEXPATR, a species-specific expansin, enhances cotton fiber elongation through cell wall restructuring. *Plant Biotechnol. J.* [Epub ahead of print].10.1111/pbi.12450
60. Liang G., He H., Li Y., Yu D. (2012). A new strategy for construction of artificial miRNA vectors in*Arabidopsis*. *Planta* 235 1421–1429. 10.1007/s00425-012-1610-5
61. Liu G. J., Xiao G. H., Liu N. J., Liu D., Chen P. S., Qin Y. M., et al. (2015b). Targeted lipidomics studies reveal that linolenic acid promotes cotton fiber elongation by activating phosphatidylinositol and phosphatidylinositol monophosphate biosynthesis. *Mol. Plant* 8 911–921. 10.1016/j.molp.2015.02.010
62. Liu N., Tu L., Tang W., Gao W., Lindsey K., Zhang X. (2014). Small RNA and degradome profiling reveals a role for miRNAs and their targets in the developing fibers of *Gossypium barbadense*. *Plant J.*80 331–344. 10.1111/tpj.12636
63. Liu Q., Singh S., Chapman K., Green A. (2009). "Bridging traditional and molecular genetics in modifying cottonseed oil," in *Genetics and Genomics of Cotton. Plant Genetics and genomics, Crops and Models 3*, ed. Paterson A. H., editor. (New York, NY: Springer Science+Business Media;), 353–382. 10.1007/978-0-387-70810-2_1541
64. Liu Q., Singh S., Green A. (2000). Genetic modification of cotton seed oil using inverted-repeat gene-silencing techniques. *Biochem. Soc. Trans.* 28 927–929. 10.1042/bst0280927
65. Liu Q., Singh S. P., Green A. G. (2002). High-stearic and High-oleic cottonseed oils produced by hairpin RNA-mediated post-transcriptional gene silencing. *Plant Physiol.* 129 1732–1743. 10.1104/pp.001933
66. Liu X., Zhao B., Zheng H. J., Hu Y., Lu G., Yang C. Q., et al. (2015a). *Gossypium barbadense*genome sequence provides insight into the evolution of extra-long staple fiber and specialized metabolites. *Sci. Rep.* 5:14139 10.1038/srep14139
67. Luo M., Xiao Y., Li X., Lu X., Deng W., Li D., et al. (2007). GhDET2, a steroid 5alpha-reductase, plays an important role in cotton fiber cell initiation and elongation. *Plant J.* 51 419–430. 10.1111/j.1365-313X.2007.03144.x
68. Machado A., Wu Y., Yang Y., Llewellyn D. J., Dennis E. S. (2009). The MYB transcription factor GhMYB25 regulates early fiber and trichome development. *Plant J.* 59 52–62. 10.1111/j.1365-313X.2009.03847
69. Macron D. (2014). *New Data Published on RNAi-based Cotton Plant Improvement.* Available at:https://www.genomeweb.com/rnai/new-data-published-rnai-based-cotton-plant-improvement
70. Mao Y. B., Cai W. J., Wang J. W., Hong G. J., Tao X. Y., Wang L. J., et al. (2007). Silencing a cotton bollworm P450 monooxygenase gene by plant-mediated RNAi impairs larval tolerance of gossypol.*Nat. Biotechnol.* 25 1307–1313. 10.1038/nbt1352
71. Mao Y. B., Tao X. Y., Xue X. Y., Wang L. J., Chen X. Y. (2011). Cotton plants expressing CYP6AE14 double-stranded RNA show enhanced resistance to bollworms. *Transgenic Res.* 20 665–673. 10.1007/s11248-010-9450-1
72. Mao Y. B., Xue X. Y., Tao X. Y., Yang C. Q., Wang L. J., Chen X. Y. (2013). Cysteine protease enhances plant-mediated bollworm RNA interference. *Plant Mol. Biol.* 83 119–129. 10.1007/s11103-013-0030-7
73. Martin G. S., Liu J., Benedict C. R., Stipanovic R. D., Magill C. W. (2003). Reduced levels of cadinane sesquiterpenoids in cotton plants expressing antisense (+)-delta-cadinene synthase.*Phytochemistry* 62 31–38. 10.1016/S0031-9422(02)00432-6
74. Miller P. A., Rawlings J. O. (1967). Selection for increased lint yield and correlated responses in upland cotton, *Gossypium hirsutum* L. *Crop Sci.* 7 637–640. 10.2135/cropsci1967.0011183X000700060024x
75. Napoli C., Lemieux C., Jorgensen R. (1990). Introduction of a chimeric chalcone synthase gene into petunia results in reversible co-suppression of homologous genes in trans. *Plant Cell* 2 279–289. 10.1105/tpc.2.4.279
76. Palle S. R., Campbell L. M., Pandeya D., Puckhaber L., Tollack L. K., Marcel S., et al. (2013).RNAi-mediated Ultra-low gossypol cottonseed trait: performance of transgenic lines under field conditions. *Plant Biotechnol. J.* 11 296–304. 10.1111/pbi.12013
77. Pang J., Zhu Y., Li Q., Liu J., Tian Y., Liu Y., et al. (2013). Development of *Agrobacterium*-mediated virus-induced gene silencing and performance evaluation of

four marker genes in *Gossypium barbadense*. *PLoS ONE* 8:e73211 10.1371/journal.pone.0073211

78. Pang M., Woodward A. W., Agarwal V., Guan X., Ha M., Ramachandran V., et al. (2009). Genome-wide analysis reveals rapid and dynamic changes in miRNA and siRNA sequence and expression during ovule and fiber development in allotetraploid cotton (*Gossypium hirsutum* L.). *Genome Biol.*10:R122 10.1186/gb-2009-10-11-r122

79. Paterson A. H., Wendel J. F., Gundlach H., Guo H., Jenkins J., Jin D., et al. (2012). Repeated polyploidization of *Gossypium* genomes and the evolution of spinnable cotton fibers. *Nature* 492 423–427. 10.1038/nature11798

80. Potikha T. S., Collins C. C., Johnson D. I., Delmer D. P., Levine A. (1999). The involvement of hydrogen peroxide in the differentiation of secondary walls in cotton fibers. *Plant Physiol.* 119 849–858. 10.1104/pp.119.3.849

81. Pu L., Li Q., Fan X., Yang W., Xue Y. (2008). The R2R3 MYB transcription factor GhMYB109 is required for cotton fiber development. *Genetics* 180 811–820. 10.1534/genetics.108.093070

82. Qiu L., Hou L., Zhang B., Liu L., Li B., Deng P., et al. (2015). Cadherin is involved in the action of*Bacillus thuringiensis* toxins Cry1Ac and Cry2Aa in the beet armyworm, *Spodoptera exigua*. *J. Invertebr. Pathol.* 127 47–53. 10.1016/j.jip.2015.02.009

83. Qu J., Ye J., Geng Y. F., Sun Y. W., Gao S. Q., Zhang B. P., et al. (2012). Dissecting functions of KATANIN and WRINKLED1 in cotton fiber development by virus-induced gene silencing. *Plant Physiol.* 160 738–748. 10.1104/pp.112.198564

84. Ramon M., Devos Y., Lanzoni A., Liu Y., Gomes A., Gennaro A., et al. (2014). RNAi-based GM plants: food for thought for risk assessors. *Plant Biotechnol. J.* 12 1271–1273. 10.1111/pbi.12305

85. Rathore K. S., Sundaram S., Sunilkumar G., Campbell L. M., Puckhaber L., Marcel S., et al. (2012).Ultra-low gossypol cottonseed: generational stability of the seed-specific, RNAi-mediated phenotype and resumption of terpenoid profile following seed germination. *Plant Biotechnol. J.* 10 174–183. 10.1111/j.1467-7652.2011.00652.x

86. Renny-Byfield S., Gallagher J. P., Grover C. E., Szadkowski E., Page J. T., Udall J. A., et al. (2014). Ancient gene duplicates in *Gossypium* (cotton) exhibit near-complete expression divergence. *Genome Biol. Evol.* 6 559–571. 10.1093/gbe/evu037

87. Ricroch A. E., Bergé J. B., Kuntz M. (2011). Evaluation of genetically engineered crops using transcriptomic, proteomic, and metabolomic profiling techniques. *Plant Physiol.* 155 1752–1761. 10.1104/pp.111.173609

88. Roberts A. F., Devos Y., Lemgo G. N., Zhou X. (2015). Biosafety research for non-target organism risk assessment of RNAi-based GE plants. *Front. Plant Sci.* 6:958 10.3389/fpls.2015.00958

89. Romano N., Macino G. (1992). Quelling: transient inactivation of gene expression in *Neurospora crassa* by transformation with homologous sequences. *Mol. Microbiol.* 6 3343–3353. 10.1111/j.1365-2958.1992.tb02202.x

90. Ruan M. B., Zhao Y. T., Meng Z. H., Wang X. J., Yang W. C. (2009). Conserved miRNA analysis in*Gossypium hirsutum* through small RNA sequencing. *Genomics* 94 263–268. 10.1016/j.ygeno.2009.07.002

91. Ruan Y. L., Llewellyn D. J., Furbank R. T. (2003). Suppression of sucrose synthase gene expression represses cotton fiber cell initiation, elongation, and seed development. *Plant Cell* 15 952–964. 10.1105/tpc.010108

92. Saha S., Stelly D. M., Raska D. A., Wu J., Jenkins J. N., McCarty J. C., et al. (2012). "Chromosome substitution lines: concept, development and utilization in the genetic improvement of upland cotton," in *Plant Breeding*, ed. Abdurakhmonov I. Y., editor. (Rijeka: InTech Press;), 107–128. 10.5772/35585

93. Sen G., Blau H. (2005). Argonaute 2/RISC resides in sites of mammalian mRNA decay known as cytoplasmic bodies. *Nat. Cell Biol.* 7 633–636. 10.1038/ncb1265

94. Shan C. M., Shangguan X. X., Zhao B., Zhang X. F., Chao L. M., Yang C. Q., et al. (2014). Control of cotton fiber elongation by a homeodomain transcription factor GhHOX3. *Nat Comm.* 5:551910.1038/ncomms6519

95. Shang X., Chai Q., Zhang Q., Jiang J., Zhang T., Guo W., et al. (2015). Down-regulation of the cotton endo-1,4-β-glucanase gene KOR1 disrupts endosperm cellularization, delays embryo development, and reduces early seedling vigour. *J. Exp. Bot.* 66 3071–3083. 10.1093/jxb/erv111

96. Shapulatov U. M., Buriev Z. T., Ulloa M., Saha S., Devor E. J., Ayubov M., et al. (2015).Characterization of small RNAs and their targets from *Fusarium* oxysporum infected and non-infected cotton root tissues. *Plant Mol. Biol. Rep.* 1–9. 10.1007/s11105-015-0945-z

97. Shi Y. L., Guo S. D., Zhang R., Meng Z. G., Ren M. Z. (2014). The role of Somatic embryogenesis receptor-like kinase 1 in controlling pollen production of the *Gossypium* anther. *Mol. Biol. Rep.* 41411–422. 10.1007/s11033-013-2875-x

98. Shweta, Khan J. A. (2014). In silico prediction of cotton (*Gossypium hirsutum*) encoded microRNAs targets in the genome of *Cotton leaf* curl *Allahabad* virus. *Bioinformation* 10 251–255. 10.6026/97320630010251

99. Simó C., Ibáñez C., Valdés A., Cifuentes A., García-Cañas V. (2014). Metabolomics of genetically modified crops. *Inter. J. Mol. Sci.* 15 18941–18966. 10.3390/ijms151018941

100. Stokstad E. (2006). *Taking the Toxin Out of Cotton. Science News.* Available at:http://news.sciencemag.org/2006/11/taking-toxin-out-cotton

101. Su X., Qi X., Cheng H. (2014). Molecular cloning and characterization of enhanced disease susceptibility 1 (EDS1) from *Gossypium barbadense*. *Mol. Biol. Rep.* 41 3821–3828. 10.1007/s11033-014-3248-9

102. Sunilkumar G., Campbell L. M., Hossen M., Connell J. P., Hernandez E., Reddy A. S., et al. (2005).A comprehensive study of the use of a homologous

promoter in antisense cotton lines exhibiting a high seed oleic acid phenotype. *Plant Biotechnol. J.* 3 319–330. 10.1111/j.1467-7652.2005.00126.x

103. Sunilkumar G., Campbell L. M., Puckhaber L., Stipanovic R. D., Rathore K. S. (2006). Engineering cottonseed for use in human nutrition by tissue-specific reduction of toxic gossypol. *Proc. Natl. Acad. Sci. U.S.A.* 103 18054–18059. 10.1073/pnas.0605389103
104. Tan J., Tu L., Deng F., Hu H., Nie Y., Zhang X. (2013). A genetic and metabolic analysis revealed that cotton fiber cell development was retarded by flavonoid naringenin. *Plant Physiol.* 162 86–95. 10.1104/pp.112.212142
105. Tang W., He Y., Tu L., Wang M., Li Y., Ruan Y. L., et al. (2014). Down-regulating annexin gene GhAnn2 inhibits cotton fiber elongation and decreases Ca2+ influx at the cell apex. *Plant Mol. Biol.*85 613–625. 10.1007/s11103-014-0208-7
106. Tang W., Samuels V., Whitley N., Bloom N., DeLa Garza T., Newton R. J. (2004). Post-transcriptional gene silencing induced by short interfering RNAs in cultured transgenic plant cells.*Genomics Proteomics Bioinformatics* 2 97–108.
107. Tang W., Weidner D. A., Hu B. Y., Newton R. J., Hu X. H. (2006). Efficient delivery of small interfering RNA to plant cells by a nanosecond pulsed laser-induced stress wave for posttranscriptional gene silencing. *Plant Sci.* 171 375–381. 10.1016/j.plantsci.2006.04.005
108. Tao X. Y., Xue X. Y., Huang Y. P., Chen X. Y., Mao Y. B. (2012). Gossypol-enhanced P450 gene pool contributes to cotton bollworm tolerance to a pyrethroid insecticide. *Mol. Ecol.* 21 4371–4385. 10.1111/j.1365-294X.2012.05548.x
109. Townsend B. J., Poole A., Blake C. J., Llewellyn D. J. (2005). Antisense suppression of a (+)-delta-cadinene synthase gene in cotton prevents the induction of this defense response gene during bacterial blight infection but not its constitutive expression. *Plant Physiol.* 138 516–528. 10.1104/pp.104.056010
110. United States Department of Agriculture [USDA] Portal (2015). *Animal and Plant Health Inspection Service, Biotechnology Regulatory Services (BRS) Permits, Notifications, and Petitions.* Available at:https://www.aphis.usda.gov/wps/portal/aphis/ourfocus/biotechnology/sa_permits_notifications_and_petitions
111. United States Department of Agriculture [USDA] Report (2013). *Cotton and Wool Outlook.* Available at: http://www.ers.usda.gov/publications/cws-cotton-and-wool-outlook/cws-13e.aspx#.UjhSRRzCBmg
112. Walford S. A., Wu Y., Llewellyn D. J., Dennis E. S. (2011). GhMYB25-like: a key factor in early cotton fiber development. *Plant J.* 65 785–797. 10.1111/j.1365-313X.2010.04464.x
113. Walford S. A., Wu Y., Llewellyn D. J., Dennis E. S. (2012). Epidermal cell differentiation in cotton mediated by the homeodomain leucine zipper gene, GhHD-1. *Plant J.* 71 464–478. 10.1111/j.1365-313X.2012.05003.x
114. Wang H., Guo Y., Lv F., Zhu H., Wu S., Jiang Y., et al. (2010a). The essential role of GhPEL gene, encoding a pectate lyase, in cell wall loosening by depolymerization of the de-esterified pectin during fiber elongation in cotton. *Plant Mol. Biol.* 72 397–406. 10.1007/s11103-009-9578-7
115. Wang L., Li X. R., Lian H., Ni D. A., He Y. K., Chen X. Y., et al. (2010b). Evidence that high activity of vacuolar invertase is required for cotton fiber and *Arabidopsis* root elongation through osmotic dependent and independent pathways, respectively. *Plant Physiol.* 154 744–756. 10.1104/pp.110.162487
116. Wang K., Wang Z., Li F., Ye W., Wang J., Song G., et al. (2012). The draft genome of a diploid cotton*Gossypium raimondii*. *Nat. Genet.* 44 1098–1103. 10.1038/ng.2371
117. Wang L., Cook A., Patrick J. W., Chen X. Y., Ruan Y. L. (2014). Silencing the vacuolar invertase gene GhVIN1 blocks cotton fiber initiation from the ovule epidermis, probably by suppressing a cohort of regulatory genes via sugar signaling. *Plant J.* 78 686–696. 10.1111/tpj.12512
118. Wang L., Wang X., Jin X., Jia R., Huang Q., Tan Y., et al. (2015b). Comparative proteomics of Bt-transgenic and non-transgenic cotton leaves. *Proteome Sci.* 13:15 10.1186/s12953-015-0071-8
119. Wang M., Yuan D., Tu L., Gao W., He Y., Hu H., et al. (2015a). Long noncoding RNAs and their proposed functions in fiber development of cotton (*Gossypium* spp.). *New Phytol.* 207 1181–1197. 10.1111/nph.13429
120. Wang Q., Zhang B. (2015). MicroRNAs in cotton: an open world needs more exploration. *Planta* 2411303–1312. 10.1007/s00425-015-2282-8
121. Wang Z., Dong Y., Desneux N., Niu C. (2013). RNAi silencing of the HaHMG-CoA reductase gene inhibits oviposition in the Helicoverpa armigera cotton bollworm. *PLoS ONE* 8:e6773210.1371/journal.pone.0067732
122. Xu L., Zhang W., He X., Liu M., Zhang K., Shaban M., et al. (2014). Functional characterization of cotton genes responsive to *Verticillium dahliae* through bioinformatics and reverse genetics strategies.*J. Exp. Bot.* 65 6679–6692. 10.1093/jxb/eru393
123. Xu W. L., Zhang D. J., Wu Y. F., Qin L. X., Huang G. Q., Li J., et al. (2013). Cotton PRP5 gene encoding a proline-rich protein is involved in fiber development. *Plant Mol. Biol.* 82 353–365. 10.1007/s11103-013-0066-8
124. Yin Z., Li Y., Yu J., Liu Y., Li C., Han X., et al. (2012). Difference in miRNA expression profiles between two cotton cultivars with distinct salt sensitivity. *Mol. Biol. Rep.* 39 4961–4970. 10.1007/s11033-011-1292-2
125. Yoo M. J., Szadkowski E., Wendel J. F. (2013). Homoeolog expression bias and expression level dominance in allopolyploid cotton. *Heredity (Edinburgh)* 110 171–180. 10.1038/hdy.2012.94
126. Zhang B., Yang Y., Chen T., Yu W., Liu T., Li H., et al. (2012). Island cotton Gbve1 gene encoding a receptor-like protein confers resistance to both defoliating and non-defoliating isolates of *Verticillium dahliae*. *PLoS ONE* 7:e51091 10.1371/journal.pone.0051091
127. Zhang H., Wan Q., Ye W., Lv Y., Wu H., Zhang T. (2013a). Genome-wide analysis of small RNA and novel microRNA discovery during fiber and seed initial development in *Gossypium hirsutum. L.PLoS ONE.* 8:e69743 10.1371/journal.pone.0069743

128. Zhang Y., Zhang F., Li X., Baller J. A., Qi Y., Starker C. G., et al. (2013b). Transcription activator-like effector nucleases enable efficient plant genome engineering. *Plant Physiol.* 161 20–27. 10.1104/pp.112.205179

129. Zhang H. B., Li Y., Wang B., Chee P. W. (2008). Recent advances in cotton genomics. *Int. J. Plant Genom.* 2008:742304 10.1155/2008/742304

130. Zhang M., Zheng X., Song S., Zeng Q., Hou L., Li D., et al. (2011). Spatiotemporal manipulation of auxin biosynthesis in cotton ovule epidermal cells enhances fibre yield and quality. *Nat. Biotechnol.* 29453–458. 10.1038/nbt.1843

131. Zhang T., Hu Y., Jiang W., Fang L., Guan X., Chen J., et al. (2015). Sequencing of allotetraploid cotton (*Gossypium hirsutum* L. *acc. TM*-1) provides a resource for fiber improvement. *Nat. Biotechnol.* 33 531–537. 10.1038/nbt.3207

132. Zhang X., Wang L., Xu X., Cai C., Guo W. (2014). Genome-wide identification of mitogen-activated protein kinase gene family in *Gossypium raimondii* and the function of their corresponding orthologs in tetraploid cultivated cotton. *BMC Plant Biol.* 14:345 10.1186/s12870-014-0345-9

133. Zhu Q. H., Fan L., Liu Y., Xu H., Llewellyn D., Wilson I. (2013). miR482 regulation of NBS-LRR defense genes during fungal pathogen infection in cotton. *PLoS ONE* 8:e8439010.1371/journal.pone.0084390

134. Zhu Y., Wang H., Peng Q., Tang Y., Xia G., Wu J., et al. (2015). Functional characterization of an anthocyanidin reductase gene from the fibers of upland cotton (*Gossypium hirsutum*). *Planta* 2411075–1089. 10.1007/s00425-014-2238-4

CHAPTER 16

REVIEW OF RECENT TRANSGENIC STUDIES ON ABIOTIC STRESS TOLERANCE AND FUTURE MOLECULAR BREEDING IN POTATO

Akira Kikuchi, Huu Duc Huynh, Tsukasa Endo and Kazuo Watanabe

Faculty of Life and Environmental Sciences, Gene Research Center, University of Tsukuba, 1-1-1 Ten-noudai, Tsukuba, Ibaraki 305-8752, Japan

ABSTRACT

Global warming has become a major issue within the last decade. Traditional breeding programs for potato have focused on increasing productivity and quality and disease resistance, thus, modern cultivars have limited tolerance of abiotic stresses. The introgression of abiotic stress tolerance into modern cultivars is essential work for the future. Recently, many studies have investigated abiotic stress using transgenic techniques. This manuscript focuses on the study of abiotic stress, in particular drought, salinity and low temperature, during this century. Dividing studies into these three stress categories for this review was difficult. Thus, based on the study title and the transgene property, transgenic studies were classified into five categories in this review; oxidative scavengers, transcriptional factors, and above three abiotic categories. The review focuses on studies that investigate confer of stress tolerance and the identification of responsible factors, including wild relatives. From a practical application perspective, further evaluation of transgenic potato with abiotic stress tolerance is required. Although potato plants, including wild species, have a large potential for abiotic stress tolerance, exploration of the factors responsible for conferring this tolerance is still developing. Molecular breeding, including genetic engineering and conventional breeding using DNA markers, is expected to develop in the future.

INTRODUCTION

Since the 19th century, the CO_2 concentration in the atmosphere has been gradually increasing and recently reached 400 ppm at some observation locations (Japan Meteorological Agency 2014). Global warming has become a major issue in the last decade. Although it may be difficult to associate raising CO_2 concentrations with extreme weather conditions, such as heavy rain, drought, or low (high) temperatures, these conditions appear to be occurring with increased frequency (Japan Meteorological Agency 2013). Agriculture was established by adapting plants to grow in certain climatic conditions, thus, productivity depends on the weather conditions and typically decreases during, or following, periods of extreme weather (Boyer 1982). In addition, various human activities have resulted in saline or drought conditions, thereby decreasing the available area of arable land. There are reports that over 800 million ha of land throughout the world are saline, covering over 6% of the total land area (Munns 2005). Further, over 20% of arable land has salinity damage (Munns 2005). Irrigation has contributed to the availability of arable land, with recent estimates suggesting that around 20% of the world's arable land is irrigated and provides 40% of food and feed (Kondrak *et al.* 2012). However, 50% of irrigated land suffers from salinity damage (Geilfus *et al.* 2010). Some studies predict that 30% of arable land will be lost within 25 years and around

Citation: Akira Kikuchi, Huu Duc Huynh, Tsukasa Endo and Kazuo Watanabe, Review of recent transgenic studies on abiotic stress tolerance and future molecular breeding in potato, doi:10.1270/jsbbs.65.85.

half of arable land will be unavailable by 2050 (Wang *et al.* 2003). Currently, 70% of the world's fresh water is consumed by agriculture (Kondrak *et al.* 2012). Thus, effective utilisation of limited water resources is an important aspect to consider in the near future.

Potato is the fourth highest produced crop in the world and the most important non-cereal food crop (Pino *et al.* 2007), with world production estimated to be over 350 million tons in 2013 (http://faostat.fao.org). Potato is a versatile crop, and is utilized as not only food and feed but also in industrial materials and fuels (Ortiz and Watanabe 2004). Potato is a temperate or continental zone crop and is cultivated worldwide in areas such as Northern China, Europe, North America and the highlands of South America (Pino *et al.* 2007). Although detailed descriptions about the origin and history of the potato will not be discussed here, there are various types of wild relative species present in its original region of origin and their habitat ranges from cool highland zones to tropical lowlands (Hawkes 1994). The genetic variation of these wild relatives is expected to be high. Potato has limited abiotic stress tolerance (Shin *et al.* 2011, Vasquez-Robinet *et al.* 2008). In particular, drought stress is inimical for potato due to their shallow root system (Bouaziz *et al.* 2012), which has a rooting zone between 50 cm to 80 cm depth (Dalla Costa *et al.* 1997, Ekanayaka and Midmore 1992). In addition to drought stress, potato is vulnerable to salt stress. In an experimental evaluation, production decreased by half under 50 mM NaCl conditions, and was negligible under 150 mM NaCl conditions (Sherraf *et al.* 1994). In a field evaluation, production of potato was also reduced by half under 5.9 mS/m electric conductivity (EC) (Kotuby-amacher *et al.* 2000). Furthermore, potato suffers damage at −3°C and has no ability to acclimate to cold conditions (Chen and Li 1980). Therefore, of the temperate crops, potato is susceptible to low temperatures. From an agricultural point of view, frost is one of the most harmful low temperature conditions that causes damage to crops (Barrientos *et al.* 1994, Chen and Li 1980, Costa and Li 1993, Vega and Bamberg 1995).

Traditional breeding of crops has focused on increasing productivity and quality and disease resistance, thus, modern cultivars have limited abiotic stress tolerance. The cold tolerant wild species *Solanum commersonii*was used as a genetic donor of cold tolerance and cold acclimation for *S. tuberosum*. However, this breeding effort failed to confer cold tolerance traits to *S. tuberosum* (Cardi *et al.* 1993, Estrada 1982, Estrada *et al.* 1993, Iovene *et al.* 2004, Pavek and Corsini 2001). Potato is the one of the highest crops in the Harvest Index (HI), like sweet potato, and its value is over 60% (Mazurczyk *et al.* 2009). Since HI indicates the available portion of each crop, high HI crops are important for the establishment of sustainable societies in the future. Investigation of abiotic stress tolerance in potato is essential for future cultivation efforts and molecular breeding of potato, including genetic engineering, is expected to develop in anticipation of future crop requirements.

Recently, many studies have been undertaken to investigate abiotic stress in potato using transgenic techniques. This manuscript focuses on the three types of abiotic stress, drought, salinity and low temperature, and discusses recent studies from 2000 onwards. These three abiotic stresses have specific individual characteristics, but they also overlap, and it was difficult to clearly divide them into three categories. Furthermore, these environmental stresses are known to enhance the accumulation of ROS in plant cells. Various types of ROS result in the oxidation of many cellar components and lead to oxidative destruction of the cell (Asada and Takahashi 1987, Mittler 2002). Enhancement of the ROS scavenging system is expected to confer tolerance to various types of stresses in plants. On the other hand, transcriptional factors (TFs) govern various related downstream genes. In fact, overexpression of abiotic stress related TF also confers the abiotic stress tolerance to plants. Thus, in this review, based on the study title and the transgene property, transgenic studies were classified into five categories: oxidative scavengers, transcriptional factors and the three abiotic stresses, drought, salinity and low temperature (Table 1). In addition to transgenic potato, exploring studies of responsible genes for abiotic stress tolerance also are introduced in this review. Potato is high potential crop plant that can acquire abiotic stress tolerance by molecular breeding including conventional methods with molecular markers and GE techniques.

Table 1: Details of recent transgenic potato studies on abiotic stress referred to in this manuscript

	Host cultivar	Transgene	Origin	Promoter	Treatment*	Plant type**	Evaluation***	Reference
Drought	Desiree	*anti-FDA*	Potato	*35S*	DA	PG	PC	Ambard-Bretteville *et al.* 2003
	Jowon	*YUCCA6*	*Arabidopsis*	*35S*	WW	PG	Sur	Kim *et al.* 2013
	White Lady	*TPS1*	Yeast	*DS2*	Nor	PG	Tr	Kondrak *et al.* 2011
	Gannongshu2	*BADH*	Spinach	*rd29A*	0.5 M Na/15% PEG	Inv	G	Zhang *et al.* 2011
	Superior	*codA*	Rhizobacterium	*SWPA2*	5 μM MV+0.4% Sol/WW/0.15 M Na	SC/PG/SC	G/G/G	Ahmad *et al.* 2008
	Albatros	*GgpPS*	Rhizobacterium	*rd29A/35S*	WD/0.1 M Na	PG	G	Sievers *et al.* 2013
	Desiree	*PR-10a*	Potato	*MAN*	0.5 M Sol or 0.32 M Na	Inv (callus)	G	El-Banna *et al.* 2010
	Superior	*codA*	Rhizobacterium	*SWPA2*	20% PEG	SC	MDA	Cheng *et al.* 2013a
	Desiree	*SST+FFT*	Globe artichoke	*35S*	WD	SC	TP	Knipp and Honermeier 2006
	White Lady	*TPS1*	Yeast	*DS2*	WD	PG	Tr	Kondrak *et al.* 2012
	White Lady	*TPS1*	Yeast	*DS2*	WW	SC	RWC/CF	Stiller *et al.* 2008
Salinity	Jubilei Zhukova	*NHX2*	Barley	*35S*	0~0.25 M Na	Inv	G	Bayat *et al.* 2010
	Gannong2	*NHX1*	*Arabidopsis*	*35S*	0~9 g/L Na	PG	Sur, Y	Wang *et al.* 2010
	Nicola	*P5CS*	*Arabidopsis*	*35S*	0~0.18 M Na	Inv	Y	Hmida-Sayaria *et al.* 2005
	Daejima	*GDP*	Oyster mushroom	*35S*	2 M Na	SC	Sur	Jeong *et al.* 2001
	Marfona	*mtlD*	*E. coli*	*35S*	0~0.15 M Na	H	G	Rahnama *et al.* 2011
Low temperature	Russet Burbank	*AFP*	Synthetic	*35S*	−2.5°C	PG	IL	Wallis *et al.* 1997
	Desnitsa	*Δ12*	Cyanobacterium	*35S*	−7°C /8°C	Inv	IL/Sur,VD	Amiri *et al.* 2010
	Desnitsa	*Δ12*	Cyanobacterium	*35S*	−9°C	Inv	IL, MDA	Demin *et al.* 2008
	Desiree	*ω9*	Wild potato	*35S*	−7~0°C	SC	IL	De Palma *et al.* 2008
	Desiree	*INVase*	Yeast	*B33*	−9°C	Inv	MDA	Deryabin *et al.* 2003
	Desnitsa	*Δ12*	Cyanobacterium	*35S*	−7°C	Inv	MDA	Amiri *et al.* 2007
Oxidative Scavenger	Taedong Valley	*GalURase*	Strowberry	*35S*	10 μM MV/0.1 M Na/0.02 M Zn	Inv	Y	Hemavathi *et al.* 2011
	Atlantic	*SOD+APX+NDPK2*	Cassava, Pea, *Arabidopsis*	*SWPA2*	0~400 μM MV/42°C	SC	D	Kim *et al.* 2010
	Atlantic	*2-cys Prx*	*Arabidopsis*	*SWPA2/35S*	0~300 μM MV/42°C	SC	VD/CF	Kim *et al.* 2011
	Atlantic	*SOD+APX*	Cassava/Pea	*SWPA2*	0~250 μM MV/42°C	SC	VD/RWC	Tang *et al.* 2006
	Atlantic	*NDPK2*	*Arabidopsis*	*SWPA2/35S*	0~250 μM MV/42°C/0.08 M Na	PG	VD/RWC/G	Tang *et al.* 2008
	Desiree	*SOD/DHN4/DREB1/ROB5*	Wheat/Barley/Canola/Bromegrass	*35S/COR78*	44°C/WD/10°C, N	PG, F	G, Yield	Waterer *et al.* 2010
	Superior	*SOD+APX+codA*	Cassava, Pea, Rhizobacterium	*SWPA2*	10 μM MV/0.2 M Na/WW	SC	IL/G/G	Ahmad *et al.* 2010
	Kufri Sutlej	*SOD*	Himalayan cinquefoil	*35S*	WW	SC	PR	Pal *et al.* 2013
	Taedong Valley	*GalURase*	Strowberry	*35S*	0.2 M Na	Inv	G	Upadhyaya *et al.* 2011
	Taedong Valley	*GalURase*	Strowberry	*35S*	0~0.5 M Na	PG	CF	Venkatesh *et al.* 2012
	Taedong Valley	*anti-MSP*	Potato	*35S*	Nor	PG	Y	Gururani *et al.* 2012
	Atlantic	*GR1*	*Arabidopsis*	*35S*	3 μM MV/200 μM Cd, WW	SC/Inv, SC	IL/G,VD	Eltayeb *et al.* 2010
	Atlantic	*DHAR1*	*Arabidopsis*	*35S*	3 μM MV/0.05 M Na/5% PEG	Inv	IL/G/G	Eltayeb *et al.* 2011
	Taedong Valley	*anti-MSP*	Potato	*35S*	0.2 M Na/0.02 M Zn/0.20 M Man	PG	VD	Gururani *et al.* 2013
	Taedong Valley	*GalURase*	Strowberry	*35S*	0~0.2 M Na	Inv	G	Hemavathi *et al.* 2009
	Taedong Valley	*GLOase*	Rat	*35S*	0.1 M Na	Inv	G	Hemavathi *et al.* 2010

Transcriptional factor	Umatilla	*CBF1-3*	*Arabidopsis*	*rd29A/35S*	–7~0°C	SC	IL	Pino *et al.* 2007
	Superior	*EREBP1*	Potato	*35S*	0.075 M Na/8~10°C	Inv	G/Y	Lee *et al.* 2007
	Atlantic	*PF1*	Pepper	*35S*	–9~0°C/45°C/0.4 M/0.075M MV	Inv	IL/Sur/Sur/CF	Youm *et al.* 2008
	Superior	*MYB1R-1*	Potato	*35S*	WW	SC	VD/G/Sur	Shin *et al.* 2011
	LongShu3	*MYB1*	Sweetpotato	*SWPA2*	0.4 M Na/WW	SC	VD	Cheng *et al.* 2013b
	Umatilla	*CBF1*	*Arabidopsis*	*35S*	–16~0°C	SC	IL	Pino *et al.* 2008
	Cardinal	*CBF1*	Wild potato	*35S*	PEG	Inv	G/Sur	Pino *et al.* 2013
	Desiree	*DREB1A*	*Arabidopsis*	*rd29A*	1M Na	Inv	VD	Behnam *et al.* 2006
	Desiree	*DREB1A*	*Arabidopsis*	*rd29A*	2M Na	SC	VD	Celebi-Toprak *et al.* 2005
	Desiree	*DREB1A*	*Arabidopsis*	*rd29A*	–20°C	Inv	Sur	Behnam *et al.* 2007
	Desiree	*DREB1A*	*Arabidopsis*	*rd29A*	PEG	Inv	VD	Huynh *et al.* 2014
	Belle de Fontenay	*DREB1*	Potato	*35S*	0.1M Na/WW	PG	G	Bouaziz *et al.* 2013
	Desiree	*DREB1B*	*Arabidopsis*	*35S*	WW/–6°C	SC	RWC/Sur	Movahedi *et al.* 2012
	Spunta	*DREB2*	Potato	*35S*	0.2 M Na	Inv	G	Bouaziz *et al.* 2012
	Umatilla/*S. commersonii*	*CBF1*	*Arabidopsis*	*35S*	Nor	SC	Tr	Carvallo *et al.* 2011

Plant performance, such as yield or survival rate, and evaluation in near natural conditions are weighed heavily in this list.
* Treatment abbreviations are as follows: Cd: $CdCl_2$, DA: Dry air, Man: Mannitol, MV: Methyl viologen, Nor: Normal conditions, Na: NaCl, PEG: Polyethylene glycol, Sol: Solbitol, WD: Water deficient, WW: Water withheld, Zn: $ZnCl_2$.
** Plant type abbreviations are as follows: F: Field plant, H: Plant in hydroponic culture, Inv: in vitro plant, PG: Potted plant in greenhouse or outside, SC: Plant in soil in controlled conditions.
*** Evaluation abbreviations are as follows: CF: Chlorophyll fluorescence, G: Growth, IL: Ion leakage, MDA: Malondialdehyde content, PC: Proline content, PR: Photosynthetic rate, RWC: Relative water content, Sur: Survival rate, TP: Turgor potential, Tr: Transcriptome, VD: Visible damage, Y: Yield.

DROUGHT

Under drought condition, cytoplasm is getting dehydrated and following reduction of various enzymatic activities. Furthermore, ions also accumulate along with pH change. This causes enzyme inactivation (Sinha 2004). Potato can grow under many different climatic conditions; however, its shallow root system makes it sensitive to drought (Pino *et al.* 2013) as it is thought to restrict water uptake (Dalla Costa *et al.* 1997,Ekanayake and Midmore 1992). Drought affects primary vegetative growth in potato, such as shoot length, leaf size and leaf number, and photosynthetic rate (Dalla Costa *et al.* 1997, Deblonde and Lendent 2001,Ekanayake and Midmore 1992, Kiziloglu *et al.* 2006, Weisz *et al.* 1994), and drought is a particularly harmful stress during tuber formation as it influences tuber production and quality (Deblonde and Lendent 2001, Ekanayake and Midmore 1992, Weisz *et al.* 1994). Thus, avoidance of drought stress is one of the most important issues in potato production.

One strategy used by plants to adapt to drought is the accumulation of compatible solutes. Glycine betaine (GB) is a general osmoprotective compound found in plants, animals and microorganisms. Accumulation of GB in higher plants is known to occur in response to various abiotic stresses, such as drought, high salinity and cold (Jagendorf and Takabe 2001, Rontein *et al.* 2002). Exogenous GB also enhances the tolerance of plants to various stresses (Ashraf and Foolad 2007, Chen and Murata 2008, Hoque *et al.* 2007, Mahouachi *et al.* 2012, Park *et al.* 2006). GB functions not only as an osmolyte but also as a molecular chaperon (Sakamoto and Murata 2002), and in plants it is synthesized from choline via betaine aldehyde (Sakamoto and Murata 2000). By manipulating this pathway, GB accumulation may occur in non-GB-accumulating plants. Interestingly, enhancement of this pathway using genetic engineering (GE) techniques has succeeded in conferring drought stress tolerance to a number of plant species (Cheng *et al.* 2013a). The betaine aldehyde dehydrogenase (BADH) is one of the GB bio-synthesis enzymes that catalyses betaine aldehyde in GB. A transgenic potato was generated with the *BADH* transgene from spinach driven by the stress-induced promoter *rd29A* from *Arabidopsis thaliana* (Zhang *et al.* 2011). BADH activity was abundant in all four transformants under stress. Using potted plants maintained in greenhouses, tolerance to drought and salinity stress was evaluated using growth performance and the electrical conductivity of leaves. The fresh weight of all transformants increased (17–29%) under stressful conditions compared to non-transformants (NT). Furthermore, there was a negative relationship ($r = -0.9495$) between BADH activity and the relative electrical conductivity in the leaves of the transformants (Zhang *et al.* 2011). This result demonstrates that BADH conferred drought and salinity stress tolerance to potato plants.

The choline oxidase (COD) does not exist in higher plants and it directly catalyses choline in GB. Transgenic potatoes were generated using the *codA* gene from *Arthrobacter globiformis* (rhizobacterium) driven by the oxidative stress-induced promoter *SWPA2* from sweet potato (Kim *et al.* 2003), with a transit peptide from the small subunit of Rubisco from tobacco (Ahmad *et al.* 2008). Two transformants were evaluated for the accumulation of GB and tolerance to stress (drought and salinity). GB accumulation observed in transformants was 3–5 times higher following methyl viologen (MV) treatment, and transformants showed better performance under drought and salinity stress compared to NTs. Potted plants in growth chambers were evaluated for drought tolerance by withholding water for 14 days. Transformants maintained higher water content and had over twice the biomass (dry weight) of NTs. In addition, the same transformants under in vitro conditions showed tolerance to MV based on ion leakage measurements (Ahmad *et al.* 2008). A further study has been published by the same group. Subsequently, they focused on the antioxidant system under drought conditions (Cheng *et al.* 2013a). Tolerance was evaluated using photosynthetic parameters, chlorophyll content and malonic dialdehyde (MDA) content. Potted plants in a growth chamber were subjected to 4 days drought (10% or 20% PEG) and 2 days rehydration. Transformants had more efficient photosynthetic systems, higher chlorophyll content and lower MDA content than NTs. Furthermore, the activity of antioxidant enzymes, such as superoxide dismutase (SOD), catalase (CAT) and peroxidase (POD), was observed to be higher in transformants than NTs. Transformants also showed better recovery from water deficiency than NTs. The studies detailed above indicate that accumulation of GB is a good strategy for conferring drought stress tolerance to potato.

Trehalose accumulation has been observed in bacteria, yeast and desiccation tolerant plants under stressful conditions (Iordachescu and Imai 2008); however, other plants are also known to accumulate small amounts of trehalose (Kondrak *et al.* 2011). Trehalose is also an osmolyte that has the ability to adjust osmosis and protect molecules (Jain and Roy 2009). That is synthesized from glucose-6-phosphate (G6P) and UDP-glucose via trehalose-6-phosphate (T6P), and trehalose and T6P are synthesized by T6P phosphatase (TPP) and T6P synthase (TPS), respectively. Accumulation of trehalose has been observed in a tobacco transformant containing a single yeast *TPS* (*TPS1*) gene (Holmstrom *et al.* 1996). Transgenic potato plants were generated containing the yeast *TPS1* gene driven by the stress-induced promoter *DS2* from potato (Stiller *et al.* 2008). Transgene expression was not induced by stress and expression levels were low in two independent lines. Although trehalose accumulation was not evaluated, water withholding was assessed in potted plants maintained in greenhouses. Both transformant lines showed higher relative water content and photosynthesis than NTs under drought conditions. Stomatal conductance was monitored in potted plants in the field under water withholding conditions and was observed to be lower in NTs than in transformants 3 days after drought treatment. At the beginning of the desiccation period, stomatal conductance was higher in NTs than in both transformant lines, possibly as a result of low stomatal density in transformants. Thus, trehalose accumulation was observed to confer drought tolerance to potato plants. However, transformants showed low stomatal density, which may cause low CO_2 fixation rates, leading to growth retardation under optimum conditions. An additional study was published using transcriptome analysis to investigate pleiotropy (Kondrak *et al.* 2011, 2012). A complex regulatory network, including transcriptional factors and other regulatory proteins, resulted in multiple phenotypic traits in transformants with the yeast *TPS1* gene.

Glucosylglycerol (GG) is a compatible solute typically found in moderate halotolerant cyanobacteria and some bacteria (Pocard *et al.* 1994, Reed and Stewart 1985, Roder *et al.* 2005). Of the terrestrial plant species, only *Myrothamnus flabellifolia*, one of the resurrection plants, is known to accumulate GG under drought conditions (Bianchi *et al.* 2006), and some members of Liliaceae constitutively synthesize GG (Kaneda *et al.* 1984). The GG-phosphate phosphatase/synthase (*ggpPS*) gene was transferred from *Azotobacter vinelandii*(rhizobacterium) to *Arabidopsis*. The transformant was observed to accumulate GG and exhibit stress tolerance (Klähn *et al.* 2009). To confer drought and salinity tolerance to potato, potato plants were generated expressing the *ggpPS* gene under control of the constitutively expressed cauliflower mosaic virus *35S*promoter (*35S*) or the *rd29A* promoter (Sievers *et al.* 2013). Although both transformants accumulated GG in leaves, only the *rd29A* transformants accumulated GG in tubers. In greenhouse experiments, a number of lines from both transformants exhibited improved shoot growth compared to NTs under drought and salinity conditions.

Drought tolerance has also been conferred to potato using other approaches. To understand the synthetic pathway of fructans, transformants with *sucrose: sucrose 1-fructosyltransferase* (*1-SST*) and *fructan: fructan 1-fructosyltransferase* (*1-FFT*) genes from the globe artichoke were generated and were able to synthesize the high molecular weight carbohydrate inulin (Hellwege *et al.* 2000). Inulin has been defined as a type of polysaccharide consisting of sucrose with β(2-1)-linked fructose (fructan) chains. It has been suggested that fructans contributes to the osmotic stress tolerance of organisms (Hendry 1993, Park *et al.* 1999), possibly via its stabilising effects on membranes and proteins (Fricke and Pahlich 1990, Vereyken *et al.* 2003). It has been suggested that inulin also plays a role in conferring drought tolerance (Hendry 1993, Pilon-Smits *et al.* 1995). Accumulation of proline was observed to be suppressed in *1-SST/1-FFT* transformants under water deficit conditions; however, the turgor potential in leaves was reduced (Knipp and Honermeier 2006). Fructan is thought to contribute to osmotic control in potato transformants. As reported in the other study, repression of formate dehydrogenase (FDA) enhances accumulation of proline under drought stress (Ambard-Bretteville *et al.* 2003). This transformant had reduced levels of the FDA protein and accumulated formate in leaves and tubers, and under drought stress, the transformant accumulated proline more rapidly than NTs. However, there was no description of drought tolerance in this study.

The YUCCA family is known to contribute to auxin bio-synthesis in plants. A rice mutant,*CONSTITUTIVELY WILTED1*, was deficient in the YUCCA homologue (Woo *et al.* 2007) and the activation tag line of *YUCCA7* in *Arabidopsis* demonstrated drought tolerance (Lee *et al.* 2011). Drought tolerance by over-expression of *AtYUC6* in potato was evaluated in 4-month-old potted plants in greenhouses by monitoring recovery after withholding water for 18 days (Kim *et al.* 2013). Although NTs wilted, transformants survived and exhibited higher water content in leaves and lower reactive oxygen species (ROS) content with increased expression of *CuZnSOD* compared to NTs. Based on these characteristics, this transformant may also be tolerant of other abiotic stresses.

The studies discussed above demonstrate that drought tolerance is not easily controlled and various factors are intricately involved. It has been reported that biotic stress related proteins are also associated with abiotic stresses, such as osmotic and salinity stress. Pathogenesis-related proteins (PR proteins) are induced by pathogen infection or wounding, while PR-10 proteins are also induced by abiotic stresses, such as drought, salinity or oxidative stress (Liu and Ekramoddoullah 2006). Based on the results of proteomic analyses, the PR-10a protein in potato was induced by osmotic or salinity stress, and over-expressing callus of this gene indicated osmotic and salinity tolerance in suspension culture (El-Banna *et al.* 2010). This transformed callus showed low oxidised glutathione content and high proline accumulation under 0.5 M sorbitol or 0.32 M NaCl containing medium compared to NT callus. From these studies, it is evident that various genes contribute to confer drought tolerance to potato.

SALINITY

Potato is a moderately salt-sensitive crop and is able to withstand soil salinity up to approximately 2.0 dS/m (Mass and Hoffman 1977). Dependent on the soil type, EC values reached 5.8 dS/m following 100 mM NaCl irrigation in our experiment. Salt affected soil has been shown to have a negative effect on potato crops: salinity at 2.0 g/L delays seed germination, retards growth and results in reduced tuber yield with low quality (Katerji *et al.* 2003), and salinity of 5.9 dS/m decreases potato production by half (Kotubyamacher *et al.* 2000). Irrigated land is gradually is suffering from salinity damage and combating salinity stress is a pressing issue for world agriculture, including potato production.

Salinity damage results in two different adverse effects on plant cells, ionic stress and osmotic stress. Osmotic regulation confers salinity tolerance to plants: a high solute concentration (molar concentration) causes water to remain in the cytoplasm and osmoprotectants contribute to maintaining the function of proteins. Mannitol is an osmoregulatory substance typically used in experiments; it is found in many plant species and is known as the most abundant sugar alcohol in nature (Patonnier *et al.* 1999, Prabhavathi and Rajam 2007, Shen *et al.* 1997). Potato plants do not naturally accumulate mannitol, so transformants with the *mannitol-1-phosphate dehydrogenase* (*mtlD*) gene from *E. coli* were generated to confer salinity tolerance to potato (Rahnama *et al.* 2011). Accumulation of mannitol in these transformant lines was abundant under 50 mM NaCl salinity conditions in vitro and in hydroponic culture; however, it was decreased dependent on salinity strength in high NaCl concentrations. Growth parameters were assessed in a hydroponic culture system to evaluate tolerance. In 100 mM NaCl conditions, the dry weight of NT shoots was reduced by 76.5%; however, in the elite transformant line the dry weight of shoots was only reduced by 17.3%. Thus, these results suggest that the bacterial *mt1D* gene may confer the salinity tolerance to potato. Accumulation levels of mannitol were not enough to osmotic adjustment in these transformants, mannitol may function as an osmoprotectant in this case.

Proline is known to be an important stress responsive substance and is also thought to be an osmoprotectant. The *pyrroline-5-carboxylate synthetase* (*P5CS*) gene from *Arabidopsis* was introduced into potato driven by the *35S* promoter (Hmida-Sayaria *et al.* 2005). Under 180 mM NaCl conditions in vitro, proline accumulation in leaves was over five times higher in the elite transformant line than in NTs. The elite transformant line also showed higher tuber production than NTs under 100 mM NaCl conditions using potted plants in a greenhouse. Improvements in proline bio-synthesis resulted in enhanced salinity tolerance in potato. In addition to the above studies, transfer of the *glyceraldehyde-3-phosphate dehydrogenase* (*GPD*) gene from oyster mushrooms into yeast resulted in the transfer of salinity tolerance (Jeong *et al.* 2000). This group also tried to introduce this gene, driven by the *35S* promoter, into potato (Jeong *et al.* 2001). They evaluated salinity tolerance of potato in a growth chamber by cultivating plants in fresh water for 10 days following 2 M NaCl treatment for 12 h. Although NTs were wilted, six independent transformant lines survived. The mechanism of tolerance was not mentioned; however, transfer of the *GPD* gene may confer salinity tolerance to potato.

Although one strategy for adapting to salinity is similar to that used for adapting to drought, another strategy is to remove Na^+ from the cytoplasm. Extrusion of Na^+ from cells is carried out by Na^+/H^+ antiporters in the plasma membrane and compartmentation into vacuoles is undertaken by Na^+/H^+ antiporters in the vacuolar membrane. However, exclusion of ions from cells is not sufficient to protect plants against high salinity because the excluded ions remain in the apoplast resulting in osmotic stress component (Blumwald *et al.* 2000). A transgenic potato was generated with a *vacuolar* Na^+/H^+ *antiporter* (*HvNHX2*) gene from barley driven by the *35S* promoter (Bayat *et al.* 2010). Although NTs were unable to survive 200 mM NaCl in vitro, five of the six transformant lines survived. Three elite transformant lines showed higher K^+ content compared to NTs under 100 and 150 mM NaCl conditions. On the other hand, another group reported about transformants with the *AtNHX1* gene from *Arabidopsis*, and evaluated the performance of eight lines using potted plants in a greenhouse (Wang *et al.* 2010). Salt concentrations that resulted in a 50% reduction in tuber yield (YR_{50}) and 50% plant survival (LD_{50}) were employed as tolerance indicators in this work. The elite transformant lines indicateed over twice higher in both indicators compared to NTs. Furthermore, tolerant lines maintained higher chlorophyll content and had lower proline accumulation than NTs. These results indicate that tolerant lines may lessen the salinity stress by confining Na^+ to vacuole wells. However, tolerant lines did not show a remarkable increase in Na^+ content in leaves under high salt concentrations. These studies demonstrate that both strategies work to confer salinity tolerance in potato. These above studies demonstrate that both osmotic regulation and removal of Na^+ work to confer salinity tolerance in potato.

LOW TEMPERATURE

Potato plants like cool climates, thus, the commercial cultivation of potato is suitable for high altitude or highland regions. As a consequence, potato production often suffers from low temperature stress, especially just after planting or just before harvesting. The display of a low temperature tolerant phenotype is important, and introgression of this phenotype has been an urgent requirement in potato production. Low temperature causes reduction of various enzymatic activities and of floridity of bio-membrane. Under freezing temperature, ice formation occurs in intercellular spaces. This leads to dehydration in cytoplasm (Sinha 2004).

One strategy for adapting to low temperatures is the desaturation of fatty acids in the bio-membrane. Unsaturated fatty acids are known to increase in the bio-membrane of various plants species exposed to low temperatures (Clarkson *et al.* 1980, Horvath *et al.* 1981, Kodama *et al.* 1995, Smolenska and Kuiper 1997) and they are also associated with chilling or freezing tolerance (Miquel and Browse 1992, Miquel *et al.* 1993,Quinn *et al.* 1989). Desaturation of fatty acids in constitutive lipids increases the fluidity of the membrane by depressing the solidification point and maintains the function of the bio-membrane, even at low temperatures. Desaturases involved in the desaturation of fatty acids can be classified into two types; delta (Δ) and omega (ω). Δ and ω notations indicate the position of the carbon bonds from the carbonyl carbon and the methyl end, respectively (Napier *et al.* 1999). There are many reports that transformation of desaturases confers low temperature tolerance in various plant species, such as tobacco (Kodama *et al.* 1994) and tomato (Yu *et al.* 2009). In potato, the *Δ12-desaturase* gene from cyanobacteria was introduced to plants and its ability to improve low temperature sensitivity was evaluated (Amiri *et al.* 2007). From the analysis of in vitro plants, all four lines had reduced MDA content under −7°C conditions, and three of the four lines showed increased unsaturated fatty acid content. They concluded that introduction of the *Δ12-desaturase* gene from cyanobacteria conferred low temperature tolerance to potato. Subsequently, they focused on only three lines that increased unsaturated fatty acid content. From the evaluation of a four-month-old in vitro cultivation at 8°C, all three transformants showed cold stress tolerance traits, such as viability, less visible damage and high microtuber production (Amiri *et al.* 2010). Using a part of same transformant lines (probably), generation of oxidative molecules under low temperature was evaluated in in vitro (Demin *et al.* 2008). Following exposure to low temperature, the presence of oxidative molecules, including superoxide anion and hydrogen peroxide, and subsequently a decrease in MDA content, were observed in transformant lines compared to NTs. However, activity of SOD was lower in transformant lines than that in NTs, suggesting that the generation of oxidative molecules was suppressed in the *Δ12-*

desaturase transformants. On the other hand, ω9-desaturase from *Solanum commersonii* was transformed to potato (De Palma *et al.* 2008). *S. commersonii* is a wild relative of *S. tuberosum* and has high abiotic stress tolerance, including cold, drought and salinity tolerance. The transformant lines had a C18:1 Δ11 fatty acid that was not found in the vector control (VC) transformants. The freezing tolerance of the potted plants was evaluated in a growth chamber by ion leakage. The LT_{50} of transformants decreased by approximately 2°C after cold acclimation (De Palma*et al.* 2008). Although these results are not from natural conditions, they imply that desaturation of fatty acids in membranes may confer a certain level of cold tolerance to plants.

Another strategy used by plants against low temperature stress is osmotic regulation. A high solute concentration (molar concentration) can enable plant cells to retain water or prevent ice formation. Generally, ice formation begins in the apoplastic area where solute concentration is the lowest. As ice formation expands in the apoplast, water is gradually withdrawn from the cytosol and cell desiccation progresses (Steponkus and Webb 1992). To prevent ice formation, yeast invertase was introduced to potato. Since over-expression in leaf or cytosolic localization of this gene (product) caused low productivity of tuber (Heineke *et al.* 1992,Sonnewald *et al.* 1997), the transgene was driven by the tuber specific promoter patatin *B33* and fused with the protease inhibitor II leader peptide for apoplastic enzyme localisation (Deryabin *et al.* 2003). Using in vitro plants, invertase activity, sugar content and low temperature tolerance (measured by MDA content) were investigated. Invertase activity doubled, especially that of the acidic type, and sugar content, such as sucrose, also increased in the apoplastic and the cytosolic fractions under control condition. After exposing plants to −9°C for 20 min, MDA content was 1.5 times higher than that under non-stress condition in VC plants; however, the level of MDA was stable in transformants. An extended low temperature treatment at 5°C for 5 days also revealed that MDA content was lower in transformants than that in VC plants. This work used the *B33* promoter driven transformant whose transgene is thought to be abundant in tubers. Readers may also be interested in whether the transgene contributes to the enhancement of low temperature tolerance in leaves. The author also evaluated *B33* promoter activity in leaves, stems and roots using the *B33* promoter*GUS* transformant of potato. Of the three tissues evaluated, GUS activity was highest in leaves, and was approximately 10% of the activity observed for the *35S* promoter GUS in leaves. Based on this result, some phenotypes, including bio-chemical properties, are thought to have been caused by weak expression of the transgene in potato leaves. Under natural conditions, low temperature tolerance may be more essential in aerial sections of the plant rather than the tubers. It is expected that high expression of this invertase in aerial sections of the potato plant may confer increased tolerance of low temperatures. Based on the results of the above studies, it is evident that various genes may enhance low temperature tolerance in potato. However, there are no reports conferring cold acclimation to potato using GE techniques as detailed mechanisms and the genes responsible remain to be elucidated.

In addition to the above examples, I would like to highlight a previous study that transferred the synthetic anti-freezing protein (AFP) to potato (Wallis *et al.* 1997). AFP inhibits ice growth and recrystallization, and is effective in protecting against frost damage at concentrations several orders of magnitude lower than freezing point depression by mol concentration (Carpenter and Hansen 1992, De Vries 1986). Performance of an AFP transformant was evaluated by measuring electrolyte release under freezing temperatures. The decrease in electrolyte release was observed to be dependent on the accumulation of AFP at −2.5 and −3.0°C. Synthetic AFP was regarded as conferring low temperature tolerance to potato.

OXIDATIVE SCAVENGERS

Environmental stresses such as drought, salinity and cold are known to enhance the accumulation of ROS in plant cells. Various types of ROS result in the oxidation of many cellar components and lead to oxidative destruction of the cell (Asada and Takahashi 1987, Mittler 2002). Enhancement of the ROS scavenging system is expected to confer tolerance to various types of stresses in plants. Scavenging of ROS in plant cells is carried out via enzymatic and non-enzymatic mechanisms. A network of low molecular weight antioxidants, such as ascorbate (AsA) and glutathione (GSH), and antioxidative enzymes, such as SOD and CAT, are important for preventing and controlling the dangerous effects of ROS (Noctor and Foyer 1998).

Improving the stress tolerance of plants using low molecular weight antioxidants could be performed by increasing the production rate of reduced forms of GSH and AsA, and the enlargement of either pool. AsA is a crucial antioxidant molecule that scavenges free radicals and protects from ROS generated by adverse environmental conditions. AsA can directly scavenge free radicals (Halliwell and Gutterridge 2000) and is an important electron donor for H_2O_2 via AsA peroxidase (APX) (Noctor and Foyer 1998). On the other hand, GSH is involved in detoxification of xenobiotic compounds (Noctor and Foyer 1998, Potters *et al.* 2002) and redox regulation in the cell (Noctor *et al.* 1998). GSH also reduces oxidised AsA (dehydroascorbate: DHA) to AsA. GSH is converted to glutathione disulfide (GSSG) by scavenging oxidative molecules. GSSG is catalysed to GSH by GSH reductase (GR) in the NADPH-dependent manner. DHA is reduced to AsA by dehydroascorbate reductase (DHAR). To enhance the ratio of AsA, the *DHAR1* gene from *Arabidopsis*driven by the *35S* promoter was introduced into potato (Eltayeb *et al.* 2011). The transformants showed higher MV tolerance, evaluated by ion leakage of leaf discs, and higher chlorophyll content compared to NTs. Furthermore, these transformants grew better under salinity and drought conditions compared to NTs. In addition to this work, same group also tried to enhance the ability of GSSH reduction by introducing the*GR1* gene from *Arabidopsis*, which is a cytoplasmic isoform, driven by the *35S* promoter (Eltayeb *et al.* 2010). The transformants with the *GR1* gene had 6.5 times higher GSH content, 5.8 times higher GR activity, and showed higher relative AsA content compared to NTs. Improved tolerance was observed as increased root growth in a heavy metal ion and low ion leakage from leaf discs assessed by MV treatment. Based on these results, high levels of the reduced forms of GSH and AsA may confer various abiotic stress tolerances to potato.

Furthermore, there are five recent reports from a Korean

group demonstrating enhancement of the bio-synthesis of AsA. The transformant was generated with the *d-galacturonic acid reductase* (*GalUR*) gene from strawberry driven by the *35S* promoter (Hemavathi *et al.* 2009). Three independent transformant lines showed high AsA accumulation with high GalUR activity compared to NTs. Two elite lines indicated a reduction in leaf disc breaching following MV treatment (~10 mM), NaCl (~600 mM) and mannitol (~300 mM). Furthermore, the in vitro growth performance of these two lines was better than that of NTs in NaCl (~200 mM) containing media. Similar results were observed in over-expression of the *l-gulono-γ-lactone oxidase* gene in potato (Hemavathi *et al.* 2010). Biochemical properties, such as ROS scavenging enzymatic activity and proline or MDA content, of one line of the *GalUR* transformants were also further analysed during MV, NaCl or ZnCl treatment (Hemavathi *et al.* 2011). SOD, CAT, APX, DHAR and GR activity was higher in transformants than in NTs under stressful conditions. Levels of reduced forms of AsA and GSH were higher in transformants than in NTs under three different stressful conditions and control conditions. Furthermore, under stressful conditions, microtuber production was over twice as high, with associated increased proline content and reduced MDA content, in transformants than in NTs. It is thought that enhanced accumulation of AsA causes upregulation of the antioxidant system, improving tolerance to various abiotic stresses. Single gene over-expression of *GalUR* resulted in alteration of various antioxidant systems. The *GalUR* transformant maintained photosynthetic performance and showed improved productivity under salinity conditions (Venkatesh *et al.* 2012). The same group also focused on metheylglyoxal (MG) accumulation (another cytotoxic compound) under salinity conditions (Upadhyaya *et al.* 2011). MG concentration is regulated by the glyoxalase system, which is dependent on GSH concentration (Yadav *et al.*2005a, 2005b). In contrast to NTs, transformants maintained MG levels under salinity conditions by increasing the activity of two glyoxalases.

Same Korean group tried to suppress evolving oxidative molecules from photosystem II. They focused on the manganese-stabilising protein (MSP), which is a key component of the oxygen-evolving complex (Williamson *et al.* 2011), and generated both sense and antisense transformants with the *MSP* gene from *S. tuberosum* driven by the *35S* promoter (Gururani *et al.* 2012). MSP sense plants had almost same the properties (morphology, tuber yield, photosynthetic performance) as NTs. However, higher tuberisation frequency and higher tuber yield, with increased total soluble sugars, was observed in antisense plants compared to NTs and sense plants. Furthermore, antisense plants had greater photosynthetic activity and lower oxygen evolution compared to NTs and sense plants. Proline and AsA content was also higher in antisense plants than in NTs and sense plants. Consecutively using these antisense plants, they evaluated abiotic stress tolerance, including as salinity (200 mM NaCl), heavy metal ions (20 mM $ZnCl_2$) and drought (200 mM mannitol) (Gururani *et al.* 2013). In addition to proline and AsA content, α-tocopherol and ABA levels also increased in antisense plants compared to NTs. Levels of oxidative scavenging enzymes, such as APX, CAT, SOD and DHAR, were also higher in antisense plants than in NTs under stressful conditions. Using potted plants in greenhouses, NTs showed more severe damage under the above stressful conditions compared to antisense plants. Suppression of MSP appears to confer higher performance to potato plants under stressful conditions and situations with high scavenging ability.

Antioxidative enzymes, such as SOD and CAT, are also important for direct scavenging of ROS. SOD catalyses dismutation of the superoxide radical ($O_2^{\bullet-}$) into H_2O_2 and O_2 (Asada 1999), and SOD contributes to an organism's primary defence against oxidative molecules. Cytosolic copper-zinc (Cu/Zn) SOD from *Potentilla atrosanguinea* (Himalayan cinquefoil) was transformed to potato driven by the *35S*promoter (Pal *et al.* 2013). Potted plants evaluation revealed that the highest SOD activity line of these transformants indicated higher net photosynthetic rate and stomatal conductance under control and salinity conditions compared to NTs. Drought conditions reduced water potential, photosynthetic activity and chlorophyll content, but increased proline and $O_2^{\bullet-}$ content in NTs. However, these changes were diminished in transformants. Another Korean group generated potato transformants (SSA) with the*Cu/ZnSOD* gene from *Manihot esculenta* (Cassava) and the *APX* gene from *Pisum sativum* driven by the oxidative stress inducible *SWPA2* promoter (Tang *et al.* 2004) with a transit peptide for chloroplast (Tang *et al.* 2006). APX reduces H_2O_2 to H_2O utilising ascorbate as an electron donor. These SSA lines indicated MV tolerance by bleaching of leaf discs. The MV and high temperature tolerances of two elite lines from SSA transformants were subsequently evaluated using potted plants in a growth chamber. Both lines indicated diminishing MV damage, including visible features and a reduction in dry weight, and a reduction in chlorophyll content. These two transformant lines also showed better growth and maintained photosynthetic activity under high temperature conditions. However, it is difficult to discuss simultaneous effects of SOD and APX in the SSA line as the work on a single transformation of SOD was done by another group using different cultivars and different evaluation methods. Furthermore, they introduced the*codA* gene, with a transit peptide for chloroplast, from rhizobacterium driven by the *SWPA2* promoter to the SSA transformant (Ahmad *et al.* 2010). These transformants (SSACs) had three transgene products (SOD, APX and GB) induced by oxidative stress in chloroplasts. Two lines of the SSAC transformants indicated better growth performance than NTs and SSA transformants under saline (200 mM NaCl) and drought (withholding water for 14 days) stress conditions.

The same group also tried to confer abiotic stress tolerance to potato by transferring the *nucleoside diphosphate kinase 2* (*NDPK2*) gene from *Arabidopsis* (Tang *et al.* 2008). NDPK is a housekeeping enzyme that maintains the levels of (d)NTPs. Recently, it was suggested that this enzyme is also associated with signal transduction in oxidative stress (Otero 2000). Transformants generated with the *AtNDPK2* gene, driven by the *SWPA2* or the *35S* promoter, were evaluated for their tolerance to various stressful conditions using potted plants. Following treatment with 250 mM MV in a greenhouse, transformants showed 30–50% less visual damage than NTs. Furthermore, transformants exhibited reduced damage under high temperature (42°C) and salinity (80 mM NaCl) conditions compared to NTs. It is thought that the expression of *NDPK2*caused upregulation of *APX* and consequently enhanced various abiotic stress tolerances. Subsequently, this group transformed the *SWPA2:AtNDKP2* construct into an SSA transformant (Tang *et al.* 2006) and generated new transformants that possessed three transgenes (SSAN) to evaluate the contribution of multiple genes (Kim *et al.* 2010). Although the activity of products from transgenes was increased in each transformant under oxidative stress conditions, APX may be upregulated in transformants with the *NDKP2*transgene. Furthermore, CAT activation was clearly observed in the transformant with *NDKP2*. The SSAN transformants showed high activity of three major oxidative scavenging enzymes (SOD, APX and CAT). Based on an ion leakage evaluation treatment using 3 mM MV, SSAN transformants showed significantly higher tolerance than other transformants and NTs. Other parameters also indicated high tolerance in SSAN transformants under oxidative and high temperature stress conditions. Although simultaneous effects were mentioned in this work, there were no statistically significant differences between SSA and SSAN, except for ion leakage evaluation under oxidative stress.

The studies discussed provide knowledge on oxidative stress tolerance and future breeding for abiotic stress tolerance in, not only potato, but also higher plant species. It is expected that simultaneous effects of transgenes will be evaluated in field trials. In addition to the various studies discussed above, the same group generated transformants with *2-cystein peroxiredoxin* (*2-Cys Prx*) genes from *Arabidopsis* driven by the*SWPA2* or the *35S* promoter, and evaluated their tolerance under oxidative and high temperature stress conditions using potted plants (Kim *et al.* 2011). The 2-Cys Prx is an antioxidative enzyme that removes peroxides and protects the photosynthetic membrane from oxidative damage (Dietz 2003). Under oxidative stress treatment using 300 mM MV, photochemical yield and chlorophyll content were reduced by 60% in NTs but by only 20–35% in transformants compared to control conditions. Moreover, photochemical yield after treatment at 42°C for 10 h was reduced by 31% in NTs, but by only 7% in an elite transformant line compared to before treatment. These transformants had reduced damage to photosynthetic mechanisms under stressful conditions and showed high oxidative and high temperature stress tolerance.

Another group attempted to generate six types of transformants using four types of transgenes (Waterer *et al.* 2010) driven by the stress inducible *COR78* promoter from *Arabidopsis* (Thomashow 1999) or the *35S*promoter. The four transgenes were mitochondrial *MnSOD* (*SOD3:1*) from wheat, *dehydrin 4* (*DHN 4*) from barley, a cold inducible transcriptional factor *DREB/CBF* from canola and the stress associated gene *ROB5*belonging to the LEA group 3 protein from bramegrass. The *COR78* promoter was used for all four genes, and the *35S* promoter was used for the *SOD* and *ROB5* genes (Waterer *et al.* 2010). Based on results using potted plants in a growth chamber, all transgenes were found to enhance high temperature (44°C) tolerance, especially the *COR78:DHN4* and *COR78:ROB5* transformants. Under low temperature conditions (10°C), the *35S:SOD3.1* transformant grew better than the NTs. The freezing temperatures, measured using ion leakage, were also reduced in *COR78:SOD3.1*, *35S:SOD3.1* and *35S:ROB5* transformants. These experiments were also carried out four times in the field (2001, 2003, 2005 and 2006). Evaluation in field conditions is quite different from that under artificial conditions as the weather is unstable and uncontrollable. In 2005, the weather differed from the other years, and was relatively cooler and wetter. The tuber yield of almost all the transformants indicated a reverse tendency in 2005 compared to the other years. The overall average of tuber yield taken across the four years is apt to extinguish the stress tolerance trait of transformant due to differences in weather tendency conditions (e.g. warm or cool, dry or rainy). However, *COR78* driven transformants, except DREB/CBF, indicated improved productivity compared to NTs. This work indicates that GE techniques may contribute to breeding programs using abiotic stress tolerance for practical purposes.

TRANSCRIPTIONAL FACTORS

All the examples discussed above were aimed at altering single or multiple genes. In contrast, there are some studies that use transcriptional factors (TFs), which influence the expression of various downstream genes. TFs regulate the expression of downstream genes by binding to specific elements of target genes (Lee *et al.* 1999, Ramanjulu and Bartels 2002). When plants detect stress in the environment, the responsible TF is induced, which controls the necessary gene set for the corresponding physiological responses. Since this mechanism is a native physiological response in plants, it has been suggested that transformation of TFs can mimic natural situations in which plants are exposed to a stress (Pino *et al.* 2007).

In *Arabidopsis*, identification and characterisation of abiotic stress related TFs was carried out, and of these, drought-responsive element binding factors (DREB) and C-repeat binding factors (CBF) have been well studied (DREB1A-B-C corresponds to CBF3-1-2, respectively; Gilmour *et al.* 1998, Liu *et al.* 1998,Stockinger *et al.* 1997). These TFs have the APETALA2 (AP2)/ethylene response factor (ERF) domain composed of approximately 60 amino acids (Sakuma *et al.* 2002). To confer abiotic stress tolerance to potato, the *AtDREB/CBF* gene was transferred to potato. There are several studies of *AtDREB/CBF* transformants mainly conducted by two different groups. One group employed the *rd29A* promoter (Behnam *et al.* 2006,Celebi-Toprak *et al.* 2005) as it had already been revealed that over-expression of *AtDREB1A* by the *35S*promoter resulted in growth retardation in *Arabidopsis* plants under control conditions (Kasuga *et al.* 1999). Using 4–6 week-old cuttings propagated potato on soil, transformants were able to survive 2 M NaCl (Behnam *et al.* 2006). A positive relationship was observed between tolerance levels and integrated copy numbers or expression levels of the transgenes. However, transformant lines with high integrated copy numbers did not follow this relationship. Furthermore, this group also evaluated freezing tolerance and compared this to tolerance of salinity conditions (Behnam *et al.* 2007). Evaluation of freezing tolerance was carried out at –20°C for 1 h in vitro after acclimation at 4°C. Expression levels of transgenes were positively correlated with survival rate after freezing, except one transformant line with high transgene expression (D138). Salinity and freezing tolerances were expected to show the same tendency in all transformant lines; however, some lines exhibited only salinity or freezing tolerance. D138 is a typical example and this line was categorized as a salinity tolerant line with high expression of transgene. Differences in tolerance may be due to positional effects, such as insertion mutation or alteration of epigenetic regulation by the transgene. These various lines may help to clarify downstream regulation or help identify responsible genes using microarray analyses. In addition to salinity and freezing evaluations, these transformants were also evaluated for drought tolerance in vitro (Huynh *et al.* 2014). Evaluation of drought tolerance for intact plants was carried out in liquid culture at −1.8 MPa PEG8000 for 9 days and consecutively without PEG for 3 days. The expression levels of transgenes were also positive correlated to drought tolerance levels. From these studies, AtDREB1A was observed to enhance not only freezing, but also drought and salinity tolerance in potato.

Another group has also generated transformants with all three *DREB/CBF* genes driven by the *35S* or the*rd29A* promoter independently (Pino *et al.* 2007). Among these six transformants, *AtCBF2* driven by both promoters did not show freezing tolerance. However, *AtCBF1* and *AtCBF3* driven by both promoters improved their LT_{50} values by approximately 2°C, evaluated using ion leakage analysis in potted plants in a growth chamber. While, these *35S* type transformants showed some negative phenotypes, such as plant stunting, delayed flowering and reduction of tuber production, the *rd29A* type transformants displayed improvements of these negative phenotypes. This work elucidated that *AtCBFs*, except *AtCBF2*, enhance freezing tolerance in potato, and directly indicated that in potato the *rd29A* promoter may improve defects caused by constitutive expression of the *AtCBF1* and *AtCBF3* genes. Furthermore, the *35S:AtCBF1* gene also enhanced freezing tolerance by approximately 2°C in *S. commersonii*, which is a cold tolerant wild relative of *S. tuberosum* (Pino *et al.* 2008). However, only the freezing tolerance of the transformants was evaluated. Transformant lines with the *AtDREB1B* gene driven by the *35S* promoter were evaluated by a different group from the two mentioned above (Movahedi *et al.* 2012). Using potted plants, in addition to freezing stress tolerance, transformant lines showed drought tolerance. Based on these results, it is expected that *AtDREB1B* will also enhance freezing, drought and salinity tolerance in potato, similar to *AtDREB1A*.

Other transcriptional factors have also been transformed into potato. One study has evaluated transformants with the *pathogen and freezing tolerance-related protein 1* (*CaPF1*) gene from *Capsicum annuum* (pepper) driven by the *35S* promoter (Youm *et al.* 2008). This gene encodes the AP2/ERF-type TF that was isolated by inoculation with a bacterial pathogen (*Xanthomonas axonopodis*) and was found to affect the expression of genes that contain either a GCC to a CRT/DRE box in their promoter regions in over-expression of*Arabidopsis* (Yi *et al.* 2004). Transformant lines in potato indicated enhanced tolerance to freezing, heat, drought, oxidative stress and heavy metal ions. Furthermore, tuber formation in vitro was retarded in transformant lines compared to NTs. Another study evaluated transformants with the *IbMyb1* gene from sweet potato driven by the *SWPA2* promoter, which encodes the R2R3-type Myb TF (Cheng *et al.* 2013b). The R2R3-type Myb TF is known to be involved in secondary metabolism and responds to biotic and abiotic stresses (Dubos *et al.* 2010, Stracke *et al.* 2007). The IbMyb1 regulates anthocyanin biosynthesis in tuberous roots (Mano *et al.* 2007). Using potted plants under 400 mM NaCl conditions, two independent transformant lines exhibited better root growth than NTs and contained higher levels of secondary metabolites, such as total phenols, total flavonoids and anthocyanin. Salinity, UV-B and drought stress induced the expression of flavonol and anthocyanidin biosynthesis genes, which cooperated with *IbMyb1*expression. Transformants also showed higher radical scavenging activity than NTs. Over-expression of*IbMyb1* was observed to affect secondary metabolism, and consequently may improve abiotic stress tolerance in potato.

The examples discussed above indicate that certain heterogenic TFs could enhance abiotic stress tolerance in potato. Recent studies have attempted to isolate orthologue genes for the TFs discussed above in potato, and to confirm their functions in potato using GE techniques. Almost all of these studies confirmed abiotic stress tolerance in the transformants and described properties of the transgene products as the TF, such as nuclear localisation, binding ability to specific DNA sequences or regulation of downstream genes. Here, the discussion focuses on enhancement of abiotic stress tolerance. The *Ethylene responsive element binding protein 1* (*StEFEBP1*) gene from *S. tuberosum* was isolated using a cold induced cDNA pool, and then its TF was characterized by constitutive over-expression via the *35S* promoter in potato (Lee *et al.* 2007). Under cold conditions, yield increased by approximately 50% in the elite transformant line compared to NTs. Growth performance was also better in all transformant lines than in NTs under salinity conditions (75 mM NaCl) in vitro. The same group generated transformants with the *StMyb1R-1* gene, which encodes the R-1 type MYB-like TF, previously isolated as a stress inducible gene (Lee *et al.* 2007). The performance of transformants was evaluated under drought conditions, using potted plants in a greenhouse, by withholding water for 15 days and re-watering for 7 days (Shin *et al.* 2011). During the stress treatment, transformants displayed less visual damage with reduced water loss and higher survival rate than NTs. Sensitivity for ABA also increased in transformants. Under control conditions, obvious negative traits did not find from the picture in this report (Shin *et al.* 2011). Another group attempted to isolate the *DREB* orthologue from *S. tuberosum*based on information from a database, and generated transformants to characterize the isolated genes. They generated transformant lines with the *StDREB2* gene driven by the *35S* promoter, and evaluated their salinity tolerance (Bouaziz *et al.* 2012). Transformant lines showed better growth in vitro under saline conditions (200 mM NaCl for 40 days), and had less water loss and higher proline accumulation than NTs (100 mM NaCl for 20 days). Furthermore, transformants were also generated with the *StDREB1A* gene driven by the*35S* promoter, and their tolerance of salinity and drought was evaluated (Bouaziz *et al.* 2013). Transformant lines had less water loss and higher proline accumulation than NTs under saline conditions (100 mM NaCl for 20 days). Using potted plants in a greenhouse, elite lines exhibited better growth of their aboveground portions under saline (100 mM NaCl for 20 days) and drought conditions (withholding water for 25 days). As in the examples above, identified TFs have been shown to enhance abiotic stress tolerance in potato. It is expected that these efforts will contribute to our understanding of the tolerant mechanisms used in potato against abiotic stress.

The *35S:AtCBF1* construct has also been transformed into *S. tuberosum* and *S. commersonii*, and the two transformant lines were compared to each other (Pino *et al.* 2008). While this study was partially introduced in the previous section, there were a few differences between the two species. *S. commersonii* is known to exhibit cold acclimation (Chen and Li 1980, Costa and Li 1993) and thickening of the leaves following exposure to low temperatures (Chen *et al.* 1977, Estrada 1982, Palta and Li 1979). Cold acclimation was not conferred to *S. tuberosum* by over-expression of the *AtCBF1* gene, while increased leaf thickness was observed in both species. The cold tolerant property of *S. commersonii* is thought to consist of a complex of traits that are not controlled by a single TF. Accumulation of similar studies will advance our understanding of cold tolerance and cold acclimation mechanisms in *S. commersonii.*

STUDY OF STRESS TOLERANCE IN *S. COMMERSONII*

S. tuberosum can endure temperatures of −3°C and does not have the ability to acclimate to cold conditions. Conversely, *S. commersonii* can endure temperatures of −5°C before acclimation and can survive at −11°C after acclimation (Chen and Li 1980, Costa and Li 1993). Molecular analyses of *CBF* in *S. commersonii*were performed. First, transcriptome analyses using *S. tuberosum* and *S. commersonii* were conducted under low temperature conditions (2°C) or by transfromants with *35S:AtCBF3* (Carvallo *et al.* 2011). This work revealed that a cold regulated gene induced at a similar degree and with similar kinetics exists in both *S. tuberosum* and *S. commersonii.* However, considerable differences were observed in the gene sets composed of the cold transcriptome and CBF regulons of these two species. It was suggested that these differences contribute to the differences in freezing tolerance of the two species, including cold acclimation. Additionally, the genetic structure of *CBFs* have been investigated and compared among *Solanum* species and *Arabidopsis*(Pennycooke *et al.* 2008). Gene clusters of CBF (*CBF3-CBF1-CBF2* cluster) were maintained among investigated *Solanum* species and *Arabidopsis*, with some deletion or duplication. In particular, *S. tuberosum*and *S. commersonii* had one more *CFB2* (remnant *CBF2* in *S. commersonii*) in adjacent in *CBF2*. Although all *CBFs* in *Arabidopsis* were induced by cold, only *CBF1* showed cold inducible properties in the *Solanum*species. Furthermore, cold inducible *CBF4* existed on another locus of *S. tuberosum* and *S. commersonii. S. tuberosum* had *CBF5* closely linked to *CBF4*, and it was suggested that the *CBF5-CBF4* cluster resulted from duplication of the *CBF3-CBF1-CBF2* cluster. DNA motifs in up-stream regions of *CBF1* were found in *CBF4* in *Solanum* species, which were conserved in all three *CBFs* in *Arabidopsis*. These results suggest that *CBF1* or *CBF4* are responsible for the cold response in the *Solanum* species.

Transformants of *S. tuberosum* and *S. commersonii* with the *CBF1* (*ScCBF1*) gene from *S. commersonii*driven by the *35S* promoter were generated and evaluated for drought tolerance in vitro (Pino *et al.* 2013). Both transformants indicated better overall plant growth and higher accumulation of proline compared to NTs. The expression level of *DNH10* was observed to increase during drought stress. This work did not focus on cold tolerance or cold acclimation. Although it is difficult to judge from the information presented here, simple over-expression may not be able to confer the cold acclimation trait to *S. tuberosum*. Further accumulation of studies on cold acclimation using GE techniques is necessary to clarify this mechanism and to confer cold acclimation to *S. tuberosum*.

FUTURE BREEDING FOR ABIOTIC STRESS TOLERANCE

Breeding programs have brought an eagerness to confer biotic stress tolerance to potato and to improve productivity and quality of tubers rather than to enhance abiotic stress tolerance. The majority of recent cultivars and breeding materials in potato carry the above traits. For this reason, it is thought that recent genetic resources in potato may be unfit for breeding abiotic stress tolerance. However, 'Konyu1~4' cultivars were established in 2007 by crossing 'Danshaku' (Irish Cobbler) and 'Konafubuki' (Iwama 2008), whose parental cultivars are major varieties in Japan. 'Konyu' showed drought tolerance and had a larger roots mass in comparison to the parent line. Although the practicability of 'Konyu' as a commercial cultivar is unknown, this variety indicates the potential to be a genetic donor for abiotic stress tolerance in recent genetic resources.

On the other hand, *S. tuberosum* has various relatives, including wild species that can exchange genetic traits. Various related species have been utilized as genetic donors to carry out introgression of biotic stress resistance to *S. tuberosum* in the potato pedigree database (Berloo *et al.* 2007). For example cv. 'Astarte' has the cyst nematode resistance gene from *S. vernei*. Using related species as genetic donors to confer abiotic stress tolerance has also been attempted. For example, introgression of traits from *S. commersonii* to *S. tuberosum* has been conducted. However, freezing tolerance and cold acclimation have not transferred into *S. tuberosum* from *S. commersonii* using traditional breeding efforts without the addition of agronomically undesirable wild traits (Cardi *et al.* 1993, Estrada 1982, Estrada *et al.* 1993, Ioveve *et al.* 2004, Pavek and Corsini 2001). In fact, breeding potato cultivars using conventional methods is demanding and time consuming work since *S. tuberosum* is autotetraploid and has a highly complicated quantitative inheritance pattern. Some scientists have suggested that GE represents a faster and a more reliable way to improve potato cultivars (Bayat *et al.* 2010, Waterer *et al.* 2010). However, molecular breeding using GE techniques is not an easy way to establish new abiotic stress tolerance cultivars. In some cases, inconvenient phenotypes appear by conferring the desirable traits mentioned above. Typical examples are transformants with abiotic stress related TFs driven by the *35S* promoter. During controlled conditions, such as in vitro or in growth chambers, inconvenient phenotypes are often hidden (Behnam *et al.* 2006, 2007). These phenotypes gradually become significant in uncontrolled conditions or during large scale cultivation, such as in greenhouses (Shimazaki *et al.* unpublished) and fields. Although the practicability of using transformants with abiotic stress tolerance should be evaluated in the field with some repetition, studies that use continuous field evaluation are still limited. Only one study performed a four-year evaluation in the field (Waterer *et al.* 2010), and they found that conferred traits disappeared in the wet and cool weather conditions. Furthermore, it is difficult to evaluate whether the tolerance levels under controlled conditions, such as in vitro, will reach practical levels in the field. Although similar cases should be applied, compatibility may depend on the transgene, host plant properties and the field trial conditions. For this reason, various case studies, including those that fail, should be collated for future assessment of transgenic potato with abiotic stress tolerance. From the studies introduced above, some transgenes have been shown to enhance abiotic stress tolerance without negative effects. It is expected that these transformants will be planted in fields to provide further information on their performance and tolerance.

In contrast, many wild relative species have thrived in their habitats under various levels of drought conditions. Drought tolerant genotypes were selected from various wild species using seedling growth (Watanabe *et al.* 2011) as the growth profiles differed for each genotype. Although one genotype indicated a slow growth profile in control conditions, seedling growth was not affected under drought conditions (Fig. 1). Slow growth is not good as a commercial trait, but this tolerant trait may contribute to potato breeding depending on the mating profile. To accomplish introgression of traits in relative species to *S. tuberosum*, donor plants derived from certain wild relative species have been prepared for future breeding. In the case of conferring abiotic stress tolerance via introgression of their traits, molecular markers are strong tools for screening objective genotypes. Specific makers can be used to retain the objective gene and to eliminate agronomically undesirable wild traits. Studies on transgenic potato with TFs may lead to the development of a specific maker set. It is expected that over-expression of TFs will contribute to the understanding of the mechanisms involved in abiotic stress tolerance and to identify genes directly responsible in potato or wild relatives. The whole genome sequence of *S. phureja* has already been identified (The Potato Genome Sequencing Consortium 2011). Quantitative trait locus (QTL) analysis has become a good method to identify responsible genes by generating transformants of the corresponding DNA region. Although there are many available genes from other plant species, exploring new genes from *Solanum* species is important for the potato breeding program and incorporating abiotic stress tolerance. Conventional breeding and molecular methods, including exploration of responsible genes and maker development using GE techniques, should be used together to develop cultivars tolerant of various abiotic stresses.

Figure 1: Seedling growth under drought conditions. Seeds were placed on 1/2 MS solid medium, and cultured at 25°C for 35 days (16L/8D). C indicates control conditions; D indicates drought stress conditions with 4% mannitol. All control plants grew well in control conditions, while their growth was inhibited under drought conditions. Three seeds did not germinate under drought conditions. All plants of the wild species germinated in both conditions, but wild plants grew slower than control plants. However, wild seedlings exhibited better growth under drought conditions than control seedlings.

Finally, we would like to mention some miscellaneous impressions. Descriptions such as 'T1 generation' and 'wild type (WT)' were observed in some literatures. This 'generation' increases by passing through a tuber, not by mating. Furthermore, 'WT' was used to mean the same as a non-transformed host cultivar. Potato is a vegetative propagated cultivar, but not is wild species. These terminologies may be common in some scientists, but we feel a sense of incongruity towards it.

ACKNOWLEDGEMENTS

We greatly appreciate USDA NRSP-6 for providing the seeds of wild relative species. This paper was supported in part by the joint research program "Plant Transgenic Research Design, University of Tsukuba". This paper supports the relevant background information for a CREST, JST project titled 'Fundamental research for technology development in productivity through simultaneous improvement of functions of sink and source organs in plants'. And this paper also supports the activities on JST/JICA, SATREPS (Science and Technology Research Partnership for Sustainable Development) project "Diversity Assessment and Development of Sustainable Use of Mexican Genetic Resources".

REFERENCES

1. Ahmad,R., M.Kim, K.Back, H.Kim, H.Lee, S.Kwon, N.Murata, W.Chung and S.Kwak (2008) Stress-induced expression of choline oxidase in potato plant chloroplasts confers enhanced tolerance to oxidative, salt, and drought stresses. Plant cell Rep. 27: 687–698.
2. Ahmad,R., Y.Kim, M.Kim, S.Kwon, K.Cho, H.Lee and S.Kwak (2010) Simultaneous expression of choline oxidase, superoxide dismutase and ascorbate peroxidase in potato plant chloroplasts provides synergistically enhanced protection against various abiotic stresses. Physiol. Plant. 138: 520–533.
3. Ambard-Bretteville,F., C.Sorin, F.Rebeille, C.Hourton-Cabassa and C.des Francs-Small (2003) Repression of formate dehydrogenase in Solanum tuberosum increases steady-state levels of formate and accelerates the accumulation of proline in response to osmotic stress. Plant Mol. Biol. 52: 1153–1168.
4. Amiri,R., I.Goldenkova-Pavlova, N.Yur'eva, V.Pchelkin, V. Tsydendambaev, A.Vereshchagin, A.Deryabin, T.Trunova, D.Los and A.Nosov (2007) Lipid fatty acid composition of potato plants transformed with the delta12-desaturase gene from cyanobaterium. Russian J. Plant Physiol. 54: 600–606.
5. Amiri,R., N.Yur'eva, K.Shimshilashvili, I.Goldenkova-Pavlova, V.Pchelkin, E.Kuznitsova, V.Tsydendambaev, T.Trunova, D.Los, G.Jouzani et al. (2010) Expression of acyl-lipid 12-desaturase gene in prokaryotic and eukaryotic cells and its effect on cold stress tolerance of potato. J. Integr. Plant Biol. 52: 289–297.
6. Asada,K. and M.Takahashi (1987) Production and scavenging of active oxygen in photosynthesis. In: Kyle,D., B.Osmond and C. Arntzen (eds.) Photoinhibition, Esevier, Amsterdam, pp. 227–287.
7. Asada,K. (1999) The water-water cycle in chloroplasts: scavenging of active oxygens and dissipation of excess photons. Annu. Rev. Plant Physiol. Plant Mol. Biol. 50: 601–639.
8. Ashrafa,M. and M.Foola (2007) Roles of glycine betaine and proline in improving plant abiotic stress resistance. Environ. Exp. Bot. 59: 206–216.
9. Barrientos,M., E.Mol, G.Peruzzo, A.Contreras and M.Alberdi (1994) Responses to cold of chilean wild Solanum species. Environ. Exp. Bot. 34: 47–54.

10. Bayat,F., B.Shiran, D.Belyaev, N.Yur'eva, G.Sobol'kova, H.Alizadeh, M.Khodambashi and A.Babakov (2010) Potato plants bearing a vacuolar Na+/H+ antiporter HvNHX2 from barley are characterized by improved salt tolerance. Russian J. Plant Physiol. 57: 696–706.

11. Behnam,B., A.Kikuchi, F.Celebi-Toprak, S.Yamanaka, M.Kasuga, K.Yamaguchi-Shinozaki and K.Watanabe (2006) The Arabidopsis DREB1A gene driven by the stress-inducible rd29A promoter increases salt-stress tolerance in tetrasomic tetraploid potato (Solanum tuberosum) in proportion to its copy number. Plant Biotechnol. 23: 169–177.

12. Behnam,B., A.Kikuchi, F.Celebi-Toprak, M.Kasuga, K.YamaguchiShinozaki and K.Watanabe (2007) Arabidopsis rd29A::DREB1A enhances freezing tolerance in transgenic potato. Plant Cell Rep. 26: 1275–1282.

13. Berloo,R., R.Hutten, H.Eck and R.Visser (2007) An online potato pedigree database resource. Potato Res. 50: 45–57

14. Bianchi,G., A.Gamba, R.Limiroli, N.Pozzi, R.Elster, F.Salamini and D.Bartels (2006) The unusual sugar composition in leaves of the resurrection plant Myrothamnus flabellifolia. Physiol. Plant. 87: 223–226.

15. Blumwald,E., G.Aharon and M.Apse (2000) Sodium transport in plant cells. Biochim. Biophys. Acta 1465: 140–151.

16. Bouaziz,D., J.Pirrello, H.Amor, A.Hammami, M.Charfeddine, A.Dhieb, M.Bouzayen and R.Gargouri-Bouzid (2012) Ectopic expression of dehydration responsive element binding proteins (StDREB2) confers higher tolerance to salt stress in potato. Plant Physiol. Biochem. 60: 98–108.

17. Bouaziz,D., J.Pirrello, M.Charfeddine, A.Hammami, R.Jbir, A.Dhieb, M.Bouzayen and R.Gargouri-Bouzid (2013) Overexpression of StDREB1 transcription factor increases tolerance to salt in transgenic potato plants. Mol. Biotechnol. 54: 803–817.

18. Boyer,J. (1982) Plant productivity and environment. Science 218: 443–448.

19. Cardi,T., E.D'Ambrosio, D.Consoli, K.Puite and K.Ramulu (1993) Production of somatic hybrids between frost-tolerant Solanum commersonii and S. tuberosum: characterization of hybrid plants. Theor. Appl. Genet. 87: 193–200.

20. Carpenter,J. and T.Hansen (1992) Antifreeze protein modulates cell survival during cryopreservation: mediation through influence on ice crystal growth. Proc. Natl. Acad. Sci. USA 89: 8953–8957.

21. Carvallo,M., M.Pino, Z.Jeknic, C.Zou, C.Doherty, S.Shiu, T.Chen and M.Thomashow (2011) A comparison of the low temperature transcriptomes and CBF regulons of three plant species that differ in freezing tolerance Solanum commersonii, Solanum tuberosum and Arabidopsis thaliana. J. Exp. Bot. 62: 3807–3819.

22. Celebi-Toprak,F., B.Behnam, G.Serrano, M.Kasuga, K.YamaguchiShinozaki, H.Naka, J.Watanabe, S.Yamanaka and K.Watanabe (2005) Tolerance to salt stress of the transgenic tetrasomic tetraploid potato, Solanum tuberosum cv. Desiree appears to be induced by the DREB1A gene and rd29A promoter of Arabidopsis thaliana. Breed. Sci. 55: 311–319.

23. Chen,H. and P.Li (1980) Characteristics of cold acclimation and deacclimation in tuber-bearing Solanum species. Plant Physiol. 65: 1146–1148.

24. Chen,P., P.Li and W.Cunningham (1977) Ultrastructural differences in leaf cells of some Solanum species in relation to their frost resistance. Bot. Gaz. 138: 276–285.

25. Chen,T. and N.Murata (2008) Glycinebetaine: an effective protectant against abiotic stress in plants. Trends Plant Sci. 13: 499–505.

26. Cheng,Y., X.Deng, S.Kwak, W.Chen and A.Eneji (2013a) Enhanced tolerance of transgenic potato plants expressing choline oxidase in chloroplasts against water stress. Bot. Stud. 54: 30–38.

27. Cheng,Y., M.Kim, X.Deng, S.Kwak and W.Chen (2013b) Enhanced salt stress tolerance in transgenic potato plants expressing IbMYB1, a sweet potato transcription factor. J. Microbiol. Biotechnol. 23: 1737–1746.

28. Clarkson,D., K.Hall and J.Roberts (1980) Phospholipid composition and fatty acid desaturation in the roots of rye during acclimatization of low temperature. Planta 149: 464–471.

29. Costa,A. and P.Li (1993) Development of cold hardiness in Solanum tuberosum by abscisic acid and mefluidide. In: Li,P. and L. Christersson (eds.) Advance in Plant Cold Hardiness. CRC Press, Boca Raton, FL, USA, pp. 139–140.

30. Dalla Costa,L., G.Vedove, G.Gianquintoi, R.Giovanardi and A. Peressotti (1997) Yield, water use efficiency and nitrogen uptake in potato: influence of drought stress. Potato Res. 40: 19–34.

31. Deblonde,P. and J.Ledent (2001) Effects of moderate drought conditions on green leaf number stem height, leaf length and tuber yield of potato cultivars. Eur. J. Agron. 14: 31–41.

32. Demin,I., A.Deryabin, M.Sinkevich and T.Trunova (2008) Insertion of cyanobacterial desA gene coding for delta 12-acyl-lipid desaturase increases potato resistance to oxidative stress induced by hypothermia. Russian J. Plant Physiol. 55: 639–648.

33. De Palma,M., S.Grillo, I.Massarelli, A.Costa, G.Balogh, L.Vigh and A.Leone (2008) Regulation of desaturase gene expression, changes in membrane lipid composition and freezing tolerance in potato plants. Mol. Breed. 21: 15–26.

34. Deryabin,A., T.Trunova, I.Dubinina, E.Burakhanova, E.Sabel'nikova, E.Krylova and G.Romanov (2003) Chilling tolerance of potato plants transformed with a yeast-derived invertase gene under the control of the B33 patatin promoter. Russian J. Plant Physiol. 50: 449–454.

35. De Vries,A. (1986) Antifreeze glycopeptides and peptides: interaction with ice and water. Meth. Enzymol. 127: 293–303.

36. Dietz,K. (2003) Plant peroxiredoxins. Annu. Rev. Plant Biol. 54: 93– 107.

37. Dubos,C., R.Stracke, E.Grotewold, B.Weisshaar,

C.Martin and L.Lepiniec (2010) MYB transcription factors in Arabidopsis. Trends Plant Sci. 15: 573–581.

38. Ekanayake,I. and D.Midmore (1992) Genotypic variation for root pulling resistance associated with drought. Euphytica 61: 43–53.

39. El-Banna,A., M.Hajirezaei, J.Wissing, Z.Ali, L.Vaas, E.HeineDobbernack, H.Jacobsen, H.Schumacher and H.Kiesecker (2010) Over-expression of PR-10a leads to increased salt and osmotic tolerance in potato cell cultures. J. Biotechnol. 150: 277–287.

40. Eltayeb,A., S.Yamamoto, M.Habora, Y.Matsukubo, M.Aono, H. Tsujimoto and K.Tanaka (2010) Greater protection against oxidative damages imposed by various environmental stresses in transgenic potato with higher level of reduced glutathione. Breed. Sci. 60: 101–109.

41. Eltayeb,A., S.Yamamoto, M.Habora, L.Yin, H.Tsujimoto and K.Tanaka (2011) Transgenic potato overexpressing Arabidopsis cytosolic AtDHAR1 showed higher tolerance to herbicide, drought and salt stresses. Breed. Sci. 61: 3–10.

42. Estrada,N. (1982) Breeding wild and primitive potato species to obtain frost resistant cultivated varieties. In: Li,P. and A.Sakai (eds.) Plant hardiness and freezing stress: mechanism and crop implications. Academic Press, New York, NY, USA, pp. 615–633.

43. Estrada,N., W.Garcia, J.Carrasco and E.Carrasco (1993) Breeding potato for tolerance to frost and resistance to late blight. Am. Potato J. 70: 809–810.

44. Fricke,W. and E.Pahlich (1990) The effect of water stress on the vacuole-extravacuole compartmentation of proline in potato cell suspension cultures. Physiol. Plant. 78: 374–378.

45. Geilfus,C., C.Zorb and K.Muhling (2010) Salt stress differentially affects growth-mediating β-expansins in resistant and sensitive maize (Zea mays L.). Plant Physiol. Biochem. 48: 993–998.

46. Gilmour,S., D.Zarka, E.Stockinger, M.Salazar, J.Houghton and M. Thomashow (1998) Low temperature regulation of the Arabidopsis CBF family of AP2 transcriptional activators as an early step in cold-induced COR gene expression. Plant J. 16: 433–442.

47. Gururani,M., C.Upadhyaya, R.Strasser, J.Woong and S.Park (2012) Physiological and biochemical responses of transgenic potato plants with altered expression of PSII manganese stabilizing protein. Plant Physiol. Biochem. 58: 182–194.

48. Gururani,M., C.Upadhyaya, R.Strasser, J.Yu and S.Park (2013) Evaluation of abiotic stress tolerance in transgenic potato plants with reduced expression of PSII manganese stabilizing protein. Plant Sci. 98: 7–16.

49. Halliwell,B. and J.Gutterridge (2000) Free Radicals in Biology and Medicine. Oxford University Press, New York.

50. Hawkes,J. (1994) Origins of cultivated potatoes and species relationships. In: Bradshaw,J. and G.Mackay (eds.) Potato genetics. Cab International, Cambridge, UK, pp. 3–42.

51. Heineke,D., U.Sonnewald, D.Bussis, G.Gunter, K.Leideriter, I.Wilke, K.Raschke, L.Willmitzer and W.Heldt (1992) Apoplastic expression of yeast-derived invertase in potato. Plant Physiol. 100: 301– 308

52. Hellwege,E., S.Czapla, A.Jahnke, L.Willmitzer and A.Heyer (2000) Transgenic potato (Solanum tuberosum) tubers synthesize the full spectrum of inulin molecules naturally occurring in globe artichoke (Cynara scolymus) root. Proc. Natl. Acad. Sci. USA 97: 8699– 8704.

53. Hemavathi, C.Upadhyaya, K.Young, N.Akula, H.Kim, J.Heung, O.Oh, C.Aswath, S.Chun, D.Kim et al. (2009) Over-expression of strawberry D-galacturonic acid reductase in potato leads to accumulation of vitamin C with enhanced abiotic stress tolerance. Plant Sci. 177: 659–667.

54. Hemavathi, C.Upadhyaya, N.Akula, K.Young, S.Chun, D.Kim and S.Park (2010) Enhanced ascorbic acid accumulation in transgenic potato confers tolerance to various abiotic stresses. Biotechnol. Lett. 32: 321–330.

55. Hemavathi, C.Upadhyaya, A.Nookaraju, H.Kim, J.Jeon, O.Ho, S.Chun, D.Kim and S.Park (2011) Biochemical analysis of enhanced tolerance in transgenic potato plants overexpressing Dgalacturonic acid reductase gene in response to various abiotic stresses. Mol. Breed. 28: 105–115.

56. Hendry,A. (1993) Evolutionary origins and natural functions of fructans—a climatological, biogeographic and mechanistic appraisal. New Phytol. 123: 3–14.

57. Hmida-Sayaria,A., R.Gargouri-Bouzida, A.Bidania, L.Jaouaa, A. Savourab and S.Jaouac (2005) Overexpression of Δ1-pyrroline5-carboxylate synthetase increases proline production and confers salt tolerance in transgenic potato plants. Plant Sci. 169: 746–752.

58. Holmstrom,K., E.Mentyle, B.Welin, A.Mandal, E.Tapio Palva, O.Tunnela and J.Londesborough (1996) Drought tolerance in tobacco. Nature 379: 683–684.

59. Hoque,J., R.Hirji, L.Adam, K.Rozwadowski, J.Hammerlindl, W.Keller and G.Selvaraj (2007) Genetic engineering of glycinebetaine production toward enhancing stress tolerance in plants: metabolic limitations. Plant Physiol. 122: 747–756.

60. Horvath,I., L.Vigh and T.Farkas (1981) The manipulation of polar head group composition of phospholipids in the wheat Miranovskaja 808 affect frost tolerance. Planta 151: 103–108.

61. Iordachescu,M. and R.Imai (2008) Trehalose biosyntesis in response to abiotic stresses. J. Integr. Plant Biol. 50: 1223–1229.

62. Iovene,M., A.Barone, L.Frusciante, L.Monti and D.Carputo (2004) Selection for aneuploid potato hybrids combining a low wild genome content and resistance traits from Solanum commersonii. Theor. Appl. Genet. 109: 1139–1146.

63. Iwama,K. (2008) Physiology of the potato: new insights into root system and repercussions for crop management. Potato Res. 51: 333– 353.

64. Jagendorf,A. and T.Takabe (2001) Inducers of glycinebetaine synthesis in barley. Plant Physiol. 127: 1827–1835.

65. Jain,N. and I.Roy (2009) Effect of trehalose on protein structure. Protein Sci. 18: 24–36.

66. Japan Meteorological Agency (2013) Climate Change Monitoring Report 2012.

67. Japan Meteorological Agency (2014) Northern pacific observation network data show atmospheric CO2 concentration exceeding 400 ppm. News Release. 26/ May

68. Jeong,M., S.Park, H.Kwon and M.Byun (2000) Isolation and characterization of the gene encoding glyceraldehyde-3 phosphate dehydrogenase. Biochem. Biophys. Res. Commun. 278: 192–196.

69. Jeong,M., S.Park and M.Byun (2001) Improvement of salt tolerance in transgenic potato plants by glyceraldehyde-3 phosphate dehydrogenase gene transfer. Mol. Cells 12: 185–189.

70. Kaneda,M., K.Kobayashi, K.Nishida and S.Katsuta (1984) Liliosides D and F, two glycerol glycosides from Lilium japonicum. Phytochemistry 23: 795–798.

71. Kasuga,M., Q.Liu, S.Miura, K.Yamaguchi-Shinozaki and K. Shinozaki (1999) Improving plant drought, salt, and freezing tolerance by gene transfer of a single stress-inducible transcription fac tor. Nat. Biotechnol. 7: 287–291.

72. Katerji,N., J. van Hoorn, A.Hamdy and M.Mastrorilli (2003) Salinity effect on crop development and yield, analysis of salt tolerance according to several classification methods. Agric. Water Manage. 62: 37–66

73. Kim,J., D.Baek, H.Park, H.Chun, D.Oh, M.Lee, J.Cha, W.Kim, M.Kim, W.Chung et al. (2013) Overexpression of Arabidopsis YUCCA6 in potato results in high-auxin developmental phenotypes and enhanced resistance to water deficit. Mol. Plant 6: 337–349.

74. Kim,K., S.Kwon, H.Lee, Y.Hur, J.Bang and S.Kwak (2003) A novel oxidative stress-inducible peroxidase promoter from sweetpotato: molecular cloning and characterization in transgenic tobacco plants and cultured cells. Plant Mol. Biol. 51: 831–838.

75. Kim,M., Y.Kim, S.Kwon, D.Yun, S.Kwak and H.Lee (2010) Enhance tolerance to methyl viologen-induced oxidative stress and high temperature in transgenic potato plants overexpressing the CuZnSOD, APX and NDPK2 genes. Physiol. Plant. 140: 153–162.

76. Kim,M., Y.Kim, S.Kwon, B.Jang, S.Lee, D.Yun, J.Cho, S.Kwak and H.Lee H (2011) Overexpression of 2-cysteine peroxiredoxin enhances tolerance to methyl viologen-mediated oxidative stress and high temperature in potato plants. Plant Physiol. Biochem. 49: 891–897.

77. Kiziloglu,F., M.Sahin, T.Tunc and S.Diler (2006) The effect of deficit irrigation on potato evapotranspiration and tuber yield under cool season and semiarid climatic condition. J. Agronomy 5: 284–288.

78. Klähn,S., D.Marquardt, I.Rollwitz and M.Hagemann (2009) Expression of the ggpPS gene for glucosylglycerol biosynthesis from Azotobacter vinelandii improves the salt tolerance of Arabidopsis thaliana. J. Exp. Bot. 60: 1679–1689.

79. Knipp,G. and B.Honermeier (2006) Effect of water stress on proline accumulation of genetically modified potatoes (Solanum tuberrosum L.) generating fructans. J. Plant Physiol. 163: 392–397.

80. Kodama,H., T.Hamada, G.Horiguchi, M.Nishimura and K.Iba (1994) Genetic enhancement of cold tolerance by expression of a gene for chloroplast ω-3 fatty acid desaturase in transgenic tobacco. Plant Physiol. 105: 601–605.

81. Kodama,H., G.Horiguchi, T.Nishiuchi, M.Nishimura and K.Iba (1995) Fatty acid desaturation during chilling acclimation is one of the factors involved in conferring low-temperature tolerance to young tobacco leaves. Plant Physiol. 107: 1177–1185.

82. Kondrak,M., F.Marincs, B.Kalapos, Z.Juhasz and Z.Banfalvi (2011) Transcriptome analysis of potato leaves expressing the trehalose- 6-phosphate synthase 1 gene of yeast. PLoS ONE 6:e23466.

83. Kondrak,M., F.Marincs, F.Antal, Z.Juhasz and Z.Banfalvi (2012) Effects of yeast trehalose-6-phosphate synthase 1 gene expression and carbohydrate contents of potato leaves under drought stress conditions. BMC Plant Biol. 12: 74–86.

84. Kotuby-Amacher,J., R.Koenig and B.Kitchen (2000) Salinity and Plant Tolerance. Electronic publishing. AG-SO-03.

85. Lee,H., L.Xiong, M.Ishitani, B.Stevenson and J.Zhu (1999) Coldregulated gene expression and freezing tolerance in an Arabidopsis thaliana mutant. Plant J. 17: 301–308.

86. Lee,H., D.Shin, S.Park, S.Han, M.Jeong, T.Kwon, S.Lee, S.Park, B.Yi, H.Kwon et al. (2007) Ethylene responsive element binding protein 1 (StEREBP1) from Solanum tuberosum increases tolerance to abiotic stress in transgenic potato plants. Biochem. Biophys. Res. Commun. 353: 863–868.

87. Lee,M., J.Jung, D.Han, P.Seo, W.Park and C.Park (2011) Activation of a flavin monooxygenase gene YUCCA7 enhances drought resistance in Arabidopsis. Planta 235: 923–938.

88. Liu,J. and A.Ekramoddoullah (2006) The family 10 of plant pathogenesis-related proteins: Their structure, regulation, and function in response to biotic and abiotic stresses. Physiol. Mol. Plant Pathol. 68: 3–13.

89. Liu,Q., M.Kasuga, Y.Sakuma, H.Abe, S.Miura, K.YamaguchiShinozaki and K.Shinozaki (1998) Two transcription factors, DREB1 and DREB2, with an EREBP/AP2 DNA binding domain separate two cellular signal transduction pathways in drought- and low-temperature-responsive gene expression, respectively, in Arabidopsis. Plant Cell 10: 1391–1406.

90. Maas,E. and G.Hoffman (1977) Crop salt tolerance—Current assessment. J. Irrig. Drain. Div. 103: 115–134.

91. Mahouachi,J., R.Argamasilla and A.Gomez-Cadenas (2012) Influence of exogenous glycine betaine and abscisic acid on papaya in responses to water-deficit stress. J. Plant Growth Regul. 31: 1–10.

92. Mano,H., F.Ogasawara, K.Sato, H.Higo and Y.Minobe (2007) Isolation of a regulatory gene of anthocyanin

biosynthesis in tuberous roots of purple-fleshed sweet potato. Plant Physiol. 143: 1252–1268.

93. Mazurczyk,W., A.Wierzbicka and C.Trawczynski (2009) Harvest index of potato crop grown under different nitrogen and water supply. Acta Sci. Pol. Agricultura 8: 15–21.

94. Miquel,M. and J.Browse (1992) Arabidopsis mutants deficient in polyunsaturated fatty acid synthesis. Biochemical and genetic characterization of a plant oleoyl-phosphatidylcholine desaturase. J. Biol. Chem. 267: 1502–1509.

95. Miquel,M., D.James, H.Dooner and J.Browse (1993) Arabidopsis requires polyunsaturated lipids for low-temperature survival. Proc. Natl. Acad. Sci. USA 90: 6208–6212.

96. Mittler,R. (2002) Oxidative stress, antioxidants and stress tolerance. Trends Plant Sci. 7: 405–410.

97. Movahedi,S., B.Tabatabaei, H.Alizade, C.Ghobadi, A.Yamchi and G.Khaksar (2012) Constitutive expression of Arabidopsis DREB1B in transgenic potato enhances drought and freezing tolerance. Biol. Plant. 56: 37–42.

98. Munns,R. (2005) Genes and salt tolerance: bringing them together. New Phytol. 167: 645–663.

99. Napier,J., L.Michaelson and A.Stobart (1999) Plant desaturases: harvesting the fat of the land. Cur. Opin. Plant Biol. 2: 123–127.

100. Noctor,G., A.Arisi, L.Jouanin, K.Kunert, H.Rennenberg and C.Foyer (1998) Glutathione: biosynthesis, metabolism and relationship to stress tolerance explored in transformed plants. J. Exp. Bot. 49: 623–647.

101. Noctor,G. and C.Foyer (1998) Ascorbate and glutathione: keeping active oxygen under control. Annu. Rev. Plant Physiol. Plant Mol. Biol. 49: 249–279.

102. Ortiz,R. and K.Watanabe (2004) Genetic contributions to breeding polyploid crops. In: Pandalai,S. (ed.) Recent Research Developments in Genetics & Breeding. Vol. 1, Part II. pp. 269–286.

103. Otero,A. (2000) NM23/nucleoside diphosphate kinase and signal transduction. J. Bioenerg. Biomembr. 32: 269–275.

104. Pal,A., K.Acharya, S.Vats, S.Kumar and P.Ahuja (2013) Overexpression of PaSOD in transgenic potato enhances photosynthetic performance under drought. Biol. Plant. 57: 359–364.

105. Palta,J. and P.Li (1979) Frost-hardiness in relation to leaf anatomy and natural distribution of several Solanum species. Crop Sci. 19: 665–671.

106. Park,E., Z.Jeknic and T.Chen (2006) Exogenous application of glycinebetaine increases chilling tolerance in tomato plants. Plant Cell Physiol. 47: 706–714.

107. Park,J., S.Kwon, K.Song, J.Kwak, S.Lee, Y.Nam, J.Shin, Y.Park, S.Rhee and K.Park (1999) Transgenic tobacco plants expressing the bacterial levansucrase gene show enhanced tolerance to osmotic stress. J. Microbiol. Biotechnol. 9: 213–218.

108. Patonnier,M., J.Peltier and G.Marigo (1999) Drought-induced increase in xylem malate and mannitol concentrations and closure of Fraxinus excelsior L. stomata. Exp. Bot. 50:1223–1229.

109. Pavek,J. and D.Corsini (2001) Utilization of potato genetic resources in variety development. Am. J. Potato Res. 78: 433–441.

110. Pennycooke,J., H.Cheng, S.Roberts, Q.Yang, S.Rhee and E. Stockinger (2008) The low temperature-responsive, Solanum CBF1 genes maintain high identity in their upstream regions in a genomic environment undergoing gene duplications, deletions, and rearrangements. Plant Mol. Biol. 67: 483–497.

111. Pilon-Smits,E., M.Ebskamp, M.Paul, M.Jeuken, P.Weisbeek and S.Smeekens (1995) Improved performance of transgenic fructanaccumulating tobacco under drought stress. Plant Physiol. 107: 125–130.

112. Pino,M., J.Skinner, E.Park, Z.Jeknic, P.Hayes, M.Thomashow and T.Chen (2007) Use of a stress inducible promoter to drive ectopic AtCBF expression improves potato freezing tolerance while minimizing negative effects on tuber yield. Plant Biotechnol. J. 5: 591– 604.

113. Pino,M., J.Skinner, Z.Jeknic, P.Hayes, A.Soeldner, M.Thomashow and T.Chen (2008) Ectopic AtCBF1 over-expression enhances freezing tolerance and induces cold acclimation-associated physiological modifications in potato. Plant Cell Environ. 31: 393–406.

114. Pino,M., A.Avila, A.Molina, Z.Jeknic and T.Chen (2013) Enhanced in vitro drought tolerance of Solanum tuberosum and Solanum commersonii plants overexpressing the ScCBF1 gene. Cien. Inv. Agr. 40: 171–184

115. Pocard,J., L.Smith, G.Smith and D.Le Rudulier (1994) A prominent role for glucosylglycerol in the adaptation of Pseudomonas mendocina SKB70 to osmotic stress. J. Bacteriol. 176: 6877–6884

116. Potters,G., L.De Gara, H.Asard and N.Horemans (2002) Ascorbate and glutathione: guardians of the cell cycle, partners in crime? Plant Physiol. Biochem. 40: 537–548.

117. Prabhavathi,V. and M.Rajam (2007) Mannitol-accumulating transgenic eggplants exhibit enhanced resistance to fungal wilts. Plant Sci. 173: 50–54.

118. Quinn,P., F.Joo and L.Vigh (1989) The role of unsaturated lipids in membrane structure and stability. Prog. Biophys. Mol. Biol. 53: 71–103.

119. Rahnama,H., H.Vakilian, H.Fahimi and B.Ghareyazie (2011) Enhanced salt stress tolerance in transgenic potato plants (Solanum tuberosum L.) expressing a bacterial mtlD gene. Acta Physiol. Plant. 33: 1521–1532.

120. Ramanjulu,S. and D.Bartels (2002) Drought- and desiccation-induced modulation of gene expression in plants. Plant Cell Environ. 25: 141–151.

121. Reed,R. and W.Stewart (1985) Osmotic adjustment and organic solute accumulation in unicellular cyanobacteria from freshwater and marine habitats. Marine Biology. 88: 1–9.

122. Roder,A., E.Hoffmann, M.Hagemann and G.Berg (2005) Synthesis of the compatible solutes glucosylglycerol and trehalose by saltstressed cells of Stenotrophomonas

strains. FEMS Microbiol. Lett. 243: 219–226.

123. Rontein,D., G.Basset and A.Hanson (2002) Metabolic engineering of osmoprotectant accumulation in plants. Metab. Eng. 4: 49–56.

124. Sakamoto,A. and N.Murata (2000) Genetic engineering of glycinebetaine synthesis in plants: current status and implications for enhancement of stress tolerance. J. Exp. Bot. 51: 81–88.

125. Sakamoto,A. and N.Murata (2002) The role of glycine betaine in the protection of plants from stress: clues from transgenic plants. Plant Cell Environ. 25: 163–171.

126. Sakuma,Y., Q.Liu, J.Dubouzet, H.Abe, K.Shinozaki and K.YamaguchiShinozaki (2002) DNA-binding specificity of the ERF/AP2 domain of Arabidopsis DREBs, transcription factors involved in dehydrationand cold-inducible gene expression. Biochem. Biophys. Res. Commun. 290: 998–1009.

127. Shen,B., R.Jensen and H.Bohner (1997) Increased resistance to oxidative stress in transgenic plants by targeting mannitol biosynthesis to chloroplasts. Plant Physiol. 113: 1177–1183.

128. Sherraf,I., S.Tizroutine, M.Chaput, M.Allot, I.Mussio, D.Sihachakr, L.Rossignol and G.Ducreux (1994) Production and characterization of intergeneric somatic hybrids through protoplast electrofusion between potato (Solanum tuberosum) and Lycopersicon pennellii. Plant Cell Tissue Organ Cult. 37: 137–144.

129. Shin,D., S.Moon, S.Han, B.Kim, S.Park, S.Lee, H.Yoon, H.Lee, H.Kwon, D.Baek et al. (2011) Expression of StMYB1R-1, a novel potato single MYB-like domain transcription factor, increases drought tolerance. Plant Physiol. 155: 421–432.

130. Sievers,N., K.Muders, M.Henneberg, S.Kl-hn, M.Effmert, H. Junghans and M.Hagemann (2013) Establishing glucosylglycerol synthesis in potato (Solanum tuberosum L. cv. Albatros) by expression of the ggpPS gene from Azotobacter vinelandii. J. Plant Sci. Mol. Breed. 2: 1. doi: 10.7243/2050-2389-2-1.

131. Sinha,R. (2004) Part III Growth and Developmental Physiology. 24. Stress physiology. In: Sinha,R. (ed.) Modern Plant Physiology, Alpha Science International Ltd. Pangbourne, UK, pp. 551–560.

132. Smolenska,G. and P.Kuiper (1997) Effect of low temperature upon lipid and fatty acid composition of roots and leaves in winter rape plants. Physiol. Plant. 41: 29–35.

133. Sonnewald,U., M.Hajlrezaei, J.Kossmann, A.Heyer, R.Thethewey and L.Willmitzer (1997) Increased potato tuber size resulting from apoplastic expression of a yeast invertase. Nat. Biotechnol. 15: 794–797.

134. Steponkus,P. and M.Webb (1992) Freeze-induced dehydration and membrane destabilization in plants. In: Somero et al. (eds.) Water and Life, Springer-Verlag, Berlin, pp. 338–362.

135. Stiller,I., S.Dulai, M.Kondrak, R.Tarnai, L.SzaboL, O.Toldi and Z.Banfalvi (2008) Effects of drought on water content and photosynthetic parameters in potato plants expressing the trehalose-6- phosphate synthase gene of Saccharomyces cerevisiae. Planta 227: 299–308.

136. Stockinger,E., S.Gilmour and M.Thomashow (1997) Arabidopsis thaliana CBF1 encodes an AP2 domain-containing transcriptional activator that binds to the C-repeat/DRE, a cis-acting DNA regulatory element that stimulates transcription in response to low temperature and water deficit. Proc. Natl. Acad. Sci. USA 94: 1035–1040.

137. Stracke,R., H.Ishihara, G.Huep, A.Barsch, F.Mehrtens, K.Niehaus and B.Weisshaar (2007) Differential regulation of closely related R2R3-MYB transcription factors controls flavonol accumulation in different parts of the Arabidopsis thaliana seedling. Plant J. 50: 660–667.

138. Tang,L., S.Kwon, S.Kwak, C.Sung and H.Lee (2004) Selection of transgenic potato plants expressing both CuZnSOD and APX in chloroplasts with enhanced tolerance to oxidative stress. Korean J. Plant Biothechnol. 31: 109–113.

139. Tang,L., S.Kwon, S.Kim, J.Kim, J.Choi, K.Cho, C.Sung, S.Kwak and H.Lee (2006) Enhanced tolerance of transgenic potato plants expressing both superoxide dismutase and ascorbate peroxidase in chloroplasts against oxidative stress and high temperature. Plant Cell Rep. 25: 1380–1386.

140. Tang,L., M.Kim, K.Yang, S.Kwon, S.Kim, J.Kim, D.Yun, S.Kwak and H.Lee (2008) Enhanced tolerance of transgenic potato plants overexpressing nucleoside diphosphate kinase 2 against multiple environmental stresses. Transgenic Res. 17: 705–715.

141. The Potato Genome Sequencing Consortium (2011) Genome sequence and analysis of the tuber crop potato. Nature 475: 189–195.

142. Thomashow,M. (1999) Plant cold acclimation: freezing tolerance genes and regulatory mechanisms. Annu. Rev. Plant Physiol. Plant Mol. Biol. 50: 571–599.

143. Upadhyaya,C., J.Venkatesh, M.Gururani, L.Asnin, K.Sharma, H. Ajappala and S.Park (2011) Transgenic potato overproducing Lascorbic acid resisted an increase in methylglyoxal under salinity stress via glyoxalase enzyme activity. Biotechnol. Lett. 33: 2297– 2307.

144. Vasquez-Robinet,C., S.Mane, A.Ulanov, J.Watkinson, V.Stromberg, D.De Koeyer, R.Schafleitner, D.Willmot, M.Bonierbale, H.Bohnert et al. (2004) Physiological and molecular adaptations to drought in Andean potato genotypes. J. Exp. Bot. 59: 2109–2123.

145. Vega,S. and J.Bamberg (1995) Screening the U.S. potato collection for frost hardiness. Am. Potato J. 72: 13–21.

146. Venkatesh,J., C.Upadhyaya, J.Yu, A.Hemavathi, D.Kim, R.Strasser and S.Park (2012) Chlorophyll a fluorescence transient analysis of transgenic potato overexpressing D-galacturonic acid reductase gene for salinity stress tolerance. Hortric. Environ. Biotechnol. 53: 320–328.

147. Vereyken,I., V.Chupin, F.Hoekstra, S.Smeekens and B. de Kruijff (2003) The effect of fructan on membrane lipid organization and dynamics in the dry state. Biophys. J. 84: 3759–3766.

148. Wallis,J., H.Wang and D.Guerra (1997) Expression of a

synthetic antifreeze protein in potato reduces electrolyte release at freezing temperatures. Plant Mol. Biol. 35: 323–330.

149. Wang,L., J.Zhang, D.Wang, J.Zhang, Y.Cui, Y.Liu, H.Yangac and Binyu (2010) Assessment of salt tolerance in transgenic potato carrying AtNHX1 gene. Crop Sci. 53: 2643–2651.

150. Wang,W., B.Vinocur and A.Altman (2003) Plant responses to drought, salinity and extreme temperatures: towards genetic engineering for stress tolerance. Planta 218: 1–14.

151. Watanabe,K., A.Kikuchi, T.Shimazaki and M.Asahina (2011) Salt and drought stress tolerance in transgenic potatoes and wild species. Potato Res. 54: 319–324.

152. Waterer,D., N.Benning, G.Wu, X.Luo, X.Liu, M.Gusta, A.McHughen and L.Gusta (2010) Evaluation of abiotic stress tolerance of genetically modified potatoes (Solanum tuberosum cv. Desiree). Mol. Breed. 25: 527–540

153. Weisz,R., K.Kaminski and Z.Smilowitz (1994) Water deficit effects on potato leaf growth and transpiration: utilizing fraction extractable soil water for comparison with other crops. Am. J. Potato Res. 71: 829–840.

154. Williamson,A., B.Conlan, W.Hillier and T.Wydrzynski (2011) The evolution of Photosystem II: insights into the past and future. Photosynth. Res. 107: 71–86.

155. Woo,Y., H.Park, M.Suudi, J.Yang, J.Park, K.Back, Y.Park and G.An (2007) Constitutively wilted 1, a member of the rice YUCCA gene family, is required for maintaining water homeostasis and an appropriate root to shoot ratio. Plant Mol. Biol. 65: 125–136.

156. Yadav,S., L.Singla-Pareek, M.Ray, M.Reddy and S.Sopory (2005a) Methylglyoxal levels in plants under salinity stress are dependent on glyoxalase I and glutathione. Biochem. Biophys. Res. Commun. 337: 61–67.

157. Yadav,S., L.Singla-Pareek, M.Reddy and S.Sopory (2005b) Transgenic tobacco plants overexpressing glyoxalase enzymes resist an increase in methylglyoxal and maintain higher reduced glutathione levels under salinity stress. FEBS Lett. 579: 6265–6271.

158. Yi,S., J.Kim, Y.Joung, S.Lee, W.Kim, S.Yu and D.Choi (2004) The pepper transcription factor CaPF1 confers pathogen and freezing tolerance in Arabidopsis. Plant Physiol. 136: 2862–2874.

159. Youm,J., J.Jeon, D.Choi, S.Yi, H.Joung and H.Kim (2008) Ectopic expression of pepper CaPF1 in potato enhances multiple stresses tolerance and delays initiation of in vitro tuberization. Planta 228: 701–708.

160. Yu,C., H.Wang, S.Yang, X.Tang, M.Duan and Q.Meng (2009) Overexpression of endoplasmic reticulum omega-3 fatty acid desaturase gene improves chilling tolerance in tomato. Plant Physiol. Biochem. 47: 1102–1112.

161. Zhang,N., H.Si, G.Wen, H.Du, B.Liu and D.Wang (2011) Enhanced drought and salinity tolerance in transgenic potato plants with a BADH gene from spinach. Plant Biotechnol. Rep. 5: 71–77.

CHAPTER 17

EPIGENETIC SILENCING IN TRANSGENIC PLANTS

Sarma Rajeevkumar[1], Pushpanathan Anunanthini[2], and Ramalingam Sathishkumar[2]

[1]Molecular Plant Biology and Biotechnology Division, Central Institute of Medicinal and Aromatic Plants Research Centre, Bangalore, India

[2]Plant Genetic Engineering Laboratory, Department of Biotechnology, Bharathiar University, Coimbatore, India

ABSTRACT

Epigenetic silencing is a natural phenomenon in which the expression of genes is regulated through modifications of DNA, RNA, or histone proteins. It is a mechanism for defending host genomes against the effects of transposable elements and viral infection, and acts as a modulator of expression of duplicated gene family members and as a silencer of transgenes. A major breakthrough in understanding the mechanism of epigenetic silencing was the discovery of silencing in transgenic tobacco plants due to the interaction between two homologous promoters. The molecular mechanism of epigenetic mechanism is highly complicated and it is not completely understood yet. Two different molecular routes have been proposed for this, that is, transcriptional gene silencing, which is associated with heavy methylation of promoter regions and blocks the transcription of transgenes, and post-transcriptional gene silencing (PTGS), the basic mechanism is degradation of the cytosolic mRNA of transgenes or endogenous genes. Undesired transgene silencing is of major concern in the transgenic technologies used in crop improvement. A complete understanding of this phenomenon will be very useful for transgenic applications, where silencing of specific genes is required. The current status of epigenetic silencing in transgenic technology is discussed and summarized in this mini-review.

INTRODUCTION

Conventionally closely related species are easier to breed than inter species/genus due to compatibility issues, and this is considered a major limitation. Transgenic technologies have allowed gene transfer to completely unrelated organisms. All these advances have increased the global transgenic plant cultivation to 181 million hectares (James, 2014). Transgenic plants with stacked genes are gaining more importance lately. Here, different genes are expressed in one transgenic plant from a single transformation event, or in consecutive steps either by re-transformation or by conventional genetic crosses involving different transgenic lines expressing a single transgenic event (Dietz-Pfeilstetter, 2010). To date, diverse traits such as disease resistance, stress tolerance, nutritional improvement, and the use of plants as host systems to produce economically important molecules have been successfully proven (Ahmad et al., 2012). The purpose of gene transfer to plants in all the above cases was to achieve specific desirable traits, where lines that failed to meet expectations are discarded, so that the best

Citation: Sarma Rajeevkumar, Pushpanathan Anunanthini, and Ramalingam Sathishkumar, Epigenetic silencing in transgenic plants, doi: 10.3389/fpls.2015.00693.

performers can be propagated (Kohli et al., 2006, 2010). Initial reports of unforeseen low gene-expression levels or silencing of transgenes were considered failures. Later, those minor glitches emerged as a principal factor elucidating the role of epigenetics in this emerging technology (Meyer et al., 2013).

A major prerequisite for plant expressing a transgene is stability and segregation. Several reports have documented a deviation from the Mendelian segregation ratios in transgenic plants (Shrawat et al., 2007;Weinhold et al., 2013). This revealed the existence of hitherto unknown cellular mechanisms which regulate expression of transgenes. In the last three decades, many reports on transgene instabilities as well as the reasons behind these events were the main focus (Charrier et al., 2000; Graham et al., 2011; Stroud et al., 2013). The explanation for inactivation/silencing of transgene activity was a lack of transcription due to methylation of the promoter along with condensation of chromatin, or degradation of transcripts by different mechanism (Fagard and Vaucheret, 2000; Table 1).

Table 1: Reports of epigenetic silencing in transgenic plants

Target plant	Gene (s)	Transgene effects	Reference
Arabidopsis thaliana	Selectable marker genes (*npt*/*hpt*)	Repeated sequence of target gene at same loci lead to repeat-induced gene silencing (RIGS).	Assaad et al., 1993
Nicotiana tabacum	Selectable marker gene (*npt*)	*De novo* methylation mediated silencing of *nptII*	Ingelbrecht et al., 1994
Petunia hybrida	Flavonoid hydroxylase gene, maize *A1* gene	Hypermethylation of 35S promoter directed *A1* gene expression lead to variegated flower pigmentation in transgenic *Petunia* lines	Meyer et al., 1994
Avena sativa	*bar* and *gusA*	Direct DNA–DNA interaction between multiple transgene copies resulted in silencing of *bar*/*gusA* gene to different levels.	Pawlowski and Somers, 1998
Oryza sativa	*bar* gene	Methylation of Ubi1 promoter lead to silencing of *bar* gene and bialaphos sensitivity in transgenic rice	Kumpatla and Hall, 1998
Saccharum officinarum	sorghum mosaic potyvirus strain SCH coat protein (CP) gene	Reduced transcript level lead to post-transcriptional gene silencing (PTGS) of CP gene in transgenic sugarcane.	Ingelbrecht et al., 1999
Oryza sativa	*GUS* gene	Reintroduction of GUS gene in GUS transformed rice lead to suppression of GUS expression due to PTGS	Kanno et al., 2000
N. tabacum	*GUS* gene	Gene silencing through DNA methylation lead to reduced expression of *GUS* gene in transgenic tobacco lines	Day et al., 2000
Petunia	*CHS gene*	White-flowering phenotype due to chalcone synthase transgene-induced silencing as a result of altered methylation in promoter	Kanazawa et al., 2007
A. thaliana	Phytochrome A/ DNA methyl transferase I gene	Exonic methylation can lead to chromatin modification further resulting in altered gene expression mediated through reduction in the transcription rate.	Chawla et al., 2007
N. tabacum	*nptII*	Target gene was silenced by PTGS based on the loci of intergration	Khaitova et al., 2011
Gentiana verna	CaMV35S promoter	*De novo* methylation of the enhancer region of CaMV 35S promoter silencing is triggered by histone H3 deacetylation.	Yamasaki et al., 2011
A. thaliana	*A. thaliana* repressor of silencing1 mutant	Mutants treated with sulfamethazine exhibited reduced levels of DNA methylation and released transgene silencing. Exogenous application of p-Aminobenzoic acid restored transcriptional gene silencing (TGS) in SMZ-treated mutants	Zhang et al., 2012
N. tabacum	CaMV35S promoter	DNA methylation and heterochromatic histone marks were studied in different epialleles of 35S promoter driven tobacco transgenic calli. Transient loss of euchromatin modifications lead to *de novo* DNA methylation further leading to formation of stable repressed epialleles with recovered eukaryotic marks	Křížová et al., 2013
N. tabacum	*A. thaliana* a repressor of silencing gene (ROS1)	Transgenic lines over-expressing *At ROS1*showed higher level of demethylation in promoter as well as coding region of various genes involved in flavanoid biosynthesis and antioxidant defense response	Bharti et al., 2015

EPIGENETICS

The British developmental biologist Conrad H. Waddington coined the term "epigenetics". Epigenetics deals with studies related to interactions of genes and their products, which determine the phenotype of a system (Waddington, 1942). During the course of an organism's development, cell fate is determined by genes and by other (epigenetic) factors, which underlies the notion of "epigenesis". Modern biology has redefined as a phenomenon in which a gene's activity is modulated by modifications of nucleic acids or the physical packaging of the chromatin in which it is embedded.

Two main classes of transgene-silencing phenomena have been reported to date. The first concerns position effects, in which the expression of a foreign gene is negatively regulated by flanking host DNA or chromosomal location (Matzke et al., 2000). The expression of a gene integrated into a region of euchromatin is also influenced by regulatory sequences of host genes (Kohli et al., 2006). Transgene integration into heterochromatic regions also leads to silencing (Grewal and Elgin, 2002).

The second class of silencing phenomena is based on epigenetic regulation and is a type of inactivation mechanism that can arise when multiple copies of the same or homologous sequence are introduced in a genome. Since interactions between homologous nucleic acid sequences are responsible for these silencing, it is also called homology-dependent gene silencing (HDGS) (Meyer and Saedler, 1996). Over the years, it has become clear that HDGS occurs through distinct processes, frequent one being involvement of inverted DNA repeats (IRs) and dsRNA. T-DNA integration at the same chromosomal site leads either to 'head-to-tail' direct repeats (DR) or to 'head-to-head' or 'tail to tail' inverted repeats (IR). T-DNAs that are arranged as IRs are often shown to have low basal expression (Mishiba et al., 2005). IRs have the ability to interact with homologous sequences elsewhere in the genome leading to chromatin remodeling. They can also induce a sequence-specific RNA degradation process, possibly via the formation of dsRNAs (Figure1A).

A

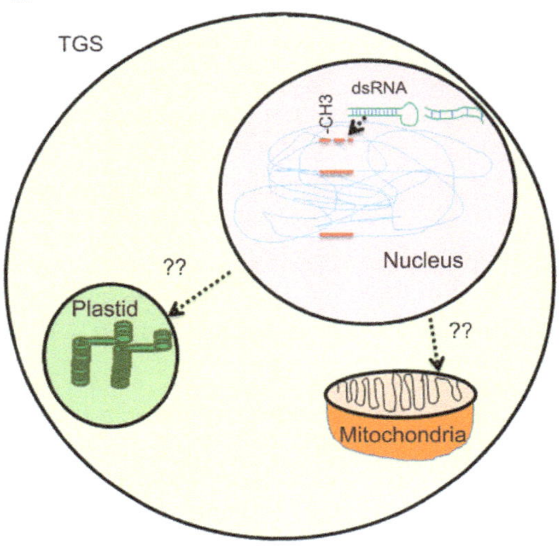

B

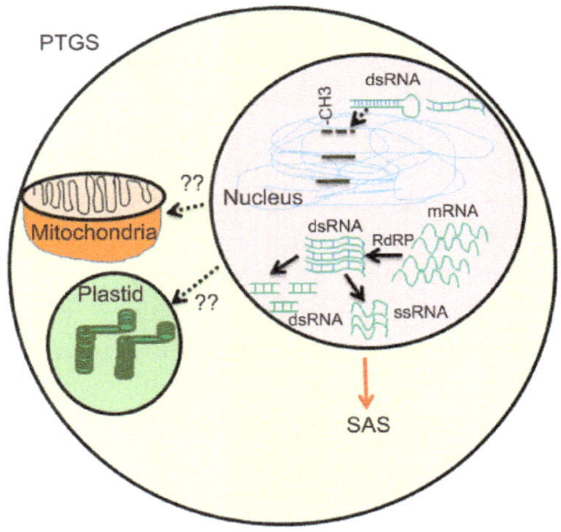

Figure 1: Overview of gene silencing in transgenic plants. (A) Transcriptional gene silencing (TGS)- DNA methylation induces dsRNA by endogenous gene or multiple copies of transgenes. Presence of multiple copies of transgene induces formation of dsRNA. Single copy .

HOMOLOGY DEPENDENT GENE SILENCING

A major breakthrough in understanding epigenetic silencing in transgenic plants was first identified in transgenic tobacco, where interaction between two homologous promoters led to DNA methylation and silencing (Matzke et al., 1989). Two types of HDGS are known based on the stage at which it occurs, called transcriptional gene silencing (TGS), which is coupled with transcription or by promoter modification, and post-transcriptional gene silencing (PTGS), which occurs after the formation of mRNA (Jauvion et al., 2012). In TGS, interacting genes that share homology in promoter regions are highly methylated. PTGS involves sequence-specific transcript turnover in the cytosol, which further requires high homology between interacting genes. Potential factors influencing HDGS are degree of homology between the transgene and endogenous gene, the complexity of the host genome, the genomic position of two transgenes, etc. A transgene locus with a complex structure with multiple scrambled T-DNAs has been reported to have strong silencing activities in tobacco, implicating transgene complexity; and vector DNA also decides the efficiency of HDGS (Fu et al., 2000). Complexity of T-DNA structure and integrated vector sequences have been shown to regulate transgene expression in grapevine (Gambino et al., 2010). An increase of endogenous transcript levels above a critical threshold induces specific degradation of homologous transcripts.

TRANSCRIPTIONAL GENE SILENCING (TGS)

Transgenes silenced at the transcriptional level acquire metastable epigenetic status that is associated with altered methylation patterns. Transgenes are frequently methylated in cytosine residues that are located within CG, CNN, or CNG sequences. *De novo* DNA methylation can be highly sequence-specific for a specific transgene (Matzke et al., 2007). Fungi or plants expressing foreign genes also exhibit non-symmetrical methylation leading to silencing of endogenous genes. Factors responsible for non-symmetrical methylation are still obscure.

Non-symmetrical methylation patterns are aided by RNA-chromatin mechanism (McGinnis et al., 2006).

Methylation in promoter regions, histones, or in coding regions influence gene expression at both the transcriptional (Huettel et al., 2006) and post transcriptional level (Regulski et al., 2013; Tsuchiya and Eulgem, 2013). Chromatin remodeling is involved in maintenance of silenced status and also in transmission of non-symmetrical methylation patterns (Meyer, 1999). Another interesting fact about TGS in transgenic plants is the association of DNA methylation along with structural changes, as methylated and silenced transgenes were less susceptible to endonucleases, reflecting an increased level of chromatin condensation (Van Blokland et al., 1997). Hence, TGS-based silencing might also involve structural changes similar to heterochromatinization, which could be the cause of these structural changes. The responsiveness of TGS of transgenes in response to environmental change was confirmed (Meyer et al., 1992; Meyer, 2015).

Transcriptional gene silencing can be further divided into two classes:

Transcriptional *cis* Inactivation

In plants, transgenes integrate into the genome at random positions by illegitimate recombination; hence, copy number, their integration site, and local arrangement differ in each transformation event. Also, an inverse relation between transgene copy number and gene expression suggests that multicopy integration can lead to silencing. Integrated foreign genes can undergo TGS in *cis* when multicopy T-DNA is integrated at a locus adjacent to hypermethylated regions of the host genome (Mishiba et al., 2005). More rarely, single copy transgene integration at a hypomethylated locus can lead to *cis* inactivation (Meyer and Heidmann, 1994;Elmayan and Vaucheret, 1996). A maize *A1* gene involved in floral pigmentation when overexpressed in*Petunia* led to silencing of *A1*; however, it was not silenced when *Gerbera dihydroflavonol-4-reductase* was over expressed in *Petunia* suggesting that the transgene also influenced the silencing process. Hence, some degree of difference in DNA composition of the transgene and surrounding host genomic sequences can be recognized by the cellular machinery as foreign non-compatible DNA, leading to specific methylation and silencing (Elomaa et al., 1995). It is believed that *cis* TGS occurs as a result of pairing between closely associated copies of transgenes or endogenous genes, which leads to the formation of secondary DNA structures which are sites for DNA methylation (Vaucheret and Fagard, 2001). Cytosine methylation at CpG and CpNpG sites of transgene and the 35S promoter were also detected in transgenic grapevine transformed with Grapevine fanleaf virus (GFLV) coat protein gene (Gambino et al., 2010).

Transcriptional Trans-Inactivation

Transcriptional gene silencing can result from unidirectional effects of one transgene on another transgene or homologous endogenous gene. A transgene can be methylated and silenced when it is crossed with a plant in which the homologous gene is in a silenced state (Meyer et al., 1993). *De novo* methylation of one transgene is mediated by a second transgene under control of the same promoter leading to TGS in trans (Fagard and Vaucheret, 2000). Experiments using dsRNA-containing promoter sequences initiated TGS and subsequently*de novo* DNA methylation of the corresponding transgene or endogenous gene, implying a role of an RNA intermediate in TGS (Meyer, 2000). Vaucheret and Fagard (2001) reported the role of different genes, including *ddm1* and *ddm2* in TGS in *Arabidopsis* transgenic lines. Yamasaki et al. (2011) reported methylation of asymmetric cytosine in the enhancer region of 35S promoter in transgenic gentian.

POST-TRANSCRIPTIONAL GENE SILENCING

Post-Transcriptional Gene silencing is a condition where transcripts do not accumulate in spite of continuous transcription (Vaucheret et al., 2001). PTGS can silence both transgenes and endogenous genes if both are homologous. An endogenous gene could be switched off, when a plant is transformed with another copy of the same gene. When genes involved in pigmentation, such as *chalcone synthase A* in *Petunia*, were overexpressed, many transgenic lines partially or completely lost activity of both transgene and endogenous gene (Napoli et al., 1990; Van der Krol et al., 1990). This was later called 'co-suppression', which was a result of degradation of mRNA of both transgene and endogenous gene. Analysis of degradation products in tobacco expressing -1,3-glucanase revealed that RNAs are first cleaved by endonucleases, which are further degraded by various exonucleases (Van Eldik et al., 1998). Silencing of two endogenous genes in*Arabidopsis thaliana* was triggered by the antisense and hpRNA transgenes, and silencing in this case was dependent on ploidy level, as it was less pronounced in 4n compared to 2n *Arabidopsis*. Studies indicated that transgenes were more methylated in 4n than 2n *Arabidopsis* suggesting transgenes are transcriptionally repressed in 4n plants, thus resulting in reduced expression levels compared to diploid plants (Finn et al., 2011).

Transgene-induced viral resistance, recovery from infection and proteins encoded by viruses that counteract PTGS suggested it as a potential defense response to check viral infections (Brigneti et al., 1998; Kasschau and Carrington, 1998; Dalmay et al., 2000). It is speculated that the concentration of specific RNAs derived from both transgene and endogenous gene is critical to activate PTGS. dsRNAs are one of the potential candidates, as they are formed between RNAs transcribed from IR and gene homologues. dsRNA is used as a template by RNA-directed RNA polymerase (RdRP) and transcription of dsRNA by RdRP would result in antisense RNAs, which ultimately could target complementary transcripts for degradation by dsRNA-specific RNases (Bond and Baulcombe, 2015; Figure1B).

Post-Transcriptional *cis*-Inactivation

Post-transcriptional gene silencing *cis*-inactivation is observed when foreign genes like β-Glucuronidase, neomycin phosphotransferase, etc., were driven under strong 35S promoter (Dehio and Schell, 1994;Ingelbrecht et al., 1994; Elmayan and Vaucheret, 1996). When a 35S promoter with a double enhancer was used, more transformants showed PTGS (Elmayan and Vaucheret, 1996; English et al., 1996). Initially, perceptions about PTGS were driven by higher transcript abundance above a threshold level, which ultimately triggered degradation of transgenic RNA. Later, it was found that the level of transcription was not always found to be significantly higher in silenced plants. The presence of IR at transgene locus of silenced lines was proposed to play a crucial role in *cis* inactivation (English et al., 1996). In same year, different models for PTGS were proposed considering RNA abundance and IRs (Baulcombe, 1996). Transgene RNA could be specifically degraded when tagged with specific molecules; these tag molecules were later named small complementary RNA (cRNA). RdRP catalyzed synthesis of cRNA using transgene RNA as template (Dougherty and Parks, 1995). They could also be internal fragments generated from transgene RNA by pairing between aberrant mRNA and normal transgene RNA due to the presence of internal sequence complementarily (Metzlaff et al., 1997). cRNA can interact with mRNA forming dsRNA, which are the target for the cellular enzymes like double-strand RNase. DNA-DNA interactions can lead to methylation, which can further interfere with transcription, ultimately producing aberrant RNA. These aberrant RNAs or higher transcript abundance were owing to the use of a strong promoter that triggered methylation of the coding sequence of the respective transgene (Wassenegger et al., 1994). Interestingly, Kanazawa et al. (2007)reported conversion of PTGS to TGS in *Petunia* transgenic lines as a consequence of the transgene homologous to an endogenous gene in host genome.

Post-Transcriptional Trans-Inactivation

Post-transcriptional gene silencing was originally reported as coordinated silencing of both transgenes as well as endogenous genes, which is generally termed 'co-suppression' (Napoli et al., 1990). Since then, several studies revealed transgenes encoding part of, or the entire transcribed sequence of, host genes have been shown to trigger co-suppression of endogenous genes (Depicker and Van Montagu, 1997). By then it was evident from studies in transgenic *Petunia* lines expressing a chalcone synthase, where efficiency of co-suppression correlated with the strength of the promoter, that there was an effect of transgene dose on co-suppression (Que et al., 1997). Besides, the efficiency of co-suppression is delayed when endogenous host genes are not expressed or when genes are transferred to a mutant devoid of functional gene homologues (Smith et al., 1990; Vaucheret et al., 1997). Hence, it can be concluded that co-suppression cannot be considered as the unidirectional silencing effect of transgenes, rather it is a synergistic phenomenon in which interaction or presence of host genes and transgenes aids aberrant RNA and/or cRNA leading to PTGS.

SYSTEMIC ACQUIRED SILENCING

A hallmark of PTGS in plants is that it systemically transmitted in a sequence-specific manner known as systemic acquired silencing (SAS). Remarkable and recurrent features in silencing patterns during developmental stages revealed propagation of a silencing message across the plant (Vaucheret et al., 1998;Kalantidis et al., 2008). Co-suppression of endogenous and transgenes of nitrate reductase, nitrite reductase and SAM synthase in tobacco led to chlorotic or necrotic phenotypes (Boerjan et al., 1994; Palauqui et al., 1996). The non-clonal patterns were observed in all transgenic lines silenced for a specific gene and a sequence-specific message was involved in the control of PTGS. Later, grafting experiments revealed that transgene-specific effector molecules were involved in propagation of *de novo* PTGS over long distances by a phenomenon called SAS (Palauqui et al., 1997). Transgenic tobacco overexpressing *A. thaliana AtMYB90*involved in anthocyanin biosynthesis showed siRNA-mediated silencing as a result of SAS (Velten et al., 2012). A SAS PTGS of transgenes in *N. benthamiana* was initiated in localized regions of the plant when a transgene-homologous DNA was introduced (Voinnet et al., 1998). The silencing signal molecules are degraded RNA, which travels through phloem across cells through plasmodesmata (Kalantidis et al., 2008). The recipient cell can also act as a source for generating secondary signals. It has been reported that sense, antisense, and ill-defined aberrant RNAs can give rise to dsRNA which can transmit signals, ultimately leading to silencing of both transgene and endogenous gene, albeit to different levels (Figure 1B).

SMALL RNAS AS SILENCING SIGNALS IN TRANSGENIC PLANTS

RNA was the driving factor for the establishment of DNA methylation patterns (Wassenegger et al., 1994) and acts a signaling agent for inducing silencing. Potato spindle tuber viroid (PSTV) in transgenic tobacco lines led to autonomous viroid RNA replication in the nucleus and induced DNA methylation in the T-DNA (Wassenegger et al., 1994). The evidence from above study clearly indicates the critical role of RNA in initiating *de novo* DNA methylation at homologous regions. Until then, DNA/RNA hybrids were believed to play a role in generating a target for *de novo* methylation. *chsA* co-suppression studies in *Petunia* led to the identification of mobile RNAs as potential candidates responsible for the induction of co-suppression (Napoli et al., 1990; Van der Krol et al., 1990). The initiation of transgene silencing has been thought to involve the generation of dsRNA. It is still under debate about factors triggering initiation of silencing even in the case of transgenes that lack unusual DNA structures.

In plants, micro RNAs (miRNAs) are produced from hairpin-like precursor RNA, which is essential for biogenesis of trans-acting siRNAs (ta-siRNAs). miRNAs are involved in regulation of gene expression by base-pairing with target RNAs further leading to their cleavage in plants. *Physcomitrella patens* transgenic lines expressing different levels of artificial miRNA (amiRNA) revealed transcript-dependent silencing of miRNA target. Thus, a crucial regulatory role of miRNAs might be conserved in other plants also, which are under investigation. siRNAs are another class of small RNAs that are involved in epigenetic modification (Miki and Shimamoto, 2008). Endogenous siRNAs can induce DNA methylation at CpG nucleotides leading to chromatin modification and silencing. Human *H1* and *Arabidopsis 7SL* RNA promoters driving*GUS* specific short hairpin RNA resulted in the efficient silencing of *GUS* at both transcript and protein level, indicating a significant role of siRNAs in epigenetic regulation. However, transgenes are generally more sensitive against RNA silencing than endogenous genes in plants.

TRANSGENE SILENCING AS PART OF THE HOST DEFENSE MECHANISM?

Silencing cannot be considered as a mechanism that evolved to regulate transgene expression; it is a part of natural plant processes. TGS and PTGS can be considered as host defense responses against 'foreign invading' viruses. Hence, transgenes or their products can be equated to cellular invaders triggering defensive reactions leading to silencing of "trans"gene. PTGS recruits cellular components acting against foreign DNA that replicates extra-chromosomally in the nucleus, or RNA in the cytoplasm. A clear connecting link between PTGS and viral resistance was established after the discovery and characterization of various viral proteins that suppress PTGS (Kasschau and Carrington, 1998; Beclin et al., 2002). TGS may use cellular components acting against invading DNA that integrates into the genome. The involvement of DNA methylation can also be considered as a part of cellular defense mechanism against transposable elements. The probable function of dsRNA in initiating methylation can be correlated to retro-elements that produce RNAs with intricate secondary structures.

STRATEGIES TO PREVENT TRANSGENE SILENCING (DEPICKER ET AL., 2005)

1. Selection of transgenic lines with single T-DNA insert
2. Organelle targeting/transformation
3. Selection of favorable/unique integration sites
4. Reactivation of silent transgenes
5. *Use of Scaffold Matrix Attachment Regions in silencing mutant host system to prevent silencing.*

CONCLUDING REMARKS

The last three decades have seen immense progress and better understanding of epigenetic effects and silencing mechanisms; transgenic technologies have played a pivotal role for these achievements. Common phenomena behind different types of silencing and recent finding of involvement of siRNAs/miRNA continue to inspire efforts of scientific community to formulate comprehensive models, which also explain the silencing mechanism from an evolutionary view point. Our understanding of the influence of various factors on stability of transgene expression is improving rapidly. We cannot control or predict integration of gene into a recipient genome, nor predict the number of copies or integrity of a transgene. Hence, a comprehensive knowledge of underlying mechanisms in integration process and the influence of chromatin remodeling leading to transgene regulation are crucial. Finally, it might be useful to keep in mind that epigenetic silencing was an unexpected phenomenon; it is still hard to foresee overcoming epigenetic related silencing in transgenic system. Nevertheless, transgenic research will continue as a platform to discover new aspects of epigenetic silencing.

AUTHOR CONTRIBUTIONS

SR and PA prepared the manuscript and RS revised it.

REFERENCES

1. Ahmad P., Ashraf M., Younis M., Hu X., Kumar A., Akram N. A., et al. (2012). Role of transgenic plants in agriculture and biopharming. *Biotechnol. Adv.* 30 524–540. 10.1016/j.biotechadv.2011.09.006
2. Assaad F. F., Tucker K. L., Signer E. R. (1993). Epigenetic repeat-induced gene silencing (RIGS) in*Arabidopsis*. *Plant Mol. Biol.* 22 1067–1085. 10.1007/BF00028978
3. Baulcombe D. C. (1996). RNA as a target and an initiator of posttranscriptional gene silencing in transgenic plants. *Plant Mol. Biol.* 32 79–88. 10.1007/BF00039378
4. Beclin C., Boutet S., Waterhouse P., Vaucheret H. (2002). A branched pathway for transgene - induced RNA silencing in plants. *Curr. Biol.* 12 684–688. 10.1016/S0960-9822(02)00792-3
5. Bharti P., Mahajan M., Vishwakarma A. K., Bhardwaj J., Yadav S. K. (2015). AtROS1 overexpression provides evidence for epigenetic regulation of genes encoding enzymes of flavonoid biosynthesis and antioxidant pathways during salt stress in transgenic tobacco. *J. Exp. Bot.* [Epub ahead of print].10.1093/jxb/erv304
6. Boerjan W., Bauw G., Van Montagu M., Inze D. (1994). Distinct phenotypes generated by overexpression and suppression of S-adenosyl-l-methionine reveal developmental patterns of gene silencing in tobacco. *Plant Cell* 6 1401–1414. 10.2307/3869977
7. Bond D. M., Baulcombe D. C. (2015). Epigenetic transitions leading to heritable, RNA-mediated de novo silencing in *Arabidopsis thaliana*. *Proc. Natl. Acad. Sci. U.S.A.* 112 917–922. 10.1073/pnas.1413053112
8. Brigneti G., Voinnet O., Li W.-X., Ding S. W., Baulcombe D. C. (1998). Viral pathogenicity determinants are suppressors of transgene silencing. *EMBO J.* 17 6739–6746. 10.1093/emboj/17.22.6739
9. Charrier B., Scollan C., Ross S., Zubko E., Meyer P. (2000). Co-silencing of homologous transgenes in tobacco. *Mol. Breed.* 6 407–419. 10.1023/A:1009672714835
10. Chawla R., Nicholson S. J., Folta K. M., Srivastava V. (2007). Transgene-induced silencing of*Arabidopsis* phytochrome A gene via exonic methylation. *Plant J.* 52 1105–1118. 10.1111/j.1365-313X.2007.03301.x
11. Dalmay T., Hamilton A., Mueller E., Baulcombe C. D. (2000). Potato virus X amplicons in*Arabidopsis* mediate genetic and epigenetic gene silencing. *Plant Cell* 12 369–380. 10.2307/3870942
12. Day C. D., Lee E., Kobayashi J., Holappa L. D., Albert H., Ow D. W. (2000). Transgene integration into the same chromosome location can produce alleles that express at a predictable level, or alleles that are differentially silenced. *Genes Dev.* 14 2869–2880. 10.1101/gad.849600
13. Dehio C., Schell J. (1994). Identification of plant genetic loci involved in a post-transcriptional mechanism for meiotically reversible transgene silencing. *Proc. Natl. Acad. Sci. U.S.A.* 91 5538–5542. 10.1073/pnas.91.12.5538

14. Depicker A., Sanders M., Meyer P. (2005). "Transgene silencing," in *Plant epigenetics. Oxford Ames*, ed. Meyer P., editor. (Iowa: Blackwell Publishing;), 1–32. 10.1002/9780470988626.ch1

15. Depicker A., Van Montagu M. (1997). Posttranscriptional gene silencing in plants. *Curr. Opin. Cell Biol.* 9 373–382. 10.1016/S0955-0674(97)80010-5

16. Dietz-Pfeilstetter A. (2010). Stability of transgene expression as a challenge for genetic engineering.*Plant Sci.* 179 164–167. 10.1016/j.plantsci.2010.04.015

17. Dougherty W. G., Parks T. D. (1995). Transgenes and gene suppression: telling us something new?*Curr. Opin. Cell Biol.* 7 399–405. 10.1016/0955-0674(95)80096-4

18. Elmayan T., Vaucheret H. (1996). Expression of single copies of a strongly expressed 35S transgene can be silenced post-transcriptionally. *Plant J.* 9 787–797. 10.1046/j.1365-313X.1996.9060787.x

19. Elomaa P., Helariutta Y., Griesbach J. R., Kotilainen M., Seppanen P., Teeri H. T. (1995). Transgene inactivation in *Petunia hybrida* is influenced by the properties of the foreign gene. *Mol. Genet. Genomics* 248 649–656. 10.1007/BF02191704

20. English J. J., Mueller E., Baulcombe D. C. (1996). Suppression of virus accumulation in transgenic plants exhibiting silencing of nuclear genes. *Plant Cell* 8 179–188. 10.1105/tpc.8.2.179

21. Fagard M., Vaucheret H. (2000). (TRANS) GENE SILENCING IN PLANTS: how many mechanisms? *Annu. Rev. Plant Physiol. Plant Mol. Biol.* 51 167–194. 10.1146/annurev.arplant.51.1.167

22. Finn E. T., Wang L., Smolilo D., Smith A. N., White R., Chaudhury A., et al. (2011). Transgene expression and transgene-induced silencing in diploid and autotetraploid *Arabidopsis*. *Genetics* 187409–423. 10.1534/genetics.110.124370

23. Fu X., Duc L. T., Fontana S., Bong B. B., Tinjuangjun P., Sudhakar D., et al. (2000). Linear transgene constructs lacking vector backbone sequences generate low-copy-number transgenic plants with simple integration patterns. *Transgenic Res.* 9 9–11. 10.1023/A:1008993730505

24. Gambino G., Perrone I., Carra A., Chitarra W., Boccacci P., Torello Marinoni D., et al. (2010).Transgene silencing in grapevines transformed with GFLV resistance genes: analysis of variable expression of transgene, siRNAs production and cytosine methylation. *Transgenic Res.* 19 17–27. 10.1007/s11248-009-9289-5

25. Graham M. W., Mudge S. R., Sternes P. R., Birch R. G. (2011). "Understanding and avoiding transgene silencing," in *Plant Transformation Technologies*, eds Stewart C. N., Touraev A., Citopvsky V., Tzfira T., editors. (West Sussex: Blackwell Publishing;). 10.1002/9780470958988.ch12

26. Grewal S. I. S., Elgin S. C. R. (2002). Heterochromatin: new possibilities for the inheritance of structure. *Curr. Opin. Gen. Dev.* 12 178–187. 10.1016/S0959-437X(02)00284-8

27. Huettel B., Kanno T., Daxinger L., Aufsatz W., Matzke A. J. M., Matzke M. (2006). Endogenous targets of RNA-directed DNA methylation and Pol IV in *Arabidopsis*. *EMBO J.* 25 2828–2836. 10.1038/sj.emboj.7601150

28. Ingelbrecht I., Van Houdt H., Van Montagu M., Depicker A. (1994). Post-transcriptional silencing of reporter transgenes in tobacco correlates with DNA methylation. *Proc. Natl. Acad. Sci. U.S.A.* 9110502–10506. 10.1073/pnas.91.22.10502

29. Ingelbrecht L. I., Irvine E. J., Mirkov T. E. (1999). Posttranscriptional gene silencing in transgenic sugarcane. dissection of homology-dependent virus resistance in a monocot that has a complex polyploid genome. *Plant Physiol.* 119 1187–1197. 10.1104/pp.119.4.1187

30. James C. (2014). *Global Status of Commercialized Biotech/GM Crops*. Ithaca, NY, USA: International Service for the Acquisition of Agri-biotech Applications (ISAAA).

31. Jauvion V., Rivard M., Bouteiller N., Elmayan T., Vaucheret H. (2012). RDR2 partially antagonizes the production of RDR6-dependent siRNA in sense transgene-mediated PTGS. *PLoS ONE* 7:e2978510.1371/journal.pone.0029785

32. Kalantidis K., Schumacher H. K., Alexiadis T., Helm J. M. (2008). RNA silencing movement in plants. *Biol. Cell* 100 13–26. 10.1042/BC20070079

33. Kanazawa A., O'Dell M., Hellens P. R. (2007). Epigenetic inactivation of chalcone synthase-a transgene transcription in petunia leads to a reversion of the post-transcriptional gene silencing phenotype. *Plant Cell Physiol.* 48 638–647. 10.1093/pcp/pcm028

34. Kanno T., Naito S., Shimamoto G. (2000). Post-transcriptional gene silencing in cultured rice cells.*Plant and Cell Physiol.* 41 321–326. 10.1093/pcp/41.3.321

35. Kasschau K. D., Carrington J. C. (1998). A counter-defensive strategy of plant viruses: suppression of posttranscriptional gene silencing. *Cell* 95 461–470. 10.1016/S0092-8674(00)81614-1

36. Khaitova C. L., Fojtova M., Krizova K., Lunerova J., Fulnecek J., Depicker A., et al. (2011).Paramutation of tobacco transgenes by small RNA-mediated transcriptional gene silencing.*Epigenetics* 6 650–660. 10.4161/epi.6.5.15764

37. Kohli A., Melendi P. G., Abranches R., Capell T., Christou P. (2006). The quest to understand the basis and mechanisms that control expression of introduced transgenes in crop plants. *Plant Signal. Behav.* 1 185–195. 10.4161/psb.1.4.3195

38. Kohli A., Miro B., Twyman R. M. (2010). "Transgene integration, expression and stability in plants: strategies for improvements," in *Transgenic Crop Plants: Principles and Development*, eds Kole C., Michler C. H., Abbott A. G., Hall T. C., editors. (Berlin: Springer-Verlag;), 201–237. 10.1007/978-3-642-04809-8_7

39. K ížová K., Depicker A., Kovarik A. (2013). Epigenetic switches of tobacco transgenes associate with transient redistribution of histone marks in callus culture. *Epigenetics* 8 666–676. 10.4161/epi.24613

40. Kumpatla S. P., Hall T. C. (1998). Recurrent onset of epigenetic silencing in rice harboring a multi-copy transgene. *Plant J.* 14 129–135. 10.1046/j.1365-313X.1998.00097.x

41. Matzke M., Kanno T., Huettel B., Daxinger L., Matzke A. J. (2007). Targets of RNA directed DNA methylation. *Curr. Opin. Plant Biol.* 10 512–519. 10.1016/j.pbi.2007.06.007

42. Matzke M. A., Mette M. F., Matzke A. J. (2000). Transgene silencing by the host genome defense: implications for the evolution of epigenetic control mechanisms in plants and vertebrates. *Plant Mol. Biol.* 43 401–415. 10.1023/A:1006484806925

43. Matzke M. A., Primig M., Trnovsky J., Matzke A. J. M. (1989). Reversible methylation and inactivation of marker genes in sequentially transformed tobacco plants. *EMBO J.* 8 643–649.

44. McGinnis K. M., Springer C., Lin Y., Carey C. C., Chandler V. (2006). Transcriptionally silenced transgenes in maize are activated by three mutations defective in paramutation. *Genetics* 173 1637–1647. 10.1534/genetics.106.058669

45. Metzlaff M., O'Dell M., Cluster P. D., Flavell R. B. (1997). RNA-mediated RNA degradation and chalcone synthase a silencing in petunia. *Cell* 88 845–854. 10.1016/S0092-8674(00)81930-3

46. Meyer P. (1999). The role of chromatin remodeling in transgene silencing and plant development. *In Vitro Cell. Dev. Biol. Plant* 35 29–36. 10.1007/s11627-999-0006-0

47. Meyer P. (2000). Transcriptional transgene silencing and chromatin components. *Plant Mol. Biol.* 43221–234. 10.1023/A:1006483428789

48. Meyer P. (2015). Epigenetic variation and environmental change. *J. Exp. Bot.* 66 3541–3548. 10.1093/jxb/eru502

49. Meyer P., Heidmann I. (1994). Epigenetic variants of a transgenic petunia line show hypermethylation in transgene DNA: an indication for specific recognition of foreign DNA in transgenic plants. *Mol. Gen. Genet.* 243 390–399. 10.1007/BF00280469

50. Meyer P., Heidmann I., Niedenhof I. (1993). Differences in DNA-methylation are associated with a paramutation phenomenon in transgenic petunia. *Plant J.* 4 89–100. 10.1046/j.1365-313X.1993.04010089.x

51. Meyer P., Linn F., Heidmann I., Heiner Meyerz A. H., Niedenhof I., Saedler H. (1992). Endogenous and environmental factors influence 35S promoter methylation of a maize A1 gene construct in transgenic petunia and its colour phenotype. *Mol. Gen. Genet.* 231 345–352. 10.1007/BF00292701

52. Meyer P., Niedenhof I., Lohuis M. T. (1994). Evidence for cytosine methylation of non-symmetrical sequences in transgenic *Petunia hybrid. EMBO J.* 13 2084–2088.

53. Meyer P., Saedler H. (1996). Homology-dependent gene silencing in plants. *Annu. Rev. Plant Physiol. Plant Mol. Biol.* 47 23–48. 10.1146/annurev.arplant.47.1.23

54. Meyer T. D., Mampaey E., Vlemmix M., Denil S., Trooskens G., Renard J. P., et al. (2013). Quality evaluation of methyl binding domain based kits for enrichment DNA-methylation sequencing. *PLoS ONE* 8:e59068 10.1371/journal.pone.0059068

55. Miki D., Shimamoto K. (2008). De novo DNA methylation induced by siRNA targeted to endogenous transcribed sequences is genespecific and OsMet1-independent in rice. *Plant J.* 56 539–549. 10.1111/j.1365-313X.2008.03624.x

56. Mishiba K., Nishihara M., Nakatsuka T., Abe Y., Hirano H., Yokoi T., et al. (2005). Consistent transcriptional silencing of 35S-driven transgenes in gentian. *Plant J.* 44 541–556. 10.1111/j.1365-313X.2005.02556.x

57. Napoli C., Lemieux C., Jorgensen R. (1990). Introduction of a chimeric chalcone synthase gene into petunia results in reversible co-suppression of homologous gene in trans. *Plant Cell* 2 279–289. 10.2307/3869076

58. Palauqui J. C., Elmayan T., de Borne F. D., Crete P., Charles C., Vaucheret H. (1996). Frequencies, timing, and spatial patterns of cosuppression of nitrate reductase and nitrite reductase in transgenic tobacco plants. *Plant Physiol.* 112 1447–1456.

59. Palauqui J. C., Elmayan T., Pollien J. M., Vaucheret H. (1997). Systemic acquired silencing: transgene specific post-transcriptional silencing is transmitted by grafting from silenced stocks to non-silenced scions. *EMBO J.* 16 4738–4745. 10.1093/emboj/16.15.4738

60. Pawlowski P. W., Somers A. D. (1998). Transgenic DNA integrated into the oat genome is frequently interspersed by host DNA. *Proc. Natl. Acad. Sci. U.S.A.* 95 12106–12110. 10.1073/pnas.95.21.12106

61. Que Q., Wang H. Y., English J. J., Jorgensen R. A. (1997). The frequency and degree of co-suppression by sense chalcone synthase transgenes are dependent on transgene promoter strength and are reduced by premature nonsense codons in the transgene coding sequence. *Plant Cell* 9 1357–1368. 10.1105/tpc.9.8.1357

62. Regulski M., Lu Z., Kendall J., Donoghue M. T., Reinders J., Llaca V., et al. (2013). The maize methylome influences mRNA splice sites and reveals widespread paramutation-like switches guided by small RNA. *Genome Res.* 23 1651–1662. 10.1101/gr.153510.112

63. Shrawat A. K., Becker D., Lorz H. (2007). Agrobacterium tumefaciens-mediated genetic transformation of barley (*Hordeum vulgare* L). *Plant Sci.* 172 281–290. 10.1016/j.plantsci.2006.09.005

64. Smith C. J., Watson C. F., Morris P. C., Bird C. R., Seymour G. B., Gray J. E., et al. (1990).Inheritance and effect on ripeing of antisense polygalacturonase genes in transgenic tomatoes. *Plant Mol. Biol.* 14 369–379. 10.1007/BF00028773

65. Stroud H., Ding B., Simon S. A., Feng S., Bellizzi M., Pellegrini M., et al. (2013). Plants regenerated from tissue culture contain stable epigenome changes in rice. *Elife* 2:e00354 10.7554/eLife.00354

66. Tsuchiya T., Eulgem T. (2013). An alternative polyadenylation mechanism coopted to the *Arabidopsis*RPP7 gene through intronic retrotransposon domestication. *Proc. Natl. Acad. Sci. U.S.A.* 110 E3535–E3543. 10.1073/pnas.1312545110

67. Van Blokland R., Ten Lohuis M., Meyer P. (1997). Condensation of chromatin in transcriptional regions of an inactivated plant transgene: evidence for an active role of transcription in gene silencing.*Mol. Genet. Genomics* 257 1–13. 10.1007/s004380050617

68. Van der Krol A. R., Mur L. A., Beld M., Mol J. N., Stuitje A. R. (1990). Flavonoid genes in petunia: addition of a limited number of genes copies may lead to a suppression of gene expression. *Plant Cell*2 291–299. 10.1105/EPC.2.4.291

69. Van Eldik G. J., Litiere K., Jacobs J. J., Van Montagu M., Cornelissen M. (1998). Silencing of beta-1,3-glucanase genes in tobacco correlates with an increased abundance of RNA degradation intermediates. *Nucleic Acids Res.* 26 5176–5181. 10.1093/nar/26.22.5176

70. Vaucheret H., Beclin C., Elmayan T., Feuerbach F., Godon C., Morel J. B., et al. (1998). Transgene-induced gene silencing in plants. *Plant J.* 16 651–659. 10.1046/j.1365-313x.1998.00337.x

71. Vaucheret H., Beclin C., Fagard M. (2001). Post-transcriptional gene silencing in plants. *J. Cell Sci.*114 3083–3091.

72. Vaucheret H., Fagard M. (2001). Transcriptional gene silencing in plants: targets, inducers and regulators. *Trends Genet.* 17 29–35. 10.1016/S0168-9525(00)02166-1

73. Vaucheret H., Nussaume L., Palauqui J. C., Quillere I., Elmayan T. (1997). A transcriptionally active state is required for post-transcriptional silencing (co-suppression) of nitrate reductase host genes and transgenes. *Plant Cell* 9 1495–1504. 10.1105/tpc.9.8.1495

74. Velten J., Cakir C., Youn E., Chen J., Cazzonelli I. C. (2012). Transgene silencing and transgene-derived siRNA production in tobacco plants homozygous for an introduced AtMYB90 construct.*PLoS ONE* 7:e30141 10.1371/journal.pone.0030141

75. Voinnet O., Vain P., Angell S., Baulcombe D. C. (1998). Systemic spread of sequence-specific transgene RNA degradation in plants is initiated by localized introduction of ectopic promoterless DNA. *Cell* 95 177–187. 10.1016/S0092-8674(00)81749-3

76. Waddington C. H. (1942). The epigenotpye. *Endeavour* 1 18–20. 10.1093/ije/dyr184

77. Wassenegger M., Heimes S., Riedel L., Sanger H. L. (1994). RNA directed de-novo methylation of genomic sequences in plants. *Cell* 76 567–576. 10.1016/0092-8674(94)90119-8

78. Weinhold A., Kallenbach M., Ian Thomas Baldwin T. I. (2013). Progressive 35S promoter methylation increases rapidly during vegetative development in transgenic *Nicotiana attenuata* plants. *BMC Plant Biol.* 13:99 10.1186/1471-2229-13-99

79. Yamasaki S., Oda M., Koizumi N., Mitsukuri K., Johkan M., Nakatsuka T., et al. (2011). *De Novo Dna Methylation of the 35s Enhancer Revealed by High-Resolution Methylation Analysis.*10.5511/plantbiotechnology.10.1222a

80. Zhang H., Deng X., Miki D., Cutler S., La H., Hou Y. J., et al. (2012). Sulfamethazine suppresses epigenetic silencing in *Arabidopsis* by impairing folate synthesis. *Plant Cell* 24 1230–1241. 10.1105/tpc.112.096149

CHAPTER 18

EXAMINING STRATEGIES TO FACILITATE VITAMIN B1 BIOFORTIFICATION OF PLANTS BY GENETIC ENGINEERING

Lucille Pourcel, Michael Moulin, and Teresa B. Fitzpatrick

Department of Botany and Plant Biology, University of Geneva, Geneva, Switzerland

ABSTRACT

Thiamin (vitamin B_1) is made by plants and microorganisms but is an essential micronutrient in the human diet. All organisms require it as a cofactor in its form as thiamin pyrophosphate (TPP) for the activity of key enzymes of central metabolism. In humans, deficiency is widespread particularly in populations where polished rice is a major component of the diet. Considerable progress has been made on the elucidation of the biosynthesis pathway within the last few years enabling concrete strategies for biofortification purposes to be devised, with a particular focus here on genetic engineering. Furthermore, the vitamin has been shown to play a role in both abiotic and biotic stress responses. The precursors for *de novo* biosynthesis of thiamin differ between microorganisms and plants. Bacteria use intermediates derived from purine and isoprenoid biosynthesis, whereas the pathway in yeast involves the use of compounds from the vitamin B_3 and B_6groups. Plants on the other hand use a combination of the bacterial and yeast pathways and there is subcellular partitioning of the biosynthesis steps. Specifically, thiamin biosynthesis occurs in the chloroplast of plants through the separate formation of the pyrimidine and thiazole moieties, which are then coupled to form thiamin monophosphate (TMP). Phosphorylation of thiamin to form TPP occurs in the cytosol. Therefore, thiamin (or TMP) must be exported from the chloroplast to the cytosol for the latter step to be executed. The regulation of biosynthesis is mediated through riboswitches, where binding of the product TPP to the pre-mRNA of a biosynthetic gene modulates expression. Here we examine and hypothesize on genetic engineering approaches attempting to increase the thiamin content employing knowledge gained with the model plant *Arabidopsis thaliana*. We will discuss the regulatory steps that need to be taken into consideration and can be used a prerequisite for devising such strategies in crop plants.

INTRODUCTION

Vitamin B_1, also known as thiamin, was the first B vitamin to be identified (Funk, 1912). In cells, it exists as three predominant vitamers, i.e., free thiamin, thiamin monophosphate (TMP), and thiamin pyrophosphate (TPP), although adenylated and triphosphorylated forms also exist (Bettendorff et al., 2007; Gangolf et al., 2010). The cofactor form of the vitamin is TPP, which is essential for the activity of enzymes involved in acetyl-CoA and amino acid biosynthesis, as well as in the Krebs and Calvin cycles. Acute deficiency in vitamin B_1 in humans leads to a disease called beriberi, which can result

Citation: Lucille Pourcel, Michael Moulin and Teresa B. Fitzpatrick, Examining strategies to facilitate vitamin B1 biofortification of plants by genetic engineering, doi: 10.3389/fpls.2013.00160.

in fatal neurological and cardiovascular disorders. Therefore, vitamin B_1 as an essential micronutrient for humans must be taken in the diet. Free thiamin is the predominant B_1 vitamer used and can be taken up *via* specific carriers in epithelial cells (Komai et al., 1974; Said et al., 1999). Paradoxically, although the main source of vitamin B_1 in the diet is plants, the edible portions of some of the most abundant crops used in human nutrition have a natural thiamin content below the recommended dietary allowance (RDA, 1.3 mg/day), e.g., rice (*Oryza sativa*, 18% RDA), wheat (*Triticum aestivum*, 25% RDA), and maize (*Zea mays*, 33% RDA; Fitzpatrick et al., 2012). In addition to a low content in thiamin, crop postharvest processing can aggravate this shortfall. Moreover, thiamin is unevenly distributed in cereal grains. The aleurone layer and germ, which are removed during the refining process, are much richer than the endosperm. Therefore, most of the thiamin is lost in the production of white flour and polished rice. To overcome this problem, wheat flour and rice are often supplemented with chemically synthesized thiamin in developed countries. However, although being common and/or compulsory (in the case of enriched wheat flour in certain developed countries; Harper et al., 1998), such postharvest supplementation remains expensive and inaccessible to people in developing countries, where crops represent the most important dietary source of vitamin B_1. An alternative approach to tackle the problem of malnutrition is to improve crop nutritional qualities through biofortification using genetic engineering or conventional breeding.

With a particular focus on genetic engineering, it is clearly imperative to know how thiamin is biosynthesized. Significant progress has been made in this area in plants over the last few years (for comprehensive reviews, see Goyer, 2010; Rapala-Kozik, 2011). Thiamin is formed by the condensation of two separately biosynthesized moieties, hydroxyethylthiazole phosphate (HET-P) and hydroxymethylpyrimidine pyrophosphate (HMP-PP) to form TMP (Figure1). The HET-P biosynthesis pathway in plants is assumed to be similar to that of yeast, in which HET-P synthase (THI4p) catalyzes the formation of the thiazole moiety from NAD^+, glycine, and a sulfur from a backbone cysteine in THI4p itself (Chatterjee et al., 2011). Homologs of *THI4*, called *THI1* in plants, have been characterized at the genetic level in maize (Belanger et al., 1995), *Arabidopsis* (Machado et al., 1996), and rice (Wang et al., 2006). The pyrimidine moiety is biosynthesized *via* an identical pathway to bacteria in which phosphomethylpyrimidine synthase (THIC) (Raschke et al., 2007) converts aminoimidazole ribonucleotide (AIR) to hydroxymethylpyrimidine phosphate (HMP-P). The latter is then phosphorylated to HMP-PP by a bifunctional protein characterized in maize as THI3 and in *Arabidopsis* as TH1 (Ajjawi et al., 2007a; Rapala-Kozik et al., 2007). The same protein catalyzes the condensation step between HET-P and HMP-PP to produce TMP (Ajjawi et al., 2007a; Rapala-Kozik et al., 2007). All of these steps have been shown to occur in the chloroplast (Chabregas et al., 2001; Ajjawi et al., 2007a; Raschke et al., 2007). As TMP is not directly phosphorylated to TPP, it is assumed to be subsequently dephosphorylated to thiamin by broad specificity phosphatases (Figure1). Indeed, a broad substrate acid phosphatase has been isolated from maize and biochemically characterized showing it can dephosphorylate TMP (Rapala-Kozik et al., 2009). However, the same phosphatase showed relatively higher specificity toward TPP (Rapala-Kozik et al., 2009). Thus, it has been suggested that TPP can also be successively dephosphorylated to thiamin. On the other hand, thiamin can be pyrophosphorylated to make the active cofactor TPP by thiamin pyrophosphokinase (TPK) (Mitsuda et al., 1979; Molin and Fites, 1980; Ajjawi et al., 2007b; Rapala-Kozik et al., 2009; Figure1). It has been suggested that the phosphatase and kinase activities are regulated such that TPP is only provided when the cell needs it (Rapala-Kozik et al., 2009). To date, the subcellular localization of the phosphatase(s) that acts on TMP (or TPP) is unknown. However, the exclusive cytosolic localization of the *Arabidopsis* TPK suggests that the pyrophosphorylation of thiamin into TPP takes place in the cytosol (Ajjawi et al., 2007b).

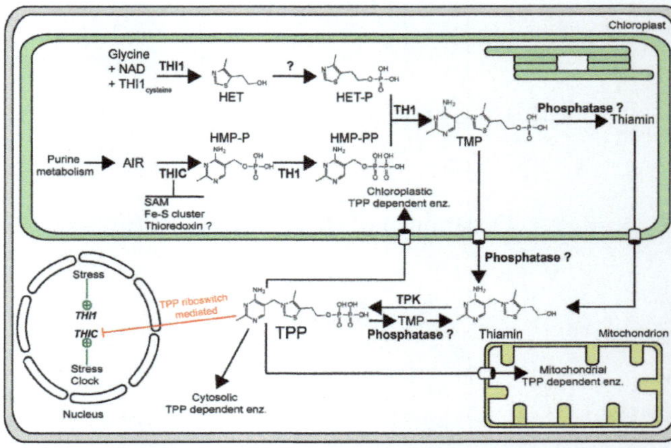

Figure 1: The thiamin biosynthesis pathway of *Arabidopsis thaliana*. As in all organisms, TMP is generated from the condensation of HET-P and HMP-PP by the action of TMP synthase (TH1). THI1 catalyzes the formation of the thiazole moiety from NAD^+, glycine, and ...

In prokaryotes, algae, plants, and certain fungi, *de novo* biosynthesis of thiamin is regulated *via* RNA sequences called riboswitches (Winkler et al., 2002; Sudarsan et al., 2003). This mode of regulation is mediated through the direct binding of the product TPP to the pre-mRNA of particular thiamin biosynthetic genes without the need for intermediary proteins. The binding causes a change in RNA secondary structure that interferes with gene expression. A TPP riboswitch sequence has been found in the 3′-UTR of the *THIC*gene of species across the plant kingdom (Sudarsan et al., 2003; Bocobza et al., 2007; Wachter et al., 2007) and in the 3′-UTR of the *THI1* gene in ancient plant *taxa* (Bocobza et al., 2007). In prokaryotes, the change in mRNA secondary structure, induced by metabolite binding, alters transcription termination or translation, whereas in eukaryotes the pre-mRNA is affected by splicing (Breaker, 2011). In contrast, no TPP riboswitch has been found to date in yeast where thiamin biosynthesis is instead regulated through the coordinate action of transcriptional regulators (for review, see Nosaka, 2006; Kowalska and Kozik, 2008).

In this report, we use the current knowledge on thiamin biosynthesis and its regulation in plants to hypothesize on approaches and pitfalls for efficient biofortification of thiamin in crop plants. The hypotheses presented are corroborated by simple feeding studies in *Arabidopsis*, which in turn reveal certain bottlenecks that should be taken into account for biofortification studies.

IS BYPASSING RIBOSWITCH CONTROL A FAVORABLE WAY TO INCREASE VITAMIN B1 CONTENT?

As mentioned above, plants have a TPP riboswitch located in the 3′-UTR of *THIC* that acts as a negative regulator of the thiamin biosynthesis pathway (Bocobza et al., 2007; Raschke et al., 2007; Wachter et al., 2007). The mechanism behind this riboswitch has been studied in *Arabidopsis*. Specifically, supplementation experiments with TPP (1 μM) or thiamin (100 μM) indicates that under such conditions the pre-mRNA of*THIC* undergoes a conformational change that exposes a splice site in the 3′-UTR of this gene (Bocobza et al., 2007; Wachter et al., 2007). As a consequence, there is splicing of intron 2 within the 3′-UTR of *THIC*eliminating the consensus polyadenylation signal and resulting in a transcript that is unstable. On the other hand, in the absence of supplementation there is strong accumulation of an intron retained variant (as the splice site is not exposed). The latter transcript has a single consensus polyadenylation signal that is more stable (Bocobza et al., 2007; Wachter et al., 2007). It is therefore assumed that when TPP reaches high enough levels, regulation of expression of the *THIC* gene occurs through this alternative splicing mechanism, i.e., high TPP levels lead to an unstable mRNA whereas the gene is stably expressed when TPP levels are low. However, the actual levels of TPP required in the cell to induce this splicing event remain to be determined. *In vitro* studies indicate a tight binding constant for TPP to the *Arabidopsis THIC* riboswitch (500 nM; Bocobza et al., 2007). As an equilibrium between the different pools of *THIC* transcript would be expected and as TPP levels can reach high micromolar quantities at the cellular level (Ajjawi et al., 2007b;Bocobza et al., 2013), we hypothesize that the TPP binding constant for the riboswitch is likely to be much larger *in vivo*.

As THIC is thought to be one of the key enzymes of the biosynthesis pathway, modulating this regulation would seem to be a tangible approach to increase thiamin content in plants. Indeed, very recently, Bocobza et al. (2013) have studied the impact of the *THIC* promoter as well as its riboswitch on vitamin B_1 levels in*Arabidopsis*. Interestingly, they could show that *THIC* is regulated in a circadian fashion through its promoter region and that there is an oscillation of the B_1 vitamer, TMP. Specifically, the *THIC* transcript is highest at the end of the light period and is lowest at the end of the dark period. In addition, the authors introduced a construct harboring the *THIC* promoter, gene and a mutated version of the riboswitch (A515G in the 3′-UTR) that can no longer bind TPP tightly (Bocobza et al., 2007) into a *THIC* knockdown mutant line (Kong et al., 2008). The expectation was that intron 2 of the 3′-UTR would be retained leading to a more stable*THIC* transcript and thereby enhancement of the vitamin B_1 content. The approach led to an up to threefold accumulation of TMP in leaves. On the other hand, the levels of the cofactor form, TPP, did not show a considerable change and surprisingly, the thiamin vitamer could not be detected. In seeds, an approximate 20% increase in thiamin content was observed in the same lines but no increase in TMP or TPP was noted (Bocobza et al., 2013). However, these transgenic lines exhibited severe chlorosis, growth retardation, and delayed flowering, especially under short day conditions (10 h of light). In another approach, the authors constitutively overexpressed *THIC* in wild-type plants under the control of the *UBIQUITIN1* promoter and terminator. While *THIC* was indeed overexpressed (up to 25-fold) in some of the transgenic lines and a moderate increase in the levels of TMP and TPP was observed (up to 1.5-fold) in leaf material, the corresponding lines were chlorotic (Bocobza et al., 2013). Based on a further comprehensive series of metabolomic and physiological experiments, the authors concluded that the alteration of riboswitch function in this way, while leading to a moderate increase in TPP levels, leads to enhanced carbohydrate oxidation as a result of higher activity of the associated TPP requiring enzymes. Clearly, this has an undesirable impact on the plant manifested as augmented flux through central metabolism and as a consequence negatively perturbing metabolic homeostasis. Therefore, this strategy appears to be unsuitable for vitamin B_1biofortification of plants.

ARE BOTH BRANCHES OF THE THIAMIN BIOSYNTHESIS PATHWAY REQUIRED FOR VITAMIN B1 BIOFORTIFICATION?

Several studies have demonstrated that thiamin is accumulated in plants during responses to abiotic and biotic stress conditions as well as oxidative stress (Ahn et al., 2005; Ribeiro et al., 2005; Rapala-Kozik et al., 2008;Tunc-Ozdemir et al., 2009). However, few studies have looked at the response of the entire core thiamin biosynthetic genes until recently. Rapala-Kozik et al. (2012) examined the response of *THIC*, *THI1*, *TH1*, and *TPK* (Figure1) under several types of abiotic stress in *Arabidopsis*. We found it interesting that under conditions where upregulation of the expression of the genes was observed, e.g., salt stress, both *THIC* and*THI1* were upregulated to similar relative levels, at least during the initial phases of the response. TPK was also upregulated under the same conditions but to a lesser extent (Rapala-Kozik et al., 2012). Notably in this study, while TPP levels did not show considerable alterations, there were significant changes in the level of the thiamin vitamer, in particular. As the key genes of both the pyrimidine and thiazole branches (i.e., *THIC*and *THI1*) were upregulated concomitant with an increase in thiamin in this study, this prompted us to raise the question whether both branches of the thiamin biosynthesis pathway need to be manipulated in order to increase vitamin B_1 levels for biofortification purposes in plants. To test this hypothesis, we grew*Arabidopsis* seedlings in the presence of the thiamin precursors from the pyrimidine branch (i.e., HMP) and/or the thiazole branch (i.e., HET). It has previously been demonstrated that both of these compounds are taken up and can restore growth in the corresponding *Arabidopsis* thiamin requiring mutants (Feenstra, 1964). However, as we had the *Arabidopsis thiC* knock-out mutant (affected in the pyrimidine branch) in hand (Raschke et al., 2007), we re-confirmed this for HMP. The *thiC* mutant is severely chlorotic and developmentally impaired when grown on basal salt medium (Figure2A, top panels). Supplementation with HMP, HMP + HET, or thiamin in the medium fully rescues the chlorotic and developmental phenotype (Figure2A, top panels). Notably, HMP supplementation alone rescues the phenotype, thus reaffirming its uptake, while HET does not, as would be expected. Neither HMP nor HET supplementation affected wild-type seedling development (Figure2A, bottom panels).

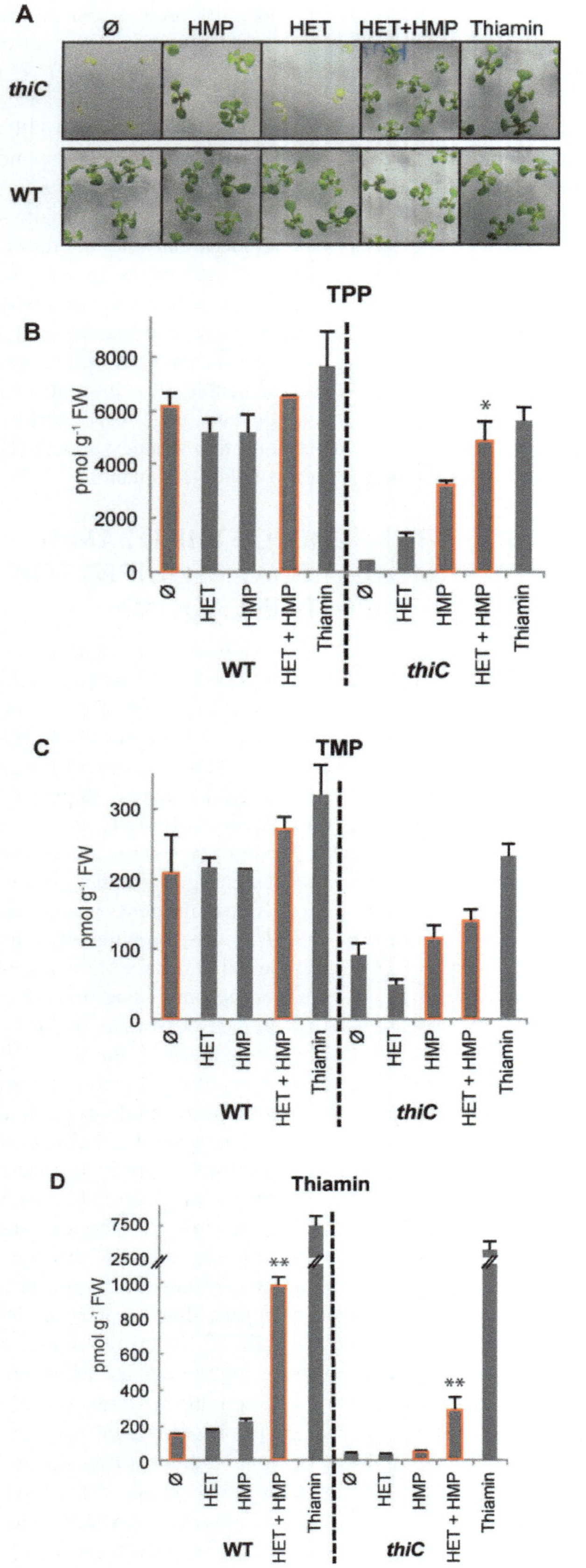

Figure 2: Effect of feeding vitamin B_1 **precursors on** *Arabidopsis*. (A) Eleven-day-old wild-type (WT) and *thiC* seedlings (SAIL_793_H10) were grown on half-strength basal salt medium (Murashige and Skoog, 1962) containing 0.5% agar, with or without HMP, HET, HMP and ...

We next quantified the individual B_1 vitamer contents (i.e., thiamin, TMP, TPP) by high-performance liquid chromatography (HPLC). Similar to other studies on *Arabidopsis* plant seedlings (Rapala-Kozik et al., 2012;Bocobza et al., 2013), TPP was found to be the most abundant vitamer in wild-type seedlings (~6151 pmol g^{-1} FW) compared to TMP and thiamin (~207 and 127 pmol g^{-1} FW, respectively; Figures2B–D). As expected, we observed that the TPP content increased in the *thiC* mutant when HMP (1 μM) was included in the medium, although the level was not completely restored to that of the wild-type (~3220 pmol g^{-1} FW) under these conditions (Figure2B). Interestingly, we observed an additive effect upon supplementation of both HMP and HET to the *thiC* mutant with the TPP levels approaching those of wild-type seedlings under the same conditions (Figure2B). This suggests that HET is indeed taken up, metabolized and converted into TPP by the seedlings. Supplementation with thiamin (1 μM) restores the TPP level in *thiC* to wild-type levels (Figure2B). In the wild-type seedlings on the other hand, supplementation with either HMP or HET alone did not significantly change the B_1 vitamer content (Figures2B–D). However, when both HMP and HET were added together to the medium, the level of thiamin, in particular, was found to be substantially higher (eightfold increase) from 127 to 973 pmol g^{-1} FW (Figure2D). On the one hand, these results suggest that if one precursor is available in excess the other becomes limiting and further implies that bypassing the *THIC*riboswitch alone will not induce B_1 vitamer accumulation sufficiently. A similar increase in the thiamin vitamer content (sevenfold increase) was observed when both HMP and HET were added to the *thiC* mutant, i.e., from 39 to 272 pmol g^{-1} FW compared to HMP alone (Figure2D). These observations provide strong evidence that the first two steps of the thiamin biosynthesis pathway (i.e., those catalyzed by THIC and THI1) are critical in order to increase the thiamin content of plants.

EXAMINING THE BALANCE BETWEEN THE PHOSPHATASE AND KINASE ENZYMES IN RELATION TO ENHANCING THIAMIN CONTENT

It is interesting that while the thiamin vitamer content was considerably increased by the dual supplementation of HET and HMP, TPP and TMP vitamer contents were not considerably perturbed in wild-type *Arabidopsis* (Figures2B–D). On the one hand, this could suggest that the protein machinery that converts the products of THIC and THI1 into thiamin (i.e., TH1 and a phosphatase) are not limiting for its enhanced production, whereas, the steps necessary for the conversion of thiamin into TPP are. However, as mentioned above, it has been suggested from a study in maize that there is a dynamic equilibrium between phosphatase and kinase actions that interconvert thiamin and its esters depending on the cellular requirement for TPP-dependent enzymes (Rapala-Kozik et al., 2009). To date, it is not known whether the product of the TH1 condensation reaction, TMP, is dephosphorylated into thiamin in the chloroplast, or in the cytosol (Figure1). However, it is known that thiamin is pyrophosphorylated to TPP in the cytosol by TPK of*Arabidopsis* (Ajjawi et al., 2007b). Therefore, we hypothesize that steps restricting the production of the TPP vitamer in particular, could be either the export of thiamin (or

TMP) from the chloroplast to the cytosol, the availability of the cytosolic kinase TPK, or dephosphorylation of TPP by a phosphatase. As the transporter(s) for thiamin or its esters out of the chloroplast have not yet been identified, we fed wild-type *Arabidopsis*seedlings with thiamin and followed the production of TPP to perform preliminary tests on the former hypothesis. Significantly, we assume that exogenous thiamin accumulates in the cytosol. Firstly, we could confirm that exogenous thiamin was indeed taken up by wild-type seedlings as shown in Figure2D. Even when the thiamin content was increased 50-fold in wild-type by supplementation (Figure2D), the TPP content only increased by 10% compared to either no supplementation or both HMP and HET supplementation (Figure2B). In a parallel fashion, thiamin addition to *thiC* led to a 92-fold increase of the vitamer (Figure2D), while the level of TPP was similar to that observed under both HMP + HET supplementation (Figure2B). This strongly suggests that a bottleneck for accumulation of the TPP vitamer is either a limitation of the cytosolic TPK enzyme itself or the enhanced action of a TPP phosphatase. It remains to be determined in future studies if TPK activity is a function of its dynamic equilibrium with the associated phosphatase and the requirement of the cell for TPP. Notwithstanding, our feeding studies reveal a fortuitous outcome in the context of biofortification. Dietary vitamin B_1 is most available to animals in its non-phosphorylated form (Komai et al., 1974; Said et al., 1999). Therefore, specifically increasing levels of TPP in the context of biofortification does not appear to be necessary. Furthermore, in contrast to the findings ofBocobza et al. (2013), where an accumulation of TMP (threefold) and to a lesser extent TPP (1.5-fold) was observed with the strategy undertaken, we do not see a significant increase in either of these vitamers and there is no observable negative effect on the plants under any of the supplementation conditions employed (Figure2A). However, although Bocobza et al. (2013) did not measure thiamin, we do not know if this vitamer was also increased in their study but it is noteworthy that TPK expression was significantly increased with the strategy that was employed. Thus, whether the accumulation of thiamin is due to a limitation in TPK activity or enhancement of a TPP phosphatase, it is a fortuitous act of nature in the context of biofortification. However, any inadvertent increase in the levels of the phosphorylated vitamers through genetic engineering strategies will need to be carefully monitored.

EXAMINING HOW STRAIGHTFORWARD IT WILL BE TO INDUCE THE PYRIMIDINE AND THIAZOLE BRANCHES

To date, increasing vitamin B_1 in the context of biofortification has never been overcome in the model plant*Arabidopsis* or in any crop plant. Based on several recent studies, some of which have been discussed above, it is now becoming clear that several regulatory steps will need to be taken into account in order to enhance the vitamin B_1 content. At its most simplistic level and as shown in this report through feeding studies, genetically engineering the overproduction of the HMP and HET precursors *in planta* should led to an overaccumulation of thiamin. Logically, this could be achieved by overexpressing the THIC and THI1 proteins, respectively. The riboswitch regulation of the expression of *THIC* can be bypassed by employing a transgene construct (i.e., promoter, gene, terminator) harboring a non-functional version of the riboswitch in the 3′-UTR (e.g., A515G) or indeed by employing an alternative terminator as has recently been shown (Bocobza et al., 2013). To counteract differential expression levels, both *THIC* and *THI1* transgenes could be inserted in tandem through a binary vector, under the control of a strong promoter, as has been shown to be successful in the β-carotene (vitamin A) biofortification of rice endosperm (Ye et al., 2000). Moreover, options to use tissue-specific promoters, thus minimizing the energy drain on the whole plant, could be exploited to enhance the process. Clearly, the chosen organ/tissue would need to express the other downstream enzymes of the biosynthetic pathway (i.e., TH1, phosphatases) and the precursors for HMP and HET biosynthesis would need to be available in sufficient amounts in the same tissue. In the context of vitamin B_1 biofortification in a crop such as rice, the endosperm would be favored but whether the aforementioned requirements are fulfilled in order to engineer such a "minipathway" therein is not yet known.

However, while we surmise that increasing the expression of *THIC* and *THI1* in plants can be achieved, it is important to remember that these are complex enzymes both of which require other key cellular components to fulfill their functions. THIC contains an iron–sulfur cluster that must be coupled with a reductant in order to produce a 5′-deoxyadenosyl radical from *S*-adenosyl methionine that is necessary for its catalytic activity (Raschke et al., 2007; Chatterjee et al., 2008; Martinez-Gomez and Downs, 2008). Moreover, affinity chromatography and proteomic approaches have revealed that THIC could be a target of the thioredoxin/ferredoxin system of oxygenic photosynthesis (Balmer et al., 2003, 2006) and therefore is likely to be also regulated by light (Raschke et al., 2007). On the other hand, while the involvement of THI1 in HET biosynthesis has been deduced from genetic studies (Belanger et al., 1995; Machado et al., 1996), the molecular nature of the substrates required for its biochemical activity in plants remain unclarified. The first study on the substrates of this enzyme claimed that deoxyxylulose 5-phosphate (DXP) is used as one of the precursors for the biosynthesis of HET (Juillard and Douce, 1991), as is the case in microorganisms (Hazra et al., 2009). However, the homolog of THI1 from *Saccharomyces cerevisiae* (called THI4p) has very recently been shown to be a suicidal enzyme (i.e., it can only perform a single turnover upon which it becomes non-catalytic). In this reaction, the thiazole is assembled from sulfur of a backbone cysteine (Cys205) of THI4p itself in addition to glycine and NAD^+ (Chatterjee et al., 2011). Several key residues including this cysteine are conserved in the plant THI1 proteins. Therefore, a similar mechanism can be anticipated and would represent yet another regulatory checkpoint *via* this enzyme. Furthermore, taken together, the involvement of NADH, the ferredoxin/thioredoxin system and the necessity of active site cysteines additionally make thiamin production highly dependent on the redox state of the cell. Therefore, engineering thiamin production in plants may not be so straightforward with several regulatory checkpoints needing to be surmounted. In this context, any genetic engineering strategy that proposes to enhance production of thiamin will need to be balanced with the physiological impact on the plant. While the branches of the biosynthetic pathway will need to be balanced, the flux through the pathway and its possible draining of other resources necessary for plant health will need to be monitored. These latter statements are corroborated by the undesirable physiology observed in *Arabidopsis* by overexpression of *THIC* alone, as a result of a negative impact on metabolic homeostasis (Bocobza et al., 2013). While the moderate increase in the pool of thiamin esters resulting from the latter strategy may negatively perturb carbohydrate metabolism as suggested by the authors (Bocobza et al., 2013), possible debilitating effects from the overuse of the complex biochemistry associated with THIC also need to be taken into account.

While the thiamin biosynthetic pathway in plants is close to being completely unraveled and could be exploited for biofortification purposes, we must also draw attention to thiamin binding proteins, specific variants of the major seed storage globulins (see Rapala-Kozik, 2011 for review). These proteins have been found and characterized in several types of seeds, e.g., rice, buckwheat, sunflower, and sesame (Golda et al., 2004). They appear to be found in the peripheral cells of the seed, accumulate during seed maturation and are degraded during seed germination (Watanabe et al., 2004). Notably, it is the thiamin vitamer that accumulates during seed maturation and serves as a reserve utilized upon seed germination (Golda et al., 2004). A closer examination of these phenomena could reveal novel strategies for thiamin biofortification of seeds in particular. For example, targeting of a thiamin binding protein to the rice endosperm is a strategy to be explored.

Although it is not the focused topic of this report, exploitation of natural variation is also likely to be a worthwhile approach that should be considered. Recently, Goyer and Sweek (2011) have reported on the identification of wild potato species and primitive cultivars that accumulate high amounts of thiamin compared to a modern potato variety (Russet Burbank). They propose to integrate these materials into breeding programs in order to enhance the nutritional value of potato. A closer investigation of the molecular basis behind the increased thiamin accumulation in these varieties may help genetic engineering strategies in other crop plants.

CONCLUSION

To date, there is no evidence of toxicity upon supplementation and accumulation of thiamin in plants. This is supported by the feeding studies provided herein. Rather, thiamin has been demonstrated to have beneficial effects enhancing tolerance to both abiotic and biotic stress (Ahn et al., 2005; Ribeiro et al., 2005; Rapala-Kozik et al., 2008; Tunc-Ozdemir et al., 2009). On the other hand, genetic engineering strategies that lead to overaccumulation of thiamin esters must proceed with caution as they may have a negative impact on metabolic homeostasis as indicated by the recent study of Bocobza et al. (2013). A simple engineering of the riboswitch associated with THIC has been discussed at the outset as a likely effective strategy to manipulate thiamin levels in plants, particularly in the context of biofortification. It now appears that the challenge of enhancing thiamin levels in crop plants for biofortification purposes will not be so straightforward. We hypothesize in this study that a balanced provision of the two key precursors (HMP and HET) will lead to the accumulation of the most bioavailable vitamer, thiamin. It will have the added benefit of not being toxic to the plant and could be anticipated to even enhance tolerance to stress. However, genetic engineering of the biosynthesis of the precursor compounds in plants will come with inherent challenges of its own. As discussed above, THI1 and THIC, in particular, are highly complex enzymes and are regulated in a highly complicated fashion, the entire mechanisms of which have not been unraveled yet.

ACKNOWLEDGMENTS

Teresa B. Fitzpatrick gratefully acknowledges the support of the Swiss National Science Foundation (grant number PP00A_119186 and VITRICA, 31003A_140911) as well as funding in the context of the SINERGIA project CRSI33_127506. Michael Moulin was funded by a Marie Curie Intra-European Fellowship within the seventh European Community Framework Program (IEF-2009-254907–RIBOREGAL).

REFERENCES

1. Ahn I.-P., Kim S., Lee Y.-H. (2005). Vitamin B1 functions as an activator of plant disease resistance.*Plant Physiol.* 138 1505–1515
2. Ajjawi I., Tsegaye Y., Shintani D. (2007a). Determination of the genetic, molecular, and biochemical basis of the *Arabidopsis thaliana* thiamin auxotroph *th1*. *Arch. Biochem. Biophys.* 459 107–114
3. Ajjawi I., Rodriguez Milla M. A., Cushman J., Shintani D. K. (2007b). Thiamin pyrophosphokinase is required for thiamin cofactor activation in *Arabidopsis*. *Plant Mol. Biol.* 65 151–162
4. Balmer Y., Koller A., del Val G., Manieri W., Schürmann P., Buchanan B. B. (2003). Proteomics gives insight into the regulatory function of chloroplast thioredoxins. *Proc. Natl. Acad. Sci. U.S.A.* 100370–375
5. Balmer Y., Vensel W. H., Cai N., Manieri W., Schürmann P., Hurkman W. J., et al. (2006). A complete ferredoxin/thioredoxin system regulates fundamental processes in amyloplasts. *Proc. Natl. Acad. Sci. U.S.A.* 103 2988–2993
6. Belanger F. C., Leustek T., Chu B., Kriz A. L. (1995). Evidence for the thiamine biosynthetic pathway in higher plant plastids and its developmental regulation. *Plant Mol. Biol.* 29 809–821
7. Bettendorff L., Wirtzfeld B., Makarchikov A. F., Mazzucchelli G., Frédérich M., Gigliobianco T., et al. (2007). Discovery of a natural thiamine adenine nucleotide. *Nat. Chem. Biol.* 3 211–212
8. Bocobza S. E., Adato A., Mandel T., Shapira M., Nudler E., Aharoni A. (2007). Riboswitch-dependent gene regulation and its evolution in the plant kingdom. *Genes Dev.* 21 2874–2879
9. Bocobza S. E., Malitsky S., Araújo W. L., Nunes-Nesi A., Meir S., Shapira M., et al. (2013).Orchestration of thiamin biosynthesis and central metabolism by combined action of the thiamin pyrophosphate riboswitch and the circadian clock in *Arabidopsis*. *Plant Cell* 25 288–307
10. Breaker R. R. (2011). Riboswitches and the RNA

world. *Cold Spring Harb. Perspect. Biol.* 1 4

11. Chabregas S. M., Luche D. D., Farias L. P., Ribeiro A. F., van Sluys M.-A., Menck C. F. M., et al. (2001). Dual targeting properties of the *N*-terminal signal sequence of *Arabidopsis thaliana* THI1 protein to mitochondria and chloroplasts. *Plant Mol. Biol.* 46 639–650
12. Chatterjee A., Abeydeera N. D., Bale S., Pai P. J., Dorrestein P. C., Russell D. H., et al. (2011). *Saccharomyces cerevisiae* THI4p is a suicide thiamine thiazole synthase. *Nature* 478 542–546
13. Chatterjee A., Li Y., Zhang Y., Grove T. L., Lee M., Krebs C., et al. (2008). Reconstitution of ThiC in thiamine pyrimidine biosynthesis expands the radical SAM superfamily. *Nat. Chem. Biol.* 4 758–765
14. Feenstra W. J. (1964). Isolation of nutritional mutants in *Arabidopsis thaliana*. *Genetica* 35 259–269
15. Fitzpatrick T. B., Basset G. J., Borel P., Carrari F., DellaPenna D., Fraser P. D., et al. (2012). Vitamin deficiencies in humans: can plant science help? *Plant Cell* 24 395–414
16. Funk C. (1912). The etiology of the deficiency diseases. *J. State Med.* 20 341–368
17. Gangolf M., Wins P., Thiry M., El Moualij B., Bettendorff L. (2010). Thiamine triphosphate synthesis in rat brain occurs in mitochondria and is coupled to the respiratory chain. *J. Biol. Chem.* 285 583–594
18. Golda A., Szyniarowski P., Ostrowska K., Kozik A., Rapala-Kozik M. (2004). Thiamine binding and metabolism in germinating seeds of selected cereals and legumes. *Plant Physiol. Biochem.* 42 187–195
19. Goyer A. (2010). Thiamine in plants: aspects of its metabolism and functions. *Phytochemistry* 711615–1624
20. Goyer A., Sweek K. (2011). Genetic diversity of thiamin and folate in primitive cultivated and wild potato (*Solanum*) species. *J. Agric. Food Chem.* 59 13072–13080
21. Harper C. G., Sheedy D. L., Lara A. I., Garrick T. M., Hilton J. M., Raisanen J. (1998). Prevalence of Wernicke–Korsakoff syndrome in Australia: has thiamine fortification made a difference? *Med. J. Aust.*168 542–545
22. Hazra A., Chatterjee A., Begley T. P. (2009). Biosynthesis of the thiamin thiazole in *Bacillus subtilis*: identification of the product of the thiazole synthase-catalyzed reaction. *J. Am. Chem. Soc.* 131 3225–3229
23. Juillard J.-H., Douce R. (1991). Biosynthesis of the thiazole moiety of thiamin (vitamin B1) in higher plant chloroplasts. *Proc. Natl. Acad. Sci. U.S.A.* 88 2042–2045
24. Komai T., Kawai K., Shindo H. (1974). Active transport of thiamine from rat small intestine. *J. Nutr. Sci. Vitaminol.* 20 163–177
25. Kong D., Zhu Y., Wu H., Cheng X., Liang H., Ling H. Q. (2008). *AtTHIC*, a gene involved in thiamine biosynthesis in *Arabidopsis thaliana*. *Cell Res.* 18 566–576
26. Kowalska E., Kozik A. (2008). The genes and enzymes involved in the biosynthesis of thiamin and thiamin diphosphate in yeasts. *Cell. Mol. Biol. Lett.* 13 271–282
27. Machado C. R., de Oliveira R. L., Boiteux S., Praekelt U. M., Meacock P. A., Menck C. F. (1996).*Thi1*, a thiamine biosynthetic gene in *Arabidopsis thaliana*, complements bacterial defects in DNA repair. *Plant Mol. Biol.* 31 585–593
28. Martinez-Gomez N. C., Downs D. M. (2008). ThiC is an [Fe-S] cluster protein that requires AdoMet to generate the 4-amino-5-hydroxymethyl-2-methylpyrimidine moiety in thiamin synthesis.*Biochemistry* 47 9054–9056
29. Mitsuda H., Takii Y., Iwami K., Yasumoto K., Nakajima K. (1979). Enzymatic formation of thiamine pyrophosphate in plants. *Methods Enzymol.* 62 107–111
30. Molin W. T., Fites R. C. (1980). Isolation and characterization of thiamin pyrophosphotransferase from*Glycine max* seedlings. *Plant Physiol.* 66 308–312
31. Murashige T., Skoog F. (1962). A revised medium for rapid growth of bioassays with tobacco tissue culture. *Physiol. Plant* 15 473–497
32. Nosaka K. (2006). Recent progress in understanding thiamin biosynthesis and its genetic regulation in*Saccharomyces cerevisiae*. *Appl. Microbiol. Biotechnol.* 72 30–40
33. Rapala-Kozik M. (2011). "Vitamin B1 (thiamine): a cofactor for enzymes involved in the main metabolic pathways and an environmental stress protectant," in *Biosynthesis of Vitamins in Plants* eds Rebeille F., Douce R., editors. (London: Elsevier) 37–91
34. Rapala-Kozik M., Go da A., Kujda M. (2009). Enzymes that control the thiamine diphosphate pool in plant tissues. Properties of thiamine pyrophosphokinase and thiamine-(di)phosphate phosphatase purified from *Zea mays* seedlings. *Plant Physiol. Biochem.* 47 237–242
35. Rapala-Kozik M., Kowalska E., Ostrowska K. (2008). Modulation of thiamine metabolism in *Zea mays* seedlings under conditions of abiotic stress. *J. Exp. Bot.* 59 4133–4143
36. Rapala-Kozik M., Olczak M., Ostrowska K., Starosta A., Kozik A. (2007). Molecular characterization of the *thi3* gene involved in thiamine biosynthesis in *Zea mays*: cDNA sequence and enzymatic and structural properties of the recombinant bifunctional protein with 4-amino-5-hydroxymethyl-2-methylpyrimidine (phosphate) kinase and thiamine monophosphate synthase activities. *Biochem. J.*408 149–159
37. Rapala-Kozik M., Wolak N., Kujda M., Banas A. K. (2012). The upregulation of thiamine (vitamin B-1) biosynthesis in *Arabidopsis thaliana* seedlings under salt and osmotic stress conditions is mediated by abscisic acid at the early stages of this stress response. *BMC Plant Biol.* 12:2 10.1186/1471-2229-12-2
38. Raschke M., Bürkle L., Müller N., Nunes-Nesi A., Fernie A. R., Arigoni D., et al. (2007). Vitamin B1 biosynthesis in plants requires the essential iron sulfur cluster protein, THIC. *Proc. Natl. Acad. Sci. U.S.A.* 104 19637–19642
39. Ribeiro D. T., Farias L. P., de Almeida J. D., Kashiwabara P. M., Ribeiro A. F., Silva-Filho M. C., et al. (2005). Functional characterization of the *thi1* promoter region from *Arabidopsis thaliana*. *J. Exp. Bot.* 56 1797–1804

40. Said H. M., Ortiz A., Kumar C. K., Chatterjee N., Dudeja P. K., Rubin S. (1999). Transport of thiamine in human intestine: mechanism and regulation in intestinal epithelial cell model Caco-2. *Am. J. Physiol.* 277 645–651

41. Sudarsan N., Barrick J. E., Breaker R. R. (2003). Metabolite-binding RNA domains are present in the genes of eukaryotes. *RNA* 9 644–647

42. Tunc-Ozdemir M., Miller G., Song L., Kim J., Sodek A., Koussevitzky S., et al. (2009). Thiamin confers enhanced tolerance to oxidative stress in *Arabidopsis*. *Plant Physiol.* 151 421–432

43. Wachter A., Tunc-Ozdemir M., Grove B. C., Green P. J., Shintani D. K., Breaker R. R. (2007).Riboswitch control of gene expression in plants by splicing and alternative 3′ end processing of mRNAs. *Plant Cell* 19 3437–3450

44. Wang G., Ding X., Yuan M., Qiu D., Li X., Xu C., et al. (2006). Dual function of rice *OsDR8* gene in disease resistance and thiamine accumulation. *Plant Mol. Biol.* 60 437–449

45. Watanabe K., Nishida N., Adachi T., Ueda M., Mitsunga T., Kawamura Y. (2004). Accumulation and degradation of thiamin-binding protein and level of thiamin in wheat seeds during seed maturation and germination. *Biosci. Biotechnol. Biochem.* 68 1243–1248

46. Winkler W., Nahvi A., Breaker R. R. (2002). Thiamine derivatives bind messenger RNAs directly to regulate bacterial gene expression. *Nature* 419 952–956

47. Ye X., Al-Babili S., Klöti A., Zhang J., Lucca P., Beyer P., et al. (2000). Engineering the provitamin A (β-carotene) biosynthetic pathway into 563 (carotenoid-free) rice endosperm. *Science* 287 303–305

CHAPTER 19

NEW TOOLS FOR CHLOROPLAST GENETIC ENGINEERING ALLOW THE SYNTHESIS OF HUMAN GROWTH HORMONE IN THE GREEN ALGA CHLAMYDOMONAS REINHARDTII

Thanyanan Wannathong[2] & Janet C. Waterhouse [1] & Rosanna E. B. Young[1] & Chloe K. Economou[1] & Saul Purton[1]

[1]Algal Research Group, Institute of Structural and Molecular Biology, University College London, Gower Street, London WC1E 6BT, UK

[2]Present address: Department of Biology, Faculty of Science, Silpakorn University, Nakornpathom 73000, Thailand

ABSTRACT

In recent years, there has been an increasing interest in the exploitation of microalgae in industrial biotechnology. Potentially, these phototrophic eukaryotes could be used for the low-cost synthesis of valuable recombinant products such as bioactive metabolites and therapeutic proteins. The algal chloroplast in particular represents an attractive target for such genetic engineering, both because it houses major metabolic pathways and because foreign genes can be targeted to specific loci within the chloroplast genome, resulting in high-level, stable expression. However, routine methods for chloroplast genetic engineering are currently available only for one species—Chlamydomonas reinhardtii—and even here, there are limitations to the existing technology, including the need for an expensive biolistic device for DNA delivery, the lack of robust expression vectors, and the undesirable use of antibiotic resistance markers. Here, we describe a new strain and vectors for targeted insertion of transgenes into a neutral chloroplast locus that (i) allow scar-less fusion of a transgenic coding sequence to the promoter/5′UTR element of the highly expressed endogenous genes psaA or atpA, (ii) employ the endogenous gene psbHas an effective but benign selectable marker, and (iii) ensure the successful integration of the transgene construct in all transformant lines. Transformation is achieved by a simple and cheap method of agitation of a DNA/cell suspension with glass beads, with selection based on the phototrophic rescue of a cell wall-deficient ΔpsbH strain. We demonstrate the utility of these tools in the creation of a transgenic line that produces high levels of functional human growth hormone.

INTRODUCTION

Microalgae represent attractive biotechnology platforms for the synthesis of recombinant therapeutic proteins and other high-value products (Gong et al. 2011; Rasala and Mayfield2015). Key advantages when compared to one or more of the traditional heterotrophic platforms such as *Escherichia coli*, yeast, or CHO cells include (i) the low-cost phototrophic cultivation of the alga in sterile, controlled photobioreactors using simple and inexpensive medium; (ii) the generally recognized as safe (GRAS) status of a number of algal species including the chlorophyte *Chlamydomonas reinhardtii*, making possible the oral delivery of bioactive products in whole algae and therefore avoiding expensive downstream purification; and (iii) the chloroplast organelle as a unique biosynthetic and

Citation: Thanyanan Wannathong & Janet C. Waterhouse & Rosanna E. B. Young & Chloe K. Economou1 & Saul Purton, New tools for chloroplast genetic engineering allow the synthesis of human growth hormone in the green alga Chlamydomonas reinhardtii, DOI 10.1007/s00253-016-7354-6.

storage compartment that contains its own minimal genetic system. Importantly, the use of the chloroplast for expression of foreign genes rather than the algal nucleus confers several benefits including precise insertion into the chloroplast genome (or "plastome") via homologous recombination, high-level expression that is not subject to any gene-silencing mechanisms, the possibility of expressing multiple transgenes as operons, and accumulation of the recombinant proteins in a benign compartment where the formation of disulfide bonds occurs readily (Tran et al. 2009) but undesirable glycosylations are avoided (Gregory et al. 2013).

Chloroplast transgenics were first reported in 1991 when a bacterial antibiotic-resistance gene was expressed in *C. reinhardtii* (Goldschmidt-Clermont 1991). Since that time, there have been many reports describing the synthesis of functional therapeutic proteins in the*C. reinhardtii* chloroplast including monoclonal antibodies (Mayfield et al. 2003; Tran et al. 2009), growth factors (Rasala et al. 2010), antigens (Dreesen et al. 2010; Michelet et al.2011; Jones et al. 2013), gut-active proteins (Manuell et al. 2007; Yoon et al. 2011), anti-bacterial proteins (Braun-Galleani et al. 2015), immunotoxins (Tran et al. 2013a, 2013b), and anti-toxins (Barrera et al. 2015). In addition, efforts are being made to manipulate chloroplast biosynthetic pathways in order to synthesize novel bioactive compounds such as diterpenoids (Gangl et al. 2015; Zedler et al. 2015). This encouraging progress in the development of the algal chloroplast as a viable platform has recently led to the establishment of start-up companies seeking to exploit the technology and the demonstration of pilot-scale production of a bioactive protein (Gimpel et al. 2015).

However, there remains a need to develop improved molecular tools that address some of the current technical limitations in the generation of *C. reinhardtii* transgenic lines (Purton et al. 2013). Specifically, there is a need for a simple and reliable method of rapidly generating homoplasmic transformant lines that also avoids the use of bacterial antibiotic-resistance genes as selectable markers. Currently, transformation typically involves bombardment of an algal lawn with DNA-coated microparticles (= biolistics) and the use of the *aadA* or *aphA6* bacterial genes as selectable markers conferring resistance to spectinomycin and kanamycin, respectively (Goldschmidt-Clermont 1991; Bateman and Purton 2000). Resistant colonies are then checked for the presence of the gene of interest (GOI) and taken through multiple rounds of single-colony selection in order to ensure that the transformant lines attain a stable, homoplasmic state in which all copies of the polyploid genome contain the marker and the GOI. A simpler alternative to microparticle bombardment involves agitating a suspension of cells and transforming DNA in the presence of glass beads, although this method requires the prior removal of the cell wall either by digestion or mutation (Kindle et al. 1991; Economou et al. 2014).

Similarly, one alternative method for selection employs non-photosynthetic mutants as recipient strains where the genetic lesion is in a key photosynthetic gene on the chloroplast genome. Selection is based on the use of a wild-type copy of the gene as the marker with successfully transformed cells able to grow phototrophically on minimal medium through replacement of the mutated gene with the wild-type version (Purton 2007; Michelet et al.2011; Chen and Melis 2013). This selection strategy therefore allows the introduction of a GOI as the only transgene, avoiding the use of any antibiotic-resistance gene. Such "marker-free" transgenic lines are appropriate for industrial cultivation since they circumvent the regulatory and environmental issues associated with the possible horizontal transfer of such resistance genes into other microorganisms. Other issues associated with antibiotic-based selection include (i) the occurrence of "false-positive" colonies due to natural resistance mutations arising in *C. reinhardtii* genes, (ii) the additional metabolic burden on the chloroplast of replicating and expressing the marker, and (iii) the challenge of using antibiotic selection to drive transformant lines efficiently to a homoplasmic state.

Our recent work on *C. reinhardtii* chloroplast engineering (Braun-Galleani et al. 2015; Young and Purton 2014, 2016) has involved the use of a new recipient strain and expression vectors that allow the generation of marker-free transformants using the glass bead transformation method. In this paper, we describe these tools in detail. The recipient strain carries the nuclear *cw15* mutation resulting in a cell wall-deficient phenotype, together with a deletion of the essential photosystem II gene *psbH* to allow selection based on phototrophy. The expression vectors carry the *psbH* marker and are designed to allow simple cloning of a GOI as a perfect translational fusion with the promoter and 5′ untranslated region (UTR) from various chloroplast genes. We find that the vector that uses the *psaA* promoter/5′UTR element gives high expression levels in the chloroplast and is able to serve as a dual expression vector, since expression is also readily detectable in *E. coli*. Finally, we describe a simple three-primer PCR strategy to confirm both the correct integration of the GOI into the plastome and the establishment of homoplasmy. To validate these new tools, we have created a transgenic line expressing a synthetic gene encoding human growth hormone (hGH) and shown that the hGH is biologically active.

MATERIALS AND METHODS

C. reinhardtii strains and culture conditions

The TN72 strain is a *psbH*::*aadA* knockout mutant of the cell wall-deficient strain*cw15*.mt+ (Davies and Plaskitt 1971) and was created as described below. TN72 has been deposited in the publicly available collection of the Chlamydomonas Resource Center (chlamycollection.org) as strain CC-5168. TN72 transformant lines were initially selected on 2 % agar plates containing high salt minimal (HSM) medium but then maintained on Tris-acetate phosphate (TAP) agar plates (Harris et al. 2009) under dim light (5–10 $\mu E/m^2/s$) at 20 °C. For liquid cultures, cells were grown in TAP medium in conical flasks at 25 °C, 120 rpm at 5 $\mu E/m^2/s$ (for TN72) or 50 $\mu E/m^2/s$ (for transformants). Spectinomycin (100 μg/ml) and erythromycin (10–50 μg/ml) were added to TAP agar medium as required.

Plasmid construction

DNA manipulations were carried out using standard protocols (Green and Sambrook2012), including transformation of chemically competent *E. coli* DH5α using ampicillin selection. Oligonucleotide primers were obtained from Eurofins Genomics (Ebersberg, Germany); restriction enzymes and T4 DNA ligase for cloning procedures were purchased from

NEB (Ipswich, MA, USA) and Thermo Scientific (Waltham, MA, USA). Plasmid p72B-aadA-BM, which was used to create TN72, was made by replacing the 0.36-kb *Mlu*I-*Bst*XI region of p72B (Bateman and Purton 2000) with the "*aadA* cassette" (comprising the *aadA* coding sequence fused to the *atpA* promoter/5′UTR and *rbcL* 3′UTR) amplified from pUC-atpX-AAD (Goldschmidt-Clermont 1991) using primers that introduced an*Mlu*I and *Bst*XI site at the ends of the cassette. Transformation vector pASapI was created from p72B-SH (Bateman and Purton 2000), by first cutting p72B-SH with *Sap*I (within the pUC8 vector) and *Eco*RI, end-filling, and religating, thereby removing a 240-bp region of the vector sequence including the unwanted *Sap*I site. Then, the *Mlu*I-*Nco*I region containing the *rbcL* promoter/5′UTR element was replaced with an equivalent region amplified from *atpA* to give pASapI in which a *Sap*I site was created immediately downstream of the start codon. Vectors pSRSapI, pPSapI, and pCSapI were created from pASapI by replacing the *Mlu*I to *Sap*I region of pASapI with the promoter/5′UTR region of*psaA* exon 1, *psbA*, and *chlL*, respectively. The *ereB* coding sequence was amplified from plasmid pAT72 (Arthur and Courvalin 1986) using primers that introduce an *Nco*I and *Sph*I site at the start and end of the coding region, respectively, and cloned into these sites in pSRSapI to create pRY129a. The synthetic gene for hGH was codon-optimized for the *C. reinhardtii* chloroplast using the Codon Usage Database (www.kazusa.or.jp/codon) with *Sap*I and *Sph*I sites added for cloning. The gene was synthesized by Life Technologies (Carlsbad, CA, USA) and cloned into the four expression vectors.

Chloroplast transformation

Chloroplast transformation was based on a previously described method (Kindle et al. 1991) and involved the agitation of an algal/DNA suspension with glass beads of 400–625 μm diameter. A 400-ml culture grown to early log phase (approx. 2×10^6 cells/ml) was concentrated by centrifugation and resuspended in TAP medium to 4 ml. Three hundred microliters of cells were added to a sterilized 5-ml test tube containing 300 mg sterile glass beads, followed by 5–10 μg circular plasmid DNA. The mixture was agitated vigorously at the maximum speed of a Vortex Genie II (Fisher Scientific, Loughborough, UK) for 15 s. The cells were spread on selective agar plates (TAP + spectinomycin at 100 μg/ml for *aadA* selection; HSM for *psbH* selection) after mixing with 0.5 % molten (42 °C) agar of the same selective medium. The plates were incubated at 25 °C in dim light (~2 $\mu E/m^2/s$) overnight then transferred to a moderate light (~50 $\mu E/m^2/s$) the next day. Transformant colonies were picked after 2–3 weeks and restreaked to single colonies several times on selective media to ensure homoplasmicity, although this was often achieved after the first streaking. Homoplasmy was determined by PCR using a combination of three primers. Total genomic DNA was extracted from a loopful of cells using the Chelex 100 method as described by Werner and Mergenhagen (1998), and PCR amplification was carried out with Phusion DNA polymerase (Thermo Scientific) according to the manufacturer's instructions. The photosystem II-deficient phenotype of TN72 was confirmed by measuring the photosynthetic capacity through chlorophyll fluorescence (Maxwell and Johnson 2000). Putative transformant lines were spotted onto TAP agar plates, incubated for 5 days and scanned using a pulse-modulated imaging fluorometer (FluorCam 700MF, Photon Systems Instruments, Czech Republic) as described by Wingler et al. (2004).

Western analysis of *C. reinhardtii* and *E. coli* transformants

Protein samples from *C. reinhardtii* were prepared, fractionated on a 15 % acrylamide SDS-PAGE gel and blotted onto nitrocellulose membrane as described by Young and Purton (2014). For immunodetection, membranes were blocked in TBS-T (20 mM Tris, 137 mM NaCl, 1 M HCl pH 7.4, 0.1 % Tween-20) with 0.5 % milk powder overnight at 4 °C. Membranes were incubated with primary antibody in TBS-T with 0.5 % milk (1 h at RT), washed 3 × 5 min in TBS-T, incubated with secondary antibody in TBS-T with 0.5 % milk (1 h at RT), and washed as before. The probed blots were scanned and quantified using an Odyssey Infrared Imaging System (LI-COR). Alternatively, for Fig. 5a, the antibody binding was detected by enhanced chemiluminescence (ECL) using SuperSignal West Pico Chemiluminescent substrate (Thermo Scientific) and Hyperfilm ECL (GE Healthcare). For hGH detection, the primary antibody was chicken polyclonal antibody to human growth hormone (Abcam product ab27223) at 1:5000 dilution and the secondary was donkey anti-chicken IgY IRDye 800CW (LI-COR) at 1:5000, or goat anti-chicken IgY HRP-linked antibody (Abcam product ab97135) at 1:5000 for ECL. For RbcL detection, the primary antibody was anti-RbcL produced in rabbit (a gift from John Gray, Cambridge University, UK) at 1:5000 followed by secondary antibody Goat anti-Rabbit IgG (H&L), Dylight™ 800 Conjugated (Thermo Scientific product 35571) at 1:10,000, or HRP-linked Goat Anti-Rabbit IgG H&L (Abcam product ab6721) at 1:5000 for ECL.

For the detection of proteins in *E. coli*, 5 ml of an overnight LB culture was harvested by centrifugation (10 min at 4500 *g*) and resuspended in *X*/4 ml solution A (0.8 M Tris-HCl pH 8.3, 0.2 M sorbitol, 1 % β-mercaptoethanol) where *X* is the absorbance of the culture at 600 nm. SDS was added to a final concentration of 1 %. Samples were boiled for 5 min and passed through a 0.5-mm needle then centrifuged (2 min at 21,000 *g*), and 25 μl of supernatant was loaded. Gel and hGH immunodetection conditions using the ODYSSEY system were as described above. For both the *Chlamydomonas* and *E. coli* analysis, pure recombinant hGH (Abcam product 51232) was used as a reference.

Nb2-11 cell proliferation assay for hGH activity

Nb2-11 cells (European Collection of Cell Cultures) were grown in suspension culture in Fischer's medium supplemented with 10 % fetal bovine serum, 10 % horse serum, 2 mM l-glutamine, 0.075 % sodium bicarbonate, and 0.05 mM β-mercaptoethanol, at 37 °C in an atmosphere containing 5 % CO_2. Cells were starved for 24 h by centrifugation (750 *g* for 5 min) to remove the culture medium, followed by washing and resuspending the cells in supplemented Fischer's medium excluding fetal bovine serum. The cells were seeded in a 12-well plate at a concentration of approximately 1×10^5 cells/ml. Recombinant hGH (Abcam) was added to 1 ml of cells in a volume of 5 μl to give a final concentration of 1, 10, or 100 ng/ml. Phosphate buffer (50 mM, pH 7.4) was added to cells as a control and was used to dilute the hGH. Samples were analyzed in triplicate. Cell proliferation was monitored every 24 h using the CASY® Cell Counter and Analyzer Model TT (Innovatis, Bielefeld, Germany), with cells >8.03 μm being defined as viable. To test *C. reinhardtii*extracts, transformant TN-hGH, and the negative control TN-C, cells were cultured in 500-ml TAP medium to a concentration of 2×10^6 cells/ml, broken by three cycles of freezing in liquid nitrogen and thawing, centrifuged at 9000 *g* for 30 min, followed by ultracentrifugation of the supernatant at 100,000 *g* for 60 min. The hGH content in the TN-hGH extract was quantified by Western blotting as above and a volume containing ~100 ng hGH used in the proliferation assay, together with an equivalent amount of TN-C extract, or phosphate buffer as controls. The experiment was repeated in triplicate with proliferation measured after 96 h.

Sequence data

The Genbank accession number of the synthetic *hGH* used in the study is KU179444.

RESULTS

The ΔpsbH strain TN72 for reliable, marker-free selection of transformant lines

Our previous research has shown that disruption of the chloroplast gene*psbH*, which encodes the 9 kDa phosphoprotein subunit of photosystem II, results in a complete destabilization of this complex. As a result, *psbH* mutants are blocked in photosynthetic function and, thus, have an acetate-dependent phenotype (O'Connor et al. 1998). These mutants can be restored to phototrophy following biolistic transformation with a plasmid carrying wild-type *psbH*, with selection on minimal medium (Bateman and Purton 2000). The *psbH* gene can therefore be used as an effective selectable marker for *C. reinhardtii*, as has been established for other photosynthetic genes including *atpB*, *tscA*, and *rbcL*(Boynton et al. 1988; Goldschmidt-Clermont 1991; Chen and Melis 2013).

However, our original recipient strain "Bst-same" (O'Connor et al. 1998) contains a cell wall and therefore must be transformed either using the expensive biolistics method or by removing the cell wall using autolysin before glass bead-mediated transformation (Economou et al. 2014). To combine the benefits of photosynthetic selection and the simple, cheap glass bead method, a new *psbH* mutant (TN72) was created that carries the*cw15* nuclear mutation and thus lacks a cell wall. In addition, we eliminated one other flaw with Bst-same, which is that *psbH* is disrupted by insertion of the *aadA* marker into the coding region while GOIs are targeted to an intergenic site between *trnE2* and *psbH* that is 0.36 kb downstream of the *aadA* insertion. This means that transformation can occasionally generate transformants in which *psbH* is restored (and *aadA* eliminated), but the GOI is not introduced because homologous recombination between vector and genome has occurred in this 0.36 kb region. To solve this, the TN72 line was constructed such that the whole region is replaced by the *aadA* marker, thereby disrupting *psbH* and ensuring that transformants can arise only by integration of both *psbH* and the GOI (Fig. 1).

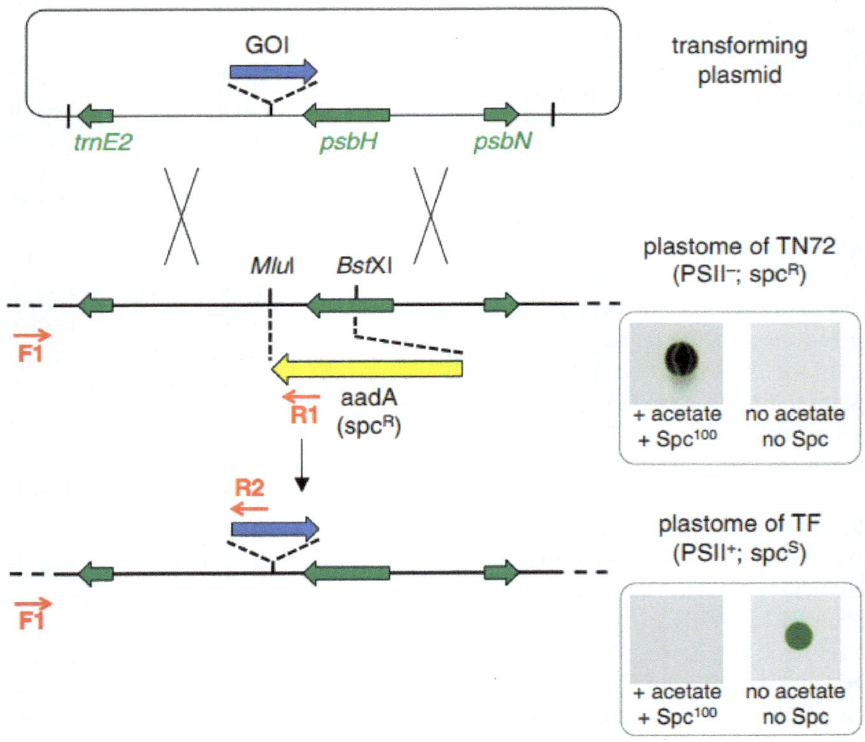

Figure 1: Strategy for generation of marker-less transformants using the TN72 recipient. The transforming plasmid contains the gene-of-interest (*GOI*) cassette which comprises a chloroplast promoter/5′UTR element from a chloroplast gene, coding sequence of the GOI and a 3′UTR from a chloroplast gene. The cassette is flanked by plastome sequence that includes *psbH* as selectable marker. Introduction into the chloroplast of strain TN72 (a knockout strain in which *psbH* and the downstream region have been replaced with the *aadA* marker) results in restoration of *psbH*, introduction of the GOI and loss of*aadA*. As a result, transformants (*TF*) are capable of phototrophic growth on acetate-free medium and are sensitive to spectinomycin (*Spc*) as shown in the *lower panel.*

New expression vectors allow a perfect fusion of the GOI to the 5′UTR and rapid establishment of chloroplast homoplasmy

Translation is a key regulatory step in gene expression in the *C. reinhardtii* chloroplast (Eberhard et al. 2002) and typically involves one or more *trans*-acting protein factors that bind to *cis* elements with the 5′UTR of the transcript and mediate efficient ribosome binding and translation initiation (Choquet and Wollman 2002; Specht and Mayfield2013). Furthermore, the bases immediately upstream of the start codon have been shown to play an important role in efficient translation initiation (Esposito et al. 2001). We therefore designed an expression vector that uses the type IIS restriction enzyme *Sap*I for cloning. As illustrated in Fig. 2, *Sap*I cuts to give a three-base 5′ overhang to one side of its recognition site. This allows the joining of a chloroplast promoter/5′UTR element to the ATG of a GOI as a perfect fusion without the incorporation of the site itself, and therefore without the need to introduce any DNA changes upstream of the start codon or add additional codons to the start of the GOI.

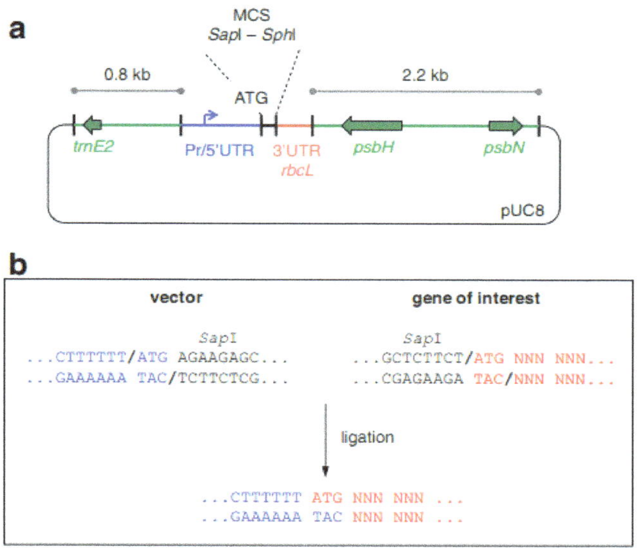

Figure 2: Design of expression vectors and *Sap*I cloning scheme. **a** Four different vectors were constructed that contained a promoter/5′untranslated region (*Pr/5′UTR*) from different chloroplast genes, namely pASapI (*atpA*), pSRSapI (*psaA*-exon 1), pPSapI (*psbA*), and pCSapI (*chlL*). **b** The positioning of the *Sap*I site in the vector and in any synthetic gene of interest to allow a perfect fusion at the ATG start codon.

In order to test different transcription and translation efficiencies, four different versions of the vector were constructed that each have the same multiple cloning site for GOI insertion followed by the 3′UTR from the endogenous *rbcL* gene, but differ in the endogenous promoter/5′UTR element used to drive expression, namely that from *psaA*exon 1 (vector pSRSapI), *atpA* (pASapI), *psbA* (pPSapI), and *chlL* (pCSapI). Each vector can be used in combination with the TN72 line to target any GOI to the neutral locus within the *psbH-trnE* intergenic region, using restoration of *psbH* as selection as illustrated in Fig. 1.

Since the *C. reinhardtii* plastome is polyploid with ~40 copies per cell under phototrophic conditions and ~100 copies under mixotrophic conditions (Lau et al. 2000), an important early step in the analysis of putative transformant lines is to confirm that the GOI has inserted into the plastome and that all copies carry the GOI—i.e., the lines are homoplasmic (Purton 2007). We have devised a PCR strategy involving three primers to confirm both integration and homoplasmy in a single step. Primers F1 and R1 generate a 0.88-kb PCR product from the untransformed TN72 plastome, whereas in the transplastomic genome, a larger band (e.g., 1.13 kb for pSRSapI transformants) is generated by primers F1 and R2, with primer R2 designed to the promoter/5′UTR element used in each vector and primer F1 outside of the left-hand region of homology carried on each vector (thereby preventing amplification from any free plasmid or plasmid integrated into the nuclear genome). Heteroplasmic lines containing some untransformed plastome copies would be expected to give rise to both PCR products.

Initial transformation experiments using pSRSapI to express a test GOI (the *E. coli ereB*gene encoding erythromycin resistance (Arthur and Courvalin 1986)) revealed that all nine colonies that were tested contained the transgene (Fig. 3a), supporting the idea that the use of the TN72 recipient ensures acquisition of the GOI in all transformants. Furthermore, it was found that selection based on restoration of phototrophy very rapidly established homoplasmy without repeated rounds of single colony isolation. This is in contrast to that observed with the commonly used *aadA* marker for spectinomycin resistance, where several rounds of colony isolation on selective medium are required to drive the transformants to a homoplasmic state (Day and Goldschmidt-Clermont 2011). The ease with which homoplasmy is obtained is probably a reflection of both the strong selective pressure to restore the photosynthetic phenotype, and the reduced plastome copy number in colonies growing on minimal medium rather than the acetate-containing medium used for *aadA* selection (Lau et al. 2000). Growth tests of three of the transformant lines on plates containing erythromycin confirmed the functional expression of *ereB* (Fig. 3b).

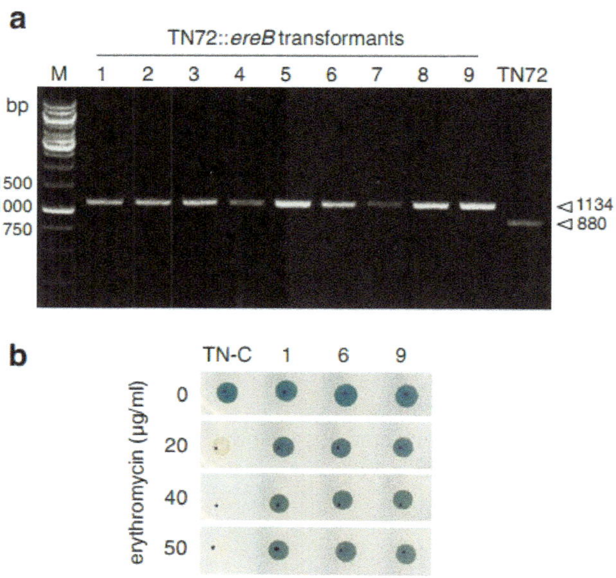

Figure 3: PCR confirmation of GOI integration and homoplasmy of pSRSapI-ereB transformants, and demonstration of erythromycin resistance. **a** PCR was performed using genomic DNA from transformants that had been streaked once only on HSM agar after

colony picking. Primers were F1, R1, and R2 (*psaA*). The presence of a 1134-bp band confirms the successful integration of *ereB*, whereas the absence of the 880-bp band that arises from the untransformed copies of the TN72 plastome indicates that the transformant lines are homoplasmic. *M* = DNA ladder of size markers. **b** Growth of the control transformant lacking *ereB* (*TN-C*) and three representative *ereB* transformants on TAP agar plates containing increasing concentrations of erythromycin.

The pSRSapI vector allows efficient expression in both C. reinhardtii and E. coli

To compare the expression levels obtained with the four different vectors, a synthetic gene encoding human growth hormone was designed based on the codon preferences of *C. reinhardtii* chloroplast genes and cloned into each vector. Transformants were generated using TN72, and homoplasmy was established by PCR as above (Fig. 4a). A representative clone for each vector was used to assess the level of recombinant hGH in both the *C. reinhardtii* chloroplast and the *E. coli* DH5α used for the cloning. As shown in Fig. 4b, c, Western blot analysis of cell extracts from the *C. reinhardtii* and *E. coli* transformant lines using anti-hGH antibodies revealed markedly differing levels of hGH for the four different promoter/5′UTR elements. In the chloroplast, both the *psaA* and *atpA* elements gave detectable levels of hGH, with *psaA* showing a higher level as has been reported previously (Michelet et al. 2011). The protein is not detected in the *psbA* transformant lines. This is not surprising given that Rasala et al. (2011) have reported that transgenes linked to the*psbA* 5′UTR are poorly expressed due to the process of auto-attenuation of *psbA*translation which is mediated via the 5′UTR. Furthermore, Kasai et al. (2003) have reported that additional *cis* elements required for efficient *psbA* translation are located downstream of the start codon and are therefore missing in the vector. There is also no detectable hGH in the *chlL* transformant lines. This probably reflects the very low level of*chlL* expression in *C. reinhardtii*, particularly in cells grown in the light (Cahoon and Timko 2000). Furthermore, work by Cheng et al. (2005) showed that the product of a transgene (*nifH*) linked to the *chlL* promoter/5′UTR element was detectable only under conditions of anaerobic growth, suggesting that use of the *chlL* element to drive transgene expression is applicable only under such cultivation conditions. The hGH signals were normalized to the signal obtained using a control antibody raised to the large subunit of Rubisco and expressed as relative values of hGH accumulation. The hGH level from the*psaA* exon 1 element was ~7.6 times higher than from the *atpA* element (Fig. 4b). In *E. coli*, the highest level of expression is obtained also from the *psaA* construct, with levels significantly higher than those from the *psbA* and *chlL* elements. The level from the *atpA*element was barely detectable (Fig. 4c). The pSRSapI vector therefore represents a useful dual vector, able to drive transgene expression in both the *C. reinhardtii* chloroplast and in *E. coli*.

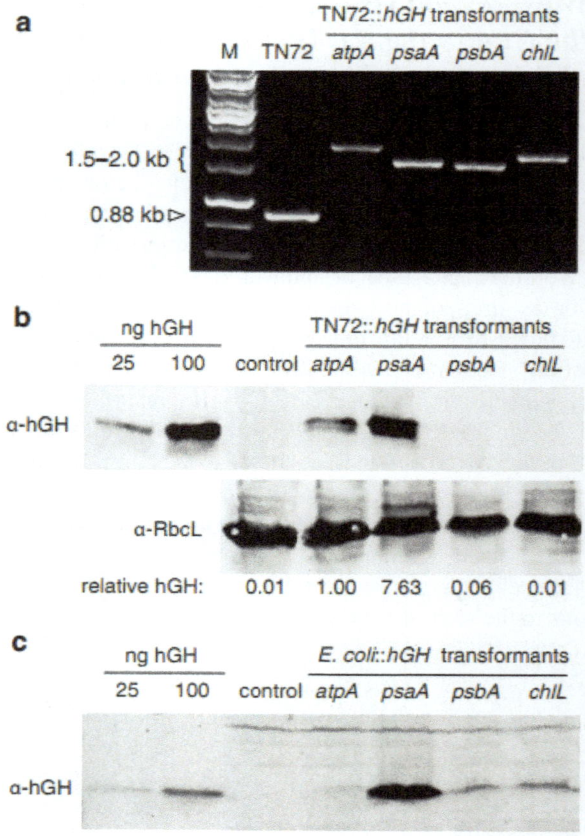

Figure 4: Analysis of hGH expression under the control of the different promoter/5′UTR elements.**a** PCR confirmation of gene integration and homoplasmy of *hGH* transformants. Three-primer PCR demonstrating the homoplasmic integration of hGH expression cassettes into the *C. reinhardtii* chloroplast genome. The TN72 cell line was transformed with plasmids that introduced the *hGH* into the plastome downstream of *psbH* as illustrated in Fig. 1. PCR of genomic DNA using primers F1, R1, and R3 (*hGH.R*) give a 0.88-kb product for the TN72 plastome, and a 1.5–2.0-kb product for the transgenic plastomes depending on the size of the promoter/5′UTR element. The absence of the 0.88-kb band indicates that the transformant lines are homoplasmic. *M* = DNA ladder of size markers.**b** Western blot analysis of the *C. reinhardtii* transformants. Crude protein extracts, equalized according to the optical density of the cultures, were fractionated by SDS-PAGE and probed with antibodies to hGH. *Control* = a TN72 transformant generated using the empty pSRSapI vector. Pure recombinant hGH was run alongside to aid quantification of the 22-kDa protein. An identical gel was blotted and probed with antibodies to the large subunit (*RbcL*) of the endogenous Rubisco enzyme as a loading control. Relative band intensities of hGH were quantified using an Odyssey Infrared Imaging System and scaled to give 1.000 for the *atpA* element. **c** Western blot analysis of the *E. coli* transformants.*Control* = an *E. coli* transformant generated using the empty pSRSapI vector.

A functional hGH is produced in the C. reinhardtii chloroplast

The transgenic line (named TN-hGH) that was generated using the pSRSapI-hGH construct was characterized further in order to assess the level of recombinant hGH produced and to determine whether the protein was functional. Figure 5a shows a Western blot in which the level of hGH in this line is compared with known amounts of a commercial sample of purified recombinant hGH. In order to mitigate against any signal differences caused by the presence of other proteins in the crude cell lysate of TN-hGH, the purified hGH was added to an equivalent crude lysate from a control transformant line (TN-C) generated using the pSRSapI vector only (Fig. 5a). Comparison of the level of hGH in the TN-hGH and spiked TN-C samples indicates that the hGH is present at ~25 μg/ml in the final TN-hGH lysate. This represents approximately 0.5 mg hGH per liter of *C. reinhardtii* culture.

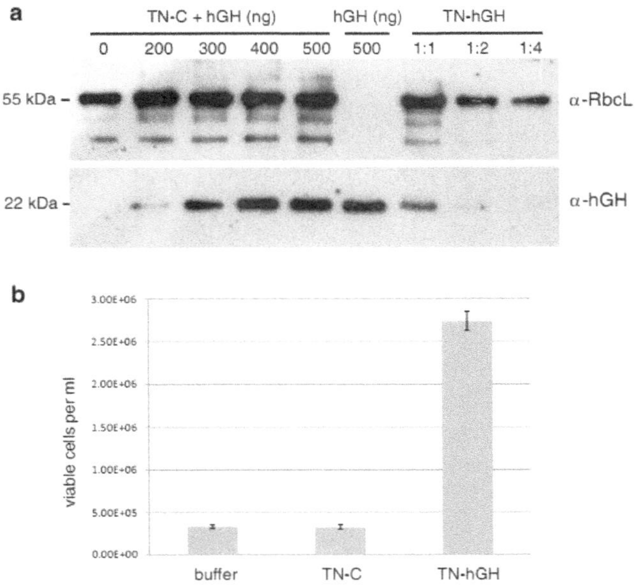

Figure 5: hGH quantification and activity in the TN72::pSRSapI-hGH transformant line. **a** Western blot analysis in which a cell extract from TN72::pSRSapI-hGH (*TN-hGH*) was compared with a control extract from TN72::pSRSapI (*TN-C*) spiked with known amounts of recombinant hGH. The lower part of the blot was probed with antibodies to hGH, and the upper part with antibodies to the large subunit (*RbcL*) of Rubisco as a loading control. Antibody binding was detected by ECL. **b** Determination of hGH activity using an Nb2-11 cell proliferation assay. Cell extract from the TN-hGH line (containing ~100 ng hGH) and the control line TN-C lacking hGH was added to serum-starved Nb2-11 cells at 1×10^5 cells/ml and proliferation determined after 96 h. An equivalent volume of phosphate buffer was also used as a negative control. *Error bars* represent one standard deviation, $n = 3$.

To determine whether the hormone is biologically active, a proliferation assay was carried out using the Nb2-11 rat lymphoma cell line. The replication of Nb2-11 cells in suspension culture is specifically stimulated by lactogenic hormones, and by hGH which is known to have lactogenic activity (Tanaka et al. 1980; Bulatov et al. 1996). This was confirmed using the commercial hGH which showed a dose-dependent effect of the hormone on the proliferation of the cell line. This proliferation was also seen when a cell extract from TN-hGH containing approximately 100 ng of hGH was added to the Nb2-11 cells, thereby confirming its biological activity. As shown in Fig. 5b, an increase in cell number of approximately eightfold was obtained after 96 h using the TN-hGH extract when compared to an equivalent extract from the control strain TN-C, or to buffer only.

DISCUSSION

Currently, the industrial biotechnology sector is almost exclusively based around the use of heterotrophic platforms (bacteria, yeasts, mammalian, and insect cells) for the biosynthesis of pharmaceutical proteins, bioactive metabolites, or other high-value products (Murphy 2012). Nevertheless, the ever-increasing growth of the global bioeconomy and the need for sustainable alternatives to petrochemical-based products is catalyzing interest in the exploitation of alternative cell factories, including photosynthetic microalgae such as *C. reinhardtii* (Wijffels et al. 2013). Recent successes in the biosynthesis of recombinant therapeutics in the *C. reinhardtii* chloroplast have underlined the potential of this alga as a commercial platform (Rasala and Mayfield 2015), and highlight the need for further development of new tools for rapid and reliable strain engineering (Scaife et al. 2015).

The recipient strain and vectors described here provide a low-cost and simple system for creating a transgenic line within a few weeks. Transformation of the TN72 strain is achieved by agitation with glass beads rather than the more technically complex particle bombardment method, and the strong selection for restoration of photosynthesis means that stable, homoplasmic lines are obtained without repeated rounds of selection. Importantly, the system guarantees that every selected colony contains the GOI since the deletion of the essential PSII gene in TN72 prevents spontaneous mutation to phototrophic growth, and the deletion of the genomic region between *psbH* and the site of GOI insertion ensures that restoration of *psbH* is always accompanied by insertion of the GOI. Our finding that the *psaA* promoter/5′UTR gives high levels of expression in the chloroplast is in agreement with the findings of Michelet et al. (2011), who reported that the same *psaA* element gave higher levels of their two recombinant proteins than elements from *atpA* and *psbA*. This identifies pSRSapI as the best of our expression vectors. Furthermore, the observation that the *psaA* promoter/5′UTR element used in pSRSapI is able to drive expression in *E. coli* provides a useful tool for testing the synthesis and function of the recombinant protein prior to generation of *C. reinhardtii* transformants. We speculate that the difference we see in *E. coli* expression levels for the different chloroplast gene elements is, in part, due to differences in the efficiency of translation initiation in *E. coli*. Of the four genes, the *psaA* sequence immediately upstream of the translational start most closely matches the *E. coli* Shine-Dalgarno consensus sequence for ribosome binding. In contrast, a weaker match is seen for *chlL* and *psbA*, and no match is seen in the *atpA* 5′UTR. Furthermore, efficient transcription from the *psaA*promoter probably also contributes to the overall expression in *E. coli*. The 5′ end of the*psaA* exon 1 mRNA was determined by Choquet et al. (1988), and Turmel et al. (1995) predicted that a TATAAT sequence within the *psaA* exon 1 promoter region corresponded to the –10 element; this exact consensus sequence is in fact preceded by TGn, correlating to the consensus extended –10 element that is thought to contribute to promoter strength (Hook-Barnard and Hinton 2007). Promoter prediction software, softberry BPROM, indicates a possible –35 element (TTGTAA) upstream of the –10 element. These data predict that the *psaA* exon 1 promoter would give strong transcription initiation in *E. coli*. While expression in *E. coli* is a desirable feature of pSRSapI, it can also present problems when cloning genes that are deleterious to the bacterium. To address this, we have recently developed a modified version of the vector carrying an engineered chloroplast tRNA gene. The cloning of synthetic genes carrying internal UGA stop codons results in premature translation termination in *E. coli*, but in the chloroplast, the tRNA recognizes the UGA as a tryptophan codon (Young and Purton 2016).

As a test of our system, we demonstrated the functional expression of a native bacterial gene (*ereB*) and of a codon-optimized synthetic gene encoding human growth hormone (hGH). Recombinant hGH is used to treat growth hormone deficiency, Turner syndrome, and several other conditions (Kirk 2012). Glycosylation of hGH is not required for biological activity, and therefore recombinant production in prokaryotic expression systems such as *E. coli* is preferred (Sockolosky and Szoka 2013). The hormone has also been expressed successfully in the tobacco chloroplast where it was shown to fold correctly, forming the two disulfide bonds within the 22-kDa molecule (Staub et al. 2000). This is in contrast to expression in the *E. coli* cytoplasm where hGH tends to accumulate within inclusion bodies that require solubilization and refolding (Sockolosky and Szoka2013). The *C. reinhardtii* chloroplast also possesses the machinery for the correct folding of recombinant proteins and the formation of multiple disulfide bonds (Tran et al. 2013b), and our finding that hGH is recovered in a soluble and active form further supports this. However, it should be noted that previous studies have indicated that the presence of the disulfide bonds in hGH is not specifically required for biological activity (Kim et al. 2013). Further improvement in the yield of hGH could be achieved by introducing into the transgenic line the nuclear mutation *raa-L121G*. This mutation blocks the *trans*-splicing of*psaA-exon1* mRNA and has been shown to greatly increase expression of transgenes fused to the *psaA* promoter/5′UTR (Michelet et al. 2011). Additionally, Ferreira-Camargo et al. (2015) have reported that the inclusion of selenocystamine in the growth medium improves the accumulation of proteins containing disulfide bonds and could therefore also increase hGH yield.

ACKNOWLEDGMENTS

This research was supported by funding from the UK's Biotechnology and Biological Sciences Research Council (BBSRC grant BB/I007660) and by European Council funding through the Framework Program 7 project GIAVAP (KBBE 2010-3: GA266401). TW was supported by a scholarship from the Royal Thai Government. We thank Andrea Townsend-Nicholson (UCL) for the assistance with the cell proliferation assay and Patrice Courvalin (Institut Pasteur,

REFERENCES

1. Arthur M, Courvalin P (1986) Contribution of two different mechanisms to erythromycin resistance in *Escherichia coli*. Antimicrob Agents Chemother 30:694–700. doi:10.1128/AAC.30.5.694
2. Barrera DJ, Rosenberg JN, Chiu JG, Chang YN, Debatis M, Ngoi SM, Chang JT, Shoemaker CB, Oyler GA, Mayfield SP (2015) Algal chloroplast produced camelid VH H antitoxins are capable of neutralizing botulinum neurotoxin. Plant Biotechnol J 13:117–124. doi:10.1111/pbi.12244
3. Bateman JM, Purton S (2000) Tools for chloroplast transformation in *Chlamydomonas*: expression vectors and a new dominant selectable marker. Mol Gen Genet 263:404–410. doi:10.1007/s004380051184
4. Boynton JE, Gillham NW, Harris EH, Hosler JP, Johnson AM, Jones AR, Randolph-Anderson BL, Robertson D, Klein TM, Shark KB (1988) Chloroplast transformation in*Chlamydomonas* with high velocity microprojectiles. Science 240:1534–1538. doi:10.1126/science.2897716
5. Braun-Galleani S, Baganz F, Purton S (2015) Improving recombinant protein production in the *Chlamydomonas reinhardtii* chloroplast using vivid Verde fluorescent protein as a reporter. Biotechnol J 10:1289–1297. doi:10.1002/biot.201400566
6. Bulatov AA, Osipova TA, Raevskaya GV (1996) Proliferative response of cultured Nb2 rat lymphoma cells to human serum prolactin and growth hormone. Bull Exp Biol Med 11:1135–1138. doi:10.1007/BF02447668
7. Cahoon AB, Timko MP (2000) *Yellow-in-the-dark* mutants of *Chlamydomonas* lack the CHLL subunit of light-independent protochlorophyllide reductase. Plant Cell 12:559–568. doi:10.1105/tpc.12.4.559
8. Chen HC, Melis A (2013) Marker-free genetic engineering of the chloroplast in the green microalga *Chlamydomonas reinhardtii*. Plant Biotechnol J 11:818–828. doi:10.1111/pbi.12073
9. Cheng Q, Day A, Dowson-Day M, Shen GF, Dixon R (2005) The *Klebsiella pneumoniae*nitrogenase Fe protein gene (*nifH*) functionally substitutes for the *chlL* gene in*Chlamydomonas reinhardtii*. Biochem Biophys Res Commun 329:966–975. doi:10.1016/j.bbrc.2005.02.064
10. Choquet Y, Goldschmidt-Clermont M, Girard-Bascou J, Kuck U, Bennoun P, Rochaix J-D (1988) Mutant phenotypes support a *trans*-splicing mechanism for the expression of the tripartite *psaA* gene in the *C. reinhardtii* chloroplast. Cell 52:903–913. doi:10.1016/0092-8674(88)90432-1
11. Choquet Y, Wollman FA (2002) Translational regulations as specific traits of chloroplast gene expression. FEBS Lett 529:39–42. doi:10.1016/S0014-5793(02)03260-X
12. Davies DR, Plaskitt A (1971) Genetical and structural analyses of cell-wall formation in*Chlamydomonas reinhardi*. Genetical Res 17:33–43. doi:10.1017/S0016672300012015
13. Day A, Goldschmidt-Clermont M (2011) The chloroplast transformation toolbox: selectable markers and marker removal. Plant Biotechnol J 9:540–553. doi:10.1111/j.1467-7652.2011.00604.x
14. Dreesen IA, Charpin-El Hamri G, Fussenegger M (2010) Heat-stable oral alga-based vaccine protects mice from *Staphylococcus aureus* infection. J Biotechnol 145:273–280. doi:10.1016/j.jbiotec.2009.12.006
15. Eberhard S, Drapier D, Wollman FA (2002) Searching limiting steps in the expression of chloroplast-encoded proteins: relations between gene copy number, transcription, transcript abundance and translation rate in the chloroplast of *Chlamydomonas reinhardtii*. Plant J 31:149–160. doi:10.1046/j.1365-313X.2002.01340.x

16. Economou C, Wannathong T, Szaub J, Purton S (2014) A simple, low cost method for chloroplast transformation of the green alga *Chlamydomonas reinhardtii*. Methods Mol Biol 1132:401–411. doi:10.1007/978-1-62703-995-6_27
17. Esposito D, Hicks AJ, Stern DB (2001) A role for initiation codon context in chloroplast translation. Plant Cell 13:2373–2384. doi:10.1105/tpc.010236
18. Ferreira-Camargo LS, Tran M, Beld J, Burkart MD, Mayfield SP (2015) Selenocystamine improves protein accumulation in chloroplasts of eukaryotic green algae. AMB Express 5:39. doi:10.1186/s13568-015-0126-3Central
19. Gangl D, Zedler JA, Włodarczyk A, Jensen PE, Purton S, Robinson C (2015) Expression and membrane-targeting of an active plant cytochrome P450 in the chloroplast of the green alga *Chlamydomonas reinhardtii*. Phytochemistry 110:22–28. doi:10.1016/j.phytochem.2014.12.006
20. Gimpel JA, Hyun JS, Schoepp NG, Mayfield SP (2015) Production of recombinant proteins in microalgae at pilot greenhouse scale. Biotechnol Bioeng 112:339–345. doi:10.1002/bit.25357
21. Goldschmidt-Clermont M (1991) Transgenic expression of aminoglycoside adenine transferase in the chloroplast: a selectable marker of site-directed transformation of chlamydomonas. Nucleic Acids Res 19:4083–4089. doi:10.1093/nar/19.15.4083
22. Gong Y, Hu H, Gao Y, Xu X, Gao H (2011) Microalgae as platforms for production of recombinant proteins and valuable compounds: progress and prospects. J Ind Microbiol Biotechnol 38:1879–1890. doi:10.1007/s10295-011-1032-6
23. Green MR, Sambrook J (2012) Molecular cloning: a laboratory manual, Fourth edn. Cold Spring Harbor Laboratory Press, Cold Spring Harbor
24. Gregory JA, Topol AB, Doerner DZ, Mayfield S (2013) Alga-produced cholera toxin-Pfs25 fusion proteins as oral vaccines. Appl Environ Microbiol 79:3917–3925. doi:10.1128/AEM.00714-13
25. Harris EH, Stern DB, Witman GB (2009) The Chlamydomonas sourcebook, Second edn. Elsevier Inc., Oxford
26. Hook-Barnard IG, Hinton DM (2007) Transcription initiation by mix and match elements: flexibility for polymerase binding to bacterial promoters. Gene Regul Syst Bio 1:275–293Central
27. Jones CS, Luong T, Hannon M, Tran M, Gregory JA, Shen Z, Briggs SP, Mayfield SP (2013) Heterologous expression of the C-terminal antigenic domain of the malaria vaccine candidate Pfs48/45 in the green algae *Chlamydomonas reinhardtii*. Appl Microbiol Biotechnol 97:1987–1995. doi:10.1007/s00253-012-4071-7
28. Kasai S, Yoshimura S, Ishikura K, Takaoka Y, Kobayashi K, Kato K, Shinmyo A (2003) Effect of coding regions on chloroplast gene expression in *Chlamydomonas reinhardtii*. J Biosci Bioeng 95:276–282. doi:10.1016/S1389-1723(03)80029-4
29. Kim MJ, Park HS, Seo KH, Yang HJ, Kim SK, Choi JH (2013) Complete solubilization and purification of recombinant human growth hormone produced in *Escherichia coli*. PLoS ONE 8(2):e56168. doi:10.1371/journal.pone.0056168
30. Kindle KL, Richards KL, Stern DB (1991) Engineering the chloroplast genome: techniques and capabilities for chloroplast transformation in *Chlamydomonas reinhardtii*. Proc Natl Acad Sci U S A 88:1721–1725
31. Kirk J (2012) Indications for growth hormone therapy in children. Arch Dis Child 97:63–68. doi:10.1136/adc.2010.186205
32. Lau KW, Ren J, Wu M (2000) Redox modulation of chloroplast DNA replication in*Chlamydomonas reinhardtii*. Antioxid Redox Signal 2:529–535. doi:10.1089/15230860050192305
33. Manuell AL, Beligni MV, Elder JH, Siefker DT, Tran M, Weber A, McDonald TL, Mayfield SP (2007) Robust expression of a bioactive mammalian protein in *Chlamydomonas*chloroplast. Plant Biotechnol J 5:402–412. doi:10.1111/j.1467-7652.2007.00249.xn
34. Maxwell K, Johnson GN (2000) Chlorophyll fluorescence—a practical guide. J Exp Bot 51:659–668. doi:10.1093/jexbot/51.345.659
35. Mayfield SP, Franklin SE, Lerner RA (2003) Expression and assembly of a fully active antibody in algae. Proc Natl Acad Sci U S A 100:438–442. doi:10.1073/pnas.0237108100
36. Michelet L, Lefebvre-Legendre L, Burr SE, Rochaix JD, Goldschmidt-Clermont M (2011) Enhanced chloroplast transgene expression in a nuclear mutant of *Chlamydomonas*. Plant Biotechnol J 9:565–574. doi:10.1111/j.1467-7652.2010.00564.x
37. Murphy CD (2012) The microbial cell factory. Org Biomol Chem 10:1949–1957. doi:10.1039/c2ob06903b
38. O'Connor HE, Ruffle SV, Cain AJ, Deak Z, Vass I, Nugent JH, Purton S (1998) The 9-kDa phosphoprotein of photosystem II. Generation and characterisation of *Chlamydomonas*mutants lacking PSII-H and a site-directed mutant lacking the phosphorylation site. Biochim Biophys Acta 1364:63–72. doi:10.1016/S0005-2728(98)00013-9
39. Purton S (2007) Tools and techniques for chloroplast transformation of *Chlamydomonas*. Adv Exp Med Biol 616:34–45. doi:10.1007/978-0-387-75532-8_4
40. Purton S, Szaub JB, Wannathong T, Young R, Economou CK (2013) Genetic engineering of algal chloroplasts: progress and prospects. Rus J Plant Physiol 60:521–528. doi:10.1134/S1021443713040146
41. Rasala BA, Mayfield SP (2015) Photosynthetic biomanufacturing in green algae; production of recombinant proteins for industrial, nutritional, and medical uses. Photosynth Res 123:227–239. doi:10.1007/s11120-014-9994-7
42. Rasala BA, Muto M, Lee PA, Jager M, Cardoso RM, Behnke CA, Kirk P, Hokanson CA, Crea R, Mendez M, Mayfield SP (2010) Production of therapeutic proteins in algae, analysis of expression of seven human proteins

in the chloroplast of *Chlamydomonas reinhardtii*. Plant Biotechnol J 8:719–733. doi:10.1111/j.1467-7652.2010.00503.x

43. Rasala BA, Muto M, Sullivan J, Mayfield SP (2011) Improved heterologous protein expression in the chloroplast of *Chlamydomonas reinhardtii* through promoter and 5′ untranslated region optimization. Plant Biotechnol J 9:674–683. doi:10.1111/j.1467-7652.2011.00620.x
44. Scaife MA, Nguyen GT, Rico J, Lambert D, Helliwell KE, Smith AG (2015) Establishing*Chlamydomonas reinhardtii* as an industrial biotechnology host. Plant J 82:532–546. doi:10.1111/tpj.12781
45. Sockolosky JT, Szoka FC (2013) Periplasmic production via the pET expression system of soluble, bioactive human growth hormone. Protein Expr Purif 87:129–135. doi:10.1016/j.pep.2012.11.002
46. Specht EA, Mayfield SP (2013) Synthetic oligonucleotide libraries reveal novel regulatory elements in *Chlamydomonas* chloroplast mRNAs. ACS Synth Biol 2:34–46. doi:10.1021/sb300069k
47. Staub JM, Garcia B, Graves J, Hajdukiewicz PT, Hunter P, Nehra N, Paradkar V, Schlittler M, Carroll JA, Spatola L, Ward D, Ye G, Russell DA (2000) High-yield production of a human therapeutic protein in tobacco chloroplasts. Nat Biotechnol 18:333–338. doi:10.1038/73796
48. Tanaka T, Shiu RP, Gout PW, Beer CT, Noble RL, Friesen HG (1980) A new sensitive and specific bioassay for lactogenic hormones: measurement of prolactin and growth hormone in human serum. J Clin Endocrinol Metab 51:1058–1063. doi:10.1210/jcem-51-5-1058
49. Tran M, Henry RE, Siefker D, Van C, Newkirk G, Kim J, Bui J, Mayfield SP (2013a) Production of anti-cancer immunotoxins in algae: ribosome inactivating proteins as fusion partners. Biotechnol Bioeng 110:2826–2835. doi:10.1002/bit.24966
50. Tran M, Van C, Barrera DJ, Pettersson PL, Peinado CD, Bui J, Mayfield SP (2013b) Production of unique immunotoxin cancer therapeutics in algal chloroplasts. Proc Natl Acad Sci U S A 110:E15–E22. doi:10.1073/pnas.1214638110
51. Tran M, Zhou B, Pettersson PL, Gonzalez MJ, Mayfield SP (2009) Synthesis and assembly of a full-length human monoclonal antibody in algal chloroplasts. Biotechnol Bioeng 104:663–673. doi:10.1002/bit.22446
52. Turmel M, Choquet Y, Goldschmidt-Clermont M, Rochaix J-D, Otis C, Lemieux C (1995) The *trans*-spliced intron 1 in the *psaA* gene of the *Chlamydomonas* chloroplast: a comparative analysis. Curr Genet 27:270–279. doi:10.1007/BF00326160
53. Werner R, Mergenhagen D (1998) Mating type determination of *Chlamydomonas reinhardtii* by PCR. Plant Mol Biol Rep 16:295–299. doi:10.1023/A:1007583508766
54. Wijffels RH, Kruse O, Hellingwerf KJ (2013) Potential of industrial biotechnology with cyanobacteria and eukaryotic microalgae. Curr Opin Biotechnol 24:405–413. doi:10.1016/j.copbio.2013.04.004
55. Wingler A, Mares M, Pourtau N (2004) Spatial patterns and metabolic regulation of photosynthetic parameters during leaf senescence. New Phytol 161:781–789. doi:10.1111/j.1469-8137.2004.00996.x
56. Yoon SM, Kim SY, Li KF, Yoon BH, Choe S, Kuo MM (2011) Transgenic microalgae expressing *Escherichia coli* AppA phytase as feed additive to reduce phytate excretion in the manure of young broiler chicks. Appl Microbiol Biotechnol 91:553–563. doi:10.1007/s00253-011-3279-2
57. Young REB, Purton S (2014) Cytosine deaminase as a negative selectable marker for the microalgal chloroplast: a strategy for the isolation of nuclear mutations that affect chloroplast gene expression. Plant J 80:915–925. doi:10.1111/tpj.12675
58. Young REB, Purton S (2016) Codon reassignment to facilitate genetic engineering and biocontainment in the chloroplast of *Chlamydomonas reinhardtii*. Plant Biotechnol J [Epub ahead of print]. doi:10.1111/pbi.12490
59. Zedler JAZ, Gangl D, Hamberger B, Purton S, Robinson C (2015) Stable expression of a bifunctional diterpene synthase in the chloroplast of *Chlamydomonas reinhardtii*. J Appl Phycol 27:2271–2277. doi:10.1007/s10811-014-0504-2

INDEX